ADVANCES
IN ENERGY PRODUCTIVITY

ADVANCES IN

ENERGY

PRODUCTIVITY

ADVANCES IN ENERGY PRODUCTIVITY

Published by
THE FAIRMONT PRESS, INC.
P.O. Box 14227
Atlanta, Georgia 30324

PREFACE – CONFERENCE SPONSOR'S STATEMENT

The papers which are compiled herein are based upon presentations made at the 5th World Energy Engineering Congress, September 14-17, 1982. The Congress is sponsored by the Association of Energy Engineers and supported by the United States Department of Energy, Office of Industrial Programs.

The sharing of information is essential to the continued growth of the energy profession. AEE is proud to have played a major role in sponsoring this important conference.

Albert Thumann, PE
Executive Director
Association of Energy Engineers

CONTENTS

CONTRIBUTORS

Allen, Robert D., P.E., *Bovay Engineers, Inc.*
Apa, R. P., *Energy, Incorporated*
Baird, George, Ph.D., *Victoria University*
Baker, James M., *Logistics Management Institute*
Beanland, Michael D., *City of Palo Alto*
Benator, Barry I., P.E., CEM, *Benatech, Inc.*
Brander, W.D.S., *Auckland Regional Authority*
Briscoe, Harry H., P.E., *Northwest Georgia Engineering Co., Inc.*
Brown, Michael, P.E., *Georgia Tech Engineering Experiment Station*
Buford, Donald R., P.E., *McDonnell Douglas Automation Company*
Bulpitt, William, P.E., *Georgia Tech Engineering Experiment Station*
Burger, Robert, *Burger Associates, Inc.*
Burrows, David L., *Tennessee Valley Authority*
Cairns, J. Robert, Ph.D., P.E., *University of Michigan*
Caratti, Giancarlo, *Istituto di Energetica*
Castonguay, Michael, *Xenergy, Inc.*
Chandrashekar, S., P.E., *United Engineers & Constructors, Inc.*
Cooper, Dale S., P.E., *Dale S. Cooper & Associates*
Cummings, Craig R., *Science Applications, Inc.*
Cutler, Robert R., P.E., *Penn Jersey Boiler and Construction, Inc.*
Davis, Richard E., *Holmes & Narver, Inc.*
Dinse, David R., P.E., *Tennessee Valley Authority*
Dougherty, N. N., *Control Energy, Inc.*
Edwards, Herbert J., Jr., *Ebasco Business Consulting Company*
Elder, Larry A., *Western Electric Company*
Elms, L. D., *L. D. Elms & Associates*
Farnham, Charles P., *Lighting Technology, Inc.*
Feledy, Charles F., CEM, *United Technologies Corporation*
Fischer, R. D., *Batelle, Columbus Laboratories*
Fisher, W. S., P.E., *General Electric Co.*
Foley, D., *Control Energy, Inc.*
Fowler, Delbert M., P.E., *Fowler/Blum Energy Consultants, Inc.*
Gagliardi, Ric, *Ford Aerospace & Communications Corporation*
Galletti, Alessandro, *ENEI, CRTN Research Center*
Gallo, Anthony J., *Corning Glass Works*
Gifford, Neil R., *United Technologies Corporation*
Gilbert, Joel S., P.E., *Mechanical Technology Incorporated*
Goetz, Donald F., P.E., *A.T. & T. Long Lines*
Gould, Barry J., *CoSyCo, Inc.*
Gupta, Vijay K., Ph.D., *Central State University*
Harris, David, *Georgia Tech Engineering Experiment Station*
Haviland, Jon R., P.E., *Ralphs Grocery Company*
Henry, Walter E., P.E., *Xenergy, Inc.*
Hodge, B. K., Ph.D., P.E., *Mississippi State University*
Janicke, Lawrence A., *Media East Communications Corp.*
Jhaveri, Arun G., *John Graham and Company*
Johnston, Walter E., *General Host Corporation*
Kardan, Cevat, P.E., *University of Houston*
Kettler, John P., P.E., *Wattmaster Company*

Kibert, Charles, J., P.E., *University of South Florida*
Kraemer, T. C., *Little General Corporation*
Krochmal, Roman S., *Harvey Morris Associates, Inc.*
LaRue, Mervin, *Rencon, Inc.*
Leonov, Abram, *Energy Management Consultants*
Limaye, Dilip R., *Synergic Resources Corporation*
Loyless, Elliott M., *W. R. Grace & Co.*
Lux, John J., Jr., *Batelle, Columbus Laboratories*
Marquette, C. James, *Baltimore Aircoil Company, Inc.*
Martelli, Francesco, *Istituto di Energetica*
McFee, John N., P.E., *Energy Incorporated*
McIntyre, Julia M., *Oak Ridge National Laboratory*
Mehta, D. Paul, Ph.D., *Bradley University*
Merat, Francis L., Ph.D., *Case Western Reserve University*
Miconi, Dante, *Istituto di Energetica*
Morris, Harvey N., *Harvey Morris & Associates*
Mozzo, Martin A., Jr., P.E., *American Standard, Inc.*
Mulhern, Thomas A., P.E., *Western Electric Company*
Needham, Victor A., III, *Tennessee Valley Authority*
Norelli, Patrick, P.E., *Westinghouse R & D Center*
Norian, Bruce D., *Energyworks, Inc.*
Ollman, Philip J., *International Harvester Company*
Ormston, Doug, *Consulting Engineer*
Parate, Nath S., Ph.D., *Tennessee State University*
Parisi, Dominick, *Mass Save, Inc.*
Payne, Pat, P.E., CEM, *Texas Energy Engineers, Inc.*
Pearson, Harry, P.E., *Johnson & Johnson Products, Inc.*
Piccolo, Frank A., *Polaroid Corporation*
Pincelli, Richard D., P.E., *ESI, Inc.*
Pollock, Edward O., *Vitro Laboratories*
Riker, Thomas H., *Polaroid Corporation*
Rolka, H., P.E., *Dow Corning*
Roy, V., *Westinghouse R & D Center*
Ruston, Michael J., *Corning Process Systems*
Sain, Stephen P., *Baltimore Aircoil Company, Inc.*
Sherry, Edward V., *Air Products and Chemicals*
Soma, John, P.E., CEM, *Consultant*
Sperberg, Richard T., *Onsite Energy*
Stecco, Sergio, *Istituto di Energetica*
Steele, W. G., Ph.D., P.E., *Mississippi State University*
Stein, David E., *Florida Solar Energy Center*
Stenhouse, Douglas, *Energy Management Consultants, Inc.*
Stovall, John P., *Oak Ridge National Laboratory*
Stubblefield, N. D., *Claude Terry & Associates, Inc.*
Swetish, John D., *Deere & Company*
Sworden, P.G., *Dow Corning*
Teji, Darshan S., P.E., CEM, *City of Phoenix*
Thompson, Ronald P., *Pacific Gas and Electric Co.*
Trieste, Richard J., P.E., *The Brooklyn Union Gas Company*
Trimm, T. Gary, *American Technical Services*
Vance, Richard F., P.E., *Energy Incorporated*
Viar, William L., P.E., *Waterland, Viar & Associates, Inc.*
Walker, R. D., Ed.D., *Mississippi Energy Ext. Center*
Westcott, William W., *Consultant*
White, Rolf B., *Radiant Tube Systems*
Wright, Richard H., P.E., *Armour and Cape, Inc.*
Yarosh, Marvin M., P.E., *Florida Solar Energy Center*
Zeman, Suzanne, *Resource Management Associates*

ix

ACKNOWLEDGMENTS

Appreciation is expressed to all those who have contributed their expertise to this volume, to the conference chairmen for their outstanding contribution to the 5th World Energy Engineering Congress, and to the officers of the Association of Energy Engineers for their help in bringing about this important conference.

INTRODUCTION

The deepening concern of energy managers to reduce energy costs and improve productivity is reflected in this comprehensive reference book, "Advances In Energy Productivity."

This text represents the latest thinking with regard to subjects such as cogeneration, fluidized bed combustion, waste heat recovery, combustion controls and energy management. Sections deal with improving energy productivity, lighting utilization, energy accounting systems and water management.

It is hoped that this comprehensive collection of state of the art applications will serve as a valuable tool for the dissemination of information to the professional community.

SECTION 1
BOILER OPTIMIZATION

Chapter 1

BOILER EFFICIENCY IMPROVEMENT PROJECTS RESULTING FROM ENGINEERING AUDITS

C. F. Feledy

INTRODUCTION

The purpose of this paper is to present a case study of a successful corporate program for conducting engineering audits of boilers to define efficiency improvement projects, and thus, reduce fuel costs and improve corporate profitability. It is not intended to be a technical discussion of the mechanics of implementing each type of project; rather, it stresses an analysis of the audit results from a management viewpoint.

United Technologies Corporation conducted engineering audits in 1979 and 1980 of 22 boilers located in 12 major facilities in order to define boiler efficiency improvement projects. This effort represented one of the major elements in UTC's successful on-going energy conservation program that had received awards from two national professional organizations (the Association of Energy Engineers and the National Energy Resources Organization) for having the best industrial conservation program in the United States. Through 1980, UTC has reduced the energy content of its products, as measured in Btu's per unit of output by 55% versus a 1972 baseline.

The boilers audited collectively represented a rated steam generation capacity of 2,113,000 pounds per hour. They ranged in size from 17,500 to 250,000 pounds per hour. UTC is not an energy intensive company like the steel or chemical industries, since it spends only about 1.3 cents out of every sales dollar on energy. However, in absolute terms, it does have a large energy bill; in 1981, UTC spent over $176 million for all forms of energy. To put this large number in perspective, these energy dollars were 49% larger than UTC's 1981 dividend payments on its common stock. When measured in energy units, UTC's 1981 energy consumption was 22.2 trillion Btu's. It is estimated that boiler fuel represents about 25% of UTC's energy costs, or namely, about $44 million in 1981. Although our divisions and groups had already done an excellent job in improving the thermal efficiency of their boilers, in view of this large fuel bill, corporate management felt that it was desirable to use an engineering consultant to conduct formal audits of boiler operations at selected major facilities in order to "fine-tune" boiler performance, upgrade operator skills, and thus, lower fuel costs and improve corporate profitability.

The Boiler Efficiency Institute of Auburn, Alabama, was selected to conduct the audits because of its excellent national reputation and extensive auditing experience. Through 1981, the Institute has audited over 800 boilers and conducted about 225 seminars and training sessions (with 15,000 attendees) for boiler operators and powerhouse engineers and supervisors. All of the UTC audits were performed by Thomas E. Burch, who holds a Master's degree in Mechanical Engineering from Auburn University. Mr. Burch worked under the supervision of David F. Dyer, Professor of Mechanical Engineering at Auburn University and President of the Boiler Efficiency Institute. Mr. Burch has extensive experience in testing boilers inasmuch as he personally has audited over 200 units.

Nine of the twelve facilities audited are located in Connecticut since the bulk of UTC's machining and electronics manufacturing operations are found there. One facility is located in New York, another in Florida and the last one is in Canada.

DESCRIPTION OF PROJECTS

As previously stated, the mechanics of implementing the eight basic types of boiler improvement projects identified will be only briefly described. A more detailed description is contained in the referenced manual and other conservation texts.[1]

A brief description of the eight types of improvement projects follows:

Load Management

This technique is applicable to only plants having two or more boilers and is very simple in concept: each boiler should be operated at the firing rate that yields the optimum point on its own thermal efficiency versus load curve. Operation of multiple boilers in this fashion will maximize efficiency of the overall system.

Air Preheating

Waste heat from the boiler and its exhaust stack is transferred to the boiler room and rises to its ceiling. This waste heat can be recovered by relocating the boiler air inlet ducts near the ceiling, thus utilizing this preheated air as combustion air.

Blowdown Recovery

Most of the waste heat generated by boiler blowdown can be recovered by installing a simple system that is basically comprised of a steam flash tank and a counter-flow heat exchanger. The recovered heat is used to preheat boiler feedwater.

Fix Air Leaks

Infiltration air will enter the combustion chamber through cracks or ports in the boiler, causing excess air conditions which in turn will reduce combustion efficiency. It is relatively easy, using conventional repair methods, to seal these openings.

Add Improved Combustion Controls

This project is largely self-explanatory. Aging control systems can be readily replaced with modern versions that will do a much better job of controlling the fuel-air ratio at the burner.

Air Atomization

Since compressed air costs less than steam to generate, it is more economical to use compressed air to atomize fuel oil than steam.

Economizer/Fouling Removal

If stack temperatures are excessive, an economizer can be installed in the stack to recover the waste heat and use it to heat boiler feedwater. Another cause of excessive stack temperatures is in the fouling of heat transfer surfaces in the boiler itself. Removal of fouling will reduce stack temperatures, and thus, improve thermal efficiency.

Reduce Excess Air

Excess air results in reduced thermal efficiency. There are numerous ways to reduce excess air, the major ones being: (1) "tuning" boilers, (2) using oxygen analyzers, (3) installing advanced fuel-air control systems, etc. Other methods are described in the referenced manual.[1]

PLANT IMPROVEMENT PROJECTS

In this section, boiler efficiency projects for each of the 12 plants are described. More detailed information on project costs and savings is contained in Exhibit I.

BOILER EFFICIENCY IMPROVEMENT PROJECTS - BY PLANT   EXHIBIT I

EFFICIENCY IMPROVEMENT PROJECT	ANNUALIZED SAVINGS (1979 $)	ESTIMATED IMPLEMENTATION COST (1979 $)	SIMPLE PAYBACK PERIOD (YEARS)
Plant A (New York - machining and assembly operation)			
Excess Air Control	44,718	50,000	1.12
Economizer/ Fouling Removal	9,438	20,000	2.12
Air Preheating	5,346	5,000	0.94
Other: Fix Leaks	40,000	90,000	2.25
Subtotals:	99,502	165,000	1.66
Plant B (Connecticut - development lab)			
Excess Air Control	370,834	80,000	0.22
Economizer/ Fouling Removal	110,623	190,000	1.72
Load Management	77,000	–	immediate
Air Atomization	58,200	25,000	0.43
Subtotals:	616,657	295,000	0.48

Plant C (Connecticut - electronics manufacturing)			
Load Management	21,000	–	immediate
Air Atomization	16,443	1,000	0.06
Subtotals:	37,443	1,000	0.03
Plant D (Canada - machining and assembly operation)			
Excess Air Control	42,788	40,000	0.93
Subtotals:	42,788	40,000	0.93
Plant E (Connecticut - machining and assembly oper.)			
Excess Air Control	92,035	100,000	1.09
Economizer/ Fouling Removal	21,396	40,000	1.87
Load Management	22,000	–	immediate
Air Preheating	6,000	5,000	0.83
Air Atomization	16,800	10,000	0.60
Subtotals.	158,231	155,000	0.98
Plant F (Connecticut - machining and assembly oper.)			
Excess Air Control	159,114	120,000	0.75
Economizer/ Fouling Removal	83,784	80,000	0.96
Blowdown Recovery	10,612	10,000	0.94
Air Preheating	8,250	5,000	0.61
Air Atomization	12,000	10,000	0.83
Subtotals:	273,760	225,000	0.82
Plant G (Connecticut - machining and assembly oper.)			
Economizer/ Fouling Removal	54,846	127,446	2.32
Air Atomization	16,800	10,000	0.60
Other: Combustion Controls	62,420	100,000	1.60
Subtotals:	134,066	237,446	1.77

Plant H (Connecticut - machining and assembly oper.)		
Load Management	30,000	-	immediate
Air Preheating	9,500	5,000	0.53
Air Atomization	23,558	15,000	0.64
Subtotals:	63,058	20,000	0.32

Plant I (Connecticut - machining operation)		
Excess Air Control	17,468	15,000	0.86
Air Preheating	5,400	5,000	0.93
Air Atomization	16,404	10,000	0.61
Subtotals:	39,272	30,000	0.76

Plant J (Connecticut - machining operation)		
Excess Air Control	12,000	10,000	0.83
Air Atomization	5,116	10,000	1.95
Subtotals:	17,116	20,000	1.17

Plant K (Florida - development center)		
Excess Air Control	205,000	710,000	3.5
Air Atomization	10,000	10,000	1.0
Subtotals:	215,000	720,000	3.3

Plant L (Connecticut - electronics manufacturing)		
Economizer/ Fouling Removal	26,250	25,000	0.95
Blowdown Recovery	5,968	6,000	1.00
Subtotals:	32,218	31,000	0.96
TOTALS:	1,729,111	1,939,446	1.12

Plant A

Reducing excess air represented the project having the greatest savings ($44,718); its payback period was 1.12 years. It was followed by the fixing air leaks project that had savings of $40,000 and a payback period of 2.25 years. However, the project having the best payback period (0.94 years) was air preheating, but it had the least savings, namely, $5,346. In total, the four projects identified had cumulative annualized savings of $99,502 and an associated implementation cost of $165,000, thus yielding an overall payback period of 1.66 years.

Plant B

Reducing excess air was also the project having the largest savings ($370,834); its payback period was 0.22 years. The economizer/fouling removal project ranked second with savings of $110,623; however, its payback was considerably longer at 1.72 years. Improved load management had the best payback of all projects since it required no implementation cost. The four projects had cumulative annualized cost savings of $616,657, while implementation costs were $295,000, representing a payback period of 0.48 years.

Plant C

With $21,000 in annual savings, improved load management represented the largest savings. Since it required no implementation costs, this project had an immediate payback. The only other project, air atomization, had savings of $16,443 and a payback period of .06 years. The two projects had total savings of $37,443, but implementation costs were only $1,000, thus yielding a payback of .03 years, that is the smallest period of all 12 plants.

Plant D

The only significant project identified was reducing excess air that had savings of $42,788. Its associated payback period is 0.93 years.

Plant E

Reducing excess air was the project having the greatest savings ($92,035); its payback period was 1.09 years. It was followed by the improved load management project that had savings of $22,000 and an immediate payback. In total, the five projects identified had savings of $158,231, with an overall payback period of 0.98 years.

Plant F

The leading project for savings was reducing excess air, worth $159,114; its payback period was 0.75 years. It was followed by the economizer/fouling removal project that had savings of $83,784 and a payback period of 0.96 years. Five projects were defined and had total savings of $273,760 and an overall payback period of 0.82 years.

Plant G

Installing improved combustion controls produced the greatest savings ($62,420) and had a payback period of 1.60 years. The economizer/fouling removal project ranked second with savings of $54,846 with a payback period of 2.32 years. In total, the three projects had savings of $134,066 with an overall payback period of 1.77 years.

Plant H

The leading project for savings was optimizing load management, worth $30,000, and its payback was immediate. It was followed by the air atomization project that had savings of $23,558 and a payback period of 0.64 years. Three projects were defined and had total savings of $63,058 and an overall payback period of 0.32 years.

Plant I

Reducing excess air produced the greatest savings ($17,468) and had a payback period of 0.86 years. The air atomization project ranked second with savings of $16,404 with a payback period of 0.61 years. In total, three projects had savings of $39,272 and an overall payback period of 0.76 years.

Plant J

Reducing excess air was the project having the largest savings ($12,000); its payback period was 0.83 years. The other project, air atomization, had savings of $5,116 and a payback period of 1.95 years. These two projects had total savings of $17,116 and an overall payback period of 1.17 years.

Plant K

The leading project for savings was reducing excess air, worth $205,000, and its payback was 3.5 years. It was followed by the air atomization project that had savings of $10,000 and a payback period of 1.0 years. These were the only two projects identified. They had total savings of $215,000 and an overall payback of 3.3 years that was the longest period of all 12 plants.

Plant L

The project having the largest savings was economizer/fouling removal with $26,250; its payback period was 0.95 years. The other project, blowdown recovery, had savings of $5,968 and a payback of 1.0 years. These two projects had total savings of $32,218 and an overall payback of 0.96 years.

RESULTS OF AUDITS

The consultant identified numerous projects for improving boiler efficiency at each of the 12 plants audited. The number of projects per plant ranged from a high of five to a low of one. A total of 36 efficiency improvement projects were identified, and they are described by plant, in Exhibit I, also.

Total annual savings resulting from the installation of these projects is $1,729,111, while implementation costs (capital plus expense dollars) are estimated to be $1,939,446. Thus, the overall simple payback period for UTC's investment in these projects is a very attractive 1.1 years. It should be noted that UTC uses the simple payback period as one of the criteria used to preliminarily rank proposed capital projects to establish relative priorities. When it is finally decided to prepare a capital appropriation request, discounted-cash-flow, return-on-investment analysis calculations are used to make the final appropriation decision.

The improvement projects have been ranked in descending order of estimated savings in Exhibit II.

ANALYSIS OF PROJECT SAVINGS EXHIBIT II

Rank	Project Description	Number of Projects	Savings ($)	% of Total	Cum. % of Total
1	Reduce Excess Air	8	$ 943,957	54.6%	54.6%
2	Add Economizer/ Fouling Removal	6	306,337	17.8	72.3
3	Air Atomization	9	175,321	10.0	82.4
4	Load Management	4	150,000	8.7	91.1
5	Other: Add Combustion Controls	1	62,420	3.6	94.7

6	Other: Fix Air Leaks	1	40,000	2.3	97.0
7	Air Preheating	5	34,496	2.0	99.0
8	Blowdown Recovery	2	16,580	1.0	100.0
	Totals:	36	$1,729,111	100.0	-----

Reducing excess air clearly ranks first in terms of estimated savings of $943,957, which represents 54.6% of total savings. Ranking second is add economizer/fouling removal with savings of $306,337, or 17.8% of total savings. Air atomization is third with savings of $175,321 (10.0%) of total savings.

Together, the first and second projects constitute a cumulative 72.3% of total savings. Also, the top-three ranked projects account for 82.4% of total savings. The remaining five projects account for only 17.6% of total savings.

Of the 36 projects defined, the length of the payback period ranged from "immediate" for four load management projects (at Plants B, C, E, and H) to 3.5 years for a reducing excess air project at Plant K. A frequency distribution of the 36 projects with respect to payback period is shown in Exhibit III.

FREQUENCY DISTRIBUTION OF PROJECT PAYBACK PERIODS EXHIBIT III

Simple Payback Period (Years)	Projects		
	Number	% of Total Number	Cum. % of Total Number
0-0.5	13	36.1	36.1
0.6-1.0	7	19.4	55.5
1.1-2.0	15	41.7	97.2
>2	1	2.8	100.0
Totals:	36	100.0	-----

Thus, it is seen that 20 of the 36 projects, or 55.5% of the total, have payback periods not exceeding one year. Also, only one project exceeds a two-year payback. It can be concluded that almost all of these projects are cost-effective and would meet the capital appropriation "hurdle" rates of most companies for proposed new investments. An "immediate" payback means that since the project requires no investment cost to be implemented, the desired results can be obtained immediately.

An analysis of the average length of the payback period by type of project is shown in Exhibit IV.

ANALYSIS OF PROJECTS BY AVERAGE PAYBACK PERIOD EXHIBIT IV

Rank	Type of Project	Average Simple Payback Period (Years)
1	Load Management	immediate
2	Air Atomization	0.58
3	Air Preheating	0.72
4	Blowdown Recovery	0.97
5	Reduce Excess Air	1.19
6	Add Economizer/Fouling Removal	1.57
7	Other: Add Combustion Controls	1.60
8	Other: Fix Air Leaks	2.25

Thus, load management ranks first with the shortest
payback period and fixing air leaks ranks last. Also,
air atomization, air preheating and blowdown recovery
each have average payback periods less than one year.

An analysis of the 36 projects for implementation
costs by type of project is shown in Exhibit V.

ANALYSIS OF IMPLEMENTATION COSTS
BY TYPE OF PROJECT   EXHIBIT V

| | | Implementation Costs | | |
| | | | % of | Cum. % of |
Rank	Type of Project	Cost	Total Cost	Total Cost
1	Load Management	$ 0	0%	0%
2	Blowdown Recovery	16,000	0.8	0.8
3	Air Preheating	25,000	1.3	2.1
4	Other: Fix Leaks	90,000	4.6	6.7
5	Other: Add Combustion Controls	100,000	5.2	11.9
6	Air Atomization	101,000	5.2	17.1
7	Economizer/Fouling Removal	482,446	24.9	42.0
8	Reduce Excess Air	1,125,000	58.0	100.0
	Totals:	$1,939,446	100.0%	-----

Load management has zero implementation cost, and
hence, ranks first. Also, blowdown recovery and air
preheating are relatively inexpensive and rank second
and third. Reducing excess air is the single most
expensive type of project; this type costs $1,125,000
and represents 58.0% of total implementation costs.

SUMMARY

This case study has demonstrated that the use of well-
qualified engineering consultants to conduct audits
of boiler performance is a very cost-effective way to
reduce fuel costs and improve corporate profitability.
Thus, it is recommended that boiler audits be included
as one of the major elements in a corporate conservation program for those companies having significant
boiler facilities.

REFERENCES

1. Measuring and Improving the Efficiency of Boilers
    (Operator Manual), by David F. Dyer and Glennon
    Maples, Boiler Efficiency Institute, P.O. Box
    2255, Auburn, Alabama.

Chapter 2

IMPROVING SEASONAL EFFICIENCY OF COMMERCIAL GAS FIRED HEATING BOILERS

J. R. Cairns

ABSTRACT

The paper evaluates the effectiveness of a boiler retro-fit procedure designed to improve the efficiency of gas fired heating boilers. The procedure is shown to produce significant improvement in seasonal efficiency by increasing combustion efficiency in the burner on-cycle and by decreasing standby losses in the burner off-cycle.

The major causes of energy loss in gas fired heating boilers—excess combustion air and convective flue losses—are described. Typical values of on-period and off-period losses are given for commercial heating boilers. The installation procedure for a unique retro-fit is described, including the method of determining the optimum volume of excess combustion air. Experimental data are given from field measurements after retrofit which show the increase achieved in steady state combustion efficiency, and the decrease in off-period flue losses.

The pre- and post-retrofit data are used in a simulation developed by the National Bureau of Standards for estimating the seasonal performance of heating systems. This simulation procedure involves 1) measuring flue gas temperature and O_2 concentration during steady-state boiler operation; 2) obtaining information on the shape of the flue gas temperature-time curves as the appliance heats up and cools down; 3) assigning appropriate values to various factors which describe the off-cycle air flow rate through the flue and stack, the degree of boiler oversizing, and the effect of boiler operation on infiltration; and 4) carrying out a calculation procedure which predicts the seasonal efficiency.

INTRODUCTION

The objectives of this study were to (1) analyze experimentally the operating effects of a retrofit procedure designed to improve the efficiency of commercial gas fired heating boilers and (2) use a computer simulation to estimate the percentage improvement in seasonal efficiency produced by the retrofit.

The subject of improving the efficiency of gas fired heating appliances has received attention recently by a number of researchers whose efforts were directed primarily at residential furnaces and boilers. Their efforts were aimed at identifying and quantifying all of the causes of heating system inefficiency and determining the effects of various heating system modifications on the steady state and annual efficiencies of the systems [1-10].[1] One of the significant results of this research was the identification of typical values of energy losses and typical system parameters which affect efficiency, as determined by extensive laboratory and field tests [4]. Recommended procedures were developed

[1]Numbers in brackets designate references at the end of this paper.

for standard test methods by which manufacturers of residential furnaces and boilers would calculate the annual fuel utilization efficiencies (AFUE) of their consumer products [9, 10]. (The Annual Fuel Utilization Efficiency is the fraction of total annual energy input that is delivered as useful heat during the heating season.) The final test procedures adopted by the federal government are those developed at the National Bureau of Standards (NBS) [9] which appeared in the Federal Register [11]. All residential furnaces and boilers now sold in the United States have AFUE's computed by the NBS computer simulation. This requirement does not extend to commercial or industrial furnaces or boilers, designated as those having inputs greater than 300,000 Btu/hr.

These developments are now having a significant impact on the market for residential heating appliances. New residential furnace and boiler models are appearing regularly now, some with annual heating efficiencies as high as 95%, an increase of over 50% above the typical annual efficiencies of just a few years ago. It is unfortunate that federal law does not require an efficiency label to be placed on the appliance, only that the dealer have the efficiency information available should the customer inquire about it. But, even with this omission, informed consumers will eventually produce market pressures to force even the most reluctant manufacturer to bring high efficiency residential appliances to the market.

There has been less progress in the retrofit market for residential appliances. The notable exception is the vent damper which reduces heat loss up the chimney when the appliance is in the off-period (after combustion has ceased). There has been little effort to develop retrofit modifications to the heating appliance itself because of limitations posed by manufacturers' warrantees and certification requirements of the American Gas Association (AGA). It appears now that replacement of an existing appliance, even a relatively recent furnace or boiler in good working condition, with a high efficiency model is a better investment than most retrofit procedures.

There has been very little effect of the research findings on the market for new commercial heating appliances, most of which are boilers. There are serious impediments to achieving large effects in this market. Most important, federal law exempts manufacturers of appliances over 300,000 Btu/hr input from conducting the AFUE test and providing the results to customers. In current sales brochures for their new commercial sized boilers, most manufacturers thus display an industry-accepted efficiency "rating" of 80% for every product, regardless of the units' steady state efficiency. Keeping in mind that the purchasing decision for commercial heating appliances is usually based on an impersonal and competitive bidding procedure, the use of an identical performance "rating" for all prospective equipment will do little to stimulate the development of high efficiency commercial appliances. Even when higher efficiency models do become available, there will be substantial resistance to replacement because of the high costs involved, particularly the high labor costs for dismantling and reinstallation in large size piping systems.

Finally, and very unfortunately, there has been virtually no application of new research results to retrofitting commercial boilers to improve efficiency. This is clearly an important area for emphasis by energy engineers: There are hundreds of thousands of these appliances now operating with low efficiency. The per-unit consumption of energy is often in the millions of Btu/hr range so savings potential is very high. The expected resistance to replacement because of high cost makes retrofitting a more likely source of energy savings than purchase of energy efficient new models when they become available, at least for the next decade.

This paper describes some of the author's results in (1) comparing energy losses of commercial heating boilers to those of residential furnaces and boilers, (2) retrofitting an atmospheric boiler to achieve higher steady-state efficiency and reduced off-period losses, and (3) predicting the improvement in seasonal efficiency using the NBS computer simulation.

TYPICAL LOSSES IN ATMOSPHERIC GAS FIRED BOILERS

Figures 1 and 2 show, respectively, schematic views of an atmospheric-fired boiler with an integral draft diverter and an external draft hood.

Definitions

Atmospheric Burner - a gas burner in which air for combustion is supplied entirely by natural draft and the inspirating force created by gas velocity through an orifice. (A power burner supplies combustion air by an electrically powered blower.)

Flue Gases - products of combustion including the inert gases and excess air which exit from the combustion chamber.

Flue - a conduit through which flue gases pass from the combustion chamber to the point of draft relief provided by a draft hood or draft diverter opening.

Draft Relief - an opening in the flue, provided by the draft diverter or draft hood, through which dilution air enters.

Draft Diverter - an integral part of a furnace or boiler which provides a side opening in the venting system such that it (1) provides for the exhaust of the products of combustion in the event of no draft, back draft, or stoppage beyond the draft diverter, (2) prevents a back draft from entering the combustion chamber, and (3) neutralizes the effect of stack action of the chimney upon the operation of the furnace or boiler, particularly so that factory settings of excess air can be maintained in the field.

Draft Hood - a device which performs the same functions as a draft diverter but is not an integral part of the furnace or boiler, but which is connected to the furnace or boiler by a short length of flue pipe.

Stack - the portion of the exhaust system downstream of the draft diverter or draft hood.

Stack Gases - the flue gases plus the dilution air which enters the draft diverter or draft hood through the side opening.

Vent Damper (Stack Damper) - a device installed downstream of the draft diverter or draft hood and which is designed to automatically open the venting system (stack) when the appliance is in operation and to automatically close the venting system when the appliance is in a standby condition.

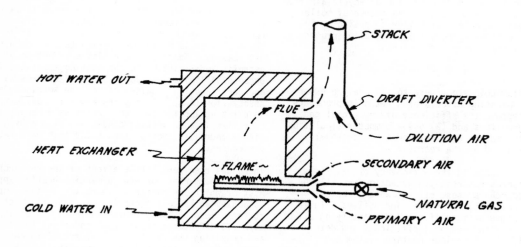

FIGURE 1. SCHEMATIC VIEW OF GAS-FIRED ATMOSPHERIC BOILER WITH DRAFT DIVERTER.

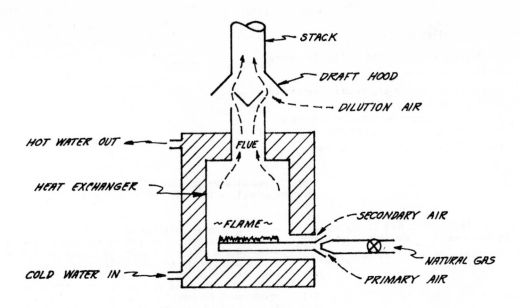

FIGURE 2. SCHEMATIC VIEW OF GAS-FIRED ATMOSPHERIC BOILER WITH DRAFT HOOD.

TABLE 1

DEFINITION OF THERMAL LOSSES ASSOCIATED WITH COMBUSTION HEATING
SYSTEMS AND TYPICAL LOSS VALUES FOR GAS FIRED ATMOSPHERIC BOILERS

(T_F = flue temp., T_o = room temp., T_1 = outside temp.)

	Typical Values of Losses for Boilers Located Inside
Group I. Burner On-period Flue Losses	
L_W - latent heat in uncondensed water	10
L_N - enthalpy of the stoichiometric products of combustion (CO_2, H_2O, N_2) between T_F and T_o	6
L_E - enthalpy of the excess air (O_2, N_2) between T_F and T_o	2-6
L_C - unburned combustibles in the flue	0
L_F - enthalpy of fuel inerts between T_F and T_o	0
Group II. Burner Off-period Flue Losses	
L_D - enthalpy of draft air flowing through the combustion chamber between T_F and T_o	11
Group III. Boiler Contribution to Building Infiltration Losses	
L_A - enthalpy of stoichiometric combustion air between T_o and T_1	1
L_{EA} - enthalpy of excess air between T_o and T_1	1
L_{DA} - enthalpy of draft air (which eventually flows through the combustion chamber) between T_o and T_1	2
$L_{H,ON}$ - enthalpy of dilution air between T_o and T_1 during the burner on-period	3
$L_{H,OFF}$ - enthalpy of dilution air (stack air) between T_o and T_1 during the burner off-period	1

TABLE 1 (cont'd)

Group IV. Miscellaneous Losses and Gains	Typical Values of Losses for Boilers Located Inside
L_L - pilot flame energy wasted during the Summer	1-2
L_J - boiler jacket losses to unheated space	0
L_V - energy wasted by the distribution system to unheated space	(0-15)
L_M - enthalpy of evaporated water in a furnace humidifier system (not applicable to boilers)	0
G_S - heat gain from stack back to building	(0-2)
Total, typical average annual loss	40%
Average annual system efficiency	60%

It should be noted that the conventional steady state efficiency, η_{SS}, as measured in a laboratory test using ANSI standards [12] includes only the losses L_W, L_N, L_E, L_C, or

$$\eta_{SS} = 1 - L_W - L_N - L_E - L_C \qquad (1)$$

Similarly, on Orsat or similar apparatus which uses CO_2 (or O_2) content of the flue gases plus net stack temperature ($T_F - T_O$) to determine efficiency in field measurements also determines the same η_{SS}, or combustion efficiency as it is usually called. (Some of the newer types of digital combustion efficiency computers calculate η_{SS} but then subtract a fixed amount, usually 3%, for jacket loss.)

From a retrofitting point of view, η_{SS} is important but not adequate information; the objective is to increase efficiency in every way possible, so an estimate is

needed of the overall, annual fuel utilization efficiency, including the losses due to cooldown of the boiler water in the off-period under part load conditions, wasted pilot heat, and all infiltration losses, before and after the retrofit. It would be impossible, of course, to determine such an overall efficiency by measurement because the value would change from day to day as outside temperature varies. The NBS computer simulation satisfies this need. It provides a means of estimating annual fuel utilization efficiency by dynamically simulating the start-up and cooldown fuel losses, and by using laboratory and field tested ratios to predict other off cycle losses such as the enthalpy of dilution air and of the draft air flowing through the combustion chamber under a typical part load operating cycle. The simulation also predicts the effect on annual efficiency losses by such factors as summer pilot loss, using electronic ignition, use of vent dampers, etc.

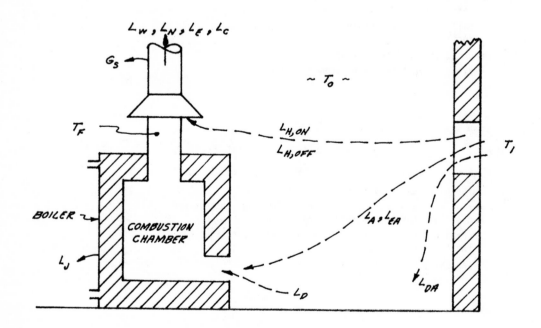

FIGURE 3. THERMAL LOSSES ASSOCIATED WITH HEATING SYSTEM.

RETROFITTING TO REDUCE LOSSES

Any retrofitting procedure must be aimed specifically at reducing one or more of the system losses described earlier. A number of these losses are, of course, not reducible, such as L_A, the enthalpy of stoichiometric combustion air from T_0 and T_1. The following list identifies those items which are not useful targets for achieving efficiency improvements by retrofitting commercial atmospheric boilers.

L_W, latent heat loss, is unrecoverable because any condensation produced will be acidic and will attack both metal parts and masonry.

L_C, unburned combustibles, is of negligible importance in properly tuned gas boilers because there will be complete combustion of hydrocarbons and there is zero or almost zero carbon monoxide in the flue gases.

L_A, enthalpy of stoichiometric combustion air from T_0 and T_1 is unavoidable.

$L_{H,ON}$, enthalpy of on-period dilution air, is unadvoidable because draft relief openings are required on atmospheric gas appliances.

L_L, summer pilot flame loss, is a very small quantity on large boilers so the cost of an electric ignition retrofit is usually not justified.

L_J, jacket losses, are already low enough to make further boiler insulation not economically feasible, and for interior boiler rooms this loss can be used to maintain the room at a comfortable level without supplementary heat.

L_V, heat loss from pipes, is usually small in commercial systems because effective pipe insulation is almost always in place, especially where pipes pass through unheated areas.

L_F, inert loss, is negligible in natural gas.

The remaining losses can be categorized as either on-period or off-period losses. A rapidly closing and tight sealing vent damper (installed downstream of the relief opening) provides an effective solution to most off-period losses. It stops $L_{H,OFF}$ by simply closing the venting system. It also stops the draft air loss, L_D, from leaving the boiler room up the stack, which in turn eliminates the infiltration loss, L_{DA}. The heat in L_D will spill out the relief opening with such a damper in place, thus warming the boiler room. It would be useful also to install an automatic flue damper (upstream of the relief opening) to trap the residual heat in the boiler and prevent L_D from spilling into the room. Such dampers are not, in general, permitted by local heating inspectors and, to the author's knowledge, no such dampers have been certified by AGA.

The remaining losses, L_N, L_E, L_{EA}, are on-period losses. The first, L_N, is the sensible heat lost in the form of high temperature flue gases and is often, mistakenly, considered unavoidable because it is thought to be caused only by poor boiler design. Poor heat exchanger design, or excessive input will, of course, result in high flue temperatures as will dirty conditions on both the fire side and water side of the boiler, but excess air also contributes to high flue temperatures. Thus, L_N is subject to partial reduction by limiting excess

air, and both L_E and L_{EA} are in their entirely caused by excess air.

Losses Caused by Excess Combustion Air

Reducing *excess air* is the first priority for reducing lost energy and thus increasing boiler efficiency. Reducing excess air has been shown to reduce losses in both the on-period and off-period through the following five effects:

1. It reduces the energy wasted to heat excess air, L_E, by decreasing the amount of excess air to be heated from T_0 and T_F.

2. It reduces the amount of infiltration air loss, L_{EA}, which must be heated during the on-period from T_1 to T_0.

3. Since excess air reduction is caused by flue baffling or other restriction to air flow through the boiler, it reduces the amount of off-period draft loss, L_D, caused by air flow through the combustion chamber.

4. The reduction in L_D causes a reduction in infiltration loss, L_{DA}.

5. Finally, a reduction in excess air increases the flame temperature, which increases heat transfer to the boiler water, which results in a decrease in the temperature of the flue products.

The last effect is less obvious than the others. A first impression is that less excess air means less combustion heat is wasted on heating excess air, so that amount of heat is available to heat the other flue products, and this might cause an *increase* in flue temperature. It is true that the flame temperature is higher with reduced excess air but an increase in flame temperature also means greater heat transfer to the water, thus reducing flue temperature. The magnitude of the temperature changes is illustrated by the following representative values of adiabatic flame temperature of methane in air (neglecting dissociation):

TABLE 2

EFFECT OF EXCESS AIR REDUCTION ON ADIABATIC FLAME TEMPERATURE AND % OXYGEN IN DRY FLUE PRODUCTS

% Excess Air	Adiabatic Flame Temperature	% O_2 in Dry Products
0	3727°F	0
50	2759°F	7.5
100	2205°F	11.1
150	1843°F	13.2

Since all of the heat transfer mechanisms from flame to boiler walls depend on flame temperature, the effect of excessive air in reducing heat transfer is seen to be an important cause of inefficiency.

THE CAUSE OF EXCESSIVE EXCESS AIR

Almost all of the air flow through the combustion chamber of an atmospheric appliance is caused by natural draft which is due principally to the bouyant effect of hot flue gases rising through a chimney. The boiler manufacturer provides for some excess air to be drawn into the combustion chamber by this draft because excess air gives assurance of complete fuel combustion and no carbon monoxide in the flue products. The manufacturer's factory tests and adjustment procedure are often done while venting the boiler into an open room, not into a chimney. Factory settings of excess air are thus based on a nominal draft produced by the short section of flue pipe used during this factory test.

Field measurements of identical boilers indicate how installation conditions can cause excessive air. In one test the author measured excess air levels in three identical boilers, each having the same 2,600,000 Btu/hr. input, same integral draft diverters, located in similar boiler room configurations, all operating in the same commercial neighborhood on the same day, but each connected to chimneys of varying materials, diameters, and heights (from 16' to 140' high). Over a prolonged operating period on the same day the boilers registered 7.6%, 9.7% and 13% flue oxygen concentrations, corresponding to approximately 50%, 75%, and 145% excess air. The manufacturer reported that the boilers had been factory adjusted for 4% oxygen level, or approximately 20% excess air. (It is interesting to review the adiabatic flame temperature data of Table 2 to appreciate the effects of these excess air variations.)

The author's measurements of excess air variations in a variety of boiler installations lead to the important conclusion that the draft diverter or draft hood does not fulfill its objective of *neutralizing the effect of stack action of the chimney upon the operation of the furnace or boiler, particularly so that factory settings of excess air can be maintained in the field.*

Failure of the draft hood or diverter to isolate the appliance is simply a result of its design, as is clear from the typical configurations shown in Figures 1 and 2.

There is abundant evidence from simulations and field measurements that there is a need for redesign of the draft hoods and draft diverters currently used in both new and retrofit installations. The present study involved a reconstruction of the draft diverter in one field installation to determine the effectiveness of one new draft hood design.

THE RETROFIT PROCEDURE AND RESULTS

Preliminary experience with the retrofit procedure and some preliminary data were described in an earlier report [13] covering a 2,600,000 Btu/hr. boiler with integral draft diverter, similar to Figure 1, which opened into a 26" diameter stack. The results that follow were obtained on that same boiler.

The following data describe the steady state operating characteristics prior to retrofit:

TABLE 3

STEADY STATE OPERATING CHARACTERISTICS PRIOR TO RETROFIT

Flue temperature: 670°F (gross temperature)

Flue oxygen content: 7.6% (no carbon monoxide)

Excess air: 50%

Flue gas velocity: not measurable because of diverter construction

Stack temperature: 310°F

Stack oxygen content: 15.3%

Stack gas velocity: 650 fpm

η_{ss}, steady state efficiency: 72.1%

The NBS simulation requires transient temperature measurements from which a dynamic operating model is constructed. The following temperature-time data were taken from this boiler installation to provide the necessary input to the simulation. (The stack velocity data is not used in the simulation. It is given here only for reference.)

TABLE 4

TRANSIENT OPERATING CHARACTERISTICS PRIOR TO RETROFIT

Flue Temperature (Gross)		Stack Velocity (fpm)
Steady state: 670°F		650
3.75 minutes after shutdown:	213°F	440
22.5 minutes after shutdown:	173°F	425
45 minutes after shutdown:	162°F	420
Start-up at 45 minutes after shutdown		
1 minute after start-up:	650°F	650
5.5 minutes after start-up:	670°F	650

The NBS simulation for this boiler, which uses a year-around standing pilot and has no vent damper, predicted the annual fuel utilization efficiency prior to retrofit to be 58.7%.

The retrofit procedure consists of sealing off the integral draft diverter and installing a new type of "extended draft hood" as shown schematically in Figure 4.

There are two principal design features of this new hood:

1) the flue gases follow an inverted path which ends at a new relief opening. The height of the relief opening above the floor is adjustable by varying the length of the inlet duct. By adjusting the height of the relief opening one can adjust the combustion chamber draft and thereby control the amount of excess air drawn into the combustion chamber.

2) the inverted flue path forms a trap for hot flue gases during the off-periods. Shortly after the burner shuts off the flue gas velocity drops rapidly to zero and the hotter gases remain trapped in the highest part of the flue passage.

The primary objective of the new hood is to better isolate the combustion chamber from the variable draft action of the stack. The installation procedure involves adjusting the height of the diverter opening, with the burner operating, so as to obtain the desired excess air before connecting the hood to the chimney. An excess air level established under this condition is that produced by the minimum draft that the boiler will experience because the boiler room temperature is higher than will ever be experienced outside during the heating season. A proper adjustment is one where excess air is at the minimum level possible without producing carbon monoxide. Both continuous display O_2 and CO detectors are used to set the hood opening at the desired height.

Whether the hood is effective in isolating the appliance from varying stack action is determined by connecting the hood to the chimney and monitoring excess air levels during varying outside temperatures during the heating season.

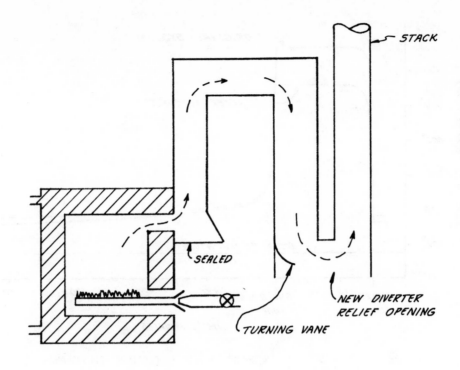

STACK

SEALED

NEW DIVERTER
RELIEF OPENING

TURNING VANE

FIGURE 4. SCHEMATIC VIEW OF RETROFIT INSTALLATION.

For this retrofit, the following data describe steady state operation before and after stack connection:

TABLE 5

STEADY STATE OPERATING CHARACTERISTICS
AFTER RETROFIT

Flue Conditions with Stack Not Connected	Flue Conditions with Stack Connected
Flue temperature: 560°F (gross)	565°F (gross)
Flue oxygen content: 2% (no carbon monoxide)	3% (no carbon monoxide)
Excess air: 8-10%	13-15%
Flue gas velocity: 280 fpm	300 fpm
η_{SS}, steady state efficiency: 79%	78.7%

These results were obtained on a day when the boiler room temperature was approximately 75-80°F and the outdoor temperature was 10-15°F. It is clear that there is a small effect of stack action on the excess air flow, producing an increase in oxygen from 2% to 3%, but this is negligibly small compared to the original installation. Periodic checks made throughout the heating season resulted in oxygen levels between 2-3% at all times. The 6.6% improvement in steady state efficiency from the original 72.1% to 78.7% is an increase of 9.2%.

The transient test was repeated after retrofit and provided the following data:

TABLE 6

TRANSIENT OPERATING CHARACTERISTICS
AFTER RETROFIT

Flue Temperature (Gross)	Flue Velocity (fpm)	Stack Velocity (fpm)
Steady state: 565°F	300	700
3.75 minutes after shutdown: 218°F	0-50	650
22.5 minutes after shutdown: 189°F	0-50	650
45 minutes after shutdown: 180°F	0-50	650
Start-up at 45 minutes after shutdown		
1 minute after start-up: 530°F	300	700
5.5 minutes after start-up: 550°F	300	700

A comparison between the transient operating characteristics is shown in Figure 5.

The NBS simulation for the retrofitted boiler predicts an annual fuel utilization efficiency of 67.7%, an increase of 15.3% over the original 58.7%.

Although no vent damper was installed as part of this retrofit, the simulation was run for the retrofitted boiler with a vent damper of the fast closing electric type which closes off 97% of the stack area. The simulation predicted an annual efficiency of 76.7%, an increase of 31% over the 58.7% annual efficiency of the original installation.

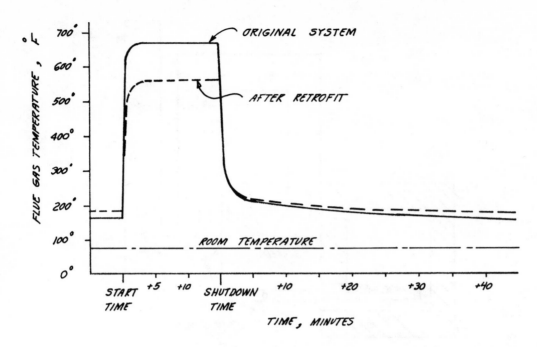

FIGURE 5. COMPARISON BETWEEN TRANSIENT FLUE TEMPERATURES BEFORE AND
AFTER RETROFIT.

CONCLUSIONS

It is clear that the retrofit did suceed in controlling excess air better than the original diverter. It is clear also that an excess air level of approximately 10% can be maintained in this boiler without creating adverse combustion conditions. It seems reasonable to think of the field adjustment of excess air done in this retrofit as a fine tuning of the boiler which should never be done at the time of manufacture because the boiler manufacturer cannot anticipate the installation circumstances of each of his boilers. Because the excess air input is trimmed as much as possible without producing carbon monoxide, it is necessary to seal the burner pressure regulator to avoid increasing the input, and a caution tag should be attached to the boiler describing how oxygen and carbon monoxide levels must be monitored during any subsequent adjustments of the burner or venting systems.

Figure 5 shows that two desired effects were achieved. There is a reduction of 105°F in on-period flue temperature that results from a reduction in excess air, thus decreasing losses L_N and L_E. After shutdown, the boiler water temperature cools slower after retrofit because of the heat trap formed in the top of the inverted flue path. In the NBS simulation, the shape of the cooldown and heat-up curves are used to determine time constants for the boiler from which a dynamic model is created that cycles on and off. The particular assumptions of this model are that the boiler is oversized by 70% and operating at an outside temperature equivalent to half load. The projected increase of 9% in annual efficiency, from 58.7% to 67.7% for this retrofit, consists of a 6.6% increase in steady state efficiency and a 2.7% increase attributed to the off-period retention of boiler heat by the trap.

The actual increase in efficiency caused by a retrofit procedure of this type will, of course, depend on the initial conditions of installation, i.e., whether the unit is running with great amounts of excess air. The actual savings achieved by the retrofit will depend on the duty cycle of the actual boiler which, in turn, depends on climate, amount of oversizing, conditions before retrofit, etc.

The reduction of 105°F in steady state flue gas temperature should not lead to significant additional condensation in the heat exchanger, particular considering the higher internal temperatures that will obtain because of slower cooling. There is, however, the possibility of additional condensation in the heat exchanger in the off-period because of the water vapor which will be trapped in the inverted flue path. Such condensation would occur only when the internal temperature falls below the dew point. The heat exchangers of sample units should be monitored after retrofit to determine if corrosion is increased.

REFERENCES

1. Bonne, U., Janssen, J. E., Nelson, L. W., and Torborg, R. H., "Control of Overall Thermal Efficiency of Combustion Heating Systems," 16th International Symposium on Combustion, MIT, Cambridge, Massachusetts, Aug. 15-22, 1976, Proceedings, p. 37.

2. Bonne, U. and Johnson, A. E., "Thermal Efficiency in Non-Modulating Combustion Systems," sponsored by NBS and ASHRAE, Purdue Univ., Lafayette, Indiana, Oct. 7-9, 1974, Proceedings, p. 1 and A14.

3. Chi, J., "Computer Simulation of Fossil-Fuel-Fired Boilers," Conference on Improving Efficiency and Performance of HVAC Equipment and Systems for Commercial and Industrial Buildings, Purdue University, April 12-14, 1976.

4. Bonne, U., and Patani, A., "Performance Simulation of Residential Heating Systems with HFLAME III," ASHRAE Transactions, Vol. 86, Part 1, 1980.

5. Janssen, J. E., and Bonne, U., "Improvement of Seasonal Efficiency of Residential Heating Systems," 1976 ASME Winter Annual Meeting, New York, NY, Dec. 5-10, 1976, Paper No. 76-WA/Fu-7 Trans. of ASME, J. Eng. Power 99A, p. 329 (1977).

6. Bonne, U., Patani, A., and Dewerth, D. W., "Dynamic Computer and Laboratory Simulations of Gas-Fired Central Heating Systems," Symposium on "Efficiency of Combustion Systems V," ASHRAE 1979 Semiannual Meeting, Philadelphia, Pa., January 28-February 1, 1979.

7. Hise, E. C., et al: "Seasonal Fuel Utilization Efficiency of Residential Heating Systems," Oak Ridge National Laboratory ORNL-NSF-EP-82, April 1975.

8. Hise, E. C., et al: "Heat Balance and Efficiency Measurements of Central, Forced Air Residential Gas Furnaces," Oak Ridge National Laboratory ORNL-NSF-EP-88, October 1975.

9. Kelly, G. M., Chi, J., and Kuklewicz, M., "Recommended Testing and Calculation Procedures for Determining the Seasonal Performance of Residential Central Furnaces and Boilers," NBSIR 78-1543, NBS, March 1978.

10. Bonne, U., and Langmead, J. P., "A Short Method to Determine Seasonal Efficiency of Fossil-Fuel Heating Systems for Labeling Purposes," ASHRAE Transactions, Vol. 84, Part 1, 1978.

11. Federal Energy Administration, "Energy Conservation Program for Appliances," Federal Register, Vol. 42, No. 155, July 29, 1977.

12. Mattocks, W. O., Chairman, American National Standards Committee Z21: AGA Approval Requirements, Z21.47, 1973, et. seq.

13. Cairns, J. R., "Boiler Efficiency Improvement Through Breeching Retrofit," Proceedings, Fourth Annual Energy Seminar, Gannon University, Erie, Pennsylvania, March, 1982.

Chapter 3

AN INDUSTRIAL BOILER DESIGNED FOR WOOD WASTE AND AGRICULTURAL RESIDUAL FUEL

M. Brown, W. Bulpitt, D. E. Harris

ABSTRACT

Utilization of low cost abundant fuels can vastly improve the operating economy of an industrial boiler. This paper describes a boiler installed by a large agricultural cooperative capable of operation on wood waste and agriculture residue fuels. A flexible design able to use wood and agricultural residue in any combination in addition to gas, oil, or coal was developed. The system has proven capable of meeting plant steam demand and state emission regulations on all fuel combinations. Information on boiler operation, maintenance, and economics is included. The experienced gained from the design and operation of this system is applicable to many other similar situations.

INTRODUCTION

Gold Kist, Inc., is one of the largest agricultural co-operatives in the country, owns and operates a 2,000 ton per day soybean oil extraction facility in Valdosta, Georgia. The extraction process requires 45,000 to 55,000 pounds per hour of 150 psig saturated steam. With natural gas and fuel oil continually escalating in price, Gold Kist began investigating the feasibility of using agricultural waste from nearby facilities as a fuel source for their boiler in 1979. In September 1980, Gold Kist agreed to participate in a program sponsored by DOE and managed by the Engineering Experiment Station at Georgia Tech.

The purpose of this program was to install on a cost-shared basis a wood fired boiler to serve as a showcase for other non-forest industries. This paper describes the Gold Kist boiler system, capable of utilizing wood waste and agricultural residue, installed under the DOE program.

FUELS

The soybean processing facility previously was supplied with steam from two Cleaver-Brooks 60,000 pound per hour saturated steam watertube boilers. Only one boiler was operated at any given time, with the other unit serving as backup. The boilers were normally fueled with natural gas. Fuel oil was used to fire the boilers if the natural gas supply was interrupted due to weather-related curtailments. Natural gas consumption in the plant was normally about 639,000 MCF annually, and fuel oil consumption was about 80,000 gallons.

In order to reduce steam costs, Gold Kist elected to install a waste fueled boiler, manufactured by Wellons Company of Sherwood, Oregon, capable of operation on wood waste and agricultural biomass waste products. Agricultural residue is an extremely attractive fuel to Gold Kist because of the amount generated in company-owned facilities. Also, an investigation made in the Valdosta area determined that the price and availability of wood residue favored its utilization.

The original objective of the new alternative fuel boiler project was to reduce dependence upon conventional petroleum fuels from outside sources. This objective is being met as Gold Kist burns wood and agricultural by-products such as peanut hulls and pecan shells from its own plants in the vicinity of Valdosta. Wood residue is transported to the plant from independent forest products industries in the Valdosta area. The new boiler system can also utilize small quantities of low sulfur coal, oil, and gas when necessary.

A study of the individual suppliers of wood residue within a 50-mile radius of Valdosta was conducted. The report indicated a wood residue availability in excess of 2,300 tons/week in the area. Suppliers were not quoted on their price to deliver the fuel, however, casual conversation with the suppliers indicated delivered wood residue prices in the range of $6 to $12 per ton, depending upon the hauling distance. The supplier name list generated from the study, along with the individual waste production available for purchase, was used to locate the most economical wood fuel in the area. Wood fuel types included bark, sawdust, chips, and sanderdust.

Gold Kist operates a peanut processing facility in Ashburn, Georgia, where peanuts are shelled. The Ashburn facility is about 70 miles from Valdosta. Peanut shells are primarily a waste product with limited commercial value. The peanut hulls are used in the production of commercial animal litter, residential fire logs, and as a filler for bulk in cattle diets. In the past, teepee-type burners were used to dispose of excess hulls that could not be sold. Some hulls were also disposed in land fills. Environmental restrictions have limited the use of the teepee burners, and as a result, a considerable quantity of excess peanut hulls are available seasonally.

Peanut hulls exhibit good properties for a solid fuel.

The heating value of peanut hulls is approximately 8,000 Btu's per pound. The hulls are relatively dry, because the peanuts must be dried to below 10% moisture content to satisfy USDA requirements for safe storage before shelling. The ash content of peanut hulls is normally between 3% and 4%, with the initial deformation temperature occurring at 2160°F, softening temperature occurring at 2210°F, and fluid temperature occurring at 2310°F. The heating value and ash content of the peanut hulls are similar to other cellulosic materials used for fuel, such as wood waste and bark. The moisture content is generally lower than other typical biomass fuels, due to the drying.

The Ashburn facility processes a large quantity of peanuts every year. Although the quantity varies every year, there are approximately 20,000 tons of peanut hulls available for fuel use after local feed sales during the processing season. The season starts in early fall when the first peanuts are harvested and continues for about nine months.

Peanut hulls are also available from a Gold Kist facility in Tifton, Georgia, about 50 miles from Valdosta. This facility has approximately 4,700 tons of peanut hulls available for fuel use after local feed sales.

Pecan shells are also available as a biomass boiler fuel from a Gold Kist facility in Waycross, Georgia. This facility generates approximately 1,000 tons per year of pecan shells after local sales. The shells also have a value as an abrasive cleaning agent in some blast cleaning operations, as a packing in oil field drilling operations, and as a garden mulch. Pecan shells have a heating value of about 8,500 Btu's per pound, and a moisture content of about 7.5%. The ash content is 2%, with an initial deformation temperature of 2660°F, a softening temperature of 2700°F, and a fluid temperatures of 2740°F.

The peanut and pecan hulls will provide approximately 4.3×10^{11} Btu's per year. A boiler efficiency of 70% translates into 3.0×10^{11} Btu's per year of steam production available from agricultural residue. The boiler system will require a total fuel input of approximately 6.6×10^{11} Btu's per year. If the annual peanut and pecan harvest goes as well as planned, the agricultural biomass waste available within the Gold Kist organization will supply the waste fuel boiler with approximately 65% of its fuel requirements. Forest products industries in the area will be called upon to supply the remaining biomass fuel to the plant since wood harvesting is less seasonal than agricultural harvesting.

BOILER SYSTEM

Once the fuel was identified, selection of a suitable boiler system could begin. The system decided on was a 60,000 lb/hr Wellons boiler. One factor influencing the selection was the cell type burners used on the Wellons system (Figure 1). These burners accommodate both high and low moisture content material and, therefore, can successfully utilize green wood and dry agricultural residue.

Feed screws deliver fuel into the "fuel cell" where combustion takes place. There are three fuel cells, each about six feet in diameter and about twelve feet high. The cells are equipped with water-cooled grates. Preheated combustion air is introduced into each cell in three locations; undergrate air, circumferential air, and overfire air. Undergrate air enters the cell through slots in the grates and assist in drying any high moisture content fuel. Circumferential air enters the cell tangentially, creating a cyclonic action within the cell. Overfire air is introduced near the top of the cell, and

provides additional air required for complete combustion of any pyrolysis products generated in the cells.

The amount of combustion air introduced into each portion of the cell can be ratioed as desired by adjusting the forced air dampers. Boiler fuel with a high moisture content may require more undergrate air, for example, than a very dry fuel such as planer shavings, or peanut hulls. The dry fuels may require little or no undergrate air for proper combustion. Of course, the correct total quantity of combustion air must be supplied in all cases.

A large furnace section, ten feet high, ten feet wide, and thirty-seven feet long, is located directly above the fuel cells. The volume of the furnace section is large enough to complete all combustion of pyrolysis gases or "sparklers" that come out of the cells. The large volume of the furnace section also allows the velocity of the combustion gases to decrease so that any heavy ash carried out of the fuel cells drop out. A large drop-out chamber at one end of the furnace section catches the ash that may have been entrained in the combustion gases.

The boiler section consists of an A-type, watertube Nebraska unit. The boiler is shop-assembled, with two mud drums and one steam drum, typical of the A-type boiler. Design pressure is 700 psig, and operating pressure is 600 psig. The boiler is equipped with a superheater section that will provide superheated steam at 750°F. It is designed to produce 80,000 pounds of steam per hour when fired with natural gas or fuel oil, and is derated to 60,000 pounds per hour for the waste fuel application. Currently, the boiler is operated at 190 psig, but will be operated at the higher pressure when cogeneration equipment is installed.

The boiler section is mounted on top of the furnace. Due to the arrangement of the A-type boiler, additional furnace volume is available between the boiler drums. The boiler tubes are located far above the combustion section, so that the possibility of clinkers and slag accumulation on them is almost non-existent.

Total heating surface in the unit is 7,928 square feet, and the superheater has a heating surface of 850 square feet. The radiant section boiler tubes are finned while the convective tubes are bare.

The evaporator section of the boiler is equipped with four fixed Diamond soot blowers. Each soot blower is mounted in the bank of tubes connecting the mud drums to the steam drum.

There are ash hoppers mounted on each mud drum to collect ash that drops out in the convective section of the boiler.

The economizer is a Tranter Kentube model with 1,418 square feet of effective surface area. The economizer is equipped with two Diamond soot blowers. It is designed to preheat boiler feedwater from 225°F to 285°F while decreasing the combustion gas temperature from 675°F to 498°F, at design conditions. The economizer has a design pressure of 900 psig.

The combustion air heater is installed in the combustion gas stream after the economizer. Combustion gases flow inside three-inch diameter tubes, while combustion air flows around the outside of the tubes. Access doors are provided for cleanout purposes.

The air heater can increase boiler efficiency by as much as ten percent and allows the use of boiler fuels with moisture contents as high as fifty percent.

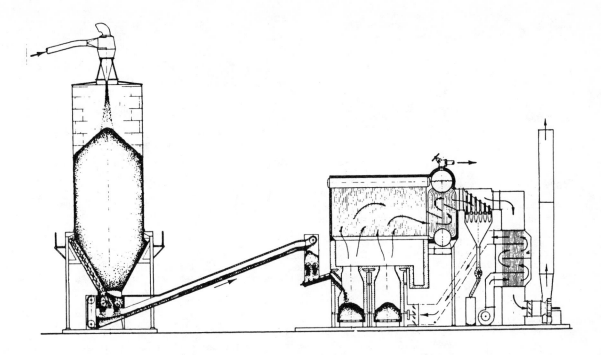

FIGURE 1. WOOD FIRED PACKAGE BOILER SYSTEM (Courtesy Wellons Inc.)

The boiler is equipped with a Coen gas-oil burner rated at 86 MMBtu/hr. The gas-oil burner is mounted on the Nebraska boiler section.

The boiler can be fired solely by the gas-oil burner, or the boiler can be fired by base-loading with wood waste or peanut hull fuel and using the gas-oil burner to take the boiler swings. If base loaded on biomass, the forced draft dampers and fuel metering screws are set manually at the desired firing rate and the firing rate controller governs the gas-oil boiler. It may be desirable to operate the boiler in this mode in case of a shortage of wood waste products and peanut hulls.

Emissions from the boiler are controlled using a mechanical "multiclone" type collector and fabric filter baghouse in series.

The dust collector was manufactured and supplied by Wellons. In order to minimize its size, the dust collector is located after the air heater. It is an effective device, removing nearly all of the large size particles reaching it, with a relatively low pressure drop. The collector's low pressure drop corresponds to low energy requirements for its operation. The Wellons boiler system normally meets all emissions requirements while burning wood waste with just the multiclone dust collector installed.

Unfortunately, the combustion of peanut hulls requires additional pollution control equipment. This is due to several factors, including the chemical composition, the moisture content, the size, and the ash content of the hulls. After considerable investigation by Gold Kist, including test firing of peanut hulls in a Wellons boiler, it was determined that a fabric filter was the best solution to the pollution control problem.

The fabric filter baghouse is from Eastern Controls, Incorporated. Prior to baghouse installation, the boiler was fired only with wood waste products, so that emissions limitations could be met with the dust collector. The fabric filter is a reverse air type, with four sections of filters. Each section contains 182 bags. The design air to cloth ratio is less than 3.0. Fiberglass bags are used in the filter.

OPERATIONAL DATA AND MAINTENANCE PROBLEMS

The boiler plant began supplying steam to the soy complex in early May 1981. Since then, the boiler operators have kept records of boiler operating data and maintenance problems that occurred. This section summarizes the information collected thus far on boiler operation and maintenance problems.

Areas where significant problems occurred with the boiler were:

o Screw conveyors in truck dump hopper
o Silo drag chain conveyor
o Level indicators in the silos
o Steam nozzles in the boiler
o Boiler shell external insulation
o Baghouse

The prominent areas are discussed in depth in the following paragraphs.

The two screw conveyors at the bottom of the truck dump hopper jammed when large pieces of wood and other foreign objects were transferred. To alleviate this problem, Gold Kist set standards for the wood waste that they will purchase. Also, they are presently working on a screening device for the hopper at the truck dump.

The level indicators, commonly called "bindicators", located in the upper portion of the Piedmont storage silos serve as sensors to activate the drag chain. However, due to poor reliability and difficult maintenance, the indicator signals have been discarded in favor of manual inspection. Fuel bridging in the silo can cause false signals to be sent to the conveyor controls and eventually shut down the boiler due to lack of fuel.

In the reclaim section, at the bottom of the silo, level indicators are also used and were not reliable. The problem was solved by simply placing the sensor in the vertical down position inside the reclaim bin. This modification allowed fuel to slide down along the sensor, rather than be held up by it and give a false signal.

Level indicators in the metering bins also gave false signals. They were replaced by rotary type fuel sensors which have given good service since installation.

In October 1981, the boiler was shut down and allowed to cool off for a manual inspection. This was done because the boiler efficiency had dropped and the flue gas temperature was higher than normal. When the inside of the unit was inspected, several problems were discovered. First, some of the steam nozzles, on two of the soot blower lines installed to clean the boiler tubes, were turned in the wrong direction. These were realigned. The pressure testing caps were removed from the other two line inlets at the steam drum. Second, ash was found around the superheaters and in the floor of the boiler just below the exhaust port. This ash was removed, but it is not yet known if any slag has returned because the boiler has not been inspected since. If the corrections made to the steam nozzles do not prevent this ash from forming in the future, more nozzles may have to be added. This is especially true in the superheater section where none are present. The boiler has operated well after this inspection.

The baghouse experienced numerous problems. The most significant one thus far has been due to condensation from the flue gas. Gold Kist engineers believe that moisture is forming in the breeching which carries flue gas from the induced draft fan to the baghouse. This moisture creates problems by causing dust caking on the bags. The caking makes it difficult for the reverse air cleanup system to knock the dust loose from the bags. Due to moisture condensation, the rotary valves must remain active or else rust will form between the rotor and the valve casing. Rust formation will lock the valves. If the rotary lock valve is stopped while peanut hulls are being burned, the system may have to be shut down for cleaning. If another fuel is being burned, the baghouse can be temporarily bypassed. To control this problem, Gold Kist insulated the breeching located under each of the baghouse sections.

The combustion cells have required very little maintenance. Regularly scheduled raking of the ash and sand off the grates is credited for the low maintenance. Slag buildup on the cell walls is acceptable as long as all of the air jets are free of obstruction. It was discovered that burning low moisture peanut hulls will raise the temperature of the cell such that slag becomes unattached from the walls, and thus falls to the cell floor. This is an acceptable way of reducing excessive slag buildup on the walls.

To reduce the high temperature that dry hulls produce, Gold Kist either wets the hulls with water or mixes them with at least 75% green wood. The mixing of wood and hulls is believed to be a better solution than wetting because water is already present in the wood. Adding water to the hulls only reduces the boiler efficiency proportionally and increases the demand for fuel.

Maintaining a small pile in each combustion cell is of primary importance. Three small piles in each cell have an advantage over a single large pile because they work together quickly to respond to sudden changes in steam demand. The use of three piles is also one of the reasons the boiler efficiency is higher than average. The cell depends upon heat from the pile for a large proportion of its energy to fully combust the biomass fuel. Underfire air must be turned off when 100% peanut hulls are fired due to their light weight.

PERFORMANCE ANALYSIS

The ideal method of analyzing boiler efficiency would be to first collect all of the data necessary and then apply the data the analytical formulas. Given a consistent heat input, the molar constituents per mole of fuel, a complete Orsat analysis of the dry flue gas, air flow rate, temperature, moisture content, steam flow rate, water flow rate and temperature, and the electrical energy consumed by the pumps, drives, and blowers, it is possible to conduct a fairly precise analysis of the boiler efficiency. However, all of this information would have to be recorded at the same instant of time because each parameter is constantly changing. Not only would this be impractical from an engineering standpoint, but also from a manpower and economic view.

The potential benefits in a continuous boiler efficiency monitoring system would not justify its costs to operate. This is why a simplified version of analyzing the boiler efficiency is used. The Department of Mechanical Engineering and the Engineering Extension Service of Auburn University has simplified the complex combustion formulas into table form. The tables allow a boiler operator to check the boiler efficiency given the type fuel, the temperature of the flue gas and one of three (excess air, percent O_2, percent CO_2) Orsat parameters. The tables assume there is always adequate combustion air available.

The data collected over the six-month monitoring period represents the average of each parameter (except fuel input) over each eight-hour shift. Any number of data averages can be used to describe the boiler efficiency over a specific period of time. However, it will still be only an estimate. The average flue temperature is approximately 350^0F and the average CO_2 reading is 12%. When this data applied to the wood fuel tables, the results are 74% excess air, 9% O_2, and 71% boiler efficiency. These results are very close to the results calculated using the formulas and educated guesses of the unknown parameters.

Some error may exist with the 71% boiler efficiency because some heat is lost at the air pollution section of the boiler. Also, the tables assume that the wood fuel is consistent in moisture content and heating value.

Most single pile wood fired boilers in operation today operate with a lower efficiency, near 65%, because of the desired ability to respond quickly to unpredictable steam demands. In other words, large burning wood piles must be maintained even if it means blowing off the excess steam until it is needed. The only other alternative with single pile burners is to allow the pile size to fluctuate with demand. Hence, the boilers' ability to quickly respond is hindered. This is why the three cell burner by Wellons is believed to have the higher efficiency. The three small piles are easier to control and can respond faster to peak steam demands than can one large pile burner.

When burning wood, an emission rate of .083 lb/MMBtu was measured. This is approximately 50% of the allowable emission rate for a unit of this size. Peanut hulls were not burned at the time these tests were run because the baghouse had not been completed. However, it is believed that with the baghouse operating, satisfactory

emission levels will also be achieved on peanut hull
fuel.

Georgia Tech engineers conducted an economic analysis of
the project. The cash flow method of analyzing the
feasibility was chosen because it approximates the
future incomes and costs according to the data given.
It also discounts future cash flow to present values by
the use of an inflation factor called "Rate for Net
Present Value." The cash flow analysis included a 20%
investment tax credit of $800,000 in year zero. A
simple payback of 3.9 years and a discounted payback of
5.8 years were calculated.

SECTION 2
ENERGY MANAGEMENT FORUM

Chapter 4

ENERGY - A BUYER'S GUIDE

E. V. Sherry

INTRODUCTION

Profit Margin is the difference between income and cost. Cost is a function of price and quantity, which is in turn affected by conservation. Price is related to capital. Energy users should always keep these factors in mind. Sometimes it is best to use more energy if it reduces price or conserves capital. We should talk of the productivity of energy, capital and labor instead of simply energy conservation.

Energy productivity is vastly improved by close cooperation between energy producers and consuming industries. Unfortunately, such cooperation is often frustrated by institutional problems or regulatory misunderstanding. Despite the fact that since 1970 much more has been said in favor of such cooperation, in the form of cogeneration and interruptible power, utilities in the United States are less disposed to be cooperative than they were in the '50s and '60s.

The Reagan administration has taken the position that the responsibility for developing and implementing energy efficient technologies resides with the private sector.

We support the theme that government action should be as neutral and permissive as possible and that Congress can promote conservation best by repealing laws and regulations rather than by passing new ones. This national policy places a big responsibility on industry to get moving and prove that we can innovate and cooperate to achieve a resurgance of productivity.

The Role of the Federal Government

While there is not complete agreement within the industrial sector, it is our opinion that there is real danger in providing explicit governmental incentives for energy conservation. Conservation investments, just like any other investment, must compete with capital available to the firm based on a number of factors including all the inherent risks associated with the project. The value judgements made by different managements and different industries to decide on competing investment opportunities are much too complex to be written into laws and regulations.

There is only so much that can be achieved by law or regulation. I think that today almost everyone realizes that we cannot substitute regulations for the infinite complexity of the market. In addition, when we try to influence the market by incentives, we run into funding restraints. As homo-economicus, we at Air Products would not oppose legislation that would give us tax or other incentives to promote energy conservation. We do, however, believe that much more could be gained by using the equivalent funds for general tax relief for industry to promote investment in new and more efficient production rather than to favor improving older plants and existing industries as is bound to be the case in any targeted incentive type legislation.

There are conservation projects on our own plate that would be sped up and given more consideration if there were additional tax incentives. There is no doubt that, with such tax incentives, additional conservation will be achieved. But would it be as economic a result as would alternatively have been obtained with the equivalent across-the-board reduction in taxes?

A Decade of Factions

It has been a fascinating experience to have been professionally involved with energy affairs through the recent period of rapid changes, when energy suddenly jumped from a function for specialists to one of national public interest.

Prior to 1970 energy prices were declining and it was possible to negotiate special rates with utilities. Natural gas was in surplus. The gas utilities fought with the electric suppliers to get new business.

It was fun to be on the buying side.

It was possible to negotiate long-term contracts for natural gas in states like Texas, at less than one tenth the price we would have to pay currently if we could get such a contract, which we cannot.

Since then we have been through the energy decade.

The period of time during which energy replaced the environment, civil rights, and Indochina as the great national debate.

The main scene of direct energy issues, that is shortage, price, balance of payments, and defense considerations, was interesting enough, but there are more interesting, more philosophical issues, at the margin of the debate.

Energy Factions

There were many active energy factions with whom we have debated in the past 10 years. All of them have a patriotic concern for the energy problem and want a solution, but each wants the solution shaped in a way that changes our social system to one more to its liking.

None is isolated from the others. As I list them you may find yourself slightly or fully within one or two or almost all of these camps. All have important views to add to the debate.

Their distinction is that they appear to put their special issue ahead of or at least equal to the energy problem itself and therefore appear to be using the energy problem to promote that issue.

Among these special interest groups would be included:

- ° The soft energy lobby,
- ° The wilderness groups,
- ° The professional regulator self-interest,
- ° The anti-nuclear movement,
- ° The social welfare activists,
- ° The anti-oil and anti-utility lobbies,
- ° Various consumer groups,
- ° Lobbies for the oil and utility industries, and, last but not least,
- ° The technocrats.

Surprisingly, consumers, except as they are included in the above categories, are not well represented. An exception is large industrial consumers who have organized and have had an impact on energy matters at the federal, state and local levels. We, at Air Products, helped organize several of these groups, notably the Industrial Energy Group (IEG) and the Electricity Consumers Resource Council (ELCON).

Conflicting Interests and Aspirations

It is an interesting and useful exercise for each of you to review your own inclinations and see how much you would support each of the groups. When each presents its own program without detailing the compromises necessary to achieve its goals they all sound rather good - each presents its own utopia:

- ° Most of us are somehow uncomfortable about the enormous institutions that affect our lives, but on the other hand, we take for granted the good service we get from utilities and integrated energy suppliers.

- ° We all support the idea of large national parks undisturbed by energy development, but we want coal and minerals available to our manufacturers at low cost and from U.S. sources which often means from other government owned land.

- ° We are tempted to think in terms of grand plans to solve all of national problems but we ridicule the government regulations that result from these plans and we hate to fill out the forms required to collect the data and administer the rules.

- ° We are uneasy about the problems associated with storage of nuclear waste and the increased radioactive materials that we must contend with as a result of nuclear power, but we lament the fact that the era of cheap energy -- too cheap to meter -- that was once touted as a future possibility, did not occur.

- ° We are all concerned with the effect of energy price increases on the needy but we want the cost of this kept under control and spread evenly.

- ° There is a general public concern for the influence of international oil companies but we can see the great benefit of U.S. dominance of this industry.

- ° Almost all of us are infected with the optimism of the technocrat. In our own lifetime we have observed the fantastic results of industrialization and industrial development, but we are reluctant to forego consumption today in order to make the investments that technology requires.

- ° And of course, we are all consumers and want low priced, readily available fuels.

The Democratic Approach

It may look like we have no national energy policy and each group could, and periodically does, present a major book arguing that no such policy exists. In fact our democracy results in a policy that includes parts of all viewpoints and therefore is perceived by many to be no policy at all.

If any one viewpoint were absolutely the best then we suffer by not following it exclusively: But none of the groups claim omniscience.

The Industrial User's Viewpoint

How do industrial buyers fit in?

Price is the element which separates sellers from buyers; the haves from the have nots and increasingly, environmentalists from average folk. The owners of energy sources and the sellers of energy want high prices. The buyers, not surprisingly, want low prices. The environmentalists line up on the side of the sellers -- high prices to them equate to less consumption and less consumption means less environmental trade-offs.

Buyers (or consumers) are, therefore, often faced with the coalition of energy marketers and environmentalists. For example, some natural gas pipeline companies have proposed that all retail natural gas prices be increased, with the excess funds being utilized by them as they see fit to develop new and in some cases novel energy sources.

Certain environmentalists call for the same higher prices with the excess funds used to promote conservation; to reverse the effects of past pollution or for other favorite projects.

No matter what disposition is made of the excess cash, it is the buyer -- the consumer -- who pays.

In relation to energy, the bottom line results from the margin between energy costs and the prices that can be obtained on the open market for energy intensive products. In the current market, such prices are weaker than energy prices. The energy costs of a product are made up of two factors: the consumption of energy per unit of product and the unit price of that energy. These two variables are not fully independent. As you know, in most energy price schedules, the unit price both for standing ready to supply and for the actual supply varies with the timing and pattern of use.

Often it is cheaper to use more energy inefficiently, if this can be done at the time and in a way that sufficiently lowers the unit price. As is the case in most of load management or curtailable energy rate schedules, there is some waste in restarting a unit, but we are compensated for that waste by a vast price improvement. This happens because our inefficiency is more than compensated for by an improved efficiency of the supplier.

This is an obvious example where the route to improvement in the bottom line is not just via conservation but is rather through a combination of conservation and price considerations.

In keeping our eye on costs, we need to closely examine hurdle rates for investment. Each project needs to be looked at separately. Based on risk, some should have low hurdle rates and others high. By risk I do not mean merely technical and cost estimating risk, but more importantly the risk that changed market conditions may change operations and reduce return.

Frequently, an energy conservation investment has little technical or cost uncertainty, and if used on a project which is already sold and thus has low market risk, deserves a low hurdle rate. In other cases, however, such projects are attached to plants with high market risk. Also, some only give a return when the plant is being pushed to capacity and do not save any energy if the plant is running at turndown. Clearly, such a project requires a higher than normal hurdle rate.

Gas Policy Needed

A specific energy legislative imperative for industry lies in the field of natural gas. The period of debate over the National Gas Policy Act was extremely divisive within industry and the political pressure from the White House for support of this bill in the final days of congressional action was beyond what should occur in a democracy.

Natural gas is a difficult commodity. It has only recently come into its own as a separate item of commerce, free from its prior stigma as a by-product of exploration and development of oil. Natural gas remains as the dominant source of energy and feedstock for the chemical industry.

Industry in the United States must again be able to contract for long term reliable supplies of this commodity if our economy is to continue to be successful. Natural gas laws and regulations must be changed to permit industry to contract for gas supplies from anywhere in North America and have these supplies transported to plant sites at minimum cost and regulatory hassle. The assurance of natural gas supply at competitive prices is essential to the health of industry.

Electric Rate Trends in the United States

Electric rate policy also needs to be re-examined.

In the United States, prior to 1970, the forces of competition brought about economical rate flexibility. Utilities were forced by competition to mold their rate structures to consumers needs. Competition came from neighboring utilities; from natural gas companies who were anxious to supply steam or gas turbine drives; and from manufacturers who could consider a variety of fuels for self-generation.

In the past 10 years, fuels for self-generation have become more difficult to obtain and recent high interest rates have dampened self-generation schemes. In addition, in the U.S., many utilities have abdicated, or have been forced to yield, their management responsibility for rate design to the local utility regulatory authority. The regulator is even more bureaucratically rule bound and therefore much more rigid in rate making and the division of authority between regulators and management makes it increasingly difficult to hold either responsible for overall results.

In the next 10 years, fuels will become more readily available, interest rates low enough to be accepted and manageable and competition will increase. The institutional structure will, however, probably have a strong hysteresis effect, with the result that U.S. electricity suppliers may lose some of their industrial sales.

Rate Trends in Europe

In contrast, in Europe, there remains considerable competition between neighboring countries for industrial business. The Common Market rules, which are designed to ensure fair competition, cannot cope with differing electricity prices, since they involve so much joint cost. Considering the differences in generating mix, indigenous fuels, residential patterns of use and exchange rates, it is next to impossible to judge how fairly each country matches costs to prices. In the United Kingdom and France in particular, the electricity suppliers are national undertakings and therefore their rates are subject to management, that is, political review, in a climate where the electricity supplier can survive even if not making a consistently adequate return.

In Germany, the sheer number of electricity suppliers, the degree of ownership by local governments and the ingenuity of the managements makes for competition within the country and an opportunity to meet competition from France. In addition, there are numerous special contracts in Germany which are not readily available for competitive inspection.

These conditions are a fertile seed bed for load management and tariff innovation. The trend is already visible.

Interruptible Rates in the U.K.

Under pressure from the industrial community and the government, the Central Electricity Generating Board (CEGB) of England and Wales has recently made available a new type of interruptible service available to customers who have at least 6 megawatts of normal winter (peak season) working day load; have agreed to be subject to load management of at least 25% of the normal winter day load or 3 mw minimum on 2 hours notice. Curtailments are usually at least 2 hours each and are limited to 60 hours total per year. The reduction in annual cost compared to firm power exceeds 20%.

Interruptible Rates in France

Electricite de France (EdF), which without a doubt has the best rate management and probably the best overall management of any electric utility in the world, is experimentally offering a new tariff option, "Effacement Jours de Pointe" (EJP), interruptible on peak days. The philosophy behind this schedule is different from the older interruptible tariff where interruptions were called upon when there were specific difficulties with the ability to supply. These supply shortages were a result of the cancellation of fossil fuel generation and the relatively longer lead time to build nuclear generation.

The logic of the new EJP option is to inform customers that a high cost situation exists, due to insufficient hydro or nuclear generation to meet normal or unusually high load or due to a shortage of transport from low cost generators to the load. The new EJP option calls for curtailment of 18 hours on 22 days during a five-month winter period. The notice period is normally 14 hours, but on occasion could be 2 hours. The reduction in price is in the range of 30%.

The new EJP option is much less supple than the type of interruptible tariff which EdF was considering last year. At that time we understand that EdF was studying a framework of pricing that has flexible pricing to match the abilities of different industrials to curtail. Under such a program, there would be a greater discount and less total hours of curtailment as the notice requirements of the customer became smaller and approached zero.

Interruptible Rates in the Common Market

Belgium has interruptible tariffs based on zero warning time and 100 hours per year maximum curtailment with an average of 61% reduction among customers who represent 10% of peak demand. The discount is about 30% of the demand charge.

The Netherlands has differing terms for curtailable service to curtail about 6% of peak demand.

As we stated previously, there are a vast variety of differing conditions for electricity supply in Germany and most are confidential.

International Arbitrage

This vast range of differing tariff structures presents an opportunity to industrial firms to practice arbitrage by moving their production from one country to another at various times depending on the variable power costs in the different countries.

In 1974 the tail block kilowatthour rates in neighboring areas of Belgium, France, Germany and Holland were about 1.3 to 2.0 U.S. cents per kwh. Today, for the same suppliers, the incremental kwh rates range from 0.8 U.S. cents per kwh for the lowest cost hours on the French Tariff, to 11.4 cents per kwh on the highest cost hours of the French Tariff. The other countries are in between at 2.5 to 5 cents per kwh. This spread will continue to grow since the running costs in the low cost hours are predicted to decrease in real terms while peak hour running costs are anticipated to increase dramatically.

Interruptible Rates in the United States

For most of the U.S. utilities, the peak demand occurs in the summer. This makes the peaking problem more difficult since the most severe weather condition occurs at the same time as the highest commercial activity. Also since storage cooling technology is not as advanced as storage heating technology, there is less opportunity to control this equipment which is actually causing the peak. These conditions should give real time price schedules greater prevalence in the United States: If we can get out of our regulatory bog. Europe is presenting us with a challenge; will we rise to meet it?

The Responsibility of Industry

Industry has a major vested interest in active participation toward solving the problems of nuclear waste disposal, breeder reactor development, reprocessing of nuclear fuel, financing of future nuclear power plants and resolution of the TMI problem. It cannot absent itself from this debate as though it were exclusively someone else's problem. The future of the electric utility industry and manufacturers are inexcorably inter-twined.

The entire basic concept of energy regulation must be reassessed. The slow process of five or more years of political debate followed by passage of massive complex legislation, the writing of rules, and the attempt to force fit myriad situations into these strict rules, is the basic flaw undermining the five parts of the National Energy Act.

We must work to bring about more teamwork between the various interests in our society, including industry, labor, consumers and environmentalists. The National Coal Policy Project of the Center for Strategic and International Studies of Georgetown University helps point the way toward development in this area. Everyone must realize that we are competing as a nation in a relatively free and open world market. Each of us, not just business, has a vested interest in improving our national productivity.

The Economy

The best means of promoting energy conservation by industry is to promote a healthy economic climate, control inflation, stop bickering, and promote public confidence in U.S. industry.

An unfortunate by-product of the "energy crisis" has been the promotion of no growth concepts. Only to no growth or low growth proponents is energy conservation a goal unto itself; their discussion centers on the theoretical optimal allocation of resources rather than full utilization of resources. The purpose of our economic system is to provide the greatest total benefit to the greatest number of our people, not just to achieve a theoretically optimum allocation of resources -- an exercise which is an intellectual nicety but if taken as an end in itself could damage the overall economy -- all other things being equal, low prices are better than high prices.

The economic goal is to curb inflation and reduce the oil payments deficit while maintaining a satisfactory level of the overall economy and reversing environmental degradation. The achievement of this goal requires that we be optimistic about our ability to solve environmental concerns and to develop the required technology through cooperative efforts.

SECTION 3
BUILDING ENERGY
UTILIZATION

Chapter 5

A REALISTIC ENGINEERING APPROACH TO SIGNIFICANT RETROFIT ENERGY CONSERVATION IN TYPICAL MEDIUM-SIZED HOSPITAL FACILITIES

A. G. Jhaveri

I. ABSTRACT

In a general-purpose, typical, medium-sized (200-plus beds) central Washington hospital, constructed during 1960's, a comprehensive energy conservation retrofit program resulted in nearly 25 percent overall savings, majority of which was provided by mechanical/HVAC, electrical, and architectural design modifications and/or systems replacement.

During the winter of 1980, Paddock & Hollingberry architects of Yakima, Washington, and John Graham and Company, consulting engineers of Seattle, Washington, conducted a Technical Assistance (TA) energy analysis for the St. Elizabeth Hospital in Yakima, Washington.

Based on detailed Technical Assistance (TA) energy analysis of the hospital's major energy consuming systems and equipment, the hospital was awarded one of the largest single grant (nearly one-half million dollars as 50 percent match) by the U.S. Department of Energy (DOE) under its Institutional Buildings Grant Program (IBGP). In addition to the cost-effective building envelope (e.g., insulation, storm windows, solar control film, etc.) and electrical power (e.g., power factor controller, energy-efficient lighting replacement) system retrofit modifications, the hospital's two existing 300-ton (each) natural gas-fired units were replaced with similar electric centrifugal chillers; existing heat recovery system was modified for improved energy efficiency; zone optimization controls were installed; and new boiler stack heat recovery system was equipped on the largest of the three boilers.

The objective of this paper is to provide relevant engineering and design approaches used in this highly effective energy conservation retrofit project.

II. EXECUTIVE SUMMARY

Over 25 percent energy savings were achieved in a 255,518-square-foot, 214-bed general hospital located in Yakima, Washington.

A 27.5% reduction in energy consumption, from 1979 to 1982, was a result of decrease in the Energy Use Index (EUI) from 513,870 Btu/SF/year to 372,447 Btu/SF/year. (This is based, partly, on the electrical conversion factor of 1 kWh = 3,413 Btu's at end use.)

In terms of current costs for energy, the reduction in electrical energy consumption of 238,000 kWh amounts to an annual cost savings of $11,323.

The reduction in natural gas energy consumption of 352,250 therms, amounts to an annual cost savings of $190,215.

Energy savings were achieved through improvements to the building envelope, lighting, power and HVAC systems.

Improvements to the building envelope included adding insulation to ceiling spaces to reduce heat loss and installing window films to reduce cooling loads.

Savings in lighting were achieved by changing out fluorescent tubes and disconnecting ballast. Mercury vapor lighting fixtures (400-watt) in parking areas were replaced with 250-watt high-pressure sodium fixtures.

A new power factor capacitors unit was installed to improve the electrical power factor, thus reducing penalty charges from the utility.

Savings in the energy consumption of HVAC sytems were achieved through the implementation of the following energy conservation measures:

1. Two existing 300-ton absorption chillers were removed and two new 300-ton centrifugal chillers were installed in a new chiller room.

2. The piping between the existing three separate chilled water systems was connected to produce a common chilled water system. These chillers ranged in size from 140 tons of capacity to 300 tons of capacity.

   Connecting the piping allowed the hospital to take maximum advantage of cooling load diversity. Chiller operation can now be sequenced according to load, assuring that the chillers are always operating in an efficient range. At times, the smallest chiller is operating to satisfy the entire hospital cooling load. Cooling tower, pumps and chillers that are not required for cooling, are deactivated.

3. A boiler stack heat recovery unit was installed on the no. 1 boiler to preheat outside air during the period November through March.

4. Several supply and exhaust fans were connected to an existing Energy Management Control System (EMCS) unit.

5. An existing heat recovery system was modified to increase heat recovery capability, thus significantly reducing the energy consumption.

III. INTRODUCTION

Attainable energy efficiency improvements of existing commercial and institutional buildings, within the United States, have unfortunately lagged behind other sectors, in spite of the significant energy conservation potential that exists, for example, in schools, hospitals and office buildings. Hospitals are notoriously high energy using/wasting facilities, partly because they operate on a 24-hour schedule and some of their activities can be classified as energy-intensive. In the state of Washington, for example, energy use in the hospitals account for some 10 percent of the total energy consumed by the commercial sector. This might not appear to be a large fraction, nonetheless, it's an important target for energy reduction activities because proportionately, energy use in each existing hospital building/facility corresponds to, on the average, nearly five times that of its counterparts (e.g. schools, office buildings, etc.). It is not uncommon to find many medium-sized, general-purpose public or private hospitals with current energy use indexes (in Btus per square feet per year) ranging from 500,000 to 750,000 -- a staggering total consumption of rapidly escalating (costs), unstable and deteriorating supply of conventional fossil fuel resources (e.g. oil, natural gas, electriity, etc.).

Within the framework of U.S. Department of Energy's Institutional Buildings Grants Program (IBGP), a comprehensive energy/engineering analysis of the St. Elizabeth Hospital in Yakima, Washington, revealed that over 25 percent energy savings can be realized by implementing many of the recommended, cost-effective energy conservation measures.

IV. DISCUSSION

St. Elizabeth Hospital, located in the Central Washington town of Yakima (population approximately 50,000), is a general purpose, 214-bed (current capacity) medical facility, serving the health care needs of greater Yakima valley. This hospital, managed by the Sisters of Providence, was originally built in the 1930's, with major structural additions and/or modifications completed during 1945, 1965 and 1975 construction renovations. Figure 1 is a photograph of the hospital building.

In general, the masonry/brick building construction of the hospital is considered good in terms of structural strength and energy insulation. The total gross area of each of the major building components is given below:

Floor:	255,518 square feet
Exterior Wall:	111,300 square feet
Roof:	70,852 square feet
Exterior Glass (windows & doors):	8,722 square feet

The typical energy cost and consumption data for the hospital during the 1979 calendar year were as follows:

Fuel Type	Consumption	Cost
a. Electricity	kWh = 7,783,600 kW demand = 13,349 kVAR power factor = 10,964	$118,714 (total)
b. Natural Gas	1,069,506 therms	$324,182

The energy use index, thus, corresponds to roughly 765,000 Btus/sq ft/ year, while the corresponding energy cost index was $1.50/sq ft/year. (Here, the electrical conversion factor used was, 1 kWh = 11,600 Btu's, at the generation point.)

Because of the escalating fuel costs, inflation and hospital operating costs, the utility expenditures at St. Elizabeth hospital have been increasing rapidly through the middle of 1981. However, because of the major energy retrofit construction projects that have been implemented during the 1981 calendar year, the current energy consumption and costs are declining significantly. This is the result of a three-phase energy conservation program that began in late 1979/ early 1980 time frame. The comprehensive energy reduction program was divided into: A) Energy Audit phase; B) Detailed Engineering Analysis phase, and C) Retrofit Design/Construction phase. The following briefly describes the methodology/approach used and recommendations proposed in each of these phases.

V. METHODOLOGY

A. Energy Audit Phase

As required by the IBGP, the Energy Audit phase must precede and completed prior to the detailed Engineering Analysis or Technical Assistance (TA) energy analysis phase. The hospital did complete a Preliminary Energy Audit (PEA) in June, 1979, in order to be eligible for the IBGP funds administered by the Washington State Energy Office. However, because of the change of Facilities Engineering Personnel at the hospital in late 1979, the Energy Audit phase had not been completed prior to the detailed engineering analysis phase. The Architect/Engineering consultant retained by the hospital, therefore, had to help the facilities personnel complete the energy audit of the hospital in December, 1979.

The methodology used during the Energy Audit phase consisted of three major elements, namely

1. Energy inventory data of the facility
2. Energy consumption calculations
3. Energy audit checklist

1. The Energy Inventory data were divided into the following categories:

 a. General administrative information

 b. Occupancy patterns/operating conditions/activity groups

 c. Physical characteristics

 d. Annual energy consumption summary

 e. Energy systems

 • Lighting
 • HVAC
 • Heating and cooling sources
 • Water
 • Special services
 • Solar and renewable resources
 • General remarks

2. The Energy Consumption calculations provided estimates of seasonal heating and cooling loads in Btus/year for the entire hospital facility.

3. The Energy Audit checklist identified no-cost and/or low-cost operational and maintenance (O&M) procedures, measures and changes, implementation of which can provide immediate energy savings to the hospital.

The hospital was able to achieve an initial reduction in energy consumption of approximately 10 percent, by implementing a cross-section of O&M's in HVAC, lighting and dministrative control systems. The largest energy and cost savings were realized through <u>thermostat</u> <u>adjustments</u> and <u>faulty</u> <u>steam</u> <u>control</u> <u>valves</u> <u>and</u> <u>traps</u> <u>replacement</u>.

B. Detailed Engineering Analysis Phase

The methodology used in this phase closely followed the April 17, 1979, Federal Register regulations 10 CFR, Part 455, Subpart C, entitled "Technical Assistance Programs for Schools, Hospitals, Units of Local Government, and Public Care Institutions," Article 455.42, <u>Contents</u> <u>of</u> <u>Program</u>, clearly delineates the items to be covered in the detailed engineering analysis.

The approach to the analysis was based on evaluation of the individual energy consuming and/or potential energy savings sytems of the hospital. These were:

1. Building Envelope -- Insulation, infiltration, solar gain/ control, etc.

2. Heating, Ventilating and Air-Conditioning (HVAC) -- Boilers, chillers, air distribution and exhaust systems.

3. Environmental Controls -- Thermostats, time clocks, resets, economizer cycle, zone optimization, humidistats, valves, etc.

4. Lighting and Power -- Lighting load reduction, task lighting, efficient lamp fixtures, demand control, power factor correction, photoelectric switching, etc.

5. Renewable and Alternate Sources -- Active solar, geothermal, incineration, waste heat recycling, day lighting, cogeneration, etc.

Energy Conservation Measures (ECM's) in each of these systems were investigated.

Prior to identifying technically and economically feasible energy conservation measures (ECM's) at the hospital, a detailed building field survey; review of plans, drawings and logs; evaluation of operations and maintenance (O&M) procedures; and analysis of building's operating characteristics as well as energy consumption profile were undertaken.

Annual energy consumption estimates were made for each of the energy consuming systems. The estimates were compared to actual energy consumption records to arrive at potential energy conservation opportunities. Where records for a portion of the system were not available, that portion was calculated using ASHRAE procedures on catalog data. National Oceanic and Atmospheric Administration (NOAA) weather data for the year of the study (1980) was utilized for HVAC systems.

Based on the comprehensive engineering analysis of each of the major energy consuming building systems, the following ECM's were recommended:

Building Envelope

ECM	Estimated Savings		Estimated Retrofit Costs	Simple Paybck Period
	Energy	Cost		
1. Add R-19 batt insulation (6") in the ceiling below 1945 wing roof.	269 MMBtu	$1,400	$3,800	2.7 yrs
2. Install R-8 foam insulation (2") in the wall utility space of 1975 hospital addition.	74 MMBtu	385	890	2.3 yrs
3. Install solar control window film on east, west, & south exposure glass (1945, 1965 & 1975 additions)	376 MMBtu	1,954	15,690	8.0 yrs
4. **Install tinted storm sash and tinted insulated glass on large waiting room windows and corridor windows (1975 addition)	88.5 MMBtu	460	6,400	13.9 yrs

HVAC/Controls System

ECM	Estimated Savings		Estimated Retrofit Costs	Simple Paybck Period
	Energy	Cost		
1. **Install <u>new</u> exhaust air heat recovery system w/ recovered heat transferred to preheat coils.	3,663	$19,050	$283,700	14.9
2. Improve efficiency of the <u>existing</u> exhaust air heat recovery system.	1,715 MMBtu	8,900	56,100	6.3 yrs
3. Install zone optimization controls for existing HVAC system.	658 MMBtu	3,950	30,000	7.6 yrs
4. Replace two 300-ton existing absorption chillers w/ two new 300-kW centrifugal chillers.	10,852 MMBtu	56,428	459,000	8.1 yrs

*Dollar amounts shown here reflect <u>1980</u> fuel prices.

**These ECM's were later deleted by the hospital because of relatively higher simple payback periods.

ECM	Estimated Savings Energy	Cost	Estimated Retrofit Costs	Simple Payback Period
5.***Install in-line boiler stack heat recovery units on three boilers.	2,104 MMBtu	10,940	69,300	6.3 yrs

***This ECM was later replaced with a significantly more energy-efficient and costly "Heat Mizer" boiler stack heat recovery unit on boiler No. 1 (largest of three).

Lighting and Power

1.	Remove, replace & reduce existing lighting fixtures (interior & exterior w/ energy-efficient systems.	2,747 MMBtu	$4,848	$42,866	8.8 yrs
2.	Install power factor correction capacitors for oversized electric motors.	425 MMBtu	2,550	21,500	8.4 yrs

Renewable and/or Alternate Energy Resources

1.	Install flat plate solar collectors on 1945 addition roof for domestic water preheating.	1,210 MMBtu	$6,290	$93,150	14.8 yrs
2.	Utilize geothermal hot water resources in conjunction w/heat pumps, plate heat exchangers, & hot water heating coils.	14,621 MMBtu	104,440	1,373,560	13.1 yrs

The above-listed ECM's are in no way complete and may have overlooked additional relevant energy conservation opportunities in the hospital. However, because of the time and cost constraints placed at the outset of this engineering analysis which was partially funded by a grant from the U.S. Department of Energy, it was concluded that the most promising and cost-effective ECM's should be first investigated, particularly those with simple payback periods of 15 years and less as required by the DOE's IBGP.

C. Retrofit Design/Construction Phase

The recommended ECM's identified during the detailed engineering analysis phase, were submitted by the hospital to the Washington State Energy Office (WSEO) and the Region X office of the U.S. Department of Energy (DOE), as part of the IBGP funding application during Cycle 1. St. Elizabeth hospital, as a result, was awarded the largest single grant, 50 percent matching, totalling a little over half a million dollars, in the

entire region. All ECM's, except solar and geothermal, were included in the grant award. DOE indicated that it would reconsider a more realistic and comprehensive alternate/ renewable ECM during the next cycle, after the impacts of Cycle 1-funded ECM's has been evaluated. Howver, because of higher simple payback periods and tight cash flow problems, the hospital decided to postpone the proposed solar/geothermal ECM until sometime in the future.

The methodology used in this phase closely followed the standard procedures used by architect/engineers during design specifications and/or construction documents preparation. The proposed energy reduction retrofit project at St. Elizabeth hospital was classified as a major comprehensive construction involving architects, engineers, contractors and equipment suppliers. Because of the magnitude of this total energy systems retrofit construction, the project was divided into two phases, namely, 1) Phase I --Building Envelope and Electrical Energy Reduction Retrofit, and 2) Phase II -- Mechanical/HVAC Controls Energy Reduction Retrofit. The Phase I construction project was completed first in late 1980 because it constituted only 10 percent of the total project retrofit costs of roughly $1 million.

"Design drawings" and "performance specifications" for each of the DOE-funded ECM's were prepared, using the Construction Specifications Institute's (CSI) MASTER FORMAT approach, first published in February, 1979. This provided a logical methodology for developing and compiling energy retrofit construction documents into competitive bid specifications. The project Architects in Yakima were responsible for developing Building Envelope ECM's specifications, while the consulting engineers of Seattle developed both the Mechanical/HVAC and Electrical/ Power ECM's. The Yakima architects were responsible for the coordination, evaluation, and negotiation of the contract bids, in conjunction with the hospital. Schematics of three typical construction design recommendations for relevant ECM's are included in this paper as Figures 2, 3, and 4.

The entire retrofit construction work was spread over an 18-month period from mid-1980 to the end of 1981, a rather fast-track schedule that resulted in significant energy and cost savings to the hospital, without compromising the quality, comfort and health service delivery functions of this large community medical facility.

One of the basic approaches used by the A/E consultants was to periodically meet, review, inspect and evaluate the various construction phase work with the general, mechanical, electrical and special energy systems contractors on the job along with the designated facilities personnel at the hospital. Not only was this approach necessary, but it also helped identify any serious or unresolved problems as well as expedite the authorized ECM retrofit construction.

One of the major contractual changes made prior to the ECM implementation, was to interconnect the chilled water piping between the new chillers (centrifugal) and the two existing chilled water systems. DOE approved this change order because it significantly simplified the interconnection among the new and existing chillers within the hospital, as well as improved the overall chiller system efficiency, operation, controls and energy use.

VI. RESULTS AND CONCLUSIONS

It is fair to report that the estimated energy and cost savings achieved as a result of implementing the recommended, grant-funded and hospital-selected ECM's, exceeded the original project objectives. Shown below is the summary of estimated results:

Total Number of ECM's Implemented: Nine (9)

Total Energy Savings in Annual MMBtu (Btu x 10^6):* 19,309

Energy Use Index (EUI) Reduction, Btu/S.F./Yr. 75,568

Annual Cost Savings, 1980 Dollars: $91,815

Total ECM Retrofit Costs:** $762,950

Average Simple Payback (years):*** 8.3

The Energy Use Index (EUI) for 1979 was 513,870 Btus/S.F./Yr. The EUI for 1981 was 372,447 Btus/S.F./Yr. Therefore, the actual reduction = 141,412 Btus/S.F./Yr., while the estimated reduction = 75,568 Btus/S.F./Yr.

   *Includes both net electrical and natural gas savings (approximately 20 percent electrical; 80 percent natural gas)

   **Total ECM Retrofit costs include 50 percent Federal DOE Grant fund, with other half contributed by the hospital or $381,475; 12 percent AE design fees; and 5 percent state sales tax.

   ***Hospital's simple payback period share is half of the above or 4.15, corresponding to 1980 dollars and fuel prices. The simple payback period will be further reduced due to escalating energy costs.

Based on the 1980 utility costs (electricity and natural gas) at the hospital, the onservative percent cost savings resulting from the energy reduction retrofit program turns out to be roughly 20 percent in 1980 dollars. However, the actual percent energy savings to be realized is estimated to be no less than 30 percent, but more like 40 percent, considering the combination of administrative, operational and maintenance (O&M), and ECM's that have been implemented by the hospital during the 1980-82 period.

To better appreciate the realistic engineering approach used in this project during energy audit, engineering analysis and design specifications/construction document phases, it is well worth to compare the estimated retrofit costs (associated with each of the recommended and Federally-funded ECM's) with the actual contractor bid costs. The following table provides a summary of comparison for the retrofitted ECM's:

ECM's	Retrofit Const Costs ($) Estimated	Actual	Annl Energy & Cost Savings 1980 Base Yr MMBtu	$$
Phase I				
1. Insulation -- all	$ 4,690	$ 5,617.50	343	$ 1,785
2. Solar control window film	15,690	6,387.03	376	1,954
3. Lighting modifications	42,866	32,690.11	2,747	4,848
4. Power factor controller	21,500	23,653.48	425	2,550
Phase I Subtotal	84,746	68,348.12	3,891	11,137
PHASE II				
1. Existing heat recovery changes	56,100	56,757.50	1,715	8,900
2. Zone controls optimization	30,000	32,699.32	658	3,950
3. Chillers replacement	459,000	392,180.34	10,852	56,428
4. Boiler stack recovery	69,300	77,553.44	2,104	10,940
Phase II Subtotal	614,400	559,190.60	15,329	80,218

Note: Savings data on couple of other retrofit ECM's were not available at this time, and therefore not included in this table.

TOTAL (PHASES I AND II)	$699,146	$627,538.72	19,220	$91,355

It can be, therefore, fairly concluded that the Energy Retrofit Reduction program at St. Elizabeth hospital in Yakima, Wahington, has successfully met the basic objectives of improved energy efficiency and cost-effective energy retrofit construction without compromising the health, welfare, safety and quality of patient-delivered medical services.

VII. ACKNOWLEDGEMENTS

The successful completion of the energy reduction retrofit project at St. Elizabeth hospital in Yakima, Washington, is truly a reflection of the team effort involving the following personnel and/or organizations:

1. U.S.Department of Energy, Region X Office in
    Seattle (now in Richland, Washington), Insti-
    tutional Building Grants Program

    ·   Mr. Eldred Smith,
        Project Manager (now with Seattle District
        Army Corps of Engineers)

    ·   Mrs. Kathy Vega
        Current Program Manager

2. Washington State Energy Office
    Olympia, Washington

    ·   Ms. Lou Ann Donnelly
        Program Manger (currently on leave)

    ·   Mr. Charles Waugh
        Staff Engineer (now with Puget Sound Power
        and Light Company, Bellevue, Washington)

3. St. Elizabeth Hospital
    Yakima, Washington

    ·   Sister Charlotte Van Dyke
        Administrator

    ·   Mr. Ronald P. Swanson
        Assistant Administrator for General
        Services

    ·   Mr. Charles (Chuck) Dedeian
        Director of Plant Services

4. Paddock & Hollingbery, Architects
    Yakima, Washington

    ·   Mr. Donald R. Hollingbery, Sr.
        Project Archtitect

    ·   Mr. Rick R. Colver
        Director of Energy Management

5. John Graham and Company
    Seattle, Washington

    ·   Mr. Larry W. Wehmhoefer
        Project Mechanical Engineer

    ·   Mr. Gerald Schneider
        Project Electrical Engineer

    ·   Mr. Robert M. Johnson
        Energy Management Engineer

    ·   The Word Processing staff

6. Selected (competitive bidding) general, mech-
    anical, electrical and energy contractors, as
    well as suppliers of energy equipment, products
    and systems.

FIGURE 1.   ST. ELIZABETH HOSPITAL BUILDING

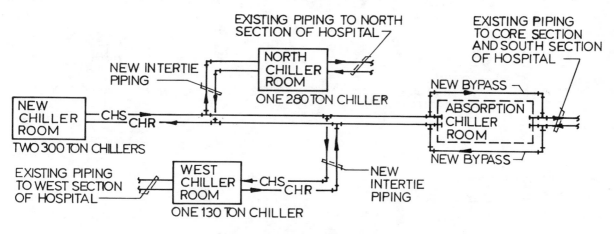

CHILLED WATER PIPING DIAGRAM

FIGURE 2.   REPLACEMENT CHILLERS SYSTEM

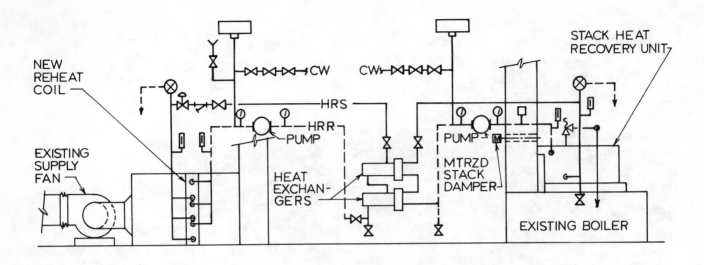

FIGURE 3. BOILER STACK HEAT RECOVERY SYSTEM

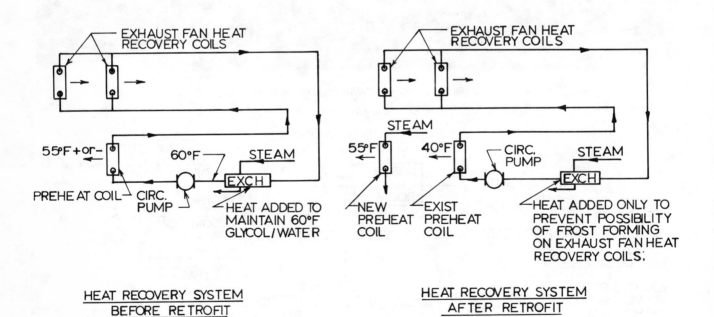

FIGURE 4. MODIFICATION TO EXISTING EXHAUST AIR HEAT RECOVERY SYSTEM

Chapter 6

ENERGY MANAGEMENT COMPUTERIZED SYSTEM

ABSTRACT

This paper describes how McDonnell Douglas Corporation was able to meet "energy crunch needs" with an 'Energy Management Computerized System' (EMCS). EMCS consists of equipment and computer software that will monitor and/or control energy use within a particular building or group of buildings. The use of EMCS as a tool for optimizing energy efficiency while maintaining desired environmental comfort standards and reducing energy cost is described. Functions vary from simple time clock schedules, to sophisticated schedules that use a computer program to monitor and control equipment operation as related to the building operation schedules, utility demands and actual outdoor weather conditions. MECS gathers data and prepares reports of energy usage and equipment operation as a function of building occupancy schedules and global weather conditions. Hardware, report capabilities, monitoring/control systems, data gathering features, utility demand shedding methods, impact on energy budgets and maintenance planning are discussed.

INTRODUCTION

This paper covers the actual successful efforts which directly relate to computerized energy efficiency approach with facts and figures. In the early 70's, this approach was considered to be a luxury, but now it has become a very important necessity for the 80's. In the energy optimization process, many building components must be accounted for in their relationship to the actual outside weather conditions, utility demands, and the building operational conditions for a full year (8760 hours). To overcome some of the possible problem areas in implementing such a program, the following steps are covered:

- o Data Gathering
- o Approved Energy Management Program
- o Evaluation and Selections of Software/Hardware
- o Engineering & Maintenance Backing of the Program
- o Results and Summary

DATA GATHERING

Data gathering was not a simple process, due to the lack of submetering for each building. In order to sell any energy efficiency improvements related to this job, it was necessary to complete an energy performance analysis before any energy retrofits to the building could be submitted. We had to provide historical data showing the total Energy Utilization Index (EUI), in BTU/SQFT/Year. This EUI is not only used for relative energy savings projections, but it is also to be used in the projected energy usage, curtailment, and cost for utility budgeting. In light of the overwhelming sequence of variables contributing to the energy behavior of a facility, it was deemed necessary to look for a means which would allow for modeling energy performance of

the building's operational schedules, HVAC Systems, energy retrofits and outside weather conditions for 8760 hours per year. The only answer found that would handle the long sequences of computations and still meet the available manhours and mandated time period was by the use of a computerized energy analysis program. This program not only provided good energy profiles for the facilities, but it supplied us with the required data for selecting the most logical operating schedules and limits to be used for the Energy Management Computerized System (EMCS).

Results of the data gathering were analyzed and selected systems were recommended for Phase I to be tied into the EMCS. (See Chart-1 and Table-1.)

We have found that the computer reduces routine data gathering tasks to manageable proportions. It's quickness and reportability gives the required management impact needed to provide real time information in a standardized format. This type of information system has aided in spotting trends, troubles, and has given us accurate forecasts of our future needs. By manually field checking the program outputs before hand we were able to provide a manual back-up system and also provide a method for debugging the program. Enhanced computer capabilities are going to be developed as the needs arise.

EMCS DATA SELECTION

ENERGY CATEGORY	Plant Winter %	Plant Summer %	Office Winter %	Office Summer %	Winter Savings %	Summer Savings %	Nite-Set Back - Dead Band	Ventilation Set Back	Zone Control & Monitoring	Improve Operations Eff.	Reduce Electrical Loads	Coordinate Systems Operations	Outside Weather Reset	Demand Limit Control	EMCS Reading	No.
SPACE HEATING	45	0	41	0	40	0	√	√	√	√	√	√	√			1
AIR CONDITIONING	0	0	0	41	0	30	√	√	√	√	√	√	√			2
LIGHTING	3	6	41	41	15	15	O	O	√	√	√	√	O			5
POWER	27	51	0	0	10	10	O	O	O	√	√	√	O	√		4
PROCESS HEATING	22	38	0	0	10	10	O	O	O	√	O	√	O	O		6
MISC. ELECTRICAL	2	3	16	16	0	0	O	O	O	O	√	O	O	O		8
KITCHENS	1	2	0	0	0	0	√	√	√	√	√	√	√			3
WATER HEATING	0	0	2	2	0	0	√	O	O	√	O	O	O	√		7

CHART-1   √ RECOMMENDED   O NOT RECOMMENDED

```
┌─────────────────────────────────────────────────────────────┐
│                        EMCS PHASE-I                          │
│                    EQUIPMENT SELECTIONS                       │
│                                                              │
├─────────────────────────────────────────────────────────────┤
```

1. The installed cost of this system is approximately $120,000.

2. The loads which are controlled by the "EMCS" Program are as follows:

    2.1 Factory Ventilating/Heating Units (39 total)

         32 - 10 HP ea., 20,000 CFM ea. = 224 KW   640,000 CFM
          2 - 15 HP ea., 24,000 CFM ea. = 20 KW   48,000 CFM
          2 - 25 HP ea., 36,000 CFM ea. = ‚34 KW   72,000 CFM
          3 - 40 HP ea., 36,000 CFM ea. = 84 KW   108,000 CFM
                                          ───────   ──────────
                           TOTALS   362 KW   868,000 CFM

    2.2 Office HVAC Units (8 systems)

          2 -167 HP ea., 99,000 CFM ea. = 222 KW   198,000 CFM
          2 -157 HP ea., 92,000 CFM ea. = 208 KW   194,000 CFM
          2 - 75 HP ea., 43,000 CFM ea. = 98 KW   86,000 CFM
          1 - 35 HP ea., 51,000 CFM ea. = 24 KW   51,000 CFM
          1 - 10 HP ea., 20,000 CFM ea. =   7 KW   20,000 CFM
                                          ───────   ──────────
                           TOTALS   559 KW   549,000 CFM

    2.3 Labs and Special HVAC Units (4 systems)

          1 - 80 HP ea., 50,000 CFM ea. = 53 KW   50,000 CFM
          1 - 53 HP ea., 43,000 CFM ea. = 36 KW   ‚47,000 CFM
          2 - 50 HP ea., 53,000 CFM ea. = 68 KW   106,000 CFM
                                          ───────   ──────────
                                          157 KW   203,000 CFM

    2.4 Cafe HVAC Units (4 systems)

          1 - 65 HP ea., 36,000 CFM ea. = 44 KW   26,000 CFM
          1 - 35 HP ea., 43,000 CFM ea. = 24 KW   43,000 CFM
          1 - 15 HP ea., 25,000 CFM ea. = 10 KW   25,000 CFM
          1 - 10 HP ea., 20,000 CFM wa. =   7 KW   20,000 CFM
                                          ───────   ──────────
                                           85 KW   124,000 CFM

    2.5 Exhaust Systems (B-106 & 107)

          2 - 5 HP ea., 17,000 CFM ea. =   6 KW   34,000 CFM

    2.6 Chilled Water System

         Winter 2 - 703 HP minimum
         Summer 3 - 703 HP maximum
                 1 - 715 HP

```
└─────────────────────────────────────────────────────────────┘
```

TABLE-1

ENERGY MANAGEMENT PROGRAM

This step in the process had to be planned out, and backed up with detailed documentation in a cost effective method that would not only sell the energy management program but would give a clear road map covering McDonnell Douglas Corporation energy plans for the next 10 years. This approved program included, but was not limited to, the following areas:

- o Energy Resources
- o Facility Energy Survey Audits
- o Objectives; Immediate, Medium, and Long Term
- o Methods To Meet Objectives
- o Monitoring Results and Actual Impact of This Program

EVALUATION GUIDE - EMCS POINTS

The following evaluation guidelines were used for suggested system tie-in points on the Energy Management Computerized System and a sample of component selection. The suggested point tie-in for the equipment such as pumps, boilers, chillers, fans, etc. must be placed in at least one of the following three classes:

Class I Mandatory Points

These points are minimum requirements to produce the estimated manpower and energy savings, and still secure equipment safety and failure detection. A start/stop indicating point for supply fans is a good example of a Class I point. Also, included in Class I are those points required for a certain utility monitoring. Another example would be utility meter input pulses required for demand limiting control.

Class II Optional Points

These points will all be useful, and their value depends on the individual situation. For example, an exhaust fan high temperature alarm might be moved up to "Mandatory" if it's failure could result in high temperature equipment failure, or left out if located in a continuously manned equipment room.

Class III Questionable Points

These points are those that should only be considered if you need to have code verification, reliability checks, or cost effectiveness. The data has no known relationship to running a building unless you plan to use it for documenting and recording information.

HARDWARE SELECTION

The initial system, referred to as "Phase I", was limited to 79 control devices in order to: (1) prove out the compatibility of hardware and software; (2) confirm estimates of energy and cost savings; and (3) determine effects on space conditions, occupants, and equipment.

Phase I has been in full operation since February 1976 and has proven to be highly successful in that (1) it produced a one year payback on energy savings; (2) we have had no major problems; (3) there has been no equipment malfunctions due to computer control; (4) it produces a daily statistical report showing electric consumption and savings in electricity for each device or group of devices; and (5) it decreased energy consumption in the related buildings by 20-30%.

Phase II expanded to a larger model of the IBM System 7 computer, which has (1) 64K memory, (2) a disc storage and (3) automatic restart in case of power failure. Phase II not only covered expansion of the IBM System 7

and tie-in of new buildings, but it also included the addition of a live weather station to provide optimization of control strategies and saving calculations.

The present EMCS consists of a central based, unmanned, control center located in one main building, with remote terminal units in over twenty (20) other St. Louis buildings occupied by McDonnell Douglas. This system is automatically controlling the heating, ventilating, and air conditioning equipment devices in remote buildings via the telephone system.

The key to the overall success of this system lies in the weather station, located on the roof of a centrally located building, which transmits analog data through a transmitter and cable to the EMCS control room where the analog data will be recorded and converted to digital form for input into the computer. The weather station selects the control strategies for cycling equipment, and provides a minute-by-minute energy saving record for each device as related to live outside weather conditions.

The IBM System 7 computer requires no operator other than to occasionally review printouts, change strategies for special holidays or special working hours, and to report device failures to the Maintenance Department.

The printer types out the following reports: (1) the daily statistical report; (2) any strategy changes which are initiated by the computer or the operator; (3) any device failure to respond to the computer or operator's command; and (4) upon special report request from operator.

All of the computer input and output data commands go through the central terminal unit (CTU) located next to the computer. The central terminal unit (CTU) interfaces remote terminal units (RTU) which are located at various remote areas. Signals to and from the CTU and RTU are over leased telephone lines or in-house phone lines. The remote terminal units properly circuit all signals to and from relay cabinets at the device being monitored and/or controlled. Each remote terminal unit is supplied with the proper circuit boards to correctly address the devices, monitor and count pulses from gas and electric meters; receive data; and in the case of the weather station, convert analog input to digital form. Through special programmed strategies, the EMCS System automatically controls and monitors all controllable devices according to the time of day, day of year, outdoor weather conditions and utility peak demand periods. If the EMCS projects that we are going to exceed the Hi-demand limit for the month, it will automatically shed (turn off) controlled devices for a minimum of 15 minutes in order to try not to exceed the Hi-demand. In the event that a new Hi-demand limit is reached, the program will automatically reset the demand limits to the new limits. If any device is shed on demand, it will automatically come back on the line with a guaranteed "ON" (running) time of 45 minutes to 2 hours before it will be available to shed again. The normal shedding cycle for each device, per month, is five times.

ENGINEERING & MAINTENANCE BACKING

As with any cost reduction plan, the success of the "EMCS" program depended upon the response at the plant level. The first challenge of this program was not in the engineering, but more of a management type, dealing with a method to create an energy awareness and to provide the tools which will increase energy efficiency on a long range, continuous basis without putting a strain on the plant maintenance, engineering, and operating personnel.

Our goals for development of such a plan covered such things as:

- Employees' interest awareness program on energy savings projects with real and lasting savings results.

- Provide operators and foremen with tools and opportunity to become an active part of the energy management team.

- Implement a vehicle which would make energy management an integral part of each department's operation.

RESULTS AND SUMMARY

Since the Phase I of EMCS System first went on the line (1976), related facilities have shown 20-30% energy savings (see Figure 1 & 2). Without losing sight of the basic tasks of providing a successful energy management program to (1) optimize energy efficiency; (2) minimize consumption use of energy; (3) use energy in it's most economical form, without breaking the first law of energy management which states that, "Plant utilities are always adequate to meet production demands". In reaching McDonnell Douglas energy goals, one of the most effective instruments which emerged thus far was the EMCS... "A sharp, double edged tool for cutting energy waste and costs".

In summary, by following a well structured approach to energy in management, centered around the EMCS, we have been able to accomplish the original energy goals and implement a long range energy program which includes, but is not limited to:

- Obtain Total Management Commitment
- Obtain Employee Cooperation
- Make Appropriate Energy Savings
- Analyze Energy Survey Results
- Set Realistic Energy Efficiency Goals
- Develop Energy Reporting Format
- Implement Energy Engineering Changes
- Optimize On Equipment
- Monitor Results

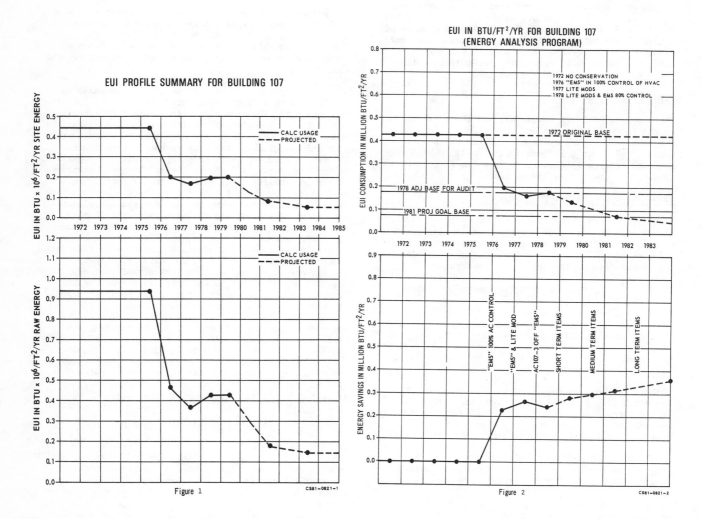

Figure 1   CS81-0821-1

Figure 2   CS81-0821-2

48

Chapter 7

THE USE OF THE ENERGY UTILIZATION INDEX IN ENERGY SURVEYS FOR GENERIC BUILDINGS

B. K. Hodge, W. G. Steele

ABSTRACT

The concept of using a target Energy Utilization Index (EUI) to determine the energy usage in efficiently operated and maintained buildings of the same generic type is investigated. The thermally light building types considered are schools and local government buildings which have daytime occupancy only. Target EUI increments for electricity and fuel usage were verified by comparison with energy usage data from efficiently run buildings. Current usage EUI's are given for eighty-four buildings, and examples of the use of the target EUI for determining energy savings are given for seven of these.

INTRODUCTION

This paper examines a methodology developed to facilitate the rapid estimate of energy and cost savings in schools and local government buildings. Within a given state, Mississippi in this case, school buildings generally have similar use patterns and some commonality in design and construction as well as similar climates. School buildings, because of these similarities, then belong to a class of generically similar buildings. This generic similarity, coupled with a low use factor (40 hours per week) and a low heating degree day climate, suggested that schools with effective energy management programs should have equivalent annual energy use on a per square foot basis. This annual energy use is called the Energy Utilization Index (EUI) and is the index about which this methodology is developed.

Other public buildings such as local government buildings, courthouses, and city halls can likewise be grouped into a generically similar category if the use patterns, climate, and construction type do not differ markedly. As a matter of fact, as Spielvogel [1] points out, the use pattern of the structure and the maintenance and operating policies for the energy using systems are more important than the building's construction type.

Such considerations led to the Building Energy Performance Standards (BEPS) program suggested by the Department of Energy. Others, namely Orlando, et al. [2], Windingland and Hittle [3], and Windingland, et al. [4] have attempted to correlate salient parameters to predict the total energy per square foot per year for a wide variety of structures in varying climates. These thrusts, however, have been predicated upon the idea that a given building's EUI could be predicted within

acceptable accuracy; but, irrespective of the building's physical characteristics, the manner in which the energy using devices are operated decisively affects the EUI. Even with sophisticated computer codes the effects of widely varying use patterns for both people and equipment are difficult to quantify. Such effects are virtually impossible to incorporate in any energy usage index which can be expressed in simple functional form. The specification of a generic class of structure in similar climates, as shown in this paper, allows a meaningful target EUI to be specified as the energy efficient operating goal.

This target EUI is generated by summing increments due to heating, cooling, cafeteria operation, and base electrical usage plus any other function which utilizes energy. These increments are derived from calculations and/or actual usage in buildings and have proven sufficiently accurate for generic buildings to be useful in specifying maintenance and operating procedure changes for more energy efficient operation. During the time period June 1980 through May 1982 over 60 school campuses and over 50 local government buildings have been examined and surveyed using the target EUI concept. In virtually every instance, the target EUI proved to be a meaningful quantity, and the actual and target EUI's, in conjunction with 12 months of utility bills and a walk-through energy audit, were sufficient to permit recommendations to be made which should save about 40 percent of the total energy used [5].

Based upon the concepts explored herein, the target EUI is meaningful if buildings are of the same generic type, have the same use factor, are used mostly during the daylight hours, and are situated in a common heating degree day climate. The only other restriction to its use is that the building structure must be such that the external weather condition is the primary factor affecting the energy usage for heating and air conditioning.

THE ENERGY UTILIZATION INDEX

The Energy Utilization Index (EUI), sometimes called the energy budget, is often used to express the energy consumed by a building or a complex of buildings. The EUI is calculated by dividing the total energy input into a building, expressed in a common unit, by the conditioned floor area. Typical units for the EUI are Btu/ft^2-year. This Index is widely used in energy auditing since it provides a common basis for comparing gross energy usage in a building. In order to generate an EUI, all sources of energy input must be converted to a common basis--usually Btu's in the United States.

Table 1 gives the commonly accepted conversion factors for electricity, natural gas, and propane--the three most extensively used energy sources in non-metropolitan areas.

This work was funded by the Mississippi Energy Extension Center as part of the DOE Energy Extension Service function.

Source	Conversion
electricity	3413 Btu/kwh
natural gas	1,030,000 Btu/MCF*
propane	91,500 Btu/gallon

* MCF = one-thousand cubic feet.

TABLE 1. BTU CONTENT OF VARIOUS ENERGY SOURCES

The electrical conversion factor of 3413 Btu/kwh is used herein since it represents the actual Btu content of a kilowatt-hour. Federal regulations for various levels of energy audits, as prescribed by the National Energy Conservation Policy Act and as reported in the April 19, 1979 Federal Register [6], require the use of the conversion factor 11,600 Btu/kwh. This larger number represents the estimated energy required to generate and deliver 1 kwh from a fossil fueled generating plant. For energy auditing purposes, the proper conversion factor is 3413 Btu/kwh since it represents the energy equivalent actually input into a structure.

The EUI is useful because it permits the rapid quantification of a building's annual energy use and provides a consistent basis with which to compare the annual total energy use of various structures. The actual value of a building's EUI is a function of many factors --the structure's size, shape, condition, orientation, climate, use, equipment, and mode of operation to formally list a few. And therein lies the greatest drawback to using the Index: depending upon the aforementioned building parameters, actual EUI's can range over nearly two orders of magnitude, from 20,000 Btu/ft^2-yr for churches to over 1,000,000 Btu/ft^2-yr for a poorly run office building. Table 2, taken from Spielvogel [7], lists approximate ranges of the EUI for various buildings. Considered in light of the information in Table 2, the calculation of an EUI for a building would permit its immediate categorization as having high, low, or typical energy use. A comparison of EUI's between buildings of markedly different sizes, shapes, and materials in widely varying climates would allow little inference about the relative efficiencies of the buildings' energy use; however, a comparison of EUI's between similar buildings with similar use patterns in similar climates would allow quantitative statements about the efficiencies of energy use to be made. In the latter case, estimates of possible EUI's for energy efficient operation of such generic categories should be possible and meaningful in the quantitative sense. This concept is explored in the next section.

	EUI (Btu/ft^2-yr)		
Type	high	low	typical
Office Buildings	400,000	40,000	150,000
Hospitals	600,000	200,000	350,000
Retail stores	300,000	70,000	120,000
Food stores	600,000	200,000	420,000
Schools	200,000	40,000	85,000
Churches	120,000	20,000	40,000
Apartments	200,000	60,000	125,000

TABLE 2. EUI RANGES FOR DIFFERENT TYPE STRUCTURES

THE TARGET EUI

If, as asserted in the previous section, generic buildings with energy-efficient operation in the same climate have similar EUI's, then a meaningful target EUI can be developed, and a building's potential for energy and cost savings can be realistically estimated with a minimum of computations. The target EUI represents the potential EUI a building would have for a given use

pattern if energy-efficient operation were achieved. The existence of a meaningful target EUI can be confirmed only by comparison with actual building energy usage data. Estimates or calculations of the endpoint energy usages should agree closely with the actual usage data from buildings with energy efficient operation. Generalizations appropriate for other climates and building types can only be validated by additional data.

As the initial step in developing a target EUI, a typical Mississippi elementary school with inefficient energy utilization was subjected to an energy audit. The results of this energy audit and the energy conservation opportunities (ECO's) examined are delineated in the report by Hodge and Steele [8]. The school possessed an EUI of 76,944 Btu/ft^2-yr--an excessive amount. Considering in order, tightening of the envelope and night set back, turning off of pilot lights in the summer, reduced hall lighting, double glazing, and an R-19 suspended ceiling, the EUI was reduced to 26,771 Btu/ft^2-yr. This represents an overall reduction in energy consumed of 62 percent. Verification of the reasonableness of this reduction was made by examining the actual data from a similar elementary school in which the ECO's had actually been carried out. Pierce St. Elementary School in Tupelo, Mississippi had undergone extensive retrofit for better energy conservation and management and had an actual EUI of 27,935 Btu/ft^2-yr. These schools which are of similar construction, and similar use and in a similar climate fit the criteria to be labeled generic and their actual and estimated EUI's indicate that the target EUI concept is a meaningful one. Subsequent energy surveys throughout the state have shown the target EUI concept to be meaningful for schools with energy efficient operation.

The target EUI is generated by adding increments for heating, air conditioning, lighting and other electrical equipment and food service. The part of the EUI attributable to each function in given in Table 3. These quantities apply only to schools with a nine month academic year and daytime operation. The values

Function	EUI Increments (Btu/ft^2-yr)
Heating	
Gas	18,000
Electric Resistance	12,600
Electric Heat Pump	6,300
Air Conditioning	4,000 (for 2 months only)
Lighting & Other Electrical	
Equipment	9,000
Food Service	9,000

Energy Conserving Features:	Reduce Target EUI By:
For Double-Glazed Windows with	
Gas Heat	3,950
Electric Resistance	2,750
Electric Heat Pump	1,375
For Drop Ceilings with R-19 Insulation with	
Gas Heat	5,900
Electric Resistance	4,100
Electric Heat Pump	2,050
For Exterior Glass Reduction	0

TABLE 3. TARGET EUI INCREMENTS FOR SCHOOL BUILDINGS

for double-glazed windows and drop ceilings with R-19 insulation are included for comparison only. They were not used in computing the target EUI's given later since the payback periods (for gas heat) are in excess of thirty years for these modifications. Also, as shown in the table, the removal of exterior glass has no energy saving effect for a building that is used primarily in the winter. The solar gain provided by the windows is beneficial so long as the windows seal tightly enough to limit infiltration. For example, the typical unairconditioned Mississippi school with natural gas heat and a cafeteria would have a target EUI of 36,000 Btu/ft^2-yr. This is composed of 18,000 Btu/ft^2-yr for heating, 9,000 Btu/ft^2-yr for lighting and 9,000 Btu/ft^2-yr for food service.

The values indicated in Table 3 were used throughout the state since the climatic variations are small; the summers are hot and humid and the winters are mild, varying from 1750 degree days (65°F base) along the Gulf Coast to 3300 degree days near the Mississippi/ Tennessee border. The validity of the target EUI concept for schools within the state is verified by Figure 1.

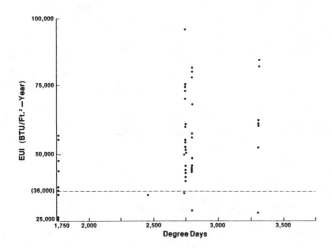

FIGURE 1. SCHOOL ENERGY SURVEY DATA

The current EUI's for the schools surveyed are shown on the Figure. Also, for comparison the nominal target EUI of 36,000 Btu/ft^2-yr is indicated. Even though there is considerable scatter in these data at each degree day location, there is still a trend of increasing EUI with number of degree days. However, it was found that schools with an effective energy management program and tight windows and doors had EUI's below the nominal target irregardless of degree days. This trend suggests that the degree day effect primarily results from nighttime operation of the HVAC system. In the calculation of the target EUI, the HVAC systems were considered to be operated only when personnel were present. The schools with EUI's falling near the 36,000 Btu/ft^2-yr value fit the generic model well and had very effective energy management programs. The four schools with EUI's well below 36,000 Btu/ft^2-yr were exceptionally tight and had either few windows or double-glazed windows and some ceiling/roof insulation.

The number of schools with EUI's in excess of the target value indicates the large potential for cost reduction in the operation of Mississippi public

schools. For the fifty school campuses surveyed, total savings of 60 billion Btu's and $417,000 were identified. (Cost savings were based on $4 per MCF of natural gas, $0.70 per gallon of propane, $0.06 per kwh of electricity and $7.00 per kw of electrical demand.) Details of the school energy data and survey results are given by Walker, Steele, and Hodge [5].

The validity of the target EUI for schools in Mississippi has been established by the combination of a detailed energy audit, the Pierce St. School data, and energy surveys of fifty school campuses. The application of the target EUI for energy surveys of this generic class, schools, is shown later.

The same logic as that used for the schools was used in developing a target EUI for local government buildings. The EUI increments for heating, lighting and air conditioning are given in Table 4.

Function	EUI Increment (Btu/ft^2-yr)
Heating	
Gas	18,000
Electrical Resistance	12,600
Electrical Heat Pump	6,300
Air Conditioning	14,000
Lighting & Other Electrical Equipment	11,000

TABLE 4. TARGET EUI INCREMENTS FOR LOCAL GOVERNMENT BUILDINGS

These quantities apply only to local government buildings with a daytime, 12-month operation. For those situations where a 24-hour operation was necessary, such as jails, the target EUI was adjusted appropriately. The variance in the construction and operation of these local government buildings made the target EUI more difficult to define than for the schools.

The compilation of data from local government building energy surveys is given in Figure 2. The nominal target EUI of 43,000 Btu/ft^2-yr is indicated in the

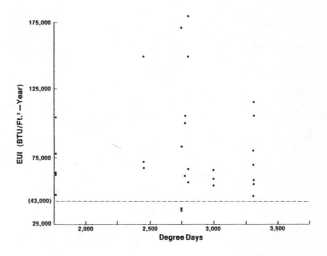

FIGURE 2. LOCAL GOVERNMENT ENERGY SURVEY DATA

Figure. The wide variability in building types and operation caused such a scatter in the EUI data that there is no degree day effect evident. Only four local government buildings, all courthouses, possessed EUI's near the target of 43,000 Btu/ft^2-yr. All of these buildings were tight and had effective energy management programs. Because of the lack of centralized authority to control the building's operation, most local government buildings did not have effective energy management programs. Many boilers and other HVAC system components were run continuously and had little or no preventive maintenance. These shortcomings are reflected in the actual EUI's of the buildings. As with schools, local government building examples using the target EUI concept are presented in the next section.

APPLICATIONS

The application of the Target EUI method for determining energy savings in generic buildings is demonstrated in this section. The example buildings are all located in Mississippi so that they all have a similar low heating degree day climate. The first four examples deal with schools which have a nine month academic year and daytime operation. The latter three examples are for local government buildings with a daytime, twelve-month operation.

School Building A

This school building was typical of most in Mississippi with walls constructed of brick veneer outside/concrete block inside, a roof of concrete and wood aggregate called TECTUM ^{R} and a continuous row of windows along the exterior wall of each classroom. The conditioned area of 88,206 ft^2 was both heated and cooled. The heating fuel had recently been converted from propane to natural gas. The energy usage for this school for the past twelve months was 7,145 Btu/ft^2-yr electricity and 27,629 Btu/ft^2-yr gas for a current usage EUI of 34,774 Btu/ft^2-yr.

Using the target EUI increments given in Table 3, the target EUI for this school building is 40,000 Btu/ft^2-yr. This target value consists of 27,000 Btu/ft^2-yr for gas and 13,000 Btu/ft^2-yr for electricity. Comparison with the current energy usage indicates that there is no potential for energy savings.

The building had windows which ranged from very tight to relatively loose. But a concentrated effort was made to keep windows and doors shut in the winter and loose windows were covered with plastic sheeting. Therefore, the building envelope was tighter than most schools. However, the significant factor in the low energy utilization was a conscientious program of turning all heaters, air conditioners and lights off during unoccupied hours. During very cold nights, a limited number of heaters were left on at a low thermostat setting to provide freeze protection. The principal and the person in charge of maintenance both verified that the systems were shut off and they also imposed strong regulations on the use of the equipment. As a result the building was being operated below the normal target energy usage.

School Building B

This building was identical in construction to Building A. The school was all electric with air conditioning and electric resistance heating. The current energy usage for the 19,498 ft^2 building was 43,799 Btu/ft^2-yr.

The target EUI for this school is 34,600 Btu/ft^2-yr. The main cause of the excess usage was the failure to turn off systems during unoccupied hours. To remedy

this situation, the installation of an energy management system with freeze protection was recommended. This system would provide both energy and demand charge savings.

The annual savings were determined by subtracting the target EUI from the current usage to yield a potential energy savings of 9,199 Btu/ft^2-yr. For this school this is equivalent to 179 x 10^6 Btu or 52,553 kwh or $3,153 at six cents per kwh.

The demand savings were calculated by first determining the base electrical demand for lighting and the cafeteria. Then it was assumed that the energy management system would be used to duty cycle the air conditioners and the heaters to reduce the monthly peak electrical demand. By cycling off in each ten minute period one out of three heater units and one out of two air conditioners the demand charge savings would be $3,502 at seven dollars per kw.

School Building C

This school was generic in construction and climate with the other examples. The primary energy usage problem in this building was the looseness of the windows. The current usage for the 60,495 ft^2 building was 7,822 Btu/ft^2-yr electricity and 53,922 Btu/ft^2-yr gas for an EUI of 61,744 Btu/ft^2-yr. The school was not air conditioned.

The target EUI in this case is 36,000 Btu/ft^2-yr (27,000 Btu/ft^2-yr gas and 9,000 Btu/ft^2-yr electricity). A comparison with the current usage shows that there is no potential for electrical savings, but that 26,922 Btu/ft^2-yr of gas can be saved. This quantity represents 1,581 MCF of natural gas or $6,325 at four dollars per MCF. This annual savings can be realized by weatherstripping the windows and by ensuring that the heating is properly controlled.

School Building D

This school was unique from most that were surveyed in the state because its windows were relatively tight-fitting and there were overhangs from the roof which shaded the windows in the summer assisting in the air conditioning of the building. However, except for these the building was of typical school construction. The current energy usage in the 75,514 ft^2 building was 22,433 Btu/ft^2-yr electricity and 38,350 Btu/ft^2-yr gas for an EUI of 60,783 Btu/ft^2-yr.

The target EUI in this case is 40,000 Btu/ft^2-yr (27,000 Btu/ft^2-yr propane gas and 13,000 Btu/ft^2-yr electricity). In this school there was no comprehensive program for shutting off equipment. Therefore, in many areas heaters and air conditioners operated continuously. By comparing the target and current usage portions of the EUI for both electricity and gas, savings are calculated of 208,709 kwh or $12,523 and 9,366 gal of propane or $6,556 at seventy cents per gallon. By duty cycling the air conditioners with an energy management system, an additional savings of $1,526 could be realized.

Local Government Building A

Local government buildings such as courthouses, city halls, etc. while different in construction from schools are still usually thermally light buildings as classified by Spielvogel [7]. The energy usage in such buildings for heating and cooling is dependent on the weather conditions. Of course, large courthouses or office buildings with several internal zones would not fall completely in the thermally light class. However, all the local government buildings in this study were essentially thermally light structures.

Building A was a courthouse which was forty years old. The building was in excellent condition with tight windows and doors. Also the building custodian was practicing an extremely conscientious program of turning off heating and cooling equipment when it was not needed. The current energy usage was 24,192 Btu/ft^2-yr electricity and 12,750 Btu/ft^2-yr gas for an EUI of 36,942 Btu/ft^2-yr.

The target EUI for this building is obtained from Table 4 as 43,000 Btu/ft^2-yr. This target value is composed of 18,000 Btu/ft^2-yr for the gas fired boiler steam heat and 25,000 Btu/ft^2-yr for the lights, air conditioning and other electrical equipment. Comparison with the current energy usage shows that there is no potential for energy savings. As noted above, this energy efficient operation is primarily due to the building custodian's control of the equipment. Due to the age and maintenance of the HVAC equipment and controls, some areas were not properly conditioned. If these deficiencies were remedied, the building energy usage would be closer to the target value.

Local Government Building B

This building was an all electric city hall of 7,776 ft^2. Heating was provided by electrical resistance. From Table 4, the target EUI for this case is 37,000 Btu/ft^2-yr.

For this city hall the current electrical usage was 149,177 Btu/ft^2-yr. The primary cause of this excessive usage was the continuous operation of the HVAC equipment. It was recommended that an energy management system be installed in this building to control equipment operating times and electrical demand. The difference in the current usage EUI and the target EUI gives a potential annual energy savings of 254,211 kwh or $15,253. Duty cycling of the equipment would provide an annual demand charge savings of $1,617.

Local Government Building C

Building C was a very old courthouse which had been renovated. The building envelope was relatively tight and the HVAC system consisted of window air conditioners and a steam boiler with radiators. The current gas usage for the 20,000 ft^2 building was 119,750 Btu/ft^2-yr and the electricity usage was 30,200 Btu/ft^2-yr.

The target EUI for this building is 43,000 Btu/ft^2-yr with 18,000 Btu/ft^2-yr for gas heating and 25,000 Btu/ft^2-yr for electrical usage. A comparison shows savings potential, especially for gas. The gas fired boiler in the building was badly in need of maintenance and it was operated continuously. Also, the steam lines in the building were uninsulated causing overheating in many areas. Proper boiler maintenance and control could save 2035 x 10^6 Btu or 1976 MCF or $7,903 per year. Replacement of the remaining incandescent lights with flourescent fixtures and the proper control of the air conditioners could save 30,472 kwh per year or $1,828.

CONCLUSION

The concept of using a target EUI to determine the energy usage in efficiently operated and maintained buildings of the same generic type was investigated. The building types considered were schools and local government buildings which had daytime occupancy only. These buildings were thermally light in that there energy usage for heating and air conditioning depended primarily on the outside weather conditions.

Comparison of the Target EUI with energy conservative examples showed the validity of this method. The difference between the target values and the current energy usage per square foot per year for electricity and fuel gave the annual energy savings when converted from Btu to amount and cost. This simple approach allowed quick estimates to be made of the energy cost savings which are necessary to determine the payback periods for energy conservation measures.

The target EUI analysis approach is not intended to replace energy surveys of buildings by qualified energy auditors. Such surveys are necessary to identify the problem areas in order to make recommendations for energy use improvements. Since in thermally light buildings with limited occupancy the primary energy savings result from the tightness of the building envelope and the proper control of the HVAC systems, the target EUI allows a quick estimate of the energy savings after these conditions are met.

The increments given in Tables 3 and 4 apply to buildings in a low heating degree day climate. However, the premise of this paper is that target EUI increments could be developed for the energy using systems in any climate. The only restriction would be that the buildings are generic with similar use patterns and some commonality in design and construction. A detailed audit of one such building would provide the analysis tool to be used for all the buildings in that class.

REFERENCES

1. Spielvogel, L. G., "Exploding Some Myths About Building Energy Use," Architectural Record, Feb. 1976, pp. 125-128.

2. Orlando, J. A., Spielvogel, L. G., and Weed, H., "Feasibility of an Index for Office Buildings," Final Report, Mathtech, Inc., Princeton, NJ, August 1976.

3. Windingland, L. M. and Hittle, D. C., "Energy Utilization Index Method for Predicting Building Energy Use," Vol. I, Report No. CERL-IR-E-105-Vol. 1, May 1977.

4. Windingland, L. M., Hittle, D. C., Hinkle, B. K., and Piper, J. E., "Energy Utilization Index Method for Predicting Building Energy Use," Vol. II, Report No. CERL-IR-E-105-Vol. 2, May 1977.

5. Walker, R. D., Steele, W. G., and Hodge, B. K., "1981 Summary Report: Energy Surveys of School and Local Government Buildings in Mississippi," MCES Report No. 1357, Mississippi State University, Mississippi State, MS, May 1982.

6. _____, "Technical Assistance and Energy Conservation Measures: Grant Programs for Schools and Hospitals and for Buildings Owned by Units of Local Government and Public Care Institutions," Federal Register, Vol. 44, No. 75, April 17, 1979, pp. 22940-22957.

7. Spielvogel, L. G., "Building Envelopes," in Energy Management Handbook, ed. W. C. Turner, Wiley-Interscience, 1982, pp. 269-302.

8. Hodge, B. K. and Steele, W. G., "Energy Conservation Opportunities in 'EFC' Schools," Mississippi Energy Extension Center Report, MCES, Mississippi State, MS, August 1981.

Chapter 8

A PROPOSED METHOD FOR THE SELECTION OF BUILDINGS WITH THE GREATEST POTENTIAL FOR ENERGY CONSERVATION

G. Baird, W. D. S. Brander

PREFACE

New Zealand, in common with most developed countries, has been forced to reappraise its energy supply and demand situation (1). Systematic studies of the various demand sectors have been carried out during the last decade. The Energy Research Group at Victoria University's School of Architecture has conducted several surveys of energy demand in the building sector (2,3). The group is now carrying out detailed studies of the energy end-use patterns of commercial buildings (4), the energy performance of schools (5), and the energy management of government buildings. The work has led to the development of several techniques related to building energy management. Possibly the most important of these techniques - how to select,from a group of buildings,those with the greatest energy conservation potential - will be described in this paper.

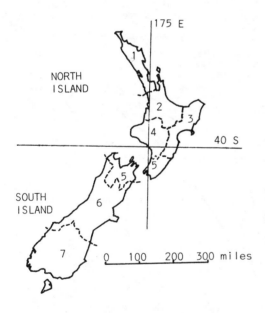

EXHIBIT 1 : ADMINISTRATIVE DISTRICTS OF THE NEW ZEALAND MINISTRY OF WORKS AND DEVELOPMENT.

MWD DISTRICT

1. Auckland
2. Hamilton
3. Napier
4. Wanganui
5. Wellington
6. Christchurch
7. Dunedin

BACKGROUND

In May of 1980, the New Zealand Government, through the Ministry of Works and Development (MWD), established a Building Energy Conservation Advisory Service (BECAS). This programme had two main elements. First, a multi-disciplinary building energy conservation advisory team in each of the Ministry's seven districts (see Exhibit 1).

Second, energy managers, appointed from the occupant government department, for each building under that department 's control.

Concurrently, and at the request of MWD, the Energy Research Group (ERG) undertook a pilot study of the nature of energy consumption and its control within the government buildings sector. The report (6) of this work presented data on existing energy use patterns, on a departmental basis; and identified the major individual consumers within the government buildings sector. The potential sources of information on building energy use were reviewed and the current building administration systems described, with respect to 'energy accountability'.

The outcome of the pilot study was a recognition of the management problems related to the effective use of resources within any multi-building programme of energy conservation. There appeared to be an abundance of technical solutions available at the building system level, but a lack of diagnostic tools to enable the manager of a group of buildings to readily pinpoint which buildings have the most energy conservation potential. Judging from the experiences of several large organisations it was apparent that in any realistic building energy conservation programme, lack of manpower, limitations of finance, and other constraints would most likely prevent all the buildings of a large group from being energy audited and having their conservation opportunities assessed. Indeed it seemed probable that such an approach would be less cost-effective than one which attempted to concentrate on those consumers with the most potential for large reductions in energy consumption.

Consequently, the next phase of the Energy Research Group's study concentrated on the information requirements, strategies and techniques which could be used to improve the effectiveness of an energy management programme. This paper deals, in turn, with the acquisition of building energy consumption information, the determination of overall priorities for building energy conservation programmes, and the selection of buildings with the greatest energy conservation potential.

ACQUIRING BUILDING ENERGY CONSUMPTION INFORMATION

For any energy conservation programme, it is essential to have knowledge of existing consumption patterns. In the case of NZ government buildings, it was found that supplier's records provided a convenient means of identifying users and their annual consumption. In a number of instances national records were kept at one location. Consumption for the 1979 calendar year was obtained from suppliers based in Wellington. Major consumers were identified and consumption was categorised on a departmental basis.

A parallel postal survey of energy consumption in buildings owned and/or occupied by government departments was also undertaken. This survey obtained annual energy consumption and cost data together with the floor area of 100 buildings.

Energy Supplier's Records

Supplier's records provided a readily usable source of information on government buildings, but from an energy management viewpoint there were a number of major gaps in the data. One drawback with these records was that they were not always categorised according to end uses. This was particularly evident in the case of oil, where there was generally no direct method of identifying whether sales to government departments were for transport use or for heat production. Similarly, as with other fossil fuels, no distinction was made between sales for process heating and space heating.

The most comprehensive data available was on coal consumption by government buildings, most users being supplied by State Coal Mines. Records of annual sales from all districts were kept by the Mines Division, Ministry of Energy. These records allowed the annual sales to all government sector delivery points to be identified.

Natural gas is supplied throughout New Zealand's North Island by nine distributors. Delivery point information and consumption for government users in each area had to be extracted from the customer files of each distributor. Manufactured gas and LPG are used in several cities in both Islands. Use of the former fuel is declining with the phasing out of government subsidies, the increased reticulation of natural gas, and supplies of LPG. Information on the consumption of government users was obtained from the suppliers.

Heating oil (comprising mainly diesel) is supplied on a regional basis, under a government contract. Delivery point statistics were kept only by companies supplying the Wellington and Christchurch regions. Other companies kept statistics based on total departmental consumption of each project. These did not indicate the quantities consumed in buildings.

The electricity consumption of government buildings was 'buried' in the customer files of over sixty electrical supply authorities. Any estimate of total government consumption would require annual extraction of the data from their records.

MWD District Records

Conventional accounting systems are not normally set up to provide adequate information on energy consumption within buildings. However, a survey of MWD Districts showed that annual energy consumption and costs, by fuel type, could be obtained from accounting records. Survey forms were sent to each district requesting the energy information together with the floor area and function of each building for which the District paid energy costs.

It was found that energy consumption information could be relatively easily extracted from accounting records, provided only the past year's consumption was required. In this respect, New Zealand is no different from other countries, where energy supply authority records and departmental accounts have provided a suitable data base for successful conservation programmes (7).

More importantly, this data can also provide sufficient information for selecting those buildings whose energy conservation potential should be assessed.

DETERMINING OVERALL PRIORITIES FOR A BUILDING ENERGY CONSERVATION PROGRAMME

With over 1800 buildings throughout New Zealand under their care, it seems certain that the MWD could reasonably apply detailed energy conservation programmes to only a fraction of their stock. However, before that fraction is determined, it is necessary to establish the effects on total sector energy use, of attempts to reduce consumption in only a proportion of buildings. Energy consumption data from several large groups of buildings will now be used to illustrate the importance of obtaining at least the annual energy consumption of each building in a target group. Energy data from a specific group of government buildings is also examined.

Using Pareto Curves

The type of information discussed previously can be used to identify the percentage of energy used by the larger consumers in a group of buildings. A convenient method of displaying and analysing this information is in the form of a pareto curve - plotting the cumulative consumption of all buildings ranked in descending order of energy use. Where energy consumption data is not available, cost data can be used to derive the curve, although this may result in an altered distribution if the effects of volume discounts on the unit cost of different fuels are significant. Pareto curves for four different groups of buildings are presented in Exhibit 2.

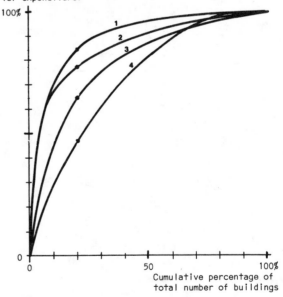

Cumulative percentage of annual energy consumption (or expenditure)

Cumulative percentage of total number of buildings

1. Wellington CBD (1057 commercial buildings)
2. Auckland Region (707 commercial buildings)
3. Gloucester County Council (330 schools)
4. Wellington Region (233 primary schools)

EXHIBIT 2 : PARETO CURVES OF ANNUAL ENERGY CONSUMPTION (OR EXPENDITURE) FOR FOUR GROUPS OF BUILDINGS

Curve 1 is for the buildings of the Wellington Central
Business District (2). The top 10% of consumers use
about 70% of this group's total energy consumption.
Curve 2, which is for a sample of commercial buildings
in the Greater Auckland Region (8) shows a similar
distribution for the top 10% of the buildings (which use
about 67% of total energy) but the curve then drops
below that for the Wellington CBD. This difference may,
along with other factors, reflect the different
sampling methods used in the two groups. The Wellington
CBD study included all the buildings in a small
geographical area in central Wellington while the
Auckland study used a 5.4 per cent sample of buildings
covering a much larger area. Hence, the dominance of
'large' buildings might not be quite so pronounced in
the Auckland study.

In contrast to the two groups of commercial buildings,
Curves 3 and 4 in Exhibit 2 show a lower fraction of
energy costs and consumption resulting from the schools
with high energy bills. The top 10% of Wellington
primary schools use only 29% of the total energy. This
probably reflects the more 'homogenous' nature of a
group of schools. The ranges of size, energy
consumption, energy intensity (e.g. energy consumption
per unit floor area) and therefore energy costs, are
generally lower than for groups of commercial buildings.

The pareto curve gives information which can provide a
useful focus for an energy management programme.
Generally, the greatest reduction in energy consumption
within a group of buildings will be achieved by
concentrating on the larger consumers. These are the
buildings which are most likely to yield greater energy
and dollar savings for a given expenditure on controls,
maintenance, etc.

Assessing the Effect of Large Consumers

The method allows assessment of the effect of energy
savings, by a fraction of the large consumers, on the
consumption by the group as a whole. For instance,
Exhibit 3, which is derived from the pareto curves of
Exhibit 2, indicates the average reductions in energy
consumption (costs) among the larger consumers, which
are required to yield an overall 10% saving in each of
the four groups of buildings.

In the Wellington CBD a 10% reduction in the total
consumption of all 1057 buildings, is equivalent to
reducing energy consumption by an average of only 14% in
the top 10% of consumers (108 buildings). Alternatively
the same result could be obtained by an average
reduction of 25% in the 26 buildings with the largest
energy consumption.

By contrast, the same strategy applied to the group of
schools might be less useful. A 27% reduction in energy
consumption among the top 10% of consumers would be
required to reduce energy consumption in the whole group
by 10%. If a maximum average reduction of 14% is
considered feasible, then this must be achieved in 28%
of the larger consumers to effect a 10% reduction over
the whole group.

The pareto curve for the 100 buildings of the MWD
Districts survey described earlier is given in Exhibit 4.

It has been based on energy costs rather than energy
consumption. This has little effect on the general
shape of the curve, which has a cumulative cost
distribution lying between those of Curve 1 and Curve 2
in Exhibit 3. The top 20% of the buildings account for
79% of the group's energy costs, while the bottom 20%
use only 2% of it.

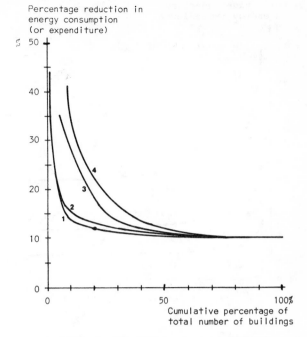

1. Wellington CBD (1057 commercial buildings)
2. Auckland Region (707 commercial buildings)[8]
3. Gloucester County Council (330 schools)[9]
4. Wellington Region (233 primary schools)

EXHIBIT 3 : REDUCTION IN ENERGY CONSUMPTION
(OR EXPENDITURE) REQUIRED TO ACHIEVE AN
OVERALL SAVING OF 10% FOR THE FOUR GROUPS OF
BUILDINGS OF EXHIBIT 2

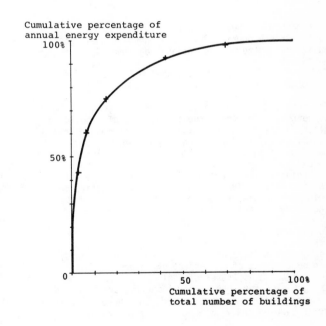

EXHIBIT 4 : PARETO CURVE OF ANNUAL ENERGY EXPENDITURE
FOR THE 100 BUILDINGS OF THE MWD SURVEY.

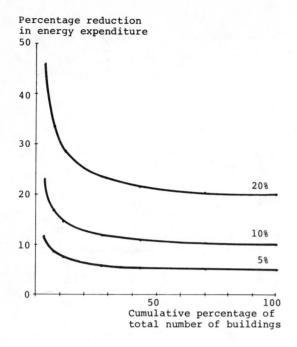

Percentage reduction
in energy expenditure

Cumulative percentage of
total number of buildings

EXHIBIT 5 : REDUCTION IN ENERGY EXPENDITURE OF LARGE
CONSUMERS REQUIRED TO ACHIEVE OVERALL
REDUCTIONS OF 5, 10 AND 20% FOR THE 100
BUILDINGS OF THE MWD SURVEY

Exhibit 5 shows the average reductions in energy costs
that are required among those buildings with high costs,
in order to reduce energy expenditure in the whole group
by 5, 10 and 20% respectively. For example, if a 20%
cost reduction could be achieved amongst the top 14% of
consumers, this would reduce total costs in the group by
10%.

This method has been applied, with considerable success,
by the UK's Property Services Agency, in their government
buildings' energy conservation programme (9).

SELECTING BUILDINGS WITH THE
GREATEST ENERGY CONSERVATION POTENTIAL

The previous section showed the importance of
concentrating on the largest consumers of a building
group in any energy conservation programme, so that a
significant fraction of the group's energy use will be
managed. This section describes the formulation of an
index, aimed specifically at ranking these buildings
according to their potential for energy conservation.
The index is designed to provide a quantitative measure
which can supplement qualitative assessment of
'conservation potential'.

Energy Use Indices

Energy use indices are usually expressed as energy input
per unit output of the object of interest. Energy
indices, such as litres of fuel per hundred kilometres or
kilograms of fuel per tonne-kilometre, are accepted
measures in the transport field, but there is no such
commonly used index of a building's energy use.

One problem with such an index is that the energy
consumption of a building is dependent on uncontrollable
factors such as sunlight and outdoor temperature. Unless
the effects of these factors are accounted for, energy
consumption in a particular building, over sequential
time periods, cannot properly be used to measure energy

performance. However, for inter-building comparisons
within the same region this problem is less significant,
since macro climatic effects will be similar.

At the present time, the Area Energy Use Index (AEUI) is
beginning to find some acceptance in the commercial
building sector. The AEUI is calculated by taking the
building's annual energy use (say, $MJ.yr^{-1}$) and dividing
by the floor area (m^2). This results in an energy
performance index expressed in $MJ.m^{-2}.yr^{-1}$. However,
particular activities often have bases other than area
which are already used as a measure of other forms of
performance, and these may frequently be more relevant.
For example, in the case of schools one may be
interested in energy use per pupil; in hospitals,
energy use per available bed; and in buildings housing
mainly clerical workers the index could be energy use
per occupant (or better still per occupant-hour).

Using the Area Energy Use Index

For the purpose of quickly assessing energy conservation
potential, a readily obtained normalising factor will
usually be preferred. In practice this will often be
floor area, leading to the use of the Area Energy Use
Index (AEUI). However, it is important to compare
AEUI's for buildings providing the same functions. A
hospital will generally have a higher AEUI than a school
but this does not imply that its energy performance is
worse. Hospitals are in continuous use and house a host
of high energy consuming services by comparison with
any conventional school.

The AEUI is an index of efficiency in terms of the first
law of thermodynamics. The greater its value the
greater the energy intensity of a given building in heat
equivalent terms. However, simply summing the energy
units of each fuel creates a thermodynamic bias in
favour of all-electric buildings, since the inherent
inefficiencies in converting fossil fuels to usable
heat, and the second law inefficiency of using a low
entropy fuel like electricity for the production of low
temperature thermal energy, are ignored.

A refinement of the energy use index which partially
accounts for these considerations is to calculate the
annual 'effective' energy use of each building by
applying a weighting factor to each fuel. The New
Zealand Ministry of Energy (10) has assumed the
following weighting factors for fuels in the industrial
and commercial sectors: electricity, 1.0; gas, 0.8;
oil, 0.7; and coal, 0.6.

The weighting factors can be applied without knowledge
of the end uses of fuels in a particular building. They
simply weight building energy consumption according to
the achievable thermal efficiencies of the fuels used.
The refined energy use index is a quick measure of how
efficiently a particular set of fuels are being used
within the building. It does not attempt to identify
inappropriate selection of fuels for a particular end
use. It is assumed that relative prices for different
types of energy will provide such an indication.

The Index of Energy Conservation Potential

A measure of energy efficiency is not sufficient in
itself to indicate the potential cost-effectiveness of
energy-reduction measures. By comparison with large
consumers, small ones tend to have less potential for
big reductions in energy use and cost, simply because
lower amounts of energy are involved. In addition, the
costs of many energy conservation opportunities are
largely independent of actual consumption. For example,
the installation of controls on a large energy flow
usually costs less per unit of energy consumed, than for
a small flow. Hence, all else being equal, energy

conservation opportunities are generally more cost-effective when applied to large users. To account for the monetary benefits of applying energy conservation opportunities to these consumers, the energy use index should be weighted by the annual cost of energy for the building.

From the foregoing discussion the following Index of Energy Conservation Potential (IECP) is proposed for assessing the energy conservation potential of buildings.

IECP = (Energy Use Index) x (Annual Total Energy Cost)

If the energy use index for an office building (say), is based on 'effective' energy and normalised by occupant-hours, then the index of energy conservation potential is given by:

$$IECP = \frac{[E + (0.8)G + (0.7)O + (0.6)C]}{\sum_{i=1}^{N} \left[\begin{array}{c} \text{hours per year worked} \\ \text{by the } i^{th} \text{ occupant} \end{array} \right]} \times \left[\begin{array}{c} \text{ANNUAL TOTAL} \\ \text{ENERGY COST} \end{array} \right]$$

where N = number of building occupants
and E, G, O and C are the annual consumptions (in consistent units) of electricity, gas, oil and coal, respectively.

The objective is to lower the value of the IECP, either by reducing annual energy requirements for a given level of occupancy, or by increasing occupancy for a given level of energy use (it will be clear that fuel substitution and tariff adjustment will also reduce the index). The use of the normalising factor 'occupancy' avoids separating the energy management function from other aspects of building management, such as building operating hours and space allocation.

Until further validation of the IECP takes place, it has been assumed that the Energy Use Index and the Annual Total Energy Cost have the same weighting in the construction of the Index of Energy Conservation Potential. A building which has either double the energy cost, or double the energy use per occupant-hour, of another building, will have double the value of IECP.

Applying the IECP

The IECP can easily be weighted to take account of other building characteristics such as HVAC type, age, etc., if experience shows these to be correlated with cost-effective energy conservation. For instance, central heating systems are likely to provide more worthwhile conservation opportunities than smaller unit systems, because of the larger energy flows involved at the central plant.

The index is not normalised for climatic effects and is only valid for comparison of buildings in the same locality over the same time period. Normalisation of total annual energy requirements to account for climatic variations is not possible, without first separating out the fraction of a building's energy requirements not related to climate. The use of the index within a given region could enable meaningful comparisons of buildings to be made over the same time period. The data on energy consumption and costs would normally be taken from suppliers invoices. Generally, these will cover different time periods for each building, and will not usually cover an exact calendar year. Therefore, unless revenue meters are read by occupants, the energy consumption for a calendar year must be interpolated from consecutive invoices.

The result of applying the IECP to the 100 buildings of

the MWD survey is shown in Exhibit 6. The IECP provides a different ranking order from that obtained using either annual total energy cost or AEUI. For example, the building ranked 6th by IECP value is ranked 25th by annual total energy cost and 4th by AEUI. The IECP can also rank a building higher than either annual total energy cost or AEUI (e.g. Buildings 1 and 10 on the IECP ranking). It is interesting to note that the building with an IECP ranking of 1 has already been the subject of a major energy conservation study.

IECP Value	RANKING ORDER		
$(MJ.m^{-2}.\$.10^6)$	IECP	Annual Total Energy Cost	AEUI
82.8	1	3	3
29.6	2	9	1
21.0	3	1	19
16.7	4	2	16
10.8	5	20	2
6.1	6	25	4
5.4	7	6	13
3.9	8	14	7
2.5	9	38	5
2.0	10	12	21

EXHIBIT 6 : RANKING ORDER, FOR THE 100 BUILDINGS OF THE MWD SURVEY - IECP, ANNUAL TOTAL ENERGY COST AND AEUI COMPARED.

For the purpose of this illustration, the IECP was calculated for the whole building group rather than for a selection of the largest consumers subdivided by building type. Normally, pareto analysis would be used to choose the size of the target group to be included in an energy conservation programme, and then the IECP ranking would be applied to that group.

CONCLUSIONS AND FURTHER ACTIVITIES

As with many human enterprises, one of the greatest obstacles to energy conservation is knowing where to start. Property owners and managers have been understandably reluctant to commit scarce resources to energy conservation, without a clear indication that such a programme is soundly based and has clear priorities.

The purpose of this paper has been to show how, with a minimum of reasonably easily obtainable data, conservation priorities can be assessed and presented in a way which can be readily grasped by any competent administrator.

The method has been illustrated with reference to New Zealand government buildings, but in principle is just as valid for any group of buildings housing similar activities in any locality. Minor adjustments to the fossil fuel weighting factors may be required so that they properly reflect attainable thermal conversion efficiencies. The IECP provides a simple tool to direct property owners and managers to buildings according to the efficiency of use of a particular fuel, the level of building energy consumption, and fuel prices. We would encourage those involved in building energy conservation to use the method, and look forward to hearing of the results of its application in a variety of circumstances.

ACKNOWLEDGEMENTS

Our main thanks must go to the Architectural Division
of the New Zealand Ministry of Works and Development
for funding this work.

The advice and encouragement of our colleagues at the
Victoria University School of Architecture are
gratefully acknowledged, as are the skills of Marilyn
McHaffie, Diana Braithwaite and Gavin Woodward in the
production of this paper.

REFERENCES

1. NZ Ministry of Energy : "Energy Research, Development
   and Demonstration in New Zealand, Volume 2 1979/80",
   Technical Publication No 10, Wellington, March 1981.

2. Baird, G; Donn, M R; Pool, F : "Energy Demand in the
   Wellington Central Business District - Final Report",
   New Zealand Energy Research and Development
   Committee, Publication No 77, Auckland (in press).

3. Donn, M R; Pool, F : "1981 Annual Building Energy Use
   Survey for the Wellington Central Business District",
   Ministry of Energy Technical Publication, Wellington
   (in press).

4. Baird, G et al : "Performance Evaluation of Air
   Conditioning and Related Energy Consuming Systems,
   by Computer". Proceedings of the International
   Congress of Air Conditioning and Computer, AICARR,
   Milan, Italy, February 1982.

5. Baird, G; Bruhns, H R : " A Preliminary Investigation
   of the Effects of Climate and Architectural Design on
   the Use of Energy in Schools in New Zealand".
   CIB-W71 Symposium on Building Climatology, Moscow,
   September 1982.

6. Baird, G; Brander, W D S : "Energy Conservation in
   Government Buildings - Report on Stage 1". Contract
   Research Paper 5, School of Architecture, Victoria
   University of Wellington, New Zealand, October 1980.

7. Baird, G; Brander, W D S; Macfarlane, J N W : "Energy
   Management in New Zealand Government Buildings".
   Proceedings of the Third CIB-W67 Symposium on Energy
   Conservation in the Built Environment, Dublin,
   Ireland, March 1982.

8. Beca Carter Hollings & Ferner and R A Shaw : "Greater
   Auckland Commercial Sector Energy Analysis". New
   Zealand Energy Research and Development Committee,
   Publication No 45, Auckland, May 1979.

9. Livesey, P M : "National Savings", Journal of the
   CIBS, Vol 3, No 8, August 1981, pp 51-53.

10.Ministry of Energy : "1981 Energy Plan",
   Wellington, 1981.

Chapter 9

A NEW ENGINEERING DESIGN APPROACH TO BUILDING ENERGY CONSERVATION

D. S. Stenhouse, A. Leonov

ABSTRACT

Various strategies have been promoted for conservation of energy in the building, industry, and transportation sectors. The authors review emerging energy sources and the subjects of actual and embodied energy. They discuss the merits of design and selection of materials and equipment based on the energy required for their manufacture, delivery, and installation.

The authors propose a system of embodied energy indices which can be used to compare existing and/or new building equipment and materials. In this respect, they recommend that priority be given to those items that are mass-produced.

They conclude that this will result in substantial energy savings since the specification of various materials and equipment in a building (and the effect this has on embodied energy) has a far greater impact on energy use and on the health of our economy than merely the regulation of actual energy required to operate a building. Establishing reasonable limits on building energy use that include both actual and embodied energy is seen as a more comprehensive approach to the energy problem and one which more properly recognizes the relationships between the building, industry, and transportation sectors. It is also proposed as a more meaningful technique by which design professionals can make legitimate decision-making tradeoffs in the total building project delivery process.

DISCUSSION

Energy and its effective use is undoubtedly one of the most important issues facing mankind today. It affects our economy and security, the industry and transportation sectors, building construction and design, and even our lifestyles.

It would not even be an exaggeration if we were to say that most people think of energy crises only in terms of frustrations experienced waiting in line or the soaring cost of a tank of gas. But the energy crisis is more pervasive. It is symptomatic of an acceleration of the general depletion of cheap conventional energy source reserves like natural gas and oil. And it should serve also as a warning about the need to use all of our natural resources wisely.

We would like to draw attention to the more timely and pertinent technical and economical aspects of energy conservation problems. So for this purpose we have tried to formulate some practical steps and measures which can be used by design engineers, industrial plant engineers, and building management personnel for effective energy conservation.

EMERGING ENERGY SOURCES

In projections of future energy use, advanced industrial nations rely to a large extent on nuclear power to meet demand and shortfalls of coventional energy sources. The International Nuclear Agency indicates that in 1975 nuclear plants (except in socialist countries) accounted for 6% of the world's installed energy production capabilities, which is estimated to be 1,111,000 MW. And it is stated that by the year 2000 nuclear power plants will generate up to 35% of the connected load, a figure approaching 4,000,000 MW.[1]

The development of nuclear energy has experienced a number of safety problems. Long-term storage of radioactive waste has not been solved. Nevertheless interest in fusion and fission will continue to grow, particularly if life safety and international security problems can be solved.

The use of coal as an alternative to nuclear has serious environmental disadvantages in terms of both extraction and combustion. Additional investments must be made to protect or restore the environment. The use of coal is seen, though, as a major solution to the energy supply problem. Coal-fired plants now supply over 40% of the electricity demand in the U.S.[2]

Synthetic liquid or gaseous fuels manufactured from our vast coal resources are attracting a great deal of attention as an alternative to depleting supplies of gasoline or natural gas.

Installations using geothermal energy are being developed for commercial and industrial applications and power generation.

A critical role will be played by co-generation to provide additional supplies of electrical energy while making use of waste heat from the generating process for space heating and/or industrial processes.

Solar energy has great promise, particularly for less exotic near term applications such as heating buildings in colder climates where there is a relatively high percentage of clear days, annual heating degree days, and where conventional sources of energy are relatively expensive. Space cooling, agriculture and industrial applications are also becoming more promising. Generating electricity with solar energy on-site or on a community or urban scale will become more feasible with economies of scale and through the experience gained from funded research and demonstrations. One interesting proposal of even greater scale involves the deployment of satellites in space to beam power from photovoltaic arrays to ground recovery stations.[3]

The combination of these various strategies for development of new or alternative energy sources is expected to increase supply and address coventional energy shortfalls.

CONSERVATION OF ACTUAL ENERGY

However it is our strong belief that energy conservation will ease the severity of conventional energy source shortfalls and give us time to decide what is the best combination of energy source options for the future. If properly implemented, energy conservation could even allow us to postpone some of the more costly options. [4] As much as half of the total energy used in the United States today is wasted. Based on projections of future energy use, eliminating waste provides an almost unexhaustible source of energy. If we could cut waste by only 40 to 50%, we would no longer have to import oil. The severity of pending energy crises and their implication on national security could thus be virtually eliminated.

The problem of implementing both qualitative and quantitive improvements to energy consumption is the central constraint to technical and economical progress. This can be illustrated by several examples.

Electrical motors use about 60% of the energy consumed by industrial and commercial sectors. [5] Increasing the efficiency of electrical motors could have the same relative effect on national energy use as implementing automobile mpg standards. Legislative action, therefore, seems as appropriate to improve performance in the building and industry sectors as it has been in the transportation sector.

Manufacturing firms have developed new models of the most commonly specified electrical motors of various capacities that have much improved power factors and efficiencies. These motors use less energy. But what is even more important, they require less embodied energy in their manufacture than other conventional motors. And the additional first cost, if any, pays for itself in a short time.

Everyone knows how important it is to reduce energy losses for electrical equipment. It is estimated that utilities could realize savings in the order of 200 million $ per year based on reduction of transformer core losses alone. [6] For instance, one new 15 KVA transformer model with a metglas core uses 56% less energy than conventionally transformers of the same rating. And the operating temperature at full load is 30% less. There are new high-intensity discharge lamps (for example, high pressure sodium (HPS) lamps) that consume 6.5 times less energy than incandescent lamps of the same rated lumen output. Conveniently shaped fluorescent lamps in popular wattages with built-in ballasts are now becoming available on the market. A general conclusion can be made that systematic modernization of energy consuming equipment produced in large quantities is one of the best ways to conserve energy.

Another effective method for conserving the actual energy consumed by industry and in buildings is the application of energy management control systems (EMCS) designed for regulating energy use and equipment operation in buildings.

The use of microprocessor-based controllers and new communication devices in lieu of hard-wired pneumatic or electric devices offers opportunities for expanded control, information gathering, and communication. For both the industry and building sectors of the economy, EMCS can have a significant impact on reduction of energy use. EMCS using computers and appropriate logic can control lighting, HVAC systems, mechanical equipment, provide life safety and security and   data for an effective preventive maintenance program, while properly managing energy use and demand and monitoring consumption over a typical day.

Total sales of EMCS were about $165 million a few years ago. But the industry predicts this figure will double over the next 10 years at an average annual rate increase of 10%. This rise will be accounted for generally through new construction. But retrofit applications are also becoming significant, and as the cost of purchased energy becomes a larger portion of operating costs. Energy conservation standards and budgets for new construction are requiring designers to think more about how buildings can be designed to save energy.

EMBODIED ENERGY CONSERVATION

But the use of actual energy used directly by buildings isn't the only way to conserve energy in the building sector. Economies in embodied energy (the energy which is expended for the manufacture of equipment or goods) provide unprecedented opportunities for improvements in energy efficiency associated with reductions in equipment weight that results in savings for transportation expense, handling, and the installation of equipment on site. The amount of energy (per unit of productivity) used in the manufacture of a given item of equipment could be used as a generalized index of embodied energy in standard industrial categories.

Though saving embodied energy is only a one time opportunity, the opportunity to conserve actual energy directly used by a building extends over the entire useful life the building or individual piece of equipment. Nevertheless, in many cases the total amount of energy that can be saved through reduction of embodied energy will exceed the actual used. It takes seven times as much energy to "make" an office building in New York than it does to operate it in one year. This is true not only of buildings but also for building products or system components that are energy-intensive yet which consume (or cause to be consumed) relatively small amounts of energy.

Engineers working in industry today, or as consultants to industry, should ask themselves what they can do to save energy. What strategies can we put in place to conserve energy while researchers try to discover (or perfect) new energy sources, or manufacturing firms improve and modernize their plants to use less energy to for the production of building materials and system components?

The answers to these questions may not be abundantly clear at first glance. However, solving them will predictably lead to the development and implementation of more appropriate analytical techniques in the design process.

One possibility is the development of standard embodied energy indices for a wide range of equipment or goods. For practical calculations the weight of equipment can be used as an approximate index of embodied energy. In many cases weight has been shown to be proportional to the energy spent in the manufacturing process to make a product. Take, for example, electrical power transformers. Data furnished by three large manufacturing firms for various transformers shows that the embodied energy for products with similar characteristics can be quite different. In making a comparison of similar transformers, one must compare not only weight, but also efficiency, power factor and impedance. For dry transformers up to 112.5 KVA, ACME transformers seem to have the least embodied energy. In the group of 150 up to 500 KVA, Gould transformers are less energy intensive; 750 up to 1000 KVA, Westinghouse transformers; and for liquid-filled transformers from 750 up to 2000 KVA, General Electric appears to manufacture transformers with minimal embodied energy.

Figures 1 & 2 show that the difference in embodied indices for transformers with similar characteristics can be as much as 25% or more. Such examples can be found through examination of other kinds of equipment (i.e., compressors, fans, pumps, blowers, chillers, etc). Indices can be used to identify the embodied energy per unit of energy consumption for compressors (KW/CFM), pumps (KW/GPM), transformers (KVA/KW), and other types of equipment.

The specification of equipment is a basic step in the design process. The use of embodied energy indices would be an incentive for manufacturing firms to produce more efficient equipment. It would encourage them to become more competitive by saving energy in the manufacturing process. National interests to save energy might thus happily coincide with industry interests to maximize profits. Overlapping interests will lead to a more viable energy conservation policy and stabile economy.

Rivers are made up of many small streams. The same is true of energy conservation. The effect of every step taken, however small, will gradually create a much larger effect, one in fact, that will increase by geometric proportions as each little success promotes favorable reactions. Due to the relative size and relationship of various sectors of the economy, these reactions will have a positive impact on the growth of industry and health of our economy.

CONCLUSION

Individual designers and firms, but also engineering associations, lawmakers, and regulatory agencies can play important roles. Project specifications and cost estimates must be expanded to include documentation of those measures incorporated in the building systems and equipment design/selection process which are expected to have energy. One method for doing this would be to use takeoffs of various building materials and equipment (that are conventionally developed for cost estimates) to sum up the embodied energy for proposed building design. The result can then be compared to some set of established standards. The proper selection of materials and equipment will have a far greater impact on energy use (and on the health of our economy) than merely the regulation of actual energy used by a building after it has been occupied. One can readily see that establishing reasonable limits on building energy use that include both actual and embodied energy is a more comprehensive approach to the energy problem because it recognizes the relationships between building, industry, and transportation sectors. We feel that annual energy budgets or building energy performance standards that incorporate actual as well as embodied energy standards are the only way to introduce rationalism into the design process and provide for legitimate tradeoffs to conserve energy in the total delivery process.

REFERENCES

1. Dr. Zigward Eclund; "Problems to be solved"; Bulletin of the International Nuclear Agency; New York; v. 19, No. 4.; Aug 1977.

2. Chauncey Starr; "Choosing Our Energy Future"; EPRI Journal; Sept. 1980.

3. Peter E. Glaser et al.; "First Step to the Solar Power Satellite"; Spectrum; May 1979.

4. Henry M. Jackson (U.S. Senator); "The Challenge. An Aggressive Energy Policy"; Consulting Engineer; Apr 1980.

5. "High-Efficiency Motors: Are they Cost Effective?"; Specifying Engineer; Aug 1980.

6. Dan Utraska; "Glossy Metals in Transformer Cores Move Toward Commercialization"; Electric Light and Power; Apr 1980.

7. A. Leonov & D. Stenhouse; "Embodied Energy in Building Equipment and Construction Materials"; Proceedings of the III National Conference of Energy Conservation; Tucson, Arizona; Jan 1979.

8. D. Stenhouse & A. Leonov; "An In-Depth Look at Embodied Energy"; Consulting Engineer; Nov 1979.

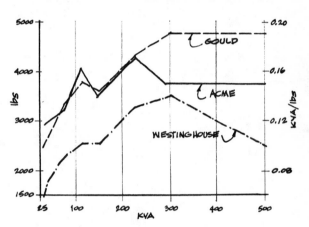

Figure 1. Comparison of Transformer Embodied Energy Indices

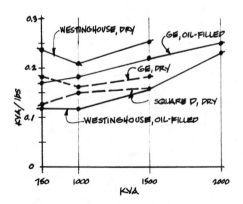

Figure 2. Comparison of Transformer Embodied Energy Indices

Chapter 10

ELECTRICAL DEMAND CONTROL - AN ESSENTIAL INGREDIENT FOR COMPLETE ENERGY MANAGEMENT

D. F. Goetz

Though emphasis today is directed toward energy conservation, electrical loads have temporal characteristics which should not be ignored. Substantial savings may result from managing when energy is used as compared to how much energy is used, particularly for the large user.

The monthly bill for large users normally contains two cost components: an energy charge and a demand charge. The energy component is related to consumption and reflects the utility's energy costs. The demand component is related to the capacity of the utility's generating plants, transmission lines, and distribution equipment which carries energy to the user. The demand component reflects the utility's capital costs.

The monthly bill prepared by the utility usually identifies the service period, the class of service (which determines the rates that apply), the energy usage in kilowatt hours, the demand in kilowatts, and the combined charges for the period. Sometimes the energy and demand charges are separated.

CONSERVATION AND LOAD MANAGEMENT

The primary approach to energy management among users is conservation. Consumption of kilowatt hours is reduced by installing more efficient loads, shutting off unused loads, and removing unnecessary loads entirely. Conservation practices are gradually introduced until a point is reached where further steps become uneconomical. But large users of electricity may obtain further savings without any compromise to existing energy conservation practices. A user may find it economical to shift deferrable loads and thereby reduce demand. However, effective management of demand requires access to the temporal characteristics of energy usage. Such details are too extensive to include in the monthly bill. Yet the monthly bill is all that some managers of electrical energy ever see!

The limit of the capacity of the generating, transmitting and distributing facilities of most utilities is determined by the ability of the equipment to dissipate heat. The amount of heat that is produced is determined by the size of the load that the equipment must carry and the length of time that this load is maintained. The demand charge is designed to distribute the cost of facility capacity in direct proportion to the influence that each user, mainly the large user, has on this thermal load. Any practice employed by the user that reduces this thermal load, and thereby makes utility capacity available to other users, is reflected in a reduction in the demand charge on the monthly bill.

Effective demand management attempts to reduce the coincidence of electrical loads which are economically deferrable in order to keep the demand charge to a minimum. Since energy consumption is not affected by deferring these loads, there is no change to the energy charge on the monthly bill. But if re-distributing these deferrable loads does appreciably reduce peak loads, a significant reduction in demand charge may result.

What users sometimes fail to realize though is that extensive research into the characteristics of the electrical load is needed before effective demand management can be achieved. First-hand experience with this data familiarizes the user with the physical and temporal characteristics of the electrical load. This knowledge is essential to make prudent decisions that result in savings in demand charge or sensible investments in demand control equipment. Since the load requirements of an operation can change, any plan for demand control should be evaluated periodically and updated.

DEMAND MEASUREMENT

The utility uses a meter to measure the user's demand. The monthly demand charge is calculated using recorded readings taken from the demand meter. The demand meter, which is usually installed on the user's premises, is therefore the final benchmark of the qualitative and quantitative aspects of user demand. A user can employ this same source of data to analyze electrical load.

If the data for the demand analysis is taken directly from the demand meter, the user is provided with the identical view the utility uses when measuring and charging for the load. A demand analysis is based upon past experience. However, if the operation does not change appreciably, the user is also provided with an idea of what the load and demand charge will look like in the future. A user may also predict quite accurately from the analysis what effect a shift of deferrable loads could have on future demand charge.

Even in most rudimentary form, demand analysis is a tedious process unless relegated to equipment data processing. Although the data generated by the demand meter is impeccably precise, it is also copious. This data is recorded by the utility in a form which is suitable for equipment processing, since it is used by the utility to prepare the monthly bill. This demand meter data record is available to the user from the utility. Any approach other than equipment processing would make demand analysis from data provided by the demand meter extremely cumbersome.

A demand meter is designed to measure the heating effect, or thermal load, placed on the utility facilities by the user's load. Therefore the demand meter is designed to measure the average rate at which energy is consumed, or average power in kilowatts, over a standard heating interval. Actually the meter generates a pulse each time a fixed number of kilowatt hours are consumed by the user. Hence the number of pulses generated by the meter in a given time is directly proportional to the average power over that interval. Pulses are accumulated for each fifteen minute interval throughout the billing period. An example of a 24 hour demand profile is shown in Figure 1.

DEMAND PROFILES

Peak demand, or the average power that establishes the capacity of the utility facility that is required by the user, is defined as the average power represented by the two contiguous fifteen

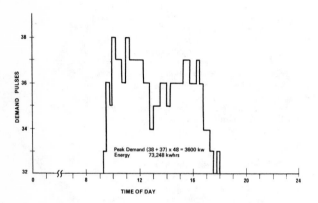

FIGURE 1. DAILY DEMAND PROFILE

minute intervals with the highest number of pulses over the billing period. Peak demand on the day represented by this data and shown in this curve occurred at 10:15 AM and reached an average of 3600 kilowatts over the half hour. Note that this same peak was reached one hour later. The second occurrence is of no consequence for demand billing purposes, only the maximum peak is significant in demand measurement. The area under this demand profile, of which only that power in excess of thirty-two pulses is shown. It happens that the peak demand for the billing period occurs on the day described by the previous demand profile. The same peak demand occurred at least a third time at 4:15 PM on another day, or days, in this billing period.

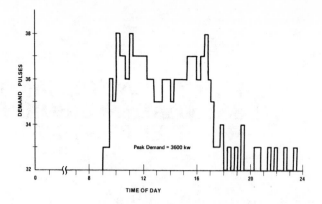

FIGURE 2. COMPOSITE MONTHLY DEMAND PROFILE

Graphic presentations such as these in slightly different format can be produced directly from the demand meter reading for any period of time using data processing facilities which are readily available to most users.

PRACTICAL DEMAND ANALYSIS SYSTEM

Demand meter readings are available from the utility as tape record, written record, or real time data. This data may be entered into the system on the user's premises and transmitted to a remotely situated time-shared data processing service. There the data is processed and the results are transmitted back to the user and appear as reports at the data terminal. The entire process is under the control of an operator and the demand analysis program. Figure 3 shows the arrangement for a time-share demand analysis system.

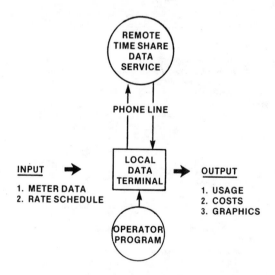

FIGURE 3. DEMAND ANALYSIS SYSTEM ARRANGEMENT

The conditions of the sale of energy and the rates that apply are contained in tariffs approved by local regulatory bodies. Tariffs are established for each class of service offered by the utility. These tariffs, like the demand meter record, are also available from the utility.

The demand analysis program can be designed to calculate the daily, weekly, and monthly consumption and demand charges using information obtained from the applicable tariff and the data provided by the demand meter. Such charges may also include or exclude additional features unique to the user such as fuel adjustments, local taxes, and special credits. The result is an exact replica of utility billing for the period.

The user may alter demand data to reflect proposed shifts in the load. The effect that this simulation of demand has on demand charge for the billing period can be calculated immediately. All details of actual and simulated load can be examined microscopically for clues to better demand management.

Because the demand meter is always in place, demand data is available for analysis at anytime. Data processing charges are incurred only while the service is being used. The data terminal and remoting line are readily accessible. This method of demand analysis is preferable to methods that employ independent metering for obvious reasons. Analysis is available by simply activating a system which always remains in place. This demand analysis system is precise, thorough, swift, simple and flexible and is an excellent tool for the progressive energy manager.

Chapter 11

INFLUENCE OF TWO-POSITION CONTROL DYNAMICS ON ENERGY UTILIZATION IN BUILDINGS

D. P. Mehta

ABSTRACT

Building envelopes, heating, ventilating and air-conditioning (HVAC) systems and building occupants interact with each other and with the exterior environments on a dynamic basis. The interactions result in a complex energy transfer between interior and exterior environments. Envelope and environmental control system dynamics have a significant influence on the energy required for comfort.

The objective of the research reported in this paper was to study the relationship between the thermostat differential, part load operation, furnace overcapacity, cycling pattern, energy consumption and occupant's thermal acceptability in a combustion-type heating system with two-position control. An analytical model was developed and applied to derive the relationships. An analysis of the relationships led to the formulation of new thermal control strategies for reducing energy consumption in buildings. It has been concluded that these energy saving strategies do not produce conditions of discomfort.

INTRODUCTION

The current and projected increases in the fuel market provide strong financial incentives for an efficient use of energy. In the United States, much attention has been paid to the consumption of energy in buildings during the past nine years. Buildings were identified as primary candidates for saving energy as early as the oil embargo of 1973 and since then, several energy saving strategies have been proposed for the design, operation and maintenance of buildings. However, very few strategies have considered the influence of control systems dynamics on the energy utilization in buildings.

In the past, research efforts of the control engineers (1-6)* have been confined to investigating analog passive circuits simulations of buildings mainly to improve the stability of environmental control systems. Nelson applied analog computer techniques to simulate a one-story house and its associated HVAC system (7). Application of an analog computer has also been reported by Magnussen to simulate an air-conditioning system in a commercial building (8). Kaya converted a thermal circuit of one zone of the multizoned commercial building simulated by Magnussen to an equivalent transfer function so that the techniques of control theory can be applied to optimize the design of the controller (9).

With the energy shortages experienced after the oil embargo of 1973, studies of the dynamics of environmental control loops began considering energy consumption. Hamilton, Leonard and Pearson investigated the dynamic responses near full load and part heating load of a discharge air temperature control system (10).

---
*Numbers in the parentheses refer to references.

Stoecker et. al. studied the dynamic characteristics of an air temperature control loop (11) and have applied these studies to investigate the effects of the throttling ranges on energy consumption (12). Room dynamics were excluded in these studies.

Very few efforts have been devoted in the past to developing models that describe dynamic interactions of weather, envelopes, internal loads, HVAC components and control systems. Harrison et. al. applied the energy balance to predict room temperature response to sudden heat disturbance inputs (13-15). The emphasis of their effort was to illustrate an approach and as such no data were taken on an actual system. Nor was any effort made to apply their studies to investigate the energy consumption patterns. Thompson et. al. reported the models for the components of a fan coil system and have studied the influence of the proportional controller on this system (16,17).

Mehta and Woods have shown that analytical models based on analog thermal circuits are valid for dynamic analysis, but too complex for energy consumption studies (18). So Mehta et. al., developed a rational model from energy balance methods for thermodynamic analysis of occupied spaces (19). Mehta has shown that the modular nature of the rational model permits its coupling to different types of HVAC systems and to the occupant to derive models which can describe dynamic interactions between the building envelopes, HVAC systems, exterior environments and the building occupants. Mehta and Woods have reported an experimental validation of the rational model coupled to a residential two-position environmental control system (21). However, their rational model included two assumptions which are not always valid. First, the storage effects of the envelope were neglected. Second, heat transfer to the occupied space by radiation was not included in the analysis.

The intent of the present paper is to report the development of a modified rational model which predicts dynamic thermal conditions in occupied spaces and includes the storage effects of the structures as well as the heat transfer by radiations. The modified rational model has been applied to derive some energy efficient strategies for two-position environmental control systems.

MODIFIED RATIONAL MODEL

The thermal processes which take place in an occupied space of a building are shown in Fig. 1. The following assumptions are made:

1. The properties of the thermal process involved in heating the occupied space are those at the location at which the occupants experience the controlled conditions.

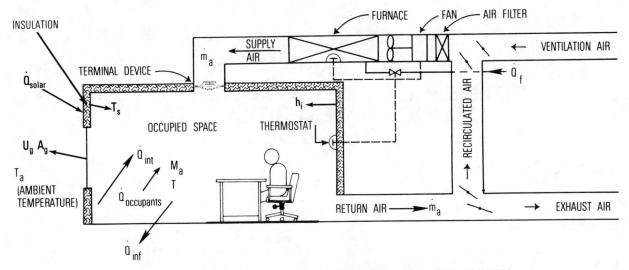

FIGURE 1. SCHEMATIC REPRESENTATION OF A BUILDING SYSTEM.

This assumption permits us to describe the process with ordinary linear differential equations instead of partial differential equations.

2. Changes in the stored energy of space air are due to changes in dry-bulb temperature; those due to humidity are considered negligible.

3. The air in the occupied space is uniformly mixed.

4. Infiltration losses are considered to be a function of supply air flow rate, occupant activity, internal heat sources, room temperature, and construction details in a given occupied space.

From an energy balance on the occupied space (Fig. 1):

$$\dot{m}_a c_{pa} T_i + \dot{Q}_{occ} + \dot{Q}_{int} + \dot{Q}_{rad} + \dot{Q}_f = \dot{m}_a c_{pa} T + \dot{Q}_{inf}$$

$$+ Ah_i(T-T_s) + A_g U_g(T-T_a) + M_{wa} c_{wa} \frac{dT^*}{dt} \quad (1)$$

If we define

$$\dot{Q} = \dot{Q}_{occ} + \dot{Q}_{int} + \dot{Q}_{rad} + \dot{Q}_f \quad (2)$$

then we may rewrite Eqn. 1, as

$$\frac{dT}{dt} = \frac{\dot{Q}}{M_{wa} c_{wa}} + \frac{\dot{m}_a c_{pa}(T_i-T)}{M_{wa} c_{wa}} - \frac{\dot{Q}_{inf}}{M_{wa} c_{wa}} - \frac{Ah_i}{M_{wa} c_{wa}}(T-T_s)$$

$$- \frac{A_g U_g}{M_{wa} c_{wa}}(T-T_a) \quad (3)$$

For most applications, relatively small fluctuations in net heat gains; space temperature and structure temperature will be experienced within the occupied space. For this reason, only small perturbations about the operating or set point will be considered. The operating point may be found by setting the time derivative of Eqn. 3 to zero:

*See list of nomenclature at the end of this article.

$$\frac{\dot{\bar{Q}}^*}{M_{wa} c_{wa}} + \frac{\dot{m}_a c_{pa}}{M_{wa} c_{wa}}(\bar{T}_i-\bar{T}) - \frac{\dot{\bar{Q}}_{inf}}{M_{wa} c_{wa}} - \frac{Ah_i}{M_{wa} c_{wa}}(\bar{T}-\bar{T}_s)$$

$$- \frac{A_g U_g}{M_{wa} c_{wa}}(\bar{T}-\bar{T}_a) = 0 \quad (4)$$

Eqs. 3 and 4 may be combined to yield an equation in terms of perturbations about the operating point.

$$\frac{d(T-\bar{T})}{dt} = \frac{\dot{Q}-\dot{\bar{Q}}}{M_{wa} c_{wa}} + \frac{\dot{m}_a c_{pa}}{M_{wa} c_{wa}}[(T_i-\bar{T}_i) - (T-\bar{T})]$$

$$- \frac{\dot{Q}_{inf} - \dot{\bar{Q}}_{inf}}{M_{wa} c_{wa}} - \frac{Ah_i}{M_{wa} c_{wa}}[(T-\bar{T}) - (T_s-\bar{T}_s)]$$

$$- \frac{A_g U_g}{M_{wa} c_{wa}}[(T-\bar{T}) - (T_a-\bar{T}_a)] \quad (5)$$

Eqn. 5 may be rewritten as:

$$\frac{d\Delta T}{dt} = \frac{\Delta \dot{Q}}{M_{wa} c_{wa}} + \frac{\dot{m}_a}{M_{wa}}(\Delta T_i-\Delta T) - \frac{\Delta \dot{Q}_{inf}}{M_{wa} c_{wa}}$$

$$- \frac{Ah_i}{M_{wa} c_{wa}}(\Delta T-\Delta T_s) - \frac{A_g U_g}{M_{wa} c_{wa}}(\Delta T-\Delta T_a) \quad (6)$$

From assumption (4):

$$\dot{Q}_{inf} = f(\dot{Q},T) \quad (7)$$

Linearizing Eqn. 7 about the operating point yields:

$$\Delta \dot{Q}_{inf} = K_1 \Delta \dot{Q} + K_2 \Delta T \quad (8)$$

The constants K_1 and K_2, are defined as follows and are evaluated at the operating point.

$$K_1 = \left.\frac{\partial \dot{Q}_{inf}}{\partial \dot{Q}}\right|_{T = constant} \quad (9)$$

$$K_2 = \left.\frac{\partial \dot{Q}_{inf}}{\partial T}\right|_{\dot{Q} = constant} \quad (10)$$

*Bar (–) notations indicate values at the operating point.

70

Substituting Eqn. 8 into Eqn. 6, taking the Laplace transform, and rearranging, we obtain for step inputs:

$$\Delta T(s) = \frac{K_{\dot{Q}} \, \Delta \dot{Q}(s)}{s(1+\tau_{os}S)} + \frac{K_{Ti} \, \Delta Ti(s)}{s(1+\tau_{os}S)} + \frac{K_{Tag} \, \Delta Tag(s)}{s(1+\tau_{os}S)}$$
$$+ \frac{K_{Tas} \, \Delta T(s)}{s(1+\tau_{os}S)} \qquad (11)$$

where $K_{\dot{Q}} = \dfrac{1 - K_1}{\dot{m}_a c_{pa} + UgAg + Ahi + K_2}$ (12)

$$K_{Ti} = \frac{\dot{m}_a c_{pa}}{\dot{m}_a c_{pa} + UgAg + Ahi + K_2} \qquad (13)$$

$$K_{Tag} = \frac{UgAg}{\dot{m}_a c_{pa} + UgAg + Ahi + K_2} \qquad (14)$$

$$K_{Tas} = \frac{Ahi}{\dot{m}_a c_{pa} + UgAg + Ahi + K_2} \qquad (15)$$

$$\tau_{os} = \frac{M_{wa} c_{wa}}{\dot{m}_a c_{pa} + UgAg + Ahi + K_2} \qquad (16)$$

A block diagram representation of Eqn. 11 is shown in Fig. 2. It may be noted that τ_{os}, the time constant of the occupied space, is a function of the enclosed mass of air, the mass of furnishings, the mass flow rate of air, the envelope characteristics (including storage effects) and infiltration. Changes in the inside surface temperatures will be envelope-sensitive and can be measured to be used as inputs to the model. Alternatively, the changes in the inside surface temperatures can be predicted from the following envelope model.

$$M_s C_s \frac{dT_s}{dt} = \Sigma hiA(T-T_s) + \left(\frac{1}{\Sigma U_s A} - \frac{1}{hiA}\right)^{-1} (Ta-Ts) \quad (17)$$

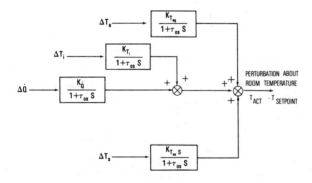

FIGURE 2. BLOCK DIAGRAM PRESENTATION OF THE MATHEMATICAL MODEL.

Eqn. 11, represented by the block diagram of Fig. 2, is in a form which can be coupled to the models of other components in an environmental control system to yield an overall transfer function for the building system. Mathematical models of components for environmental control have been presented in detail in reference (20).

CRITERIA TO EVALUATE ENERGY SAVING STRATEGIES

The criteria for the evaluation of any energy saving strategy must be based on its impact on occupant welfare because the main purpose of heating and air conditioning systems is to provide thermal comfort for the occupants. The rational model in Eqn. 11 has been developed to predict dynamic temperature perturbations about the set point in an occupied space which is in line with the definition of a true dynamic condition (22). Thermal transients can have a decided effect on a person's thermal acceptability by influences on physiological and sensory responses. Thus a review of the influences of thermal transients on human comfort, before the rational model is applied to derive energy conservation strategies, is in order.

A literature review revealed that little work has been done to study the effects of thermal transients with drifts in dry bulb temperature and humidity while the subjects stay at one place, and occupant acceptability of such procedures is unclear (23-27).

Studies by Sprague and McNall (23) showed that allowable fluctuating limits stated in ASHRAE Comfort Standard 55-66 were conservative. In a study by Griffiths and McIntyre (25) subjects were exposed to slow one-directional temperature changes of 0, 0.5, and 1.5°C/hour, centered about 23°C, over 6 hour periods. The clothing was in the 0.7 to 0.9 Clo range. From the mean thermal sensation votes of the subjects, Griffiths and McIntyre determined the corresponding Predicted Percent Dissatisfied (PPD) using the relationship developed by Fanger (28). Based on the predicted percent dissatisfied they have recommended a maximum rate of temperature change of 0.75°C/hr with a maximum deviation from the mean comfort temperature of 2.25°C.

Berglund and Gonzalez (26) have reported a study on the effects of thermal transients in which several subjective judgments of acceptability were evaluated. Dry bulb temperature was changed over a 4.5 hour period in morning experiments at rates of change in space temperature of ±0.5, 1.0, and 1.5°C/hr from a 25°C neutral point, while humidity was constant to 10 Torr. Subjects were unacclimated and wore clothing directly evaluated (29). These studies have shown that for sedentary persons, the slower temperature changes (±0.5°C/hr) from a neutral/comfort point were indistinguishable from the constant temperature conditions. The neutral point was determined by the clothing levels of the subjects. In these studies a ±.05°C/hr rate of change from a base temperature of 25°C was acceptable to 80% of the subjects.

A second study has been reported by Gonzalez and Berglund in which air temperature was allowed to rise at ±0.6°C/hr over an 8-hour working day (27). Summer clothing was worn by the subjects and two levels of elevated, but constant, humidity were employed. It has been reported from this study that there was no sign of decrements in thermal acceptability for low rates of change (≤ ±0.6°C/hr) as long as air temperature level is less than 28°C or dew point temperature is below 20° for normal summer clothing. It has also been shown in this 8 hour test study that humidity level was a less important consideration than dry bulb temperature. Table 1 summarizes the various studies discussed on the effects of temperature fluctuations on comfort.

TABLE 1   EFFECTS OF DRIFTS IN TEMPERATURE ON COMFORT, ACCEPTABILITY AND HEALTH

Year	Investigators	Conditions of test:	Findings:	Reference(s)

Temperature Drifts

Year	Investigators	Conditions of test:	Findings:	Reference(s)
1970	Sprague and McNall	1) 78 males and 78 females (age group (17.8-23.0 years) with Clo values of 0.6 were used in tests on thermal drifts. Test periods were 3 hours each. 2) In another study, 16 different tests on temperature drift of fluctuation periods varying from 1/2-1 hr, amplitudes 0.56-3.33°C, fluctuation rates 1.67-10.94°C/hr.	In practical air conditioned spaces where dry bulb air temperature fluctuates, no serious occupancy complaints should occur due to temperature fluctuations if $(\Delta T^2 \times CPH) < 4.63$ ΔT is the peak to peak amplitude of the temperature fluctuation (°C) and CPH is the cycling frequency (cycles/hour).	23
1974	Griffiths and McIntyre	Subjects were exposed to slow one-directional temperature changes of 0, 0.5, and 1.5°C/hour, centered about 23°C, over a 6 hour period. The clothing was in the 0.7 to 0.9 Clo range.	A maximum rate of temperature change of 0.75°C/hr with a maximum deviation of 2.25°C from the mean comfort temperature was recommended.	25
1978	Berglund and Gonzalez	Subjects were exposed to space temperature changes of ±0.5, 1.0, and 1.5°C/hr from a 25°C neutral point (determined from Clo values) while humidity was constant at 10 Torr. Test periods were of 4.5 hr duration.	For sedentary persons, a ±0.5°C/hr rate of change from a base temperature of 25°C was acceptable to 80% of the subjects.	26
1979	Gonzalez and Berglund	Subjects were exposed to space temperature change of ±0.6°C/hr over an 8-hour working day. Summer clothing was worn by the subjects and two levels of elevated but constant humidity were employed.	There was no sign of decrements in thermal acceptability for ±0.6°C/hr as long as air temperature is below 28°C or dew point temperature is below 25°C for normal summer clothing.	27

EFFECTS OF CONTROL DYNAMICS ON ENERGY CONSUMPTION

The rational model can be applied to determine the effects of control system dynamics on energy consumption in buildings. The residential heating system shown in Figure 1 and Figure 3 has been analyzed using the following assumptions:

1. The control mode was two position. As the boundary of the occupied space in Figure 3 is drawn across the surface of the building envelope, the dynamics of the safety switches were considered to be included as a part of the two position controller.

2. Energy was transferred to the occupied space only by the furnace, $\dot{Q}_f$.

3. Heat was transferred from the occupied space to the surroundings through the envelope. Solar radiations, infiltration and internal loads generated by occupants and lights were all considered as modifiers to a net heat load term, $\dot{Q}_{out}$.

4. Changes in stored energy were only due to changes in the dry bulb temperature of occupied space.

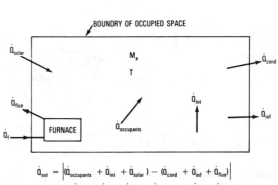

$$\dot{Q}_{out} = \left| (\dot{Q}_{occupants} + \dot{Q}_{int} + \dot{Q}_{solar}) - (\dot{Q}_{cond} + \dot{Q}_{inf} + \dot{Q}_{flue}) \right|$$
$$\text{when} (\dot{Q}_{cond} + \dot{Q}_{inf} + \dot{Q}_{flue}) > (\dot{Q}_{occupants} + \dot{Q}_{int} + \dot{Q}_{solar})$$
$$\dot{Q}_{out} = -\left[(\dot{Q}_{occupants} + \dot{Q}_{int} + \dot{Q}_{solar}) - (\dot{Q}_{cond} + \dot{Q}_{inf} + \dot{Q}_{flue}) \right]$$
$$\text{when} (\dot{Q}_{occupants} + \dot{Q}_{int} + \dot{Q}_{solar}) > (\dot{Q}_{cond} + \dot{Q}_{inf} + \dot{Q}_{flue})$$

FIGURE 3.   THERMAL PROCESSES TO DESCRIBE HEAT TRANSFERS IN EQUATION 18.

From the assumptions on the previous page, equation 1 may be simplified:

$$\dot{Q}_f = \dot{Q}_{out} + M_{wa} c_{wa} \frac{dT}{dt} \qquad (18)$$

or

$$\left(\frac{dT}{dt}\right)_{\substack{Furnace \\ ON}} = \frac{\dot{Q}_f - \dot{Q}_{out}}{M_{wa} c_{wa}} \qquad (19)$$

$$\left(\frac{dT}{dt}\right)_{\substack{Furnace \\ OFF}} = \frac{-\dot{Q}_{out}}{M_{wa} c_{wa}} \qquad (20)$$

So, for a given thermostat differential, the rate of increase or decrease of indoor air temperature depends on the heating load, $\dot{Q}_{out}$, which varies throughout the heating season. Figure 4 shows the relationship between the furnace capacity, thermostat differential, load and cycling rate. Calculations were made by using equations 18-20 and data for the Human Evaluation Center at Bradley University campus. At present, this building is being used, through the combination of professional personnel and appropriate tools for testing and evaluation, to determine performance levels, interests, aptitudes, and personality of individuals. The building has a total floor area of $223.2 m^2$. It is equipped with a gas fired forced air heating system.

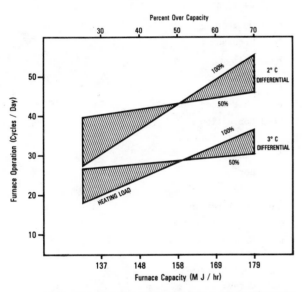

FIGURE 4. FURNACE CYCLES PER DAY FOR HUMAN EVALUATION CENTER

Figure 4 shows the relationships for two different differentials (2°C and 3°C), and two heating loads, one at the design temperature and the other at 50% of the maximum load.

EFFECTS OF OVER-CAPACITY AND PART LOAD OPERATION ON CYCLING AND ENERGY CONSUMPTION

For a thermostat differential of 2°C, it can be seen from Figure 4 that with a furnace capacity of 132 MJ/hr (25% over-capacity), the number of cycles increases from 27 cycles/day to 39 cycles/day when the heating load is reduced to 50% of the design value. Increased cycling at part load operation due to lower installed capacity (25% over-capacity) has been considered by the designers (8) to be detrimental to the system and the practice is to install a furnace with higher over-capacity (50%-

100%) so that cycling reduces at part load operation.

Certain "rules of thumb" have been devised to size furnaces with over-capacity to reduce cycling at part load operation. According to these rules, the actual installed capacity in the Human Evaluation Center is about 169 MJ/hr. At this capacity (60% over-capacity), the cycling rate reduces from 50 cycles/day at 100% load to 44 cycles/day at 50% load. However, with 25% over-capacity the number of cycles at 50% load would have been 39 cycles/day instead of 44 cycles/day. Thus, the current rules to design over-capacity furnaces fail to achieve the real objective of having fewer cycles/day at part load operation.

Bonne et. al. (30) have reported that the seasonal efficiency of a residential combustion heating system is dependent upon the cycling patterns. Based upon their model HFLAME, they have reported an efficiency for a residential furnace of 60.97% at 72 cycles/day and only 58.42% at 240 cycles/day. Thus, a properly sized furnace (having over-capacity to the left side of the crossover in Figure 4) can result in lower cycles/day (compared to the over-capacity to the right side of the cross over in Figure 4) which means higher seasonal efficiency and lower energy consumption.

EFFECTS OF THERMOSTAT DIFFERENTIAL ON CYCLING AND ENERGY CONSUMPTION

For a furnace with over-capacity to the left side of the crossover in Figure 4, the number of cycles/day increases at part load operation. Thus, the seasonal efficiency will be expected to reduce at part load operation though it is still higher than a furnace with over-capacity to the right of the crossover. It can be seen from Figure 4 that for a given over-capacity, on the left side of the crossover, the percent increase in cycles/day at part load operation reduces if the thermostat differential is increased from 2°C to 3°C. Thus, a reduction in part load efficiency on the left side of the crossover can be compensated to save fuel by increasing the differential of the thermostat. However, fuel saving strategies based on variations of thermostat differential or changes in number of cycles/hour must include considerations of the occupant's dynamic responses which were discussed above.

THERMAL CONTROL STRATEGIES FOR SAVING ENERGY

For the two-position control system, it has been shown that for certain conditions increasing the differential of the thermostat can be used to advantage to reduce the cycles per day and to increase the efficiency of a residential heating system.

Review of different studies on the effects of thermal transients has indicated that wide variations in dry bulb temperatures and their rates of change are acceptable factors to occupants at normal levels of activity and clothing. This information can be used to derive some thermal control strategies for saving energy.

New Buildings: For energy efficient design of new residential heating systems, the following procedures are suggested:

1. Calculate the design heat load and determine the crossover point as illustrated in Figure 4 for the maximum acceptable fluctuation (possibly 4.5°C – a conclusion drawn from the studies by Griffiths and McIntyre (25) Table 1).

2. Size the furnace with over-capacity less than the value at the cross-over point. The part load efficiency of the furnace so selected will be higher compared to one with over-capacity greater than the crossover point.

3. Use one of the following methods to automatically increase the differential of the thermostat (possibly up to 4.5°) in response to reduced heating loads.

   a. Redesign the anticipator heater in the thermostat.

   b. Use a microprocessor based programmable thermostat.

This will compensate for the loss of efficiency at part-load operation due to increased cycling rate.

Existing Buildings: For existing residential heating systems, the following procedures are suggested:

1. Calculate the design heat load, and for the maximum acceptable thermostat differential by the occupant without discomfort, determine the crossover-point as illustrated in Figure 4.

2. If the installed furnace capacity is greater than the crossover point, replace the existing burner with one sized to have over-capacity lower than the value at crossover point.

3. Install a thermostat with a wider differential (possibly up to 4.5°C).

CONCLUSIONS

The concept of a modified rational model has been developed and applied to study the effects of over-capacity, part load operation and thermostat differential on cycling and energy consumption in a residential combustion heating system. The effects of thermal transients on human comfort have been considered and used to derive thermal control strategies for energy conservation in buildings. Procedures to apply thermal control strategies to both existing and new buildings have been described.

ACKNOWLEDGEMENTS

Research work reported in this paper has been funded through the National Science Foundation (NSF) Research Project No. DME-8106950. Financial support from the NSF is gratefully acknowledged. However, the views presented in this paper are those of the author and not of the National Science Foundation. Special thanks are due to Mr. Joseph Wetch, graduate student at Bradley University, for his assistance. Appreciation is also extended to Mrs. Annelle Fletcher for her interest and efforts in typing this paper.

NOMENCLATURE

Variables

A   area: (m^2) or (ft^2)

C   Specific heat: (J/kg °K) or (Btu/lbm °F)
Thermal capacity: (J/°K) or (Btu/°F)

L   Length: (m) or (ft)

M   Mass: (lbm) or (kg.)

Q   Energy: (J) or (Btu)

T   Temperature: (°C) or (°F)

U   Overall heat transfer coefficient: $(J/h\ m^2\ °C)$ or $(Btu/h\ ft^2\ °F)$

c   Specific heat: (J/kg °K) or (Btu/lbm °F)

h   Heat transfer coefficient: $(J/m^2 hr\ °C)$ or $(Btu/ft^2 hr°F)$
Specific enthalpy: (J/kg) or (Btu/lbm)
Water head: (m water) or (ft water)

k   Thermal conductivity: (J/h m °C) or (Btu/h ft °F)

m   Mass: (kg) or (lbm)

s   Specific heat: (J/kg °K) or (Btu/lbm °F)

t   Time: (h)

Symbols and Subscripts

F   Floor

G   Ground; transfer function

H   Transfer function of feedback element

J   Joule

K   Proportional sensitivity
Gain

L   Load

M   Million

O   Outdoor

Occ   Occupant

W   Wall
Watts

a   Air

c   Controlled variable

Cond   Conduction

f   Fluid; furnace

g   Glass

h   Hot

i   Inlet

inf   Infiltration

int   Internal; integral

m   Manipulated variable
Metre

o   Outlet; outdoor

occ   Occupant

os   Occupied space

p   Pressure

r   Return

s   Sensor, supply
Laplace Transform variable
Surface

t   Tank

w   Water, well
Entry to plenum

wa   Weighted average

Greek Symbols

Δ   Perturbation of a variable

Θ   Temperature (°C) or (°F)

ρ Density: (kg/m^3) or (lbm/ft^3)

τ Time constant (h)

REFERENCES

1. Paschkis, V. "Periodic Heat Flow in Building Walls Determined by Electircal Analog Method." ASHVE Trans. 48(1942): 75.

2. Willcox, T.N., Dergel, C.T., Reque, S.G., Toelaer, C.M., and Brisken, W.R. "Analog Computer Analysis of Residential Cooling Loads." ASHVE Trans. 60(1954): 505.

3. Nottage, H.B. and Parmelee, G.V. "Circuit Analysis Applied to Load Estimating." ASHAE Trans. 60 (1954): 59.

4. Nottage, H.B. and Parmelee, G.V. "Circuit Analysis Applied to Load Estimating." ASHAE Trans. 61 (1955): 125.

5. Buchberg, H. "Electric Analogue Prediction of the Thermal Behavior of an Inhabitable Enclosure." ASHAE Trans. 61(1955): 339.

6. Buchberg, H. "Electric Analogue Studies of Single Walls." ASHRAE Trans. 62(1956): 177.

7. Nelson, L.W. "The Analog Computer as a Product Design Tool." ASHRAE Journal 11(1965): 37.

8. Magnussen, J.L. Analog Computer Simulation of an Air Conditioning System in a Commercial Building Incorporating Yearly Weather Data. Presented at First Symposium on the Use of Computers for Environmental Engineering Related to Buildings (supported by ASHRAE and NBS), November 30-December 2, 1970.

9. Kaya, A. "Analytical Techniques for Controller Design." ASHRAE Journal 18(April, 1976): 35.

10. Hamilton, D.C., Leonard, R.G., and Pearson, J.T. "Dynamic Response Characteristics of a Discharge Air Temperature Control System at Near Full and Part Heating Load." ASHRAE Trans. 80(1974): 181.

11. Stoecker, W.F., Rosario, A., Hadenreich, M.E., and Phelan, I.R. "Stability of an Air-Temperature Control Loop." ASHRAE Trans. 84(1978): 35.

12. Stoecker, W.F., and Daber, R.P. "Conserving Energy in Dual-Duet Systems by Reducing the Throttling Ranges of Air-Temperature Controllers." ASHRAE Trans. 84(1978): 23.

13. Zermeuhlen, R.O., and Harrison, H.L. "Room Temperature Response to Heat Disturbance Input." ASHRAE Journal 7(March, 1965): 25.

14. Zelenski, R.E., and Harrison, H.L. "Room Temperature Response to Heat Disturbance Inputs." ASHRAE Trans. 72(1966): II-2.1.

15. Harrison, H.L., Hansen, W.S., and Zelenski, R.E. "Development of a Room Transfer Function Model for Use in the Study of Short Term Transient Response." ASHRAE Trans. 75(1968): 198.

16. Thompson, J.G., and Chen, P.N.T. "Digital Simulation of the Effect of Room and Control System Dynamics on Energy Consumption." ASHRAE Trans. 85, part 2(1979): 222.

17. Thompson, J.G. "The Effect of Room and Control Systems Dynamics on Energy Consumption." ASHRAE Trans. 87, part 2: 1981: 897.

18. Mehta, D. Paul, and Woods, J.E. "Accuracy of an Analytical Model to Predict Dynamic Thermal Responses of Building Systems." Proceedings of the International Congress on Building Energy Management, Pova de Varzim, Portugal. Permagon Press. 1980.

19. Mehta, D. Paul, Woods, J.E., and Brueck, D.M. A Rational Model for Thermodynamic Analysis of Occupied Spaces. Proceedings of Energy Management Conference, Iowa State University, October 28-30, 1978.

20. Mehta, D. Paul. "Dynamic Thermal Responses of Buildings and Systems." Ph.D. Dissertation. Iowa State University, Ames, Iowa, 1980.

21. Mehta, D. Paul, and Woods, J.E. "An Experimental Validation of a Rational Model for Dynamic Responses of Buildings." ASHRAE Trans. 86(2): 1980: 497.

22. Chapman, W.P. "Research Needs in the Field of Dynamic Responses of Environmental Control Processes in Buildings." ASHRAE letter to T.C. and T.G. Chairmen. New York, July 9, 1979.

23. Sprague, C.H., and McNall, P.E. "The Effects of Fluctuating Temperature and Relative Humidity on Thermal Sensation (Thermal Comfort) of Sedentary Subjects." ASHRAE Trans. 76(1970): 34.

24. Wyon, D.P., Anderson, I., and Lundquist, G.R. "Spontaneous Magnitude Estimation of Thermal Discomfort During Changes in Ambient Temperature." J. Hygiene (Cambridge) 70(1972): 203.

25. Griffiths, I.D., and McIntyre, D.A. "Sensitivity to Temporal Conditions." Ergonomics 17(1974): 499.

26. Berglund, L.G., and Gonzalez, R.R. "Application of Acceptable Temperature Drifts to Built Environments as a Mode of Energy Conservation." ASHRAE Trans. 84(1978): 110.

27. Gonzalez, R.R., and Berglund, L.G. "Efficiacy of Temperature and Humidity Ramps in Energy Conservation." ASHRAE Journal 21(1979): 34.

28. Fanger, P.O. Thermal Comfort Analysis and Applications in Environmental Engineering. McGraw-Hill, New York, 1973.

29. Nishi, Y., Gonzalez, R.R., Nevins, R.G., and Gagge, A.P. "Field Measurements of Clothing Thermal Insulation." ASHRAE Trans. 82(1976): 248.

30. Bonne, V., Janssen, J.E., Nelson, L.W., and Torborg, R.H. "Control of Overall Thermal Efficiency of Combustion Heating Systems." Proc. of 16th International Symposium on Combustion. MIT, Cambridge, Massachusetts. August 1976.

SECTION 4
HVAC ENERGY UTILIZATION

Chapter 12

INDUSTRIAL BOILER ROOM VENTILATION

S. Chandrashekar

BUILDING DESCRIPTION

Location

The Medical Area Total Energy Plant is located in Boston. The site is a gateway to the prominent medical area complex and downtown Boston. The plant is located on the fringe of a medical institutional area and bordering residential buildings.

Impact

The plant has a visual impact that goes beyond its immediate surroundings reaching out into the residential areas of neighboring communities. The impact of plant noise to the surrounding area was another important consideration in the design. The number of openings in the walls or roof was kept minimal to reduce the visual and noise impact to the surrounding areas.

SYSTEM DESCRIPTION

Conventional System

A conventional large boiler plant ventilation design would consist of a wall louver with a damper at the outside wall, unit heaters located at strategic locations and exhaust fans or relief vents on the roof. Air intake dampers would be opened manually or automatically for ventilation air to enter the space and the hot air would exhaust through relief vents or roof exhaust fans. Boiler combustion air would be introduced through a separate air intake, silencer and preheat coil.

In this system, the ventilation air which removes the radiant heat given out by boilers, turbine generator set and other accessories equipment is controlled manually by the plant operator. This type of operation many times leads to wasting energy by operating unit heaters to keep minimum inside temperature. This is especially true if the minimum ventilation is left at summer settings. Boiler combustion air would be preheated by steam heating coils. Generally, the radiant heat given out by the equipment would not be used to preheat or mix with the combustion air to save the steam heating cost.

Energy Saving System

In the MATEP System, ventilation air is combined with combustion air and the combustion air will be part of the total ventilation air. The ventilation air will be introduced into the space at different levels. This will maintain reasonable comfort level at operating levels. The combustion air fan will drag the ventilation air over the equipment, which in turn will raise inlet air temperature without use of preheat coils.

The ventilation air is introduced into the space through multiple centrifugal fans and exhausted through multiple roof exhaust fans. Space thermostats located at different levels will average the space temperature and start the roof exhaust fans to maintain the space temperature. The space pressure controller will start the supply air and vary the inlet valve controller at the centrifugal fans in response to the amount of boiler combustion air and the space exhaust air.

The above system will be run automatically. Systems can be operated manually if the plant operator decides to bypass automatic operation.

Design Description

The ventilation system will supply combustion air for the operation of the boilers, as well as to remove the heat radiated by the equipment during operation. The volume of ventilation air forced into the plant depends on the outside air temperature and load demand on the equipment.

Major equipment which releases heat to the boiler room space are the waste heat recovery steam generators, primary steam generators, steam turbines, water chillers, motors, and forced draft fans. The heat release of the boiler area equipment is two (2) million Btu/hr for each primary steam generator, one (1) million Btu/hr for each heat recovery steam generator, 0.55 million Btu/hr for each turbine generator, 0.62 million Btu/hr for each motor-driven chiller, and 0.78 million Btu/hr for each steam-driven chiller. The total heat to be removed from the boiler-chiller area during the summer will be 5.4 million Btu/hr.

The volume of ventilation/combustion air depends on the actual equipment in operation, total heat released into the space, the outside air temperature and exhaust air requirements. During the winter months, the combustion air and exhaust air normally exceeds the ventilation air required for cooling. Ventilation supply air will be introduced into the plant at 55°F during the winter months. The minimum volume of supply air to the boiler room in winter will be 170,000 CFM. The boiler room area will employ an exhaust fan of 80,000 CFM capacity or minimum ventilation to remove odors and products of combustion in the space. 80,000 CFM represents approximately 6 air changes. As the outside air temperature rises, the volume of supply air increases.

During the summer months, the air quantity for heat removal exceeds the combustion air. The volume of supply air will be based on the temperature difference between the supply air and exhaust air. The volume of air for ventilation during summer months in the boiler room area will be 550,000 CFM with 11°F space temperature rise. Air temperature will not exceed 104°F when outside temperature is 91°F. A temperature gradient will thus be established in the space. The temperature near the operating floor will be approximately 94°F and at ceiling level, approximately 104°F. Table #1 presents a summary of air quantity analysis for the boiler room area. A positive pressure will be maintained in the area by supplying 10% excess air into the

space.
Three centrifugal fans, two 200,000 CFM and one 150,000 CFM capacity, will supply ventilation/combustion air to the boiler-chiller room. The fans are installed in the fan room located above the truck access driveway. Supply air will be drawn through weatherproof louvers and five-foot long silencers. Fans are double width, double inlet type. Supply air is ducted to the area. Distribution of supply air matches the heat release of the equipment. During the winter months, air will be transferred from the operating floor through the grating to the forced draft fans located at the basement level. Air leaving the supply air registers will be 55°F. As the air passes over the equipment, it picks up heat before entering the forced draft fan. Therefore, additional heating will not be required for combustion air. An 80,000 CFM exhaust fan will operate continuously for space ventilation. During the summer months, supply air will be transferred from the basement level to the operating level through the grating to the roof exhaust fan. Supply fans are supplied with variable inlet fan volume control. All roof exhaust fans are low-profile propellar fans. Three-foot long silencers are mounted on the roof exhaust fans.

In winter, the volume of outside air introduced into the boiler areas will be 250,000 CFM. Motor-operated dampers will open to allow the air through the heating coils. Hot water with ethylene glycol will be circulated in the heating coils.

Plant and fan noise are reduced to 49 dB at 60 feet away from the property line by installing sound attenuators, acoustic turning vanes and enclosing the fans in plenums at the air intake and exhaust.

Exhaust fans are located over the waste heat recovery boilers, primary steam generators and deaerators.

The volume of supply air into the plant will be controlled by combustion air requirements and temperature/pressure controller.

TABLE #1 - BOILER ROOM

AIR QUANTITY ANALYSIS

(Based on Load Profile)

	ITEMS	WINTER	SUMMER
1.	Number of Riley Boilers	3	--
2.	Percentage of Load	70	--
3.	Number of Heat Recovery Boilers	1	2
4.	Percentage of Load	89	85
5.	Turbine Chiller Pumps, etc., Percent Ld.	--	80 to 100
6.	Supply Air Temperature °F - Winter	55	--
7.	Supply Air Temperature °F - Summer	--	91
8.	Supply Air - CFM	190,000	550,000
9.	Combustion Air - CFM	90,000	40,000
10.	Exhaust Air - CFM	80,000	480,000
11.	Maximum Heating Air Quantity - CFM	190,000	--
12.	Number of Supply Fans Oper/CFM	1-170,000	2-200,000 1-150,000
13.	Number of Exhaust Fans Oper/CFM	1- 80,000	8- 60,000

ENERGY CONSERVATION FEATURES

Ventilation

The ventilation air was matched closely with minimum combustion air plus outside air necessary to remove the radiant heat given out by the equipment. Thermostat control in steps brings the required number of exhaust fans into the operation. The thermal load analysis indicated that only the minimum ventilation and combustion air fan will be in operation during most of the winter months. The matching of outside air to the load is important in this plant since the outside air has to go through five-foot deep silencers and high-pressure ducts.

Heating

The turbine output is increased by lowering the condensing temperature. Further use of low-pressure steam increased overall efficiency of the plant. The low-pressure steam is used to make hot water through a heat exchanger, which in turn is used to heat the plant and temper the combustion/ventilation air.

The use of ventilation air as a combustion air source eliminated the preheating of combustion air heating coils.

Electrical Power Input

Plant ventilation is automatically controlled by thermostat and static pressure controllers. The fan electric input is closely matched with the actual demand inside the plant. This type of design is important as the designed motor horsepower for the fans is in the vicinity of 300 Hp.

If a separate combustion air path was designed, additional fan energy would have been required to overcome pressure drop associated with five-foot depth of sound attenuators. This was not necessary since the ventilation air was used as a source of combustion air.

Noise

Using ventilation air as a combustion air source minimized the number of openings at outside walls, which in turn helped to control noise pollution at the surrounding area. This is critical because of the close proximity of hospitals.

Conclusion

The boiler plant ventilation system can be effectively designed to reduce the energy consumption without sacrificing the performance and operation of the plant. In this case, the simple payback period was within two years, despite the constraints that were imposed concerning location, noise and staggered construction schedule.

Chapter 13

REDUCING ENERGY COSTS OF COOLING SYSTEMS WITH ICE STORAGE

C. J. Marquette

As energy costs continue to escalate, designers of cooling systems are continually seeking new technologies to reduce operating costs. But as we often discover, old technologies in new applications can provide significant cost saving opportunities, too. Such is the case with the concept of thermal storage - an old technology in a new application.

The concept of thermal storage applied to cooling systems dates back to the 1930's when refrigeration technology was combined with thermal storage to handle infrequent, short term loads, like those common in theaters, churches, and batch process applications such as dairies. The reason for using thermal storage was to minimize the first cost of the cooling system. For example, a church may have a cooling load of 50 tons over a five hour period, occurring once a week. Rather than install a 50-ton system to operate for 5 hours, it made more sense to install a 5-ton system to generate and store cooling for 50 hours. The same total capacity of 250 ton-hours was produced in each case, and the system cost was substantially reduced, even when the cost of the storage equipment was included. This procedure was practical as long as the time available to generate cooling storage greatly exceeded the time of cooling use. Under these circumstances, the reduction in the cost of the refrigeration system would more than offset the cost of the storage equipment. Thermal storage continues to be most effective when applied to cooling systems with these characteristics.

Thermal storage is now drawing interest for broader application in comfort and process cooling systems because of major changes in the electric power industry. Many electric utility companies experience the greatest demand for electricity during the summer, largely to satisfy comfort cooling needs. The amount of power that the utility must generate peaks during daylight hours when cooling requirements are highest. As the peak demand for power increases, and approaches the limit of generating capacity, the utility must consider increasing its capacity through new plant construction or plant expansion. But with today's high construction costs, many utilities are turning to another solution - better utilization of existing capacity. Commonly called "load shifting", the idea is to cap or reduce the peak demand by moving loads to off-peak hours, thus flattening the demand curve. Utilities frequently create an incentive for load shifting by implementing a rate schedule that has higher energy and/or demand charges during peak hours.

Thermal storage is a concept that can take advantage of variable energy costs and reduce cooling system operating costs. Many comfort and process cooling loads exist for only a few hours each day, and commonly occur during hours of peak demand. Conventional cooling systems produce cooling when needed, and thus operate when power costs are highest. Thermal storage systems minimize energy costs by generating cooling capacity at off-peak times and storing it for future use. Air

conditioning applications that can benefit from thermal storage are office buildings, schools, laboratories, large retail stores, libraries, and museums; in general, any facility with a peaking load profile that is limited to certain hours of the day. The amount of cooling capacity that must be generated by the refrigeration system and stored is equal to the area under the load profile curve, usually expressed in ton-hours or Btu's.

Thermal storage can also be used for many industrial processes, such as dairies and breweries, and other types of operations with batch cooling cycles. Cooling loads of this type tend to be constant during their cycle, as shown by this load profile.

There are two types of thermal storage systems that will provide chilled water for cooling: chilled water storage and ice storage.

Chilled water storage seems to be the natural marriage between cooling system technology and the thermal storage concept. The usual approach is simply to combine a conventional chilled water system with a chilled water storage tank, as shown in Figure 1. Although this

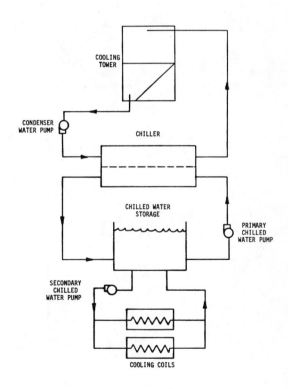

FIGURE 1: CHILLED WATER STORAGE SYSTEM

system is simple in concept, it becomes more complex when executed. This is primarily due to the limited cooling storage capability of chilled water. Refrigeration machines are commonly selected to cool water at full load through a temperature range of 15°F. Since the specific heat of water is 1 Btu per pound per °F, each pound of water circulated to the load provides 15 Btu of cooling. At 8.33 pounds per gallon, 15x8.33, or 125 Btu's are available for cooling from each gallon of water.

To put this in perspective, consider a building load profile where the instantaneous peak load is 600 tons, and the total load under the load profile curve is 5,000 ton-hours, or 60,000,000 Btu. If a chilled water storage system with a 15°F range were considered for this building, the required storage capacity would be 60,000,000 Btu ÷ 125 Btu/gal, or 480,000 gallons of water, which requires 64,000 cubic feet of space. This obviously adds considerable expense to the system and requires a great deal of space that may not be available at the jobsite.

Another major consideration in the design of chilled water storage systems is blending of the warm water returning from the system with the stored chilled water. If water is returned from the system to the same tank where the chilled water is stored, the two masses of water mix, thereby raising the temperature of the water being pumped to the cooling coils. So it becomes imperative to minimize the blending process. But current anti-blending techniques are either not completely effective, or very costly.

So while it would appear that chilled water storage is a natural combination of cooling system technology and the thermal storage concept, closer examination shows that there are substantial cost, space, and blending problems that must be solved.

Ice storage systems provide the benefits of thermal storage without these problems. Ice storage systems form ice on the surface of evaporator tubes, and store it until chilled water is needed for cooling.

A typical ice storage system shown in Figure 2, utilizes the basic components of a mechanical refrigeration system: compressor, condenser, expansion valve, and a combination evaporator/thermal storage unit, an ice builder. The ice builder is the evaporator in the system, and consists of a multiple-tube serpentine coil submerged in a tank of water, with a water agitation device included to provide uniform ice build-up and melt-down. The tank is insulated, and is usually provided with covers to minimize foreign matter infiltration.

After the refrigeration system has developed a full charge of ice, and cooling is required, a pump circulates cold water from the ice builder to the load. The warm return water is re-cooled by the melting ice.

Recall the 5000 ton-hour system that required 64,000 cubic feet of stored chilled water to handle the load. Compare this to ice storage, where cooling is based on the latent heat of fusion of water which is 144 Btu per pound. Typically, ice builder modules have volumes of about 3.2 cubic feet per ton-hour, so the 64,000 cubic feet volume required for the 5000 ton-hour load with chilled water storage would be reduced to just 16,000 cubic feet with ice storage, one-fourth the volume of chilled water.

The ice storage system also eliminates the problem of blending that is characteristic of chilled water storage systems. The return water flows into the ice builder and is cooled by the melting ice, which provides a consistent leaving temperature of about 35°F throughout most of the cycle. The availability of 35°F supply water offers opportunities to reduce system costs that are not available with chilled water storage. The temperature rise of the chilled water loop can be increased to 20°F or more, which permits a corresponding reduction in water flow. This results in smaller chilled water pipe sizes, smaller pumps, and less pump energy.

One potential disadvantage of ice storage systems is that more energy is required to make ice than to chill water because a lower evaporator temperature is required to produce ice at 32°F than to produce chilled water at 44°F. This penalty can be nearly eliminated by choosing an evaporative condenser for the system heat rejection. An evaporative condenser allows system selection and operation at a lower condensing temperature, which usually is sufficient to offset the power penalty caused by the lower evaporator temperature required by ice storage systems.

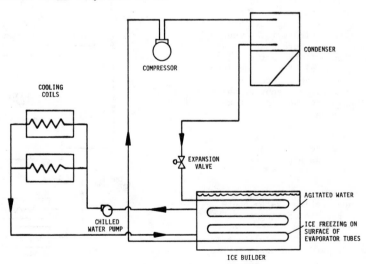

FIGURE 2: BASIC ICE STORAGE SYSTEM

So, on the basis of lower first cost, less space required, and no blending problem, with essentially no penalty in power consumption, ice storage will usually be the best choice in the selection of a thermal storage system for cooling.

The ice storage system designer has two basic operating strategies to choose from: full storage or partial storage. The best choice is determined by a number of factors associated with the specific project, such as the cost of electricity, equipment costs, load profile and space availability. A full ice storage system is one which has been selected to build the full cooling capacity for a facility of process during the hours when off-peak electrical rates are in effect, usually during the evening and early morning. During the off-peak period, the refrigeration system is operated and ice is formed on the coil surface of the ice builder until a pre-determined thickness is attained. At that point, an ice thickness sensor deactivates the refrigeration system.

During the peak period, when cooling for the building or process is required, a chilled water pump circulates water from the ice builder to the load. Return water is cooled by the melting ice, and this continues until the daily cooling or process requirement is satisfied. After electricity rates return to the off-peak schedule, a timer permits refrigeration to re-start, and the charge of ice is then rebuilt during off-peak hours for use in the next cooling cycle.

A full ice storage system makes maximum use of the thermal energy storage concept. It achieves minimum electrical power cost by avoiding peak hour demand and energy charges. However, since all of the cooling capacity is stored, the total ice storage system can be quite large and expensive to install. In some cases, operating cost savings will not be sufficient to justify the additional investment of necessary equipment. When this occurs, a partial ice storage system should be considered.

The partial ice storage concept utilizes a refrigeration system to build ice during off-peak or non-load hours, and uses the same or parallel equipment as needed to generate direct cooling during peak hours.

When cooling is needed, melting ice handles part of the load, and the refrigeration system operates in parallel to handle the remainder of the load. Therefore, the refrigeration system operates for up to 24 hours to produce a specified number of ton-hours of cooling, so the size and cost of the refrigeration plant and the ice builders are substantially less than for a full ice storage system.

There are three types of partial ice storage systems: (1) the Ice Storage/Refrigerant Coil System; (2) the Ice Storage/Parallel Evaporator System; and (3) the Compressor-Aided Storage System.

The Ice/Refrigerant Coil combination, shown in Figure 3, has refrigerant lines running directly to cooling coils for supplementary capacity. The ice building mode operates during off-peak hours as usual. During on-peak hours, the compressor circulates refrigerant directly to the cooling coils, while the remaining load is handled by melting ice. This arrangement has the lowest operating cost, since an intermediate chiller is not needed. The refrigerant coils can be operated at a relatively high evaporator temperature (approximately 45°F), which saves compressor horsepower. But, duplicate liquid refrigerant distribution equipment is required for operation at the two different evaporator temperatures, which increases the complexity and cost of the control system. Plus, some local building codes prohibit the use of direct refrigerant coils in occupied areas.

The Ice/Parallel Evaporator arrangement, Figure 4, has a chilled water evaporator instead of refrigerant coils. During the off-peak period, the ice builder develops its charge with the parallel evaporator inactive. During the peak period, refrigerant flow is routed to the parallel evaporator as it comes on-line to supplement the ice storage. This system has the advantage of using chilled water in all cooling coils for simplicity of design, and may also be attractive where use of direct refrigerant coils is restricted. However, the parallel evaporator adds substantial installation cost and since the system must operate at two different evaporator temperatures, controls are expensive and complex.

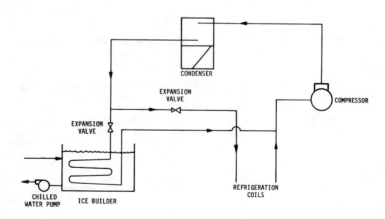

FIGURE 3: PARTIAL ICE STORAGE SYSTEM - ICE/REFRIGERANT COILS

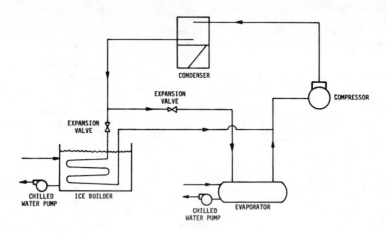

FIGURE 4: PARTIAL ICE STORAGE SYSTEM - ICE/PARALLEL EVAPORATOR

The Compressor-Aided ice storage system is shown in Figure 5. It builds ice in the usual manner during off-peak hours, but continues to operate the refrigeration system with the ice builder during peak hours, if necessary. The ice builder, aided by the compressor, acts as a chiller during the peak period. The ice charge in the ice builder becomes the evaporator coil surface, and cools the warm return water from the load. Operation of the refrigeration equipment during this time slows the melt-down process and provides more cooling capacity than would be obtainable in a conventional melt-down with idle refrigeration equipment. This concept offers simplicity of design and control of an all chilled water cooling loop, and low first cost since there is no need for a separate chilled water evaporator or duplicate refrigerant feed equipment. It also minimizes possible conflicts with codes that prohibit the presence of refrigerant piping in occupied areas, since the system uses only chilled water in the cooling coils. A disadvantage of this arrangement is that the refrigeration system operates at a low suction temperature during the ice building process and when it is necessary to operate during peak hours, resulting in slightly higher operating costs.

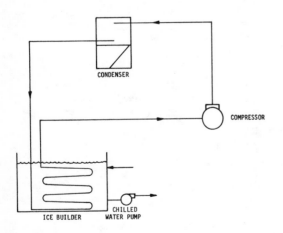

FIGURE 5: PARTIAL ICE STORAGE SYSTEM - COMPRESSOR-
AIDED BUILDER

Now that we have seen how ice storage systems operate to reduce the operating cost of cooling systems, let us illustrate the magnitude of the savings. It is difficult to generalize in this regard, because several variables, particularly the utility rate structure and the load profile, can have a dramatic effect on the results. For our illustration, we shall assume a load profile equivalent to 5000 ton-hours, occurring over a ten-hour period from 8 AM to 6 PM. This profile has a peak load of 600 tons. We shall evaluate operating cost entirely on the basis of avoided demand charge in order to simplify this example. Additional cost savings could be realized if a time-of-day energy charge were included, but we shall take a conservative approach, and work on the basis of demand charge only. We will use a demand charge of $7.00 per kilowatt, which our experience indicates is a reasonable national average, in effect from 8 AM to 6 PM. The length of the cooling season is assumed to be six months. Now let's compare a conventional chilled water cooling system with various types of thermal storage systems.

A conventional chilled water system must be sized for peak load conditions, which in this case is 600 tons. It will operate only during the daily cooling period, ten hours. Considering the energy requirements of the chiller, the cooling tower fans, and the condenser pump, the total system energy is 550 kilowatts.

The estimated average installed cost for a typical 600-ton centrifugal water chiller with cooling tower and condenser water circuit is $270,000; and since no storage equipment is necessary, the total installed cost of the refrigeration plant is $270,000. The annual demand charges at $7.00 per kilowatt are $23,100.

Now, let us consider a full chilled water storage system. Because it is a full storage system, all cooling capacity is generated and stored during the off-peak hours, so the refrigeration equipment is selected by dividing the required 5000 ton-hours by 14 hours off-peak operation, or 357 tons. Including the same components as the conventional system, but at the reduced capacity, the total system energy requirement is 337 kilowatts.

Because of the smaller cooling equipment size, the cooling equipment cost is just $161,000. But this reduction is more than offset by the enormous cost of storing chilled water. The cost of water tanks capable of storing 5000 ton-hours of cooling is approximately $440,000. As a result, first cost at $601,000 is more than twice the total first cost of the conventional system. The required storage volume is over 64,000 cubic feet.

Since the full chilled water storage system operates only during off-peak hours, it avoids the $7.00 per kilowatt demand charge, so it saves $23,100 in demand charges annually. At a first cost premium of $331,000, simple payback is calculated at 14.3 years, not a result that would motivate many building owners.

A full ice storage system requires the same capacity and operates during the same off-peak period as does the full water storage system. However, the lower suction temperature needed for building ice requires about 3% more energy. The total system energy is 348 kilowatts, which includes the compressor, evaporative condenser fans and pump, a water agitator required in the ice builder, and a small liquid refrigerant pump.

The installed cost of the refrigeration equipment is $191,000, and the installed cost of the ice builders is $277,000, resulting in a total cost of $468,000. This illustrates the lower installed cost of ice storage; it is about one-third less than the comparable chilled water storage. And the ice storage requires just 16,000 cubic feet, which is only one-fourth the volume of chilled water.

This system also operates only during off-peak hours, so it too saves $23,000 in annual demand charges. With a $198,000 first cost premium over the conventional chilled water system, the payback period is 8.6 years.

The compressor-aided system is the partial ice system that builds ice at night and also uses the ice builder, aided by the compressor, as a chiller during the day, if necessary. It requires the smallest refrigeration capacity because it operates over the longest time period. In fact, the 5000 ton-hour load divided by 24 hours of operation requires just 208 tons of refrigeration capacity. The system energy requirement is proportionately smaller, at 204 kilowatts.

Obviously, a smaller system will be less expensive; in this case, the refrigeration equipment cost is approximately $149,000. And since only a portion of the daily cooling load is stored, less ice storage is needed, and its cost ($161,000) is much less than that required for full ice storage. Therefore, the total first cost is $310,000, which is just 15% greater than the conventional chilled water system; and the required storage volume is just 9,200 cubic feet.

Finally, since the system operates at times during the demand period, a demand charge is incurred. But, at $8,585, it is much less than half the demand charge of the conventional system. Therefore, the first cost premium of $40,000 is paid back by an annual savings of $14,515....in only 2.8 years!

The details of this comparison are summarized in Table 1. It should be noted that this analysis does not include any credit for the benefits available when using an ice storage system with an increased temperature rise and a corresponding reduction in water flow--smaller chilled water pipe sizes, smaller pump, and less pump energy. When these factors are included, the first cost premium is reduced, and the payback period is even less.

TABLE 1

ALTERNATIVE SYSTEM COMPARISON

	Conventional Chilled Water	Full Chilled Water Storage	Full Ice Storage	Compressor-Aided Ice Storage
Refrigeration Capacity (Tons)	600	357	357	208
Hours of Operation	10	14	14	24
System Energy (kW)	550	337	348	204
Cooling Equipment Cost	$270,000	$161,000	$191,000	$149,000
Storage Cost	--	440,000	277.000	161,000
Total First Cost	$270,000	$601,000	$468,000	$310,000
Storage Volume (Cu.Ft.)	--	64,200	16,000	9,200
Annual Demand Charge	$23,100	--	--	$8,585
Demand Charge Savings	--	$23,100	$23,100	$14,515
First Cost Premium	--	$331,000	$198,000	$40,000
Payback Period (Years)	--	14.3	8.6	2.8

Overall, the compressor-aided system appears to be the optimum alternative for this load profile. While it has obvious merits, please realize that a compressor-aided system will not necessarily be the correct choice for every application. It and the other partial systems are usually best suited for normal day and night cycle loads, common to the HVAC applications. On the other hand, full storage systems are generally conducive to process cooling applications with longer build times and/or flexible storage space limitations.

Thermal storage systems using ice as a storage medium are an effective means of shifting cooling loads that can significantly reduce operating costs where the cost of electric power varies with the time of day. They are conceptually simple, easy to design, simple to operate, and can have first costs comparable to conventional systems that produce cooling as it is needed. For these reasons, ice storage systems deserve serious consideration by cooling system design engineers who are sensitive to present and future operating costs.

Chapter 14

HEAT PUMP STRATEGIES AND PAYOFFS

J. S. Gilbert

ABSTRACT

After evaluating numerous waste heat sources and heat pump designs for energy recovery, we have become aware that a great deal of confusion exists about the economics of heat pumps. The purpose of this article is to present some simple formulas for the design and economics of heat pump systems so that the reader may easily determine which heat pump applications are most cost effective. Commercially available equipment is considered to determine the near-term and future economic viability of each approach as a function of the source and user temperatures, and magnitude of heat flow. Generic heat pumps, including vapor recompression designs are explained, costed, estimated in performance, and evaluated as a function of basic economic parameters such as local utility and fuel rates. Steam recompression and generation of steam from moist, hot air are also investigated specifically.

INTRODUCTION

Those readers with considerable background in heat pump principles may find this presentation overly simplistic. The author's contention is that the economics are more critical than detailed design calculations in regards to the feasibility of incorporating heat pumps into plant designs. Therefore, the presentation will describe the basic heat pump cycles and then develop the energy savings achieved through their use. While equipment cost may be calculated following the general guidelines offered in this paper, installation costs must be considered individually as they are very site specific.

THE HEAT PUMP CONCEPT

Several types of heat pumps (both open and closed) are available today. All have certain basic attributes in common. Most importantly, all can be analyzed using some simple approximations of heat pump performance. The most commonly recognized type of heat pump is the conventional external refrigerant design or closed heat pump, shown schematically in Figure 1. Examples of this type are air conditioners and refrigerators. Over the past few years, this type of heat pump has also become popular in home and office comfort conditioning. These heat pumps all use relatively dense refrigerants that evaporate in a heat source and can transfer the heat to the heat user after condensing at a higher temperature.

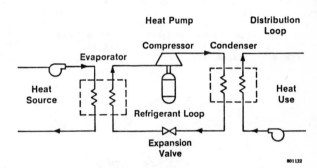

Fig. 1 A Conventional Closed-Rankine-Cycle Heat Pump

Figure 2 illustrates the closed-cycle heat pump thermodynamics. The heat from the source stream is transferred into the working fluid at the lower temperature by evaporation of the heat pump fluid. The vapors are then compressed mechanically to permit transfer of the heat by condensation to the heat user at the higher temperature. The amount of heat delivered is greater than the heat reclaimed from the waste stream by the amount of mechanical work in the heat pump. For this example, the source and use streams are assumed to stay at a constant temperature as heat is added or removed. This assumption is true only for vaporization and condensation of pure vapors. Moisture-laden air is an important exception which will be considered later in detail.

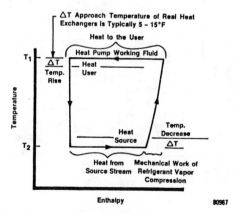

Fig. 2 Thermodynamic Description of Closed-Rankine-Cycle Heat Pumps

BASIC HEAT PUMP CATEGORIES

Before the details of heat pump performance and economics are discussed, a simplified categorical description of all heat pump types will be given. Figure 3 outlines four basic heat pump concepts ranging from a simple vapor recompression system to the waste-heat-driven, external-refrigerant system.

Type I - Simple Vapor Recompression - Common applications include steam recompression where compressing vapors for reuse in the process is less expensive than condensing and reboiling. This category of open heat pump represents a minimum capital cost but has the narrowest range of applicability, very often limited by pressure rise or steam purity. Payback periods are often less than 1 year.

Type II - Vapor Recompression Evaporator - Frequently, this type of open heat pump is an attractive alternative to multiple-effect evaporators where evaporator vapors are compressed sufficiently to transfer the latent heat of the overhead product to the incoming feed stream upon condensation. This type of heat pump is also attractive for distillation towers where overhead condenser and bottoms reboiler temperatures are within about 100°F of each other. However, polymerization or other reactions may prevent application even when condensers and reboilers are thermally close. This class of heat pumps has frequently shown payback periods of 1 to 1-1/2 years.

Type III - External Refrigerant Heat Pump - This closed-cycle heat pump is the most common form of heat pump, and is frequently applicable where refrigerated condensers are used and the evaporation load is within about 100°F of the reboiler. Although these systems can be significantly more expensive than simple vapor recompression systems, they do often produce a payback period of 1 to 2 years.

Type VI - Rankine-Cycle-Driven (Waste Heat) Heat Pump - In applications where the amount of heat to be pumped is significantly less than the amount of heat available, a Rankine-cycle drive may be feasible. This type of heat pump may be either an open or closed system. While the probability of a match between heat pump and heat source characteristics is lower than in the previous categories, and the capitalized cost is the highest, this concept of using "free" waste heat to pump heat has yielded paybacks of less than 3 years.

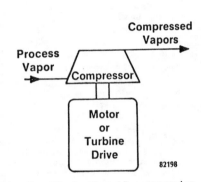

Type I - Simple Vapor Recompression

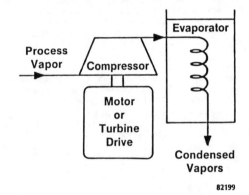

Type II - Vapor Recompression Evaporation

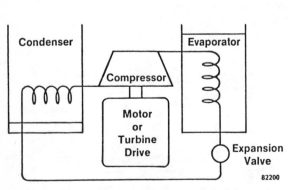

Type III - Heat Pump with External Refrigerant
(Reversed Rankine Cycle)

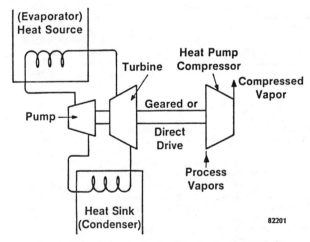

Type IV - Rankine Cycle (Waste Heat Driven)

Fig. 3 Basic Heat Pump Categories

BASIC DESIGN STRATEGIES FOR HEAT PUMPS

A heat recovery system may be designed in several ways. The approach is predicated upon providing the largest amount of heat recovery consistent with the marginal economics. That is, although recovering a given amount of heat may seem economic, the acid test is to prove that each Btu recovered was economic in it's own right.

Figure 4 was prepared to illustrate this point for a hypothetical moisture-laden exhaust. The first strategy in any heat recovery design is to look for direct exchange possibilities. The plant in this example can use 85-psig steam. Steam at this pressure can be generated directly from the exhaust using a waste heat boiler. The payback is usually less than 1 year for this type of heat recovery.

The next level of waste heat requires heat pumping (one or two stages of compression) to generate 85-psig steam. This type of equipment usually can be designed and installed to provide a less than 2 year payback in it's own right. Coupled with the direct exchange opportunity at the higher temperature, the payback will probably fall into the 1 to 1-1/2 year range.

As additional heat is reclaimed through more stages of compression, the marginal economics rapidly deteriorate. If corporate objectives can be met with 3 to 4 year paybacks, up to four stages of compression can usually be justified. The total heat recovery system will now provide about a 2 to 2-1/2 year payback.

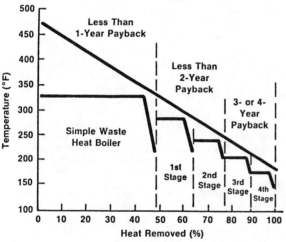

Fig. 4 Heat Recovery Strategies

82291

THE COST OF PERFORMANCE

Open heat pump designs do not require the use of a separate working fluid. Thus, the open heat pump has the potential for lower installed equipment costs. The most common design in practice is that of steam recompression service as illustrated in Figure 5. In this figure, hot condensate is shown flashed to subatmospheric steam and electrically compressed to 20 psig. The heat source is 150°F hot water which when cooled to 142°F can generate steam at 138°F or 2.74 psia. This steam is then compressed (and desuperheated if necessary) to the required delivery pressure. Virtually any delivery pressure is achievable. However, the economics will deteriorate as the required compression ratio increases.

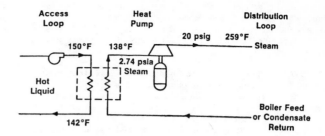

Fig. 5 Example Schematic for an Open Heat Pump Generating Steam from a Hot Liquid Stream

The performance of all heat pumps can be related to a simple expression founded on the Second Law of Thermodynamics. Using the coefficient of performance (COP) as an index of heat pumps' effectiveness, the maximum achievable COP is determined by the heat source and heat user temperatures. Figure 6 shows how heat pumped from 200°F to 300°F has a maximum achievable COP of 5.32. The factor 0.70 accounts for turbomachinery and typical thermodynamic efficiency. The COP of 5.32 means that the heat delivered to the user is over 5 times the amount of energy consumed in the electric-motor-driven compressor.

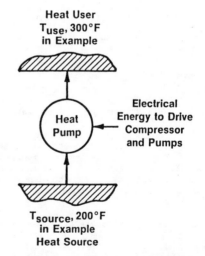

$$COP = \frac{\text{Energy Delivered to User}}{\text{Energy Used in Pumps and Compressor}}$$

$$COP = 0.70 \left(\frac{460 + T_{use}}{T_{use} - T_{source}} \right)$$

For T_{use} = 300°F and T_{source} = 200°F   COP = 5.32

82292

Fig. 6 Theoretical COP for Heat Pumps

The electrical energy consumed in the heat pump is normally quite costly. The equivalent of $0.04/kWh for electric power is $11.72/10⁶ Btu for fuel. Heat pumps would not be economic if not for the fact that realistically high COPs are commonly possible. In fact, where electricity is relatively expensive or its cost is rising more rapidly than fuel costs, heat pumps will not normally be economic.

Figure 7 was prepared to calculate the equivalent cost of fuel based upon the heat pump COP and electric rates, given that the heat source is free. Using the example from Figure 6, a COP of 5.32 will deliver heat at $2.20/10⁶ Btu, if the energy at the heat source is free. The amount of heat delivered is measured as "delivered Btus," so comparison with fuel costs should be made taking into account boiler efficiencies.

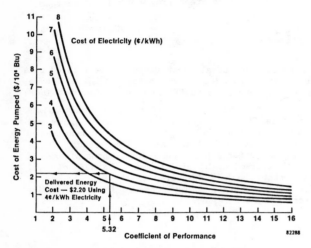

Fig. 7 Cost of Delivered Energy as a Function of Electric Costs and Heat Pump COP

BUT HOW MUCH DO YOU SAVE?

Heat pumps are purchased to save money and energy. Since energy may have significantly different values within or external to an individual process, the following discussion focuses on the savings due to pumping one million Btus from Tsource to Tuse assuming that the energy at Tsource and Tuse is valued at dollars per million source (DPMS) and dollars per million use (DPMU). The heat pump is assumed to consume electric energy valued at dollars per million Btus pumped (DPMP).

Delivering one million Btus to the heat user requires 10^6 Btu/COP in the heat pump plus $[(COP - 1)/COP] \times 10^6$ Btu from the heat source. Therefore, the energy dollar savings per million Btu through heat pumping is represented by the expression:

$$\$Savings/10^6 \text{ Btu} = DPMU - \left(\frac{COP-1}{COP}\right) DPMS - \left(\frac{DPMP}{COP}\right)$$

Quite often the heat source energy would be free or even negative (as in refrigeration heat load). As an example calculation, a number of assumptions can be made:

- The cost of energy for the 200°F source is $1/10⁶ Btu
- The value of energy delivered to the user at 300°F is $5/10⁶ Btu
- The heat pump has an electric drive
- Local electric rates are $0.04/kWh
- The COP is 5.32.

Using these values in the above formula yields:

$$\$Savings/10^6 \text{ Btu Delivered} = \$5.00 - \left[\left(\frac{5.32-1}{5.32}\right)\$1.00\right]$$

$$- \left(\frac{\$11.72}{5.32}\right)$$

$$= \$5.00 - \$0.81 - \$2.20$$

$$= \$1.99$$

or about a 40% reduction in the cost of energy delivered. However, these savings have to provide an acceptable return on investment and only reflect the technically achievable heat pump performance.

STEAM RECOMPRESSION AS HEAT PUMPING

If "waste" low-pressure steam is available, the simplest way to pump the heat is to compress the steam to a useful pressure. The horsepower required to perform this compression follows the COP relationship of Figure 6 rather closely by using saturation temperatures at the inlet and outlet conditions as the definition for source and use temperatures. Surprisingly little electrical energy is involved, especially when the pressure ratios are less than 4:1 on an absolute basis.

If an open heat pump design can be used, a condenser is unnecessary (see Figure 1). If the hot liquid happens to be water of adequate quality, both the evaporator and condenser can be eliminated since the hot water could be flashed directly to produce the water vapor.

The initial reaction to this type of open heat pump design may well be . . . "But, that's not the way a heat pump works." Nevertheless, it behaves almost identically to the typical closed-cycle designs. As a matter of fact, the compression of any vapor may be approximated by the simple expressions developed in this paper. However, certain cost and performance tradeoffs are quite dependent upon the heat source and use characteristics, and on the nature of the refrigerant. Consequently, open heat pumps are not always less expensive. On the other hand, where steam generation is desired, open heat pumps are usually less expensive than closed-cycle designs.

Table 1 provides actual performance statistics for compressing 1000 lb of steam from P1 to P2 (all pressures in psia) assuming a 70% compression efficiency. Using the latest technology single-stage compressors, this compression can be achieved even with outlet to inlet pressure ratios of 2 as indicated. The saturation enthalpies are noted as well as an "extra energy" term. This term accounts for the superheat in the steam. This energy can either be "sprayed down" to saturation, thereby increasing steam flow or can be used by the heat user directly (if superheated steam will not adversely affect the downstream heat transfer). One stage of compression (2 to 1) yields an actual COP of about 15.

TABLE 1

COEFFICIENT OF PERFORMANCE FOR 1000 LB OF STEAM AT P1

Compression		hp Required Per 1000 lb	Inlet Enthalpy (Btu/lb)	Outlet Enthalpy (Btu/lb)	"Extra" Energy (Btu/lb)	COP
Inlet P1 (psia)	Outlet P2 (psia)					
5	10	28.8	1131.1	1198.6	55.3	16.4
10	20	30.2	1143.3	1214.1	57.8	15.8
20	40	31.5	1156.3	1230.2	60.4	15.3
40	80	32.9	1169.8	1247.0	63.9	14.9
80	160	34.1	1183.1	1263.1	68.0	14.6

Figure 8 was prepared to illustrate how to determine the number of compression stages required and how they are related to delivered steam pressure. Each stage is assumed to accomplish a 2:1 pressure ratio. The stages shown were based upon delivering 80-psia (65-psig) steam. The stages are numbered from the top down since the decision to add each stage of compression has to be made on the marginal economics of recovering the heat to be used in that stage. The first stage would use a 40-psia (15-psig) inlet and could recover heat from a waste heat stream down to about 270°F. The second stage could bring steam generated at 230°F up to the point that the first stage could compress it, and so on through the next stages.

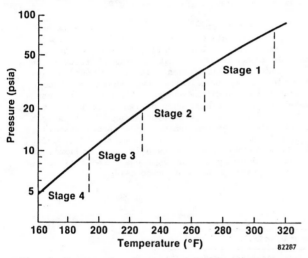

Fig. 8 Compression Stages and Usable Waste Heat Levels Assuming 65 psig Steam Is Desired

Obviously, the amount of waste heat available at a given temperature has a tremendous impact on heat pump design and feasibility. Because of this impact, external refrigerant systems may seem more effective than steam recompression since they can accomplish larger "lifts" per stage with pressure ratios of 3 or more. Actually, the external refrigerant systems do not recover any more heat per required kilowatt. That is, the savings are almost independent of the design

details. Equipment costs are dependent upon the choice of refrigerant. The COP and energy savings potentials are virtually independent of the method used in heat pumping. Comparably high turbomachinery efficiency is the only important factor.

Clearly then, pumping heat must become relatively unattractive at some point. Table 2 shows the actual COPs achievable with up to five stages of compression of steam at 5 psia. Generating 160-psia (145-psig) steam from 5-psia (subatmospheric) steam with an actual COP of 3.7 is possible. This COP is comparable to that of a home heating system with a heat pump. Unfortunately, the extra energy needed to "lift" the lowest level steam (5 psia) requires significantly more electricity than steam at 10 psia. Remember the steam brought up from 5 psia to 10 psia has to also be brought up from 10 psia to the higher pressures. Reasoning through the intricacies of this process may be confusing, so some useful guidelines have been developed.

There are no savings if:

$$\text{COP Equivalent to No Savings} = \frac{\text{DPMP} - \text{DPMS}}{\text{DPMU} - \text{DPMS}}$$

which, using the data from the previous example, works out to:

$$\text{COP Equivalent to No Savings} = \frac{\$11.72 - \$1.00}{\$5.00 - \$1.00}$$

$$= 2.68$$

The "no savings" COP is generally in the 2.5 to 3.5 range. The COPs that fall in this range merely convert thermal costs to electrical costs. If the energy delivered was only worth $3/10^6$ Btu, the "no savings" COP would have been 5.36. Systems with COPs near this point also have the largest capital investments. Our experience indicates that actual COPs less than about twice the "no savings" COP are seldom economic.

TABLE 2

OVERALL COP VERSUS STAGES

| Compression | | 1st Stage | 2nd Stage | 3rd Stage | 4th Stage | 5th Stage |
Inlet P1 (psia)	Outlet P2 (psia)					
5	10	16.4	8.5	5.8	4.5	3.7
10	20	15.8	8.2	5.7	4.4	—
20	40	15.3	8.0	5.5	—	—
40	80	14.9	7.8	—	—	—
80	160	14.6	—	—	—	—

82907

THE FRUSTRATING CASE OF INDUSTRIAL DRYING OPERATIONS

An enormous amount of energy exists in the exhausts of many industrial drying operations such as paper machine hood exhausts. Much of the energy is in the water vapor. But can it be economically recovered and how? Unfortunately, as Figure 9 illustrates, retrieving the latent heat for water-laden vapors isn't easy, even with relatively high dew points such as 170°F as shown.

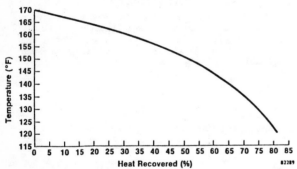

Fig. 9 Water Laden Air with a Dew Point of 170°F

As the water vapor is condensed, the dew point of the remaining air and water vapor mixture also drops. The percentage of moisture that can be removed (which approximates the recovered energy assuming about 1000 Btu/lb of water vapor condensed) from a stream at an initial dew point is summarized numerically in Table 3. As an example, the temperature would have to be brought down to 120°F to remove 81.2% of the latent heat of a stream with an initial dew point of 170°F.

In addition, a heat pump has to capture available heat by accepting it at a temperature below the lowest source temperature. Consequently, to produce steam of reasonable pressure levels, attempting to recover the latent heat of water-laden air is almost always uneconomic since the electricity consumed will cost more than the value of the energy delivered. A frequent mistake made by some heat pump designers is to conceal this uneconomic system by incorporating it with a much more economic higher temperature heat recovery system, and to show that the combination is acceptable. Going after this latent heat is usually only economic in situations where pure vapor flows are available. Contaminated vent steam is a common case.

TABLE 3

PERCENT OF MOISTURE REMOVED

| Dew Point (°F) | Saturation Humidity lb/lb D.A. | Bring Temperature Down to (°F) | | | | | | | | | |
		165	160	155	150	145	140	135	130	125	120
170	.4327	17.2	30.9	41.9	50.9	58.3	64.5	69.8	74.2	77.9	81.2
165	.3584	—	16.6	29.8	40.7	49.7	57.2	63.5	68.9	73.4	77.3
160	.2990	—	—	15.9	28.9	39.7	48.7	56.3	62.7	68.1	72.1
155	.2516	—	—	—	15.5	28.3	39.0	48.0	55.6	62.1	67.6
150	.2125	—	—	—	—	15.2	27.8	38.4	47.5	55.1	61.6
145	.1803	—	—	—	—	—	14.9	27.5	38.1	47.1	54.8
140	.1534	—	—	—	—	—	—	14.7	27.2	37.8	46.9
135	.1308	—	—	—	—	—	—	—	14.7	27.1	37.7
130	.1116	—	—	—	—	—	—	—	—	14.5	27.0
125	.0954	—	—	—	—	—	—	—	—	—	14.6
120	.0815	—	—	—	—	—	—	—	—	—	—

82908

COST GUIDELINES FOR COMMERCIALLY AVAILABLE EQUIPMENT

For the majority of heat pump applications, the heat pump components are compressors, mechanical drives, heat exchangers, piping, and controls. Typically, the user will assume responsibility for the process tie-in only, but can directly purchase the components and do the entire installation. The large heat pump applications (pumping more than 100,000 acfm* of vapors) are often the domain of A&E firms, due to the size and complexity of the equipment.

An estimate of the cost of an installed heat pump can be made by using the following guidelines. The compressor has two basic parameters that benchmark its cost: flow and horsepower. Basically, the flow is dictated by the amount of heat pumped while the horsepower is also a function of the pressure ratio required to pump the heat. A useful approximate formula for required horsepower is:

$$\text{Heat Pump Compressor hp} = \frac{\text{Energy Pumped (Btu/hr)}}{(.70)(2545 \text{ Btu/hp})}$$

$$\times \frac{T_{use} - T_{source}}{460 + T_{use}}$$

Of course, actual equipment should be selected based upon manufacturers' specifications but for early opportunity identification the cost of the compressor including controls and lube system may be approximated in 1981 dollars by the expression:

$$\text{Compressor Cost (\$)} = (\$/\text{acfm}) \times (\text{flow})^{0.5}$$

$$+ (\$/\text{hp}) \times \text{hp}$$

$$= \$2200 \times (\text{flow})^{0.5}$$

$$+ \$30 \times \text{hp}$$

where the flow is expressed in entering acfm of vapor. Corrosive-resistant materials could cost up to 1-1/2 times this amount.

Heat exchanger costs are estimated at about $6 per square foot of required surface, with typical shell costs at $25,000. Stainless generally doubles these costs. The other costs would be that of installation and process tie-in. These costs are highly variable and are difficult to estimate in general. Our experience has indicated that they are usually about as large as the equipment in question. That is, the installed cost is usually twice the cost of the compressor and heat exchangers.

CALCULATION PROCEDURE USING A DISTILLATION TOWER AS AN EXAMPLE

Figure 10 is a schematic of a distillation tower with an overhead condenser (shown dotted) and a reboiler modified to use the overhead vapors to reboil the bottoms. These vapors are compressed to the pressure corresponding to about 10°F above the bottoms so that the latent heat may be transferred. An additional trim cooler is required when the heat duty of the overhead condenser is greater than the reboiler. For this example, 20 x

10^6 Btu will be transferred from the overhead condenser at 170°F to the reboiler at 250°F. The overhead vapors in this case have a specific volume of 10.0 ft^3 lb, a latent heat of 800 Btu/lb, and are flowing at 25,000 lb/hr. Energy used in heating the reboiler is valued at $5 per million Btus, the cost of providing overhead cooling is $.50 per million Btus and the local electric rate is 4¢/kWh. The column typically runs for 8000 hours per year.

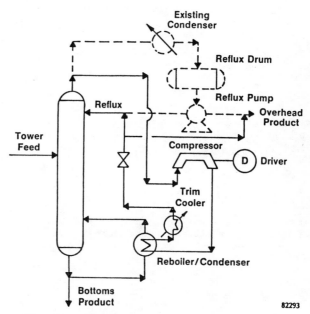

Fig. 10 Distillation Tower Heat Pump Using Overhead Vapor Recompression

$$\text{Heat Pump Work} = \left(\frac{20 \times 10^6}{.70 \times 2545}\right)\left(\frac{250 - 170}{460 + 250}\right)$$

$$= 1265 \text{ hp}$$

$$\text{Vapor Flow} = \frac{25,000 \text{ lb/hr} \times 10.0 \text{ ft}^3 \text{ lb}}{60 \text{ min/hr}}$$

$$= 4167 \text{ acfm}$$

$$\text{Compressor Cost} = 2200 \times (4167)^{0.5} + 30 \times 1265$$

$$= \$179,965$$

Heat Exchanger Area (assuming a 10°F ΔT and U = 125)

$$= \frac{20 \times 10^6}{125 \times 10}$$

$$= 16,000 \text{ ft}^2$$

95

Heat Exchanger Cost = 16,000 ft^2 x \$6/ft^2 + \$25,000

= \$121,000

Total Equipment Cost = \$179,965 + \$121,000

= \$301,000

Probable Installed Cost = \$602,000

Savings based on:

$$COP = .70 \left(\frac{460 + 250}{250 - 170} \right) = 6.21$$

Electricity Cost/10^6 Btu = \$0.04/kWh x 293 kW

= \$11.72

Savings/10^6 Btu Delivered = \$5.00 - $\left(\dfrac{6.21-1}{6.21} \right)$

x \$0.50 - $\left(\dfrac{\$11.72}{6.21} \right)$

= \$2.69

First Year Energy Savings = \$430,400

Operating Costs (assuming 7% of capital cost)
= \$42,100

Net First Year Savings = \$430,400 - \$42,100

= \$388,300

This amount would yield a simple payback period of approximately 1.5 years.

Chapter 15

VARIABLE-AIR-VOLUME SYSTEM FOR PHOENIX CITY HALL

D. S. Teji

ABSTRACT

In most old City-owned buildings, heating and cooling is accomplished by varying the temperature of a constant volume of supply air. Therefore, the supply air fans run continuously at a constant level of power input and air is supplied to all the areas regardless of heating and cooling load. In a variable air volume system a smaller total quantity of air is directed to the zones in proportion to the varying loads throughout the day. Since air distribution energy forms a substantial percentage of the total energy required for heating and air conditioning, a variable air volume system can save a considerable amount of energy.

This paper describes the Variable Air Volume System conversion for the Municipal Building (City Hall) for the City of Phoenix. This building was originally constructed with a dual duct, hot and cold, system using constant air volume with variable temperature for cooling and heating the building. Four supply air fans and two return air fans with a total of 340 horsepower installed capacity provided air circulation through the building. The purpose of this project was to retrofit the City Hall with a variable air volume system and quantify the amount of energy and cost savings.

A computer analysis which was done on the building indicated that it was possible to significantly reduce the volume of supply air. This was accomplished by turning off two out of four supply air fans and both return air fans.

Further reduction was made possible by implementing the variable air volume system. Preliminary results indicated an average of 50 percent reduction in the energy consumption of the remaining two fans was possible over the useful life of the equipment.

DESCRIPTION OF BUILDING AND EQUIPMENT

The Municipal Building (City Hall) under study has twelve stories including basement and sub-basement. The discussion is applicable to only floors one to nine above ground which are served by centralized air conditioning system. The air conditioning equipment is located on the tenth floor. The centralized system is divided into two equal air distribution sections or zones. The air handler for each zone consists of two supply air fans each driven by a 60 HP electric motor and a return air fan with a 50 HP motor. There are two 400-ton centrifugal refrigeration chillers. Two chilled water pumps and two condenser water pumps circulate chilled water and condenser water respectively. Hot water for heating is provided by a boiler and a circulation pump.

The operation schedule of the supply air and return air fans and resulting energy consumption before implement-ing VAV system are provided in Table 1.

EQUIPMENT	QTY	H.P. (EA)	"ON" HRS/DAY	DAYS/YR	KWH
Supply Air Fans (Weekdays)-Day	4	60	11	260	686,400
Supply Air Fans (Weekdays)-Night	2	60	13	260	405,600
Supply Air Fans (Weekends)	2	60	24	104	299,520
Return Air Fans	2	50	24	365	876,000
Total					2,267,520

TABLE 1 - FANS OPERATION DATA

One HP is assumed equal to one KW considering the efficiencies.

The two sets of fans for each zone are located approximately in the middle of each half of the tenth floor. The estimated volume of supply air used prior to this project was equal to 220,000 CFM at 5 inches W.C.

PREVIOUS AIR DISTRIBUTION SYSTEM

Air distribution was affected through a dual-duct air supply system. The dual-duct system consists of a cool air and a warm air supply duct. Air from these two ducts was mixed in a volume control and mixing box which supplied the mixed air to specific zones. This system is usable both for heating and cooling. The volume of mixed air supplied to space remains constant. The temperature of the mixed air varies in accordance with the heating or cooling load. Which means the air distribution power remained constant irrespective of the load in the space. During cooling season the warm air dampers in the terminal boxes are closed to allow the entire volume of air to be supplied through the cool duct. The two return air fans handled the return air, flowing to the fans through above-ceiling return air plenum and main ductwork.

The return air/outside air mixture is cooled by passing it over chilled water coils for cooling before distribution to the zones.

BUILDING ENERGY ANALYSIS

Before attempting to work on converting the building to VAV system, it was considered necessary to analyze the energy efficiency of the building. This was accomplished by making design calculations with computer aid for HVAC parameters for the building with reference to current energy conservation guidelines. The following conditions were used in the analysis:

space temperature	=	78° F
outside dry bulb (DB) temp.	=	107° F
outside wet bulb (WB) temp.	=	75° F
by-pass air factor	=	0.15
apparatus dew point	=	52° F
equipment diversity factor	=	0.4
ventilation air per person	=	10 CFM

Heat gain elements on floors one through nine were as follows:

lights	=	253.2 KW
appliances	=	88.0 KW
number of people	=	679

The calculations were based on ASHRAE Handbook of Fundamentals 1977. The computer program was designed to calculate the supply air requirement for each floor at maximum heat gain conditions. The volume of supply air thus calculated would become the maximum upper limit for the supply air volume. The actual constant volume of supply air used prior to conversion was 220,000 CFM at 5 inches W.C. The calculated volume of air for maximum design conditions worked out to 130,486 CFM. This means the volume of supply air could be varied from 130,486 CFM down to a certain minimum in accordance with the load variation.

STUDY OF FAN CURVES

The supply air fan curves are shown in Figure 1. Point "A" on the curve indicates the condition of operation prior to conversion to the VAV system. This condition represents 55,000 CFM at 5.5 inches W.C. Following the curve at point "B", a condition for supply air of 83,000 CFM at 1.1 inches of W.C. can be obtained. This means two fans at 1.1 inches W.C. are capable of supplying 166,000 CFM of air. This volume of air is more than adequate for maximum load conditions during the year as against computed volume of 130,486 CFM. This condition for the two fans correspond to point "D" on the curve. The minimum supply air condition is indicated at point "C" which corresponds to approximately 40,000 CFM for each fan at 1.0 inches W.C. static pressure with fan operating at 10 HP.

The variation of supply air volume from point "B" to "C" on the fan curve is accomplished through variable air volume system accompanied by fan speed reduction with adjustable frequency fan drive. As a result of the computer analysis and supply air fan curves the use of only two fans was tried, one in each half of the building. In addition, the return air fans, 50 HP each, were turned off.

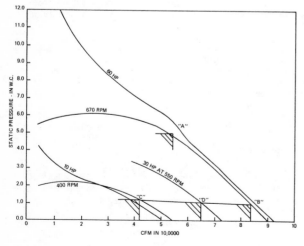

Fig. 1 - SUPPLY AIR FAN CURVE

VARIABLE AIR VOLUME RETROFIT SYSTEM

The concept of variable air volume system is based on the fan laws:

$$(1) \quad \frac{\text{air volume (CFM)}_2}{\text{air volume (CFM)}_2} \varpropto \frac{\text{fan speed (RPM)}_2}{\text{fan speed (RPM)}_1}$$

$$(2) \quad \frac{\text{fan power}_2}{\text{fan power}_1} \varpropto \left(\frac{RPM_2}{RPM_1}\right)^3 \varpropto \left(\frac{CFM_2}{CFM_1}\right)^3$$

Where suffix "1" and "2" denote conditions at full load and reduced load respectively.

As indicated, the reduction in speed (RPM) is directly proportional to the reduction of air volume (CFM) and is proportional to cube root of the fan power. Table 2 shows the relationship between RPM, HP and input KW/HP.

% Rated RPM(CFM)	% Rated H.P.	KW/Rated H.P. (out)	System Efficiency	Input KW/H.P.
100	100.0	.746	.82	.910
90	72.9	.544	.81	.672
80	51.2	.382	.80	.478
70	34.3	.256	.76	.337
60	21.6	.161	.70	.230
50	12.5	.093	.66	.141
40	6.4	.048	.64	.075
30	2.7	.020	.54	.037

TABLE 2. SAVINGS WITH ADJUSTABLE FREQUENCY DRIVE

The conversion of constant volume system to the VAV system as shown in Figure 2 was relatively simple, due to the original design of air distribution ductwork system. The terminal mixing boxes that supply air to the interior zones were converted to cooling only by disconnecting the linkage between the hot and cold dampers. The hot air damper was sealed shut. The cold damper was connected to a new pneumatic damper operator that opened as the demand for cooling was sensed by the zone thermostat. The terminal boxes that supply air to the perimeter zones were converted to heating and cooling by installing thermostats and damper motors to both hot and cold dampers with cooling only above 60°F. outdoor temperatures and heating only below 60°F.

The VAV system consists of the following components:

Thermostats

A remote sensing pneumatic thermostat bulb senses the temperature variation in space. In the perimeter zones two thermostats operate from one sensor. One of the thermostats is set at 76°F. for cooling and the other at 72°F. for heating.

Outdoor Crossover Thermostat

The outdoor sensor controls the action of the heat/cool condition. This sensor sets the cooling thermostat into action when the outdoor temperature is above 60°F. and heating thermostat at outdoor temperatures below 60°F. This control is required for perimeter zone only. The interior zones are cooling only zones, which require only one thermostat, regardless of outdoor temperatures. The thermostats work on a 2 degree temperature differential range from the room temperature set point, which in turn produces a 4 psig to 8 psig variation in pneumatic pressure.

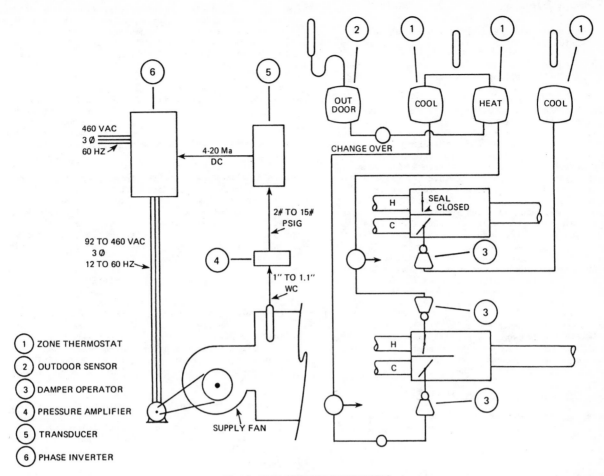

Fig. 2 - VAV SYSTEM SCHEMATIC

Legend for figure:

1. ZONE THERMOSTAT
2. OUTDOOR SENSOR
3. DAMPER OPERATOR
4. PRESSURE AMPLIFIER
5. TRANSDUCER
6. PHASE INVERTER

Pneumatic Terminal Dampers

The pressure variation produced by the thermostat, as explained in the previous paragraph, actuates the pneumatic damper operators for VAV operation. The perimeter terminal damper boxes were converted to the VAV system by installing two damper operators, one each for hot and cold. As explained before, the interior zone damper boxes were converted by completely sealing closed the hot air damper. The VAV system balances the cooling air supply volume between low cooling and high cooling demand zones and shuts down the dampers in no demand zones. These maneuvers result in a substantial savings in volume of supply air. These savings in air volume can be realized by reducing the fan speed with the help of adjustable frequency drive discussed in the following paragraph.

Adjustable Frequency Fan Drive (Inverter): The adjustable frequency drive is a current source inverter. The purpose of the drive is to provide variable speed control of the fan motor. The adjustable frequency fan drive, hereto called inverter, modulates the speed of the 60 HP, 480 volt 3-phase fan motor. An internal feedback signal is incorporated in the control system of the inverter. It adjusts and holds the speed of the fan motor and the fan to the required RPM. This process provides the exact volume of supply air at the proper static pressure that the system requires to maintain comfort zone in the building. Any small variation in duct static pressure upwards or downwards is sensed by the control system which, in turn, varies the speed of the fan to the required adjusted air volume (CFM).

Benefits of adjustable speed (frequency) fan devices are listed below:

- Lower installation cost.

- Minimum maintenance.

- Due to controlled acceleration (lengthening life of belts, pulleys, bearings and fans).

- Multi-fan capability (providing inherent matching of supply and return air fans).

- Soft starting of motors (which reduces electrical demand change by eliminating peak starting current).

● High power factor.

● Slower speed operation (reducing fan noise and vibration).

As indicated in Table 2 significant amounts of energy can be saved by operating supply air fans at reduced speeds.

The inverter takes 480 volts, 60 Hertz A.C. supply and rectifies it to direct current (D.C.) through an input bridge. The fixed voltage D.C. power is then presented to the current source chopper. The chopper converts the fixed D.C. to a pulsating D.C. then becomes a source of current for the load, which in this case is a 60 HP 3-phase fan motor. The inverter directs the chopper to feed the current and voltage required by the motor corresponding to the load at a varied cycle frequency, based on the speed requirement. The operation of this power circuitry is regulated by the control section of the inverter. The speed reference is the 4 m.a. - 20 m.a. signal that is transmitted from the transducer control panel. This reference signal sets up a modulating frequency produced by a voltage controlled oscillator which generates the signal used to control the magnitude of the motor current by turning on/off the electronic switches at the proper time. The signal from the voltage controlled oscillator is used to signal the inverter electronic switches in the proper sequence and time interval to generate the modulating 3-phase current to the motor. This is done in accordance with the requirement of speed and volume of air for the system.

Static Pressure Sensors (Transmitters)

The transmitter installed in the duct transmits the pressure rise with the VAV system to the transducer control panel for the inverter. The low static pressure of 1.1 inch of W.C. is amplified to a maximum of 15 psig. The amplification is necessary before transmitting the variation in pressure.

Transducer Control Panel

The transducer control panel, on receiving the pneumatic pressure change signal, converts it directly to a 20 milliamperes (m.a.) to 4 milliamperes (m.a.) range modulating the electric signal at 24 volt D.C. The conversion of 3 psig to 15 psig is directly proportional to electric signal of 20 m.a. to 4 m.a. The 4 to 20 m.a. modulating signal corresponds to static pressure variation from 1 inch W.C. to 1.1 inch W.C. static pressure. This signal is transmitted to the control panel of the A.C. solid state adjustable frequency fan drive (inverter).

Auxiliary Equipment

Redundancy Control: In the event of an emergency it may be necessary to connect the fan motor directly, bypassing the inverter. In order to accomplish this a second 3 pole/double throw (TPDT) center off disconnect switch with double pole double throw (DPDT) auxiliary switches was installed. This was connected between the inverter outlet power and the motor selection disconnect switch. The switch is wired so that standby power from a "live" circuit can be fed directly to the fan motor. Bypassing the inverter, the auxiliary switches were connected to the control circuits of either the "live" motor controller or the "inverter" control circuit, depending on the position of the main disconnect switch. This safety feature prevents an accidental connection of both power sources at the same time and still gives a safe redundancy system.

Fan Modifications: The fans were originally operating in parallel. These fans were equipped with inlet turning vanes. These turning vanes were locked in a position to provide balanced supply of air without surges. With the conversion to VAV system, operation of only one of the two fans was adequate for each half of the building. It was then found necessary to link these turning vanes to a pneumatic damper motor/operator. This pneumatic control will close the inlet vanes of the fan not in use to prevent back pressure. Switching on the other fan will energize the inlet vanes of the fan starting to operate, to open. As soon as the vanes are completely open a microswitch will start the inverter automatically.

Return Air Fan Bypass: After turning off the return air fan because of redundancy, the air passing through the fan was found to undergo a friction loss of approximately 0.85 inches W.C. In order to eliminate this loss, a bypass system was installed. The bypass system consists of a bypass duct, 4 ft. x 8 ft., equipped with manually controlled dampers in case use of return air is found necessary in the future.

BENEFITS OF VAV SYSTEM

1. VAV system offers an inexpensive means of temperature control in multiple zoning systems.

2. In VAV systems, advantage may be taken of shifting loads from various sources of heat gains such as lights, people, solar and equipment, permitting diversities as compared with the sum of the peak loads.

3. VAV system is virtually self-balancing.

4. It is easy and inexpensive to subdivide for new zones and to take increased loads with new usage as long as the total load does not exceed the original design simulated peak.

5. Operating cost savings are accrued from the following characteristics:

   a. Larger or smaller fan energy savings from long hour usage at reduced volumes.

   b. Refrigeration, heating and pumping energy savings resulting from diversity aspects.

   c. Because of reduced volume flow, longer filter life can be expected.

   d. Full cut off of unoccupied areas, decreasing both refrigeration and ventilation requirements.

6. The maintenance of lower than design air quantities during most of the operating time tends to reduce draft problems.

7. The VAV system operates at maximum noise level only at full load, being considerably quieter at off peak loads.

8. No zoning required in central system.

SYSTEM MONITORING

As soon as the inverters were installed and started functioning, and before all the zone terminal boxes were completely converted to VAV, a monitoring procedure was started.

In order to get a complete electrical profile of the operation, a three channel recorder was used. A digital display on the front panel gives the instantanious display of power factor or KVA. A resetable KW/hour counter display records the accumulative KW/hr.

Before starting the inverters, a chart of supply fan motor KW, KVAR and KW demand was recorded for an 8 hour period, using the "stand by" system. This provided a record of the fan operation parameters. A series of 24-hour tests were then conducted at various chart speeds in order to find the chart speed that would reflect the actual conditions. Line "A" is KW, line "B" is KVAR and line "C" is KW demand. The chart sample was run at 60 CM per hour. Figure 3 was made by taking one minute per hour from the original chart and combining them together to form a 24-hour profile. Line "A" is the KW rate with the supply fan operating in the "stand by" mode. The KW shows a constant 41.5 KW rate. Line "B" is the compressed KW and line "C" is the KVAR. The demand lines in each minute segment show what demand rise was recorded during the one minute interval. Line "D" is a graph of outdoor temperatures in DEG.F recorded at the same time that the recorder was in operation. KVAR recordings were made downstream of the inverter. The values at upstream of the inverter will be much lower.

RESULTS

During the first stage of the project after computerized analysis of the building, it was realized that the two return air fans, with a total horsepower of 100, were redundant. Also, the total supply air volume needed, even at the maximum load, could be adequately provided by two supply air fans. The final supply air volume reduction was made possible by actuating the variable air volume system associated with the adjustable frequency drive (inverter). It is now experienced that over the entire year a 50 percent reduction in energy consumption for the two remaining fans is achievable. This reduction results in an overall reduction from six air handling fans to only one fan or from 340 to 60 horsepower (H.P.).

The simple calculation for energy savings is as follows:

Estimated Energy Consumption Prior to the Project	= 2,267,520 KWH
Estimated Energy Consumption after Completion of VAV	= 525,600 KWH
Net Savings	1,741,920 KWH

At 6.6¢ per KWH rate, the amount of savings works out equal to $114,967.

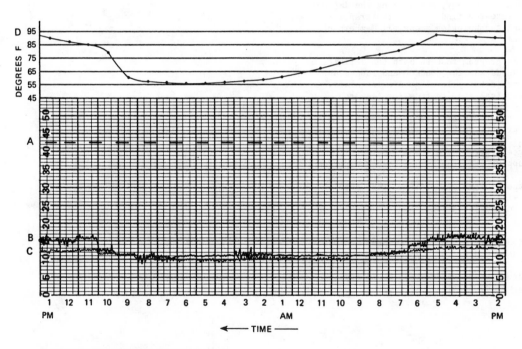

A - KW WITHOUT INVERTER SPEED CONTROL
B - KW WITH INVERTER SPEED CONTROL
C - KVAR WITH INVERTER SPEED CONTROL
D - OUTDOOR TEMPERATURE

Fig. 3 - 24 HOUR PROFILE RECORDING CHART

Table 3 shows energy consumption (billed) and peak
demand before (1978) and after (1981). Savings from
other small projects are also included.

Month	1978		1981	
	KWH	KW	KWH	KW
January	480,000	1,200	326,400	800
February	472,000	1,184	328,800	824
March	531,000	1,200	321,600	848
April	380,800	1,200	366,400	944
May	599,200	1,184	410,400	960
June	540,000	1,392	470,400	976
July	608,800	1,440	523,400	1,016
August	623,200	1,408	502,400	1,040
September	560,800	1,456	501,600	1,008
October	544,000	1,376	420,000	944
November	478,400	1,152	362,400	896
December	451,000	1,056	332,000	848
Total	6,169,400		4,865,600	

TABLE 3 . . . KWH/KW BEFORE AND AFTER

ACKNOWLEDGEMENTS

1. This project was sponsored by the Energy Task Force
   of the Urban Consortium. The Urban Consortium is
   a coalition of 37 major urban governments which
   pursues technological solutions to pressing urban
   problems. The Consortium's Energy Task Force
   supports applied research in urban energy
   management and technology, emphasizing the implementation of research findings and the transfer
   of methods and technologies among all members of
   the Consortium.

   The Task Force program fosters both the definition
   of research needs and the conduct and management
   of research projects by local government staff.
   To assure practical results from its research
   projects, the Task Force encourages the use of
   existing, state-of-the-art technology available
   through the private market.

   The Task Force program receives its primary
   financial support from the United States Department
   of Energy. The city of Chicago serves as the
   manager of the program on behalf of the Urban
   Consortium, with staff and secretariat services
   provided by Public Technology, Inc. in Washington,
   D.C.

2. Harold Sigafoos, Energy Consultant, provided project
   engineering assistance.

3. Assistance provided by other staff of Energy
   Conservation Office is also acknowledged.

Chapter 16

ENERGY CONSERVATION BY DESIGN

D. S. Cooper

GENERAL

This paper treats several ways in which Energy can be saved by "Design Considerations" which have many times been overlooked by Architects and Engineers.

It is a rather discouraging fact that plans and specifications for many new buildings do not include energy saving ideas which now form the basis for retro-fit in existing buildings.

In other words, "State of the Art" ideas in the Architectural, Mechanical and Electrical fields are not appearing, as they should, in New Building Designs.

Atrium buildings with transparent skylights, large glass areas on all exposures with no outside shading and little effective reduction in shading coefficient by reason of the glass on either internal or external surfaces, is now the rule instead of the exception.

Very few of the new buildings constructed within the past 5 years would meet A.S.H.R.A.E. 90-75 Energy requirements, which were thought to be reasonably effective in conserving Energy only a few years ago.

There are, of course, many reasons for the relaxing of our Energy Standards in the Building construction fields and it is the purpose of this paper to address these, with the hope that the trend may be reversed.

Starting with the Architectural profession, because they are the instigators of Architectural ideas which form the Envelope of a building, and once the shape, size and form has been determined to the satisfaction of the owner, there is little that the Engineer can do to make it more efficient.

The question now becomes "Who wants the building to be more energy efficient?"

Not the Architect, because it generally interfers with the esthetic features which give distinction to the overall concept of design. It also usually puts some added costs into the machine room where they cannot be seen or appreciated.

Not the Owner, because he wants to build something which incorporates ingenious and unique features which he feels may be desirable in the renting process or in the eventual sale of the building.

The Owner also intends to pass on all the operating costs, including inefficiencies, to his tenants by means of what is called a "pass-through" lease. In this way, it doesn't matter whether or not the building is energy efficient.

Not the Mortgage Banker, who loans the money for the long-term financing, because he is interested in keeping the loan at a minimum. He locks-in his loan to a standard cost per square foot which includes very little extra money for Mechanical and/or Electrical features which might save Energy.

If these people are not interested in Saving Energy, why should the designing Engineer try to save it for them?

Many Engineers have pointed out, that they have designed good, efficient systems for Architects only to have them reduced to mediocrity when the bids came in too high. The mechanical and electrical features which are not absolutely necessary for marginal function are always the first to be eliminated, if a cost reduction is necessary.

So, if the Engineer feels that he also would like to design something which he can point to with pride, there are ways and means in which he can affect the design of the Mechanical & Electrical Systems to that end.

One rather sneaky way in which this designer has been able to affect the cooling capacity of a plant is by being able to adjust operating temperatures in the more conservative direction, both on chilled water and condenser water.

This was done on Houston's ASTRODOME in which chilled water temperatures leaving the machines were specified at 42°F rather than the usual 45°F which was specified for the coils. With a 10°F rise in the coils, the return water would be at 52°F. This, in effect, made the Machine Manufacturers put in a few more tubes in their chiller, because of closer approach to freezing refrigerant temperatures. The cooling coils were then specified for a 10°F rise between 45°F and 55°F which made the coil manufacturers increase their heat transfer surfaces in order to obtain the leaving air temperatures specified.

The same thing was done on the cooling tower and condenser side of the process. The machine manufacturer was told he would have 86°F water for condenser supply with an 8°F rise to Tower, while the Tower manufacturer was specified to deliver 85°F water with an 80°F wet bulb even though Houston's usual design is for 79°F wet bulb.

The net result of the above procedure was a full 10% increase in plant capacity above design, which amounted to some 600 tons of refrigeration, for which very little extra money was spent, and which has since proved to be very desirable from the owners and operators standpoint.

ARCHITECTURAL CONSIDERATIONS

There are many ways in which architects and engineers could be very helpful in reducing the Energy Consumption of their buildings if they just knew more about the problem.

There is a very true saying which recently appeared in print, which says:

"Anyone who is not a part of the solution may be part of the PROBLEM."

1.-VAPOR BARRIER LOCATION

Take, for instance, the location of a vapor barrier in a building wall. Few people realize what it can do for the integrity of the envelope, and fewer yet know where it should be placed.

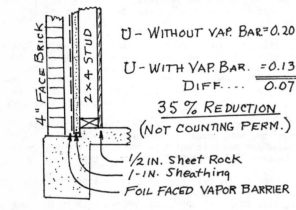

—VAPOR BARRIER LOCATION—

U - WITHOUT VAP. BAR.= 0.20

U - WITH VAP. BAR. = 0.13

DIFF.... 0.07

35 % REDUCTION
(NOT COUNTING PERM.)

½ IN. Sheet Rock
1-IN. Sheathing
FOIL FACED VAPOR BARRIER

—TYPICAL WALL SECTION—
FIG.-1

Referring to Fig. 1, the vapor barrier is shown to be located on the surface of the sheating - looking outward to the brick air space. This is the proper location for Southern climates having high moisture content air outside while trying to maintain lower humidities inside. This may <u>not</u> be the desirable place in more Northerly latitudes.

If the vapor barrier is foil faced (for reflection of radiant heat), it has the added effect of reducing the transmission factor by 35% while still reducing the travel of moisture through the wall to the inside. It is, therefore, much more effective than would be 4 inches of a blown-in insulation between the studs.

When it is considered that up to one-third of the capacity of an Air Conditioning plant is being used to wring moisture out of the air, it should be considered a major factor in the design of buildings. Residential construction is just now beginning to use the vapor barrier, but very few use the reflective type. Commercial office buildings seldom resort to such a refinement even though the cost is insignificant.

2.-DOUBLE VS. SINGLE GLAZING

There seems to be some differences of opinion as to when and where double glazing should be used. As a general rule, double glazing should <u>not</u> be used in the milder climates below latitudes 32^0-35^0. Even though

many locations below these latitudes have some cold weather in the winter, it is usually not of long duration and if double glass is used, it may be more detrimental than good.

—DOUBLE-Vs. SINGLE GLAZING—

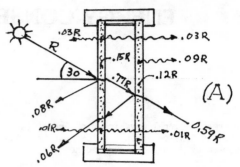

(A)

U-FACTOR = 0.5 to 0.6 (TRANSMISSION)
SHADING COEF. = 0.59 + 0.15 = 0.74

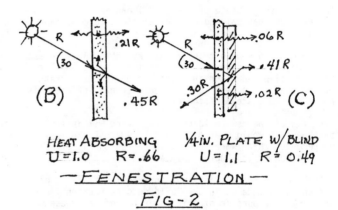

HEAT ABSORBING
U=1.0   R=.66

¼ IN. PLATE W/ BLIND
U=1.1   R= 0.49

—FENESTRATION—
FIG-2

Figure 2, shows cross sections through these types of glass in which the amounts of radiant heat and conductive heat passing through to the conditional space inside are being considered. (A)

The double glass panel in (A) shows the approximate distribution of the radiant energy as it passes thru the outside pane and then thru the inside pane with the absorption and reflection factors involved.

Due to the trapping of heat between the panes, as with the accumulation of heat in an automobile with windows closed, together with the heat absorbed by the glass itself, appreciable amounts of heat are delivered inside from re-radiation as well as direct radiation.

The transmission factor "U" is much lower than single glass, but the double glass does very little for the radiation represented by the overall shading coefficient. The heat total transmitted is then the sum of the conduction and radiation, but the temperature difference across the double glass is usually only 20^0F $\pm$, while the shading coefficient is multiplied by the sun's intensity "R" which can be of the order of 150 to 200 BTU per sq. ft. of glass.

If there are many more hours of cooling required than hours of heating, the double glass will actually admit more energy than a single pane of ¼" plate glass with a venetian blind. (C)

Heat absorbing glass is worse yet, due to the re-radiation of trapped heat in the glass as well as a normal Transmission Factor U. (B)

3.-BOXED SOFFITS

In the Gulf Coast area, Architects and Builders are using what is called "Boxed Soffits" around the outside periphery of small commercial buildings because it provides a shaded overhang for the first floor walls while allowing increased second floor area.

It also gives a massive appearance and the cavity space between first floor ceiling and the second floor or roof has ample space for duct work lighting fixtures, as well as ceiling return for air conditioning.

In general, it is not a bad arrangement, but in many cases it causes the failure of the air conditioning system or results in excessive energy usage.

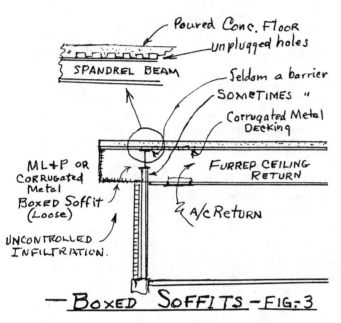

— BOXED SOFFITS -FIG-3

The reasons for the above are because the soffit overhang is seldom insulated or sealed to become a part of the conditioned interior return air space, and admits large quantities of uncontrolled outside air directly into the return air plenum space.

Sometimes the vertical wall is extended to the bottom of the spandrel beam, but seldom are the corrugated structural floor cells sealed off. Never is the soffit insulated properly to reduce the entrance of heat and humidity from outside.

Sometimes rockwool batts are shown stacked on the vertical surfaces and laying on the horizontal surfaces, but inspection reveals they are never properly spaced, placed or retained in position, to be effective for reducing conduction and certainly are not the answer to prevent the entrance of outside air loaded with moisture.

Architects should give this construction serious attention by detailing and specifying board form insulation with vapor seal on flat soffit surfaces and plug corrugated holes.

4.-ATTIC VENTILATION

The means for reducing the amount of heat from a hot attic space which passes through the ceiling construction to the conditioned space below has long been a controversial topic.

Attempts have been made to arrive at formulae which would predict the heat flow based upon temperature differences, and rate of air removal from the attic. None of these attempts have been accepted by rationally minded engineers, because the one most important factor of heat transmission has been omitted-RADIATION.

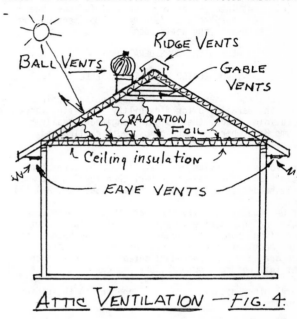

ATTIC VENTILATION — FIG. 4.

The problem has recently been substantially confirmed by tests made upon several identical houses by Houston Lighting & Power Co. All houses were new, unoccupied identically air conditioned, and together on the same side of the street for similar sun exposure.

One house used gravity eave vents with ridge vent release of hot air.

Another used gable louvres and eave vents.

Another used gravity ball vents instead of ridge vents.

And one had a hipped roof with no gable or other means of venting.

Without going into the details of the arrangements and tests, the bottom line summed up the logical conclusions, as follows:

1. The Electrical input to each house was measured for one month during the hottest summer weather. There were no significant differences in any of the meter readings. This proves that practical Attic Ventilation as incorporated into todays residential construction is ineffective in preventing heat transfer by ventilating methods.

2. The heat transfer is nearly all a result of radiation between the hot roof and the attic ceiling construction, and there is no acceptable method of attic ventilation which can prevent it.

3. The problem should therefore be attacked by the installation of the following types of insulation:

    (a) Foil faced batts between the ceiling rafters.

    (b) Foil faced batts on the underside of the roof rafters (with both foil faces looking toward the attic space).

    (c) Provide a white or light colored roof surface to reflect as much heat as possible to the outside.

    (d) Ventilate the attic by exhaust fan only when necessary to provide some air movement during access to the attic space.

MECHANICAL CONSIDERATIONS

1.-DEDICATED PUMPS IN PARALLEL

Many years ago, when large air conditioning machines were being paralleled for flexibility in meeting varying building loads, the designing engineer usually had a "Pump Bay" where all the pumps, both chilled and condenser, were placed.

Pumps for each service were paralleled (by connecting suctions and discharges together) then running a common supply header to all of the machines. (See Fig. M-1 (B))

This was done to justify the fear, that any pump could be used on any machine in the event of failure. But, a centrifuge pump with mechanical seals has become one of the most reliable components of the entire system. There are many other non-duplicated components which might fail before a pump failure.

Several other disadvantages are inherent in the piping arrangement shown in (A).

    (a) Shutting down a chiller means manually valving it off.

    (b) One pump operating on one chiller will overload the pump due to drop in system resistance.

    (c) Twice as many gate valves are needed.

    (d) This arrangement does not lend itself to automatic chiller operation, and wastes energy due to the fact that part load pump operation causes pumps to operate at points on their flow curve far beyond their design selection, thereby using more power over long periods of operating time.

— SERIES-PARALLEL PUMPS & CHILLERS —

— WRONG —
(A)
FIG. M-1

Referring to Fig. M-1 (B), each pump feeds its own chiller and can be sized to meet individual chiller size, if they are not the same make or capacity.

— DEDICATED PUMPS IN PARALLEL —

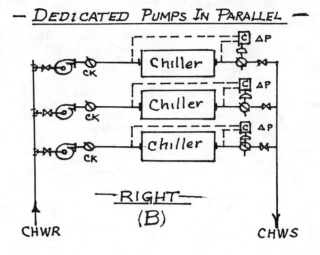

— RIGHT —
(B)

This arrangement allows any and all chillers to operate automatically without bypassing of return water through a dead chiller.

The automatic differential pressure controller (C) measures the pressure drop through the chiller and positions the downstream butterfly valve to accomplish design flow under any operating sequence, regardless of system resistance variations.

The added cost of the controls will have a short payback period from Energy Savings.

Flow control should be added in every case to limit the chilled water flow to the rated capacity of the chiller. This prevents overloading the individual pumps regardless of the sequence of operation and eliminates the need for valving off a dead chiller.

2.-PARALLEL PUMP CURVES

Pumps operating in parallel on a circulating system are always selected for maximum head and flow when all machines are developing maximum capacity. This condition actually does not exist more than about 5% of the operating time, when outside maximums are encountered. The other 95% requires some form of capacity reduction in which there may only be a single machine and pump in operation.

With the Automatic Flow Regulating Control described above, the advantages become quite apparent when the pump curves are examined in Figure M-2.

The circulating system head curve is always logrithonic or the head goes up as the square of the flow. Even though the system curve may get steeper or flatter as modulating valves respond to temperatures, the fact remains that the pump can only operate at the intersection of its curve with the system curve and establish a corresponding horsepower.

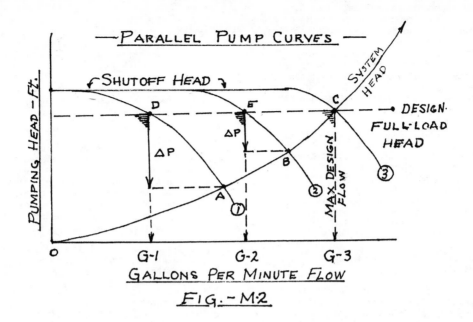

— PARALLEL PUMP CURVES —

FIG.-M·2

By reference to Fig. M-2, three pump curves are shown, indicating the expected operation when one, two and three machines are running. When the system head curve is superimposed upon the three operating modes, it becomes quite clear as to the reason for flow tion.

With only one pump in operation, the unrestricted flow would cause the pump to operate at the intersection of its curve with the system curve at point A. This would cause the pump to deliver about 75% more flow than it was designed to handle at point D, representing a serious overload.

When two pumps are operating, the point of operation would be at point B overloading both pumps by about 50% of flow.

The P shown is the pressure drop or added restrictive head necessary to make these pumps operate at design head, flow and horsepower under part load conditions.

The automatic flow controllers shown in Fig. M-1 measure the pressure drop across each chiller and modulates the motorized butterfly valve to establish design flow under any operating condition.

Mechanical Engineers should provide this control on all systems where multiple machines are to be used under varying load conditions.

3.-THREE-WAY VALVE CHARACTERISTICS

The three-way valve is the most insidious device ever invented as it is usually misapplied by unsuspecting designers and engineers.

There are in operation today, thousands of HV&AC Systems using three-way valves which have been improperly applied under the guise of being a "constant volume" means of regulating water flow through coils with bypasses.

"By placing a manual regulating valve in the bypass and setting it equal to the coil resistance, the water flow will be constant through the combination regardless of the three-way valve position." - - WRONG

The ports through the three-way valve, as with any valve, have logrithmic (square curve) characteristics. In three-way configuration, the flow through each port is the reciprocal of the other, represented by the overlapping curves. When one port opens the other closes. The bypass port (B) opens from left to right, while the

— 3-WAY VALVE CHARACTERISTICS —

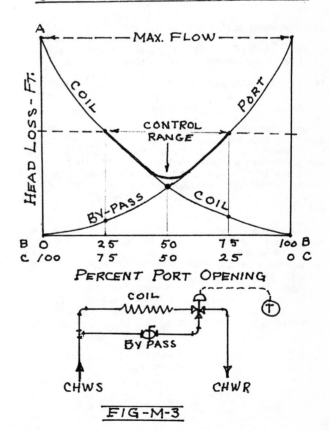

FIG-M-3

107

coil port (C) opens from right to left. The coil flow and valve size are determined at maximum, Point A, and pump head and capacity is determined to meet this condition which may occur less than 5% of the operating time.

95% of the time the combination will be operating within the control range indicated, and when the flow is established at 50-50, midway of both ports, the resistance of the combination can be as low as 25% of the head originally designed. Most of the time it is less than 50% depending upon the action of the thermostat.

Instead of being a constant flow device, it is a variable flow device, and its effect upon the hydraulics of the water circulating system and all associated components including the chiller or boiler, is devastating.

(a) Dumping large quantities of supply water back into the return system where units are located near the source, causes a destruction of pressure differential and flow to the units located further away.

(b) Mixing of supply and return water is a notorious Waster of Energy.

(c) Pumping power is proportional to flow and three-way valves increase flow without any beneficial effect.

(d) Temperature differentials at chillers and/or boilers are reduced to the point where their controls indicate a satisfied system condition which may not be true.

4.-HEAT RECOVERY SYSTEMS

Heat Recovery Systems are generally thought of as highly complex, requiring large insulated storage tanks and complicated control systems, but this is not necessarily true.

Everyone knows that a centrifugal chiller is a heat pump capable of putting into its condenser about 15,000 BTU/Ton and that this heat is usually thrown away into the atmosphere through a cooling tower.

The recovery of a large portion of this heat is possible, instantaneously, without storage, and can be Energy Efficient if there are good reasons for its use.

Many systems are designed for close control of temperature, and some times humidity, using double duct distribution, multi-zone units, or other air-to-air systems in which heating and cooling takes place simultaneously. There are still many operating systems in which terminal reheat is employed, and although they are notorious wasters of energy, it may be too costly to correct without Major Retrofit.

If the system is being designed for a New Building, the heat recovery unit can be specified with a separate heat exchanger or double-bundle condenser for direct recovery of heat into the circulating hot water loop of a 4-pipe system. If the cooling load is insufficient in cold weather, there may also be incorporated an Electric Boiler on the downstream side of the heat recovery condenser which will usually only be necessary during extremes of cold weather when perhaps no cooling is necessary.

Several simple and independent control systems are necessary to make the system automatic.

Referring to Fig. M-4, these control systems are briefly described as:

(a) System #1 - to control the condenser water by fan sequencing and dump valve bypass. (this prevents the water from getting too cold)

(b) System #2 - resets boiler water temperature from outside temperature to limit its operation to only the extremes or as required.

(c) System #3 - provides a closed loop on the main condenser, regulated from condensing temperature on the machine. (this keeps warm water in motion at all times thru the condenser to allow the auxiliary condenser to recover all possible heat from the machine sending only the excess to cooling tower.

(d) System #4 - provides regulated chilled water flow through each machine regardless of the sequence of operation as previously described in Fig. M-2.

The system shown has recently been installed in the First United Methodist Church in Houston as a part of their energy retrofit program and is now in automatic operation, without the necessity of an operating engineer.

Chillers are sequenced into operation by means of a computerized timing device which also starts and stops some 15 Air Handling units to meet the varying requirements of the church.

Since the system had previously been designed for zoned reheat, downstream from central cooling units, the temperature control of all zones was established by the use of warm water recovered from the chiller condenser of only one machine.

The electric boiler is interlocked with chiller operation to allow only enough heating elements (10 in all) to come on line to prevent establishing any new K.V.A. demands above what would be established by the two chillers operating together. Thus, in extremely cold weather, both chillers would be off, allowing full boiler capacity to be used for heat.

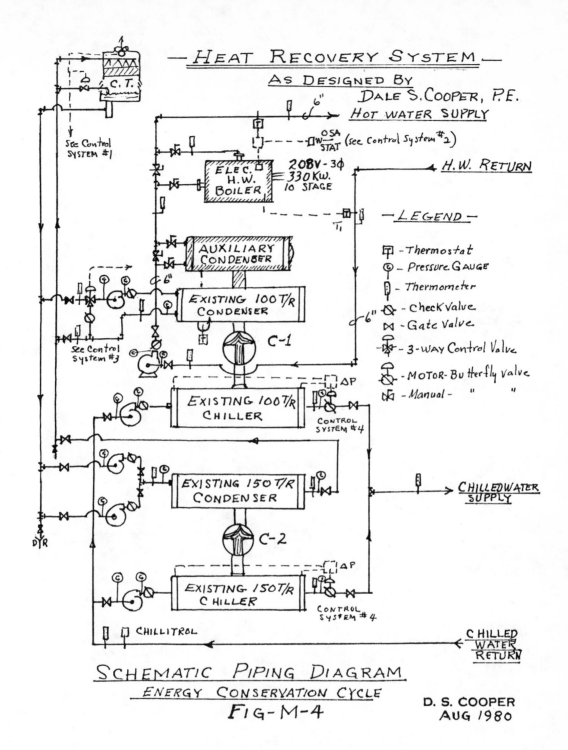

— HEAT RECOVERY SYSTEM —

AS DESIGNED BY

DALE S. COOPER, P.E.

HOT WATER SUPPLY

See Control SYSTEM #1

OSA W/STAT (see Control System #2)

H.W. RETURN

ELEC. H.W. BOILER

208V - 3∅
330 KW.
10 STAGE

— LEGEND —

- Thermostat
- Pressure GAUGE
- Thermometer
- Check Valve
- Gate Valve
- 3-WAY Control Valve
- MOTOR - Butterfly Valve
- Manual - " "

AUXILIARY CONDENSER

EXISTING 100 T/R CONDENSER   C-1

See Control System #3

ΔP

EXISTING 100 T/R CHILLER

CONTROL SYSTEM #4

EXISTING 150 T/R CONDENSER

CHILLED WATER SUPPLY

C-2

ΔP

EXISTING 150 T/R CHILLER

CONTROL SYSTEM #4

CHILLITROL

CHILLED WATER RETURN

C.T.

D/R

SCHEMATIC PIPING DIAGRAM

ENERGY CONSERVATION CYCLE

FIG-M-4

D. S. COOPER
AUG 1980

CONCLUSIONS

The features described herein are tried and proven methods for conserving Energy and represent the State-of-the-Art, although there is nothing mentioned herein that has not been known for many years.

The reasons they have seldom been incorporated into the design of new systems is because they bring about a small (15% to 20%) increase in first cost which the owner and architect cannot immediately merchandise.

In existing building retrofit, however, there is always enough money to correct the situation when the costs of operation begin to hit the pocketbook.

"The ox must be gored before the owner becomes damaged."

Chapter 17

ENERGY EFFICIENT CONTROL OF COMMON VARIABLE VOLUME SYSTEMS FOR THE SUN BELT

B. J. Gould

Variable volume systems have over a period of the past five years gained relatively universal acceptance as being the most energy efficient system for a wide range of types of building mechanical systems. The most common application has been that of commercial office building space although other facilities such as schools, public areas of hospitals, and research labs are also applying the variable volume principles.

While the development of variable volume is largely due to a search for more energy efficient systems it also provides additional zoning, versatility for modification and flexibility for on/off cycling that may not be easily obtainable in other types of systems. Much experience has been gained in the South and Western parts of the country on variable volume systems and it is a review of current control techniques that will be covered here.

TERMINAL UNITS

The design of both the equipment and the control system for the main core of a variable volume system is very often dependent on the type of terminal unit selected. The following are some of the more common systems in use at the present time.

(1) Dual duct mixing boxes for either high pressure, medium pressure or low pressure duct system. These units have independently controlled cold and hot deck dampers so that modulation and variable flow can occur in both sequences. With the incorporation of a velocity controller and the necessary relays with a separate actuator for both heating and cooling it is common to achieve a sequence which provides for sequencing modulation of both heating and cooling with a maximum flow separately selected for each. Also, a minimum flow can be maintained at all times with overlap of heating and cooling sequence occurring only during the minimum flow condition.

Figure 1 outlines a typical operating sequence over the full throttling range of a space thermostat. Velocity controller is necessary to provide both limits and will also provide velocity limiting associated with shifts in upstream static pressure. Some discussion of pressure independent operation will take place shortly in another section.

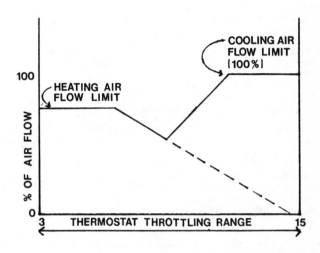

Fig. 1 DUAL DUCT BOX - DIRECT ACTING

(2) Low pressure continuous fan system. On this unit a low pressure air valve modulates to provide cold deck air to the inlet of a continuous blower which may supply air from either the cold deck or return air. The cold air valve will modulate the closed position prior to the cycling of either electric or hot water heat. The individual zone fan must run continuously during normal operating conditions.

The system does provide for modulating cooling with continuous and constant air flow to the zone, thereby eliminating one of the common hazards of variable volume in the form of variable air distribution.

(3) Modulating cold air valve with low, medium or high pressure duct supply providing the building with zone variable volume cooling but having a completely separate and independently ducted perimeter heating system. Typically the heating system in this application is sized to handle only building skinload and is arranged so that on exterior zones the cold air valve will reduce to minimum air flow prior to the cycling of the heat.

In many systems the perimeter heating zone is formed from one entire floor exposure while the cooling is broken up into multiple zones. This does present the possibility of overlap between heating and cooling depending on individual zone heating and/or cooling set points. Some of this type variable volume air valve units have self powered controls thereby requiring no instrument air for normal control.

(4) Induction type air valves have been used in some systems where after a reduction in cooling air flow a heating coil, either electric or hot water, in the downstream duct work will cycle only upon minimum air flow from the induction box. This type of box pulls a substantial percentage of its air from the return air space.

These systems were favored in the early transition to variable volume but it is evident that even though it is at reduced air flow there is still in this system a substantial amount of reheat on the heating cycle and is, therefore, no longer in favor as a design sequence.

(5) Variable cold air valve for high, medium or low pressure system with auxiliary side pocket fan operation providing constant volume heating with either electric or hot water heat. This system is a predominant system for some types of commercial office building construction. It operates fully modulating, normally open, on the cold air valve to provide cooling with individual zone control of heating arranged so that an auxiliary fan cycles on as the cold air valve approaches the closed position. This provides constant air flow for heating by pulling air from the nearby return air space through a backdraft damper upstream of the blower.

The cold air valve moves to the completely closed position prior to cycling of the electric heat or opening of the hot water valve, whichever may apply. This system has evolved over a period of time to become a factory packaged system having all major components integrated into either a single box or an air valve and a single side pocket fan assembly. Generally, the air flow of the auxiliary blower is about 65% of maximum cooling cfm.

One advantage of this arrangement for mild climate condition areas is that the auxiliary fan operates only during the heating season and therefore for a reduced number of hours. This is probably a considerable maintenance advantage compared to continuous fan operation. It should be noted that the actual power consumption of either continuous or auxiliary fans is quite small and does not have a substantial effect on total operating costs.

Tight close off of the primary air valve in this, and in fact all common heating and cooling systems, is extremely important for efficient operation. For example, if low pressure air valves are to be installed in a duct the seal between the air valve frame and the duct should be caulked. One of the more common criticisms of this type of system has revolved around installation defects due to primary cooling air valve leakage.

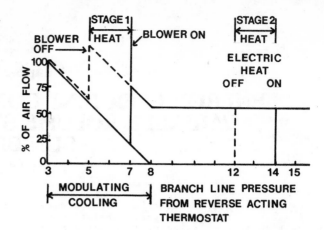

Fig. 2   SIDE POCKET FAN TERMINAL UNIT

PRESSURE INDEPENDENT OPERATION OF AIR VALVES

There are many viewpoints related to pressure independent, that is velocity sensing and/or limiting control of the VAV terminal units. It appears that in low pressure systems, providing the shift in static pressure immediately upstream of the air valve does not vary more than 1 to 1.5 WG, that under normal operating conditions velocity sensors may not be justified. This particularly applies to a building with a large percentage area of interior zone where those interior air valves are mainly open during normal operation and can also apply to systems where maximum blower static either controlled or riding the blower curve does not appreciably shift.

Air handler manufacturers have taken many steps to achieve flat blower curves with very little displacement in static pressure from the operating condition to minimum air flow condition for typical systems.

It should also be noted that pneumatic velocity controllers or limiters consume considerable instrument air and in some systems effectively quadruple the air capacity required by the compressor station. Additionally, many sensors must be specially factory installed and tested for location to insure accuracy and repeatability of the velocity sensing.

As soon as the system design shifts to either medium or high pressure, it is probably mandatory to utilize some form of velocity sensing. Air valves operating with static pressure upstream variation in the range of 1.5" WG upward are subject to both extreme variations of space air flow as well as potential air valve turbulent air noise if sound attenuation is not used.

AIR HANDLERS

(1) Single station air handlers field erected either DX or chilled water: Many of these systems provide the air source for either all or substantial sections of multiple story buildings and can be designed to operate very efficiently. The inherent size of the air handler means that either tube axial variable pitch blowers or air foil blowers with

inlet vane damper control are ideally suited to this application. Many air handlers will have two blowers operating either in unison or in sequence.

The static pressure control required for fairly uniform downstream static pressure conditions will typically be the static pressure transmitter and receiver controller driving a heavy duty actuator at the pitch control or inlet vane damper control.

It has been noticed that particularly with variable pitch control under minimum load conditions more stable operation can be achieved by using one of several receiver controllers having an integral reset sequence. This sequence implies resetting of the output signal until the system is sensing mid-throttling range thereby eliminating throttling range. Should a substantial load shift occur the controller resets on a variable time rate to that same mid-throttling range point.

Location of static pressure sensors is very much dependent on the arrangement and design of the building and has been in many cases determined by field experience.

Most systems having a single station air handler plant are readily adapted to ventilation or economizer cycle at minimum expense. There has been some contention throughout the warmer and high humidity areas of the country that this is not a justifiable sequence. However, we could take advantage of substantial free cooling during most of the heating season.

Most economizers in operation are utilizing outdoor enthalpy sensing to determine a changeover point from mechanical cooling to outdoor (free) cooling. It generally will expand the usable time period for an economizer package as compared to outdoor D.B. changeover by 10-15% of normal commercial operating time. If a building is occupied with internal loads during night time hours that percentage would, in fact, be much higher. The typical economizer system is utilizing mixed air temperature control modulating outside air and return air dampers. Either pneumatically actuated or barometric building relief dampers are required to relieve building pressure.

There has been some use of comparative enthalpy systems sensing outdoor conditions and return air conditions arranged so that should the enthalpy of outdoor be lower than that of the return air then the system will permit economizer operation along with mechanical cooling if required to maintain the necessary air handling L.A.T.

These systems are available both pneumatically and electronically and it has been found they are both fairly expensive and difficult to fine tune and maintain in calibration. This appears to be one of those control systems which theoretically would provide great energy savings but which in practice requires too much attention to be maintained at efficient operating level.

You will notice that discharge temperature control for this type system is covered

separately elsewhere.

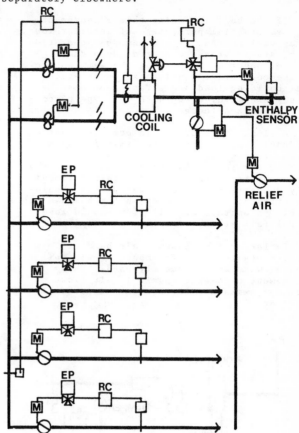

Fig.3 CENTRAL AIR HANDLER

(2) Low pressure air handlers. One very typical design in some areas of the Southwest has been to provide a single air handler for each floor of a building.

A typical unit is horizontal draw through with forward curved fan. If this air handler arrangement is set up so that the blower "rides the curve" for a substantial part of its operation, this can be an efficient system even though the concept of the equipment may not be the most efficient. If the blower selection is made so that full air flow operation is off to the right of the peak efficiency point on the blower curve so that as the load and therefore the air flow in the building decrease, the blower performance shifts up the curve to a more efficient point. This will improve both blower efficiency and reduce HP. On occasion this requires a blower section smaller than the coil section of the air handler it is conventionally matched to. Keep in mind that the peak air flow on this type of system will occur only for a very small percentage of the time.

If a static pressure controller bypass damper were used with this type of blower, it is possible to permit the blower to ride the curve back to some point near 50% of selected air flow and then arrange for a bypass damper to open to permit air flow back into the return of the air handler providing relatively efficient operation down to that 50% or 40%

capacity level.

If the building using this arrangement
has a substantial percentage of interior load,
it may well be that bypass and dump operation
of the static pressure control dump damper
will not occur except during the warm up cycle
and the cold start conditions. Here again
total building load and arrangement must be
closely examined, along with the type of control sequence selected.

A simple form of static pressure control
and/or high static pressure limit for such
systems is the use of a bypass barometric
damper mechanically balanced to insure that a
minimum static pressure is maintained immediately downstream of the blower. This should
be sufficient for all zones in the building to
be able to operate at full flow should zone
control fall at any time.

In larger low pressure air handler
arrangements inlet vane dampers on air foil
wheel blowers are frequently used. This
arrangement becomes economic in air flows
above approximately 20,000 cfm and is highly
efficient.

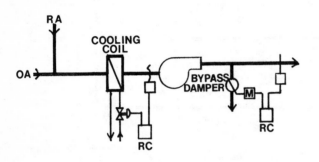

Fig. 4   LOW PRESSURE AIR HANDLER

(3) Packaged rooftop equipment which are
DX and are available from several major manufacturers. These are typically provided with
a packaged static pressure control system operating a discharge damper on the air handling
side. Temperature control here is often
achieved by an electronic temperature control
sequencer staging the DX cooling and maintaining discharge temperature. It should be noted
that these systems are ideally applied where a
building minimum air flow is perhaps rarely
less than 20% of maximum. For effective discharge damper static pressure control there
must be reasonable velocity through that discharge damper to be able to maintain downstream static, otherwise the blower can move
up the curve to what could amount to leakage
through the air valves. Thus if we have a
tight close off air valve in the terminal unit
then this type of static pressure control may
not be the ideal.

The bypass damper arrangement for a forward curved fan as referred to under individual
air handlers could be readily applied also to
rooftop units.

Several factory packaged electronic

temperature control systems have been evolved
by control manufacturers designed to maintain
discharge temperature for the typical rooftop
units. These are all beset with the normal
difficulties of maintaining discharge control
with a reasonably narrow and acceptable throttling range on any staged device. They generally have some time delay and some mode of
reset built in to avoid the problems of substantial shift in supply air temperature and
rapid short cycling under load conditions.
Effective field calibration for the individual
conditions is very important for these controllers to function as designed.

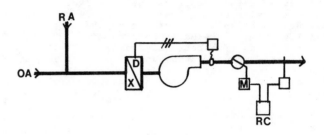

Fig. 5   ROOFTOP UNIT
COOLING COIL CONTROL

There are several schools of thought on
how important cooling coil control is on typical variable volume systems. We will cover
most of the general types of control that are
currently being used. There is one general
philosophy which says if you are going to vary
the air flow, why also vary the water flow.
In some type of basic variable volume systems
this may well apply, but when used it must be
with due caution.

(1) Wild water coils. Many systems, particularly those utilizing forward curved air
handlers with one air handler per floor of a
building are operating with wild water coils.
This is very successful, particularly where
there is substantial minimum air flow and
where some effort is being made to shift
chilled water temperatures, as we will look at
later. It has been demonstrated that upon
examining a normal 80-67 entering air WB/DB
condition, based on varying the air flow from
a nominal 100% to a minimum of 50%, and having
a 500 feet per minute nominal coil face velocity on a 6 row - 8 fin per inch coil, there
is no more than 3.5° F. dry bulb shift in
leaving air temperature.

This is quite surprising to many and
accounts for the successful operation of many
wild water coil systems in office buildings
where on low pressure duct systems the air
valves potentially see only a temperature
variation of 3.5° F. and an upstream static
pressure variation of .75" WG thereby enabling
the system to function with very few accessory
control or limiting devices.

Where used judiciously the wild water
coil is very effective. It has a disadvantage
in that should a multiple air handler building
wish to shut down some air handlers, chilled
water will continue to circulate and will
ultimately reduce air handler temperatures

114

below dewpoint and cause condensation on the air handler in the mechanical room.

Some buildings to obviate that have motorized butterfly valves to shut down when the air handler shuts down so that continued water circulation will not occur. This is substantially less cost than providing a temperature controlled globe pattern valve for each air handler in an installation.

(2) Conventionally pneumatically controlled or electronically controlled discharge temperature operating on either a straight through or three way globe pattern valve in a chilled water service has been applied in many cases. This also is a closed fast response control application where if satisfactory minimum shifts in leaving air temperature are to be achieved a controller with integral reset operating sequence should be used.

If a conventional controller is used it is very likely that a wide throttling range must be established with the inherent shift in leaving air temperature upward as load increases, thereby essentially varying cold deck air temperatures in the wrong direction.

With the use of temperature sensor and integral reset controller, either pneumatic or electronic, a constant discharge temperature can be achieved under stable load conditions which serves to stabilize the operation of the terminal air valves.

It is also practical to reset the air temperature upward within limits upon decrease in outdoor D.B. This should be viewed with caution since it implies that interior zone air valves can function at full load with a full supply air temperature higher than the peak design. It is also critical that a building having a large south solar exposure should have a sufficiently low SAT available during the highest south solar gain season which occurs in the Sunbelt in January and February.

This implies that even under very low outdoor conditions there can be sufficient solar load on the south zone to require south zone peak cooling and would, therefore, require the possibility of solar reset or override of the outdoor sensing system. Once this has been incorporated in the control system we are now again looking at a relatively complex series of components to calibrate. In the case of unattended buildings it is expected that this may be more complicated than the building justifies.

Reduced operating cost savings by resetting supply air temperature upward is not evident in most systems. Where either built up air handler or rooftop unit is using DX cooling great care must be taken to effect proper staging. This can be controlled either pneumatically or electronically. The staged DX system requires careful attention to ΔT across evaporator coil for each stage, compressor capacity, refrigerant circuiting, coil circuiting and minimum load characteristics of the refrigeration system. It is often advisable to utilize hot gas bypass on the first stage of reciprocating compressor DX systems and some type of anti-short cycling

device outside of the temperature control package should be provided.

The typical electronic temperature control system provides for integral reset with staged output and a slow reset time cycle for rapid load shift. Notwithstanding the difficulties implied by DX variable volume with varying air flow on the evaporator coil, as well as stage control, many systems are being successfully operated.

Some safety controls can be externally built in such as low suction temperature limit and compressor lock out (sensing potential liquid return to compressors) and low return air temperature. Additionally, low suction pressure limit should lock the first stage of a compressor off, pilot operated hot gas bypass as an operating system to provide protection under many very low load conditions.

START-UP CYCLES

Most buildings in the Sunbelt operate for a large percentage of each year with very low shifts between night and day time conditions. This tends to require less optimized start operation than buildings located in less temperate areas. For example, during a cooling season in the Southwest the building can probably be set up to start the same time each day with very little penalty in operating costs. However, during the heating season the same potential savings to a lesser degree are available to the colder parts of the Sunbelt as would be in a colder northern climate. The main variable here is the fact that the heating season is of short duration. There are many microprocessors available which provide the optimized start sequence and these will be discussed shortly.

We need to examine a typical side pocket fan variable volume system with a chilled water air handler per floor and separate interior to exterior pneumatic main air control loops. A fairly common start-up sequence will, for the warmup cycle during the heating season, start the air handlers and remove the instrument air from the normally open interior zone air valves so that they open 100% and apply the instrument air to the entire perimeter zones, permitting them to run into the heating cycle, thereby bringing the outer skin of the building up to temperature.

Some return air will pass back into the air handler and be returned into the interior zone through the open valves. This sequence can be initiated by a return air thermostat such that it remains in operation until a predetermined return air temperature such as 68° is achieved.

Since many buildings are shut down for extended periods, such as long weekends, it is important that some form of low temperature override be provided, such as a space thermostat independently located in the north zone area set to permit instrument air to come on to the perimeter for exterior zone units at 55° setting and to shut it off at 58° setting so that at no time will the perimeter fall below 55°. This also serves to assist the warm up sequence so that the building, if unoccupied for an extended period, is not starting from below 55° F.

CHILLER SEQUENCERS

There are many variations on chiller sequencers with some of the following being more common. It is reasonable to assume that if the building load relates to outdoor dry bulb temperature and if it is a multiple chiller installation a chiller can be shut down and locked off below a predetermined outdoor temperature. If chillers are piped in parallel it serves the purpose, at reduced building load, to elevate chilled water supply temperature as described in wild water coil sequence. There are some buildings with little interior zone area which may justifiably shut the chillers off completely below a further lower outdoor dry bulb temperature.

It is also practical to sense return water temperature on reciprocating chillers and while the chillers may continue to function on their own internal return water controls, a machine or machines may be shut off upon fall in return water temperature to insure the loading of the lead machines.

Lead/lag switching and time delay on the lag chillers during initial start-up to insure that hot system water temperature will not bring all the chiller capacity on the line under reduced load conditions are fairly common.

Chiller optimization in the form of attempting to load chillers to provide peak efficiency and performance, depending on the type of chiller, has not been commonly applied to new construction projects in this area. It is probable as utility rates increase that the techniques for this control will be more readily applied.

The measurement of BTU consumption in the building and the consequent chilling of chillers and/or dedicated pumps is probably the most advantageous energy control for centrifugal chiller plants. By this means the centrifugals can be maintained at higher load characteristics consistent with peak efficiency curves. It is apparent that this type of system, because of current initial cost, will be most commonly applied to larger buildings and substantial central plants.

VARIABLE SPEED CONTROL

Variable speed has many applications in variable volume controls. This would be the ideal arrangement for reducing air handler fan horsepower consumption, operated from static pressure control. At this moment the mechanical drives that have been made available are relatively new to the market and have not been broadly applied. The most efficient system available is that of variable frequency which unfortunately is still expensive enough for most engineers not to be able to justify the ROI.

Some systems have been installed with current clutch drives and/or DC motor drives. In all applications these drives relate to the same kind of static pressure system control as would be found with our conventional inlet vane or dump damper static pressure control.

Variable speed pumping packages are available, generally applied to the more sophisticated mechanical systems. In these, variable flow can be achieved to match either pressure differential across the system or BTU consumption of the system as measured by temperature sensors and turbine type flow meter.

Many packages have two pumps with one constant speed and one variable speed pump arranged to operate in sequence on a secondary loop so that the chillers function with constant flow and low pump horsepowers on a primary loop. The system pumps vary both input power and speed as demanded by the load on the building. In this case it is generally necessary to use straight through control valves on all cooling coils to achieve the necessary variation in flow and water temperature for the pump monitoring system operation.

ENERGY MANAGEMENT MICROPROCESSORS

The rapid evolution of building management microprocessors has changed cycling of many commercial office buildings. As a replacement for conventional time clocks it is expected that the 8, 16 or 20-point special duty building management microprocessor has just begun its substantial usage in the Sunbelt region.

These microprocessors must be examined carefully and matched to the building they are intended to operate. Either overdesign or underdesign generally results in difficult or unsatisfactory building operation. However, properly selected these packages enable building managers to provide time cycling, duty cycling of major equipment should it be desired and load sheeding for utility cost reduction. Often beedback of pertinent building operational data that can be used for improved efficiency of building operation is also available.

Generally the variable volume systems lends themselves readily to microprocessor start/stop and operation. Because of the multiple zones in a typical VAV and the individual control of those zones duty cycling and load shedding will serve mainly to let space temperatures float upward on the cooling season or downward on the heating season while the system is off. This could be outside of the design parameters for the occupied space.

A common application for office buildings recently developed involves security card access to the microprocessor override function for overtime running of the mechanical system. This uses a microprocessor and printer which will fully designate which tenants or occupants are operating the building for overtime so they can be adequately charged by the building management. This system is sometimes incorporated with a building security system but also can be applied separately with a building energy management package and is apparently very popular among unattended buildowners and operators.

Larger buildings requiring considerable transmission of switching points and data can

be very adequately handled with multiplexing
two-wire loop systems providing switching to
individual points throughout the building for
both mechanical systems and lights. The individual building and the individual microprocessor must be closely examined and matched
to expect optimum operation.

The foregoing is a review of the current
state-of-the-art in one specific region of the
country. It is hoped this will provide background and information that is useful to those
involved in system design, analysis and selection for that part of the country, as well as
other areas where system design requirements
are somewhat different.

It should be noted also that some types
of variable volume systems described here are
markedly warm climate designs and in that context are not intended for application in areas
where the heating season on a particular zone
extends for a considerable or continuous period of time.

It is apparent that variable volume
systems have found their place in commercial
system application and are continuing to
evolve and develop as engineers, contractors,
owners and suppliers gain further experience.

Chapter 18

HEAT PUMP UTILIZING GROUND WATER AS HEAT SOURCE

C. Kardan

INTRODUCTION

In a refrigeration cycle the unwanted heat from a space is pumped outside the space. In other words the space is being cooled lower than the surrounding environment. The reverse cycle of refrigeration (cooling) cycle is the heat pump which raises the temperatures of the space higher than the surrounding environment. When an actual refrigeration cycle is reversed, the evaporator becomes the condenser, and the condenser becomes the evaporator. Using the sames refrigeration equipment with modified reversing valves and controls, this conversion can easily be achieved. Based on the well known Carnot cycle the coefficient of performance of a heat pump is:

$$C.O.P. = \frac{T_1}{T_1 - T_2}$$

Where; T_1 = Condensing temperature of the refrigerant, ^{O}R

T_2 = Evaporating temperature of the refrigerant, ^{O}R

The highest coefficient of performance is obtained when $T_1 - T_2$ is lowest. In other words the highest possible evaporating temperature will give the most efficient heat pump operation. In order to achieve this, the temperature of the heat source, evaporating the refrigerant in the evaporator, must be high enough. But the temperature of the outside air during the winter season is always so low that heating is required. For this reason if a new heat source other than ambient air, which changes drastically during the winter months can be found, the heat pump will operate with a high coefficient of performance. The best heat source so far is ground water. Under certain levels below the surface, ground water temperature remains fairly constant - around 50° F. An economic analysis is made to show that the usage of ground water as a heat source for a heat pump application is very attractive, and economical when compared with a supplementary heat, e.g. gas or electric.

There are heat pumps that utilize various types of heat sources. Namely:

1. Ground to air
2. Well water or ground water to air
3. Air to air

In this paper we specifically discuss ground or well water as a heat source utilizing existing air conditioning equipment (refrigeration) to supply heat during the winter.

DISCUSSION

Since the 1950's the application of air conditioning

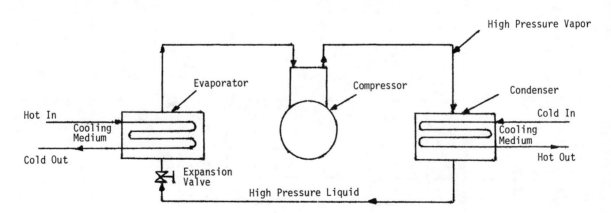

FIG. 1. BASIC COMPONENTS OF A REFRIGERATION SYSTEM

(comfort cooling) to residential and commercial places has increased tremendously. The same existing equipment could be used to supply heat to the space for comfort during winter months.

Let us look briefly at the equipment arrangement for comfort cooling (air conditioning). FIG. 1 shows the basic components of the system.

The compressor receives low pressure and low temperature refrigerant vapor and compresses it to high pressure and temperature. The condenser receives the vapor and condenses it to liquid. In this process the sensible heat and the latent heat from the refrigerant is taken by the cooling medium (air or water). Then the high pressure refrigerant is expanded through an expansion valve where its pressure is reduced. Flashing occurs, and the refrigerant at the lower pressure has a lower temperature (saturation temperature at that pressure). The vapor/liquid mixture goes through an evaporator and the liquid portion evaporates at a constant temperature extracting heat from the medium. The heat that evaporates the refrigerant comes from air as sensible heat. Then cooled air is pumped to a space for comfort. In this process not only the air is cooled, but also its relative humidity is reduced. Then the completely evaporated refrigerant is received by the compressor. The process repeats. In short, by condensation of the refrigerant at high pressure and temperature the heat is rejected, and by the evaporation of the refrigerant at low pressure and temperature the heat is received. FIG. 2 shows the cycle on a P-H diagram.

The Coefficient of Performance of a refrigeration cycle is defined as:

$$\text{C.O.P. (Carnot)} = \frac{T_2}{T_1 - T_2}$$

Where: T_2 = Evaporator temperature $^{\circ}$R

T_1 = Condenser temperature $^{\circ}$R

or

$$\text{C.O.P} = \frac{\text{Refrigeration effect}}{\text{Compressor work}}$$

The C.O.P. for a typical air conditioning system ranges form 3 to 5.

As we can see from the P-H diagram, the heat load of the condenser is composed of the heat load of the evaporator plus the heat of compression (compressor work). Now by reversing the cycle we can make the condenser operate as an evaporator and the evaporator operate as a condenser. This arrangement will provide the evaporator load plus the compressor load as a heat supply to the place where heat is needed in the winter months. The only problem is that during the winter months outside air temperature is too low (especially in northern states) to evaporate the refrigerant. Or, in other words, if the outside ambient air is too cold, the evaporator must operate at a much lower temperature. This in turn will considerably reduce the coefficient of performance for heating. In other words, a supplementary heat would be requied. This is presently the common practice in most applications.

Let us look at the coefficient of performance of the heating cycle:

$$\text{C.O.P. (heating)} = \frac{T_1}{T_1 - T_2}$$

Where: T_1 = Condensing temeprature $^{\circ}$R

T_2 = Evaporator temperature $^{\circ}$R

The coefficient of performance increases when $T_1 - T_2$ is small or the evaporator temperature is high. In most air conditioning applications, the typical evaporator temperature is about 40° F. The ambient air temperature during the winter months is much lower than 40° F. One good source of heat would be ground water. About 10 to 15 feet below the surface ground water temperature remains constant around 50 to 60° F. We can easily utilize this source to supply the necessary heat to evaporate the refrigerant. FIG. 3 shows a typical air to air heat pump operation, total heat required vs. outside ambient temperature. The balancing point occurs at about 35° F. ambient temperature. When the outside temperature drops below 35° F., then supplementary heat is needed. Also on this curve ground water as a source of heat is plotted. When ground water

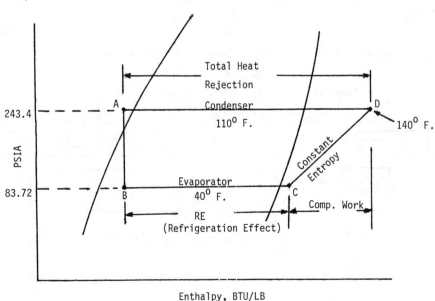

FIG. 2. P-H DIAGRAM OF A SIMPLE VAPOR COMPRESSION CYCLE. (R-22)

is used as a heat source, no supplementary heat is required even when the outdoor temperature reaches 0° F. The curve is based on a ground water temperature of 50° F. and a refrigerant evaporating temperature of 40° F. With the condenser temperature at 110° F., the theoretical coefficient of performance for heating will be:

$$\text{C.O.P. (heating)} = \frac{T_1}{T_1 - T_2} = \frac{110 + 460}{(110+460)-(40+460)} = 8$$

When we add the energy to pump the ground water, the C.O.P. falls to 7. In actual operation of the system when frictional and heat losses are accounted for, then the C.O.P. for heating drops and ranges from 3 to 4. This is still much more economical than electric heat or natural gas heating.

AVAILABILITY OF GROUND WATER

The National Water Well Association* reports that 75 percent of the U.S. has sufficient ground water for heat pump applications. Most southern states have a ground water temperature range from 60 to 70° F. The depth of a well may vary from location to location. Even northern states have ground water temperatures around 50° F., which allows a heat pump to operate efficiently. A typical four ton air conditioning unit requires about 12 GPM water flow rate. A heat pump using ground water as a heat source will pay for itself in four to six years due to energy savings. If one considers the average life of a heat pump system to be 15 to 20 years, the heat pump has a definite advantage.

CONCLUSION

The initial cost of a ground water to air heat pump is about $3000 higher than an air to air heat pump because of the wells that must be drilled. But the yearly operating cost of a ground water to air heat

* Information obtained from map published by National Water Well Association.

pump which supplies heat to a typical house requiring 70,000 BTU/hr is much less than the yearly cost of an air to air heat pump, electric resistance heat, or natural gas and oil heating. For every dollar spent for operating a ground water to air heat pump, $3 must be spent for electric heating, $1.40 for natural gas and $2.60 for oil heating. As one can see, presently the natural has heating is the closest cost wise to the ground water to air pump. But when the natural gas prices are decontrolled, the cost difference between the ground water to air heat pump and natural gas heating will be wider.

Since the oil embargo of 1973, the electrification of America is increasing. At the same time oil and gas prices are on the rise. One may safely assume that we will always have electricity in our homes regardless of what happens to natural gas and oil supplies. Then the application of the ground to air heat pump is one of the best choices for comfort cooling and heating.

REFERENCES

ASHRAE Guide and Data Book, 1961: Fundamentals and Equipment.

Dossat, Roy J. Principles of Refrigeration, second edition. John Wiley and Sons, 1978.

Federal Energy Administration, 1978. Evaluation of the air-to-air heat pump for residential space conditioning, Report No. FEA/D-76/340.

Harris, Norman C. and Conde, David F. Modern Air Conditioning Practice, second edition. MacGraw Hill, 1979.

Heat Pump Handbook, 1974: Lenox Industries, Inc. Fom No. HPH-752-L3.

Owen, Donald L. "Tapping An Energy Source for Heat Pumps," Appliance, September 1979.

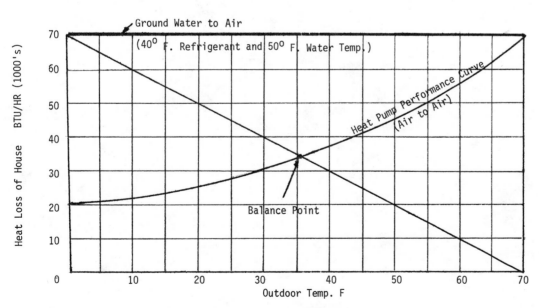

FIG.3. HEAT PUMP BALANCE POINT

Refrigeration and Air-Conditioning (Air-Conditioning
and Refrigeration Institute). Prentice-Hall 1979.

Second Annual Heat Pump Technology Conference. 1976
School of Mechanical Engineering. Oklahoma State
University.

SECTION 5
COGENERATION

Chapter 19

COGENERATION IN A COMMERCIAL ENVIRONMENT: PAPER INCINERATION WITH HEAT RECOVERY, ELECTRICITY, SPACE HEATING AND COOLING

R. Gagliardi

ABSTRACT

Besides rising energy costs, Ford Aerospace and Communications Corporation, Western Development Laboratories Division, faces the added problems of the destruction of classified and proprietary documents, and diminishing municipal sanitary landfill areas. A single solution to these immediate problems has been found in a planned incineration system with waste heat recovery and cogeneration.

The incineration system will dispose of 10,000 lbs. of paper waste daily including approved destruction of classified and proprietary material. The exhaust gas from the incinerator will produce steam from a waste heat boiler which will in turn drive a steam turbine to generate electricity. The hot water leaving the turbine will be used for space heating and cooling by absorption.

The system results in the reduction of gas and electric costs, paper waste disposal costs, and demand on local resources. The estimated cost of the project is $1.3 million and has a projected annual rate of return of 35%. Third party leasing and State financed bonds are being actively pursued.

Future plans include incineration as a service to other businesses in the area and disposal of hazardous materials.

BACKGROUND

A wholly owned subsidiary of Ford Motor Company, Ford Aerospace and Communications Corporation specializes in microwave communication satellites and ground networks. Western Development Laboratories (WDL) is a division of Ford Aerospace located in Palo Alto, California, occupying 35 buildings on 72 acres, with 4500 employes pursuing research and development of satellites.

A byproduct of this work is the generation of 10,000 lbs. of paper waste each day. A large part of this waste is made up of photocopies, preprinted forms and computer printouts. Of that 10,000 lbs., 1600 is Government classified and Company proprietary information which must be carefully and completely destroyed.

Presently, classified waste is pulverized on site by a mechanical hammermill, then transported to a local municipal landfill. Proprietary waste is shredded offsite by an independent company and recycled to become corrugated boxes. The remaining 8400 lbs. of general waste is removed from dumpsters and transported to the landfill by a disposal company.

All three methods of disposal have disadvantages which are becoming more pronounced. The hammermill which destroys classified waste is 20 years old, well beyond its expected operating life, and is highly unreliable, suffering repeated breakdowns. It produces a 'fallout' of chalky dust which finds its way into adjacent clean room areas. The vibration it generates when operating also creates a nuisance to personnel in the area. The pulverized material is highly flammable, and has to be kept moist to prevent combustion.

The offsite company which recycles WDL's proprietary waste has proven to be unreliable, being unable to accept the waste for periods of up to three months. Although there are no immediate difficulties in handling general waste, the city solid waste site is nearing capacity. Future sites are much more distant and will be more costly to users.

INCINERATION

Several alternative solutions to the waste problem have been considered, and incineration of the material provides the most distinct advantages. It is a highly reliable means of destroying classified waste which is relatively clean and quiet and meets security requirements without increased operating costs. It reduces dependency on outside sources for disposal of proprietary waste. General waste is reduced 95% by volume and 90% by weight, easing the load on municipal disposal. Perhaps the greatest advantage is that the exhaust heat can be recovered to produce steam for heating, cooling and power generation.

The application of recovered heat will result in significantly reduced energy costs. It was found that the resultant fuel value from heat recovery as related to heating, electric generation and cooling is 0.0132 $/kWh, 0.0091 $/kWh and 0.0060 $/kWh, respectively. Thus, the most cost efficient use is supplemental space heating. Of second priority, electricity can be generated for sale back to the local utility. Cooling can also be acheived through an absorption chiller, however this is the least cost effective application.

Based on this criteria, a single building (No. 3) was selected as being best suited to utilize the heat recovered from incineration. The building is the largest single user of natural gas, used for space heating, and uses an existing central hot water system rather than several package units scattered throughout the building. Energy used for heating and cooling Building 3 through 1981 is estimated at 8.3 million kWh. A monthly load profile of average heating and cooling demands was developed to relate current consumption with the use of waste heat recovery to supplement the purchased energy. Figure 1 graphically illustrates this load profile.

INCINERATION ALTERNATIVES

Four basic incineration alternatives were considered in the investigation, each with several variations which are outside the scope of this presentation.

ALTERNATIVE A:

DESCRIPTION: Incorporates small incinerator for the purpose of destroying classified and proprietary materials. No heat is recovered.

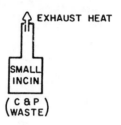

Figure 2 - Flow Diagram for Alternative A

No heating or cooling is provided for Building 3. Installation cost is estimated at $137k. Annual operating costs are $48k and annual savings are $13k from avoided transportation costs. Thus, the net operating cost the first year is $35k. Considering inflation, the estimated average net cost will be $138k over the next 20 years.

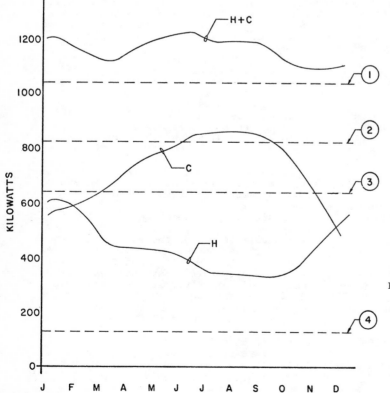

Figure 1 - Waste Thermal Capacity vs. Heating and Cooling Demand and Cogeneration.

(1) Thermal Capacity of General and Classified/Proprietary Waste

(2) Thermal Capacity After Producing Steam (Alt. C)

(3) Thermal Capacity After Cogeneration (Alt. D)

(4) Thermal Capacity of Classified & Proprietary waste only, After Producing Steam (Alt. B)

H   Average Heating Load

C   Average Cooling Load

ALTERNATIVE B:

DESCRIPTION: A boiler is added to recover heat exhausted from incinerator. Hot water from boiler is used to supplement heating of Building 3.

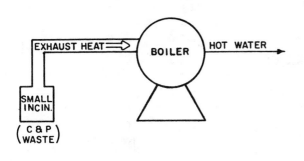

Figure 3 - Flow Diagram for Alternative B

This system would provide 31% of the space heating for Building 3. Its cost is $266k. Annual operating costs are $48k and cost avoidances are $29k per year in waste transportation and natural gas. The net operating cost the first year is then $19k, however, since energy costs are expected to escalate faster than labor costs, the system would provide an average net savings of $43k per year over the next 20 years.

ALTERNATIVE C:

DESCRIPTION: Add large incineration to burn general waste, increase size of boiler to accommodate larger incineration capacity, provide hot water storage tank and absorption chiller for heating and partial cooling of Building 3.

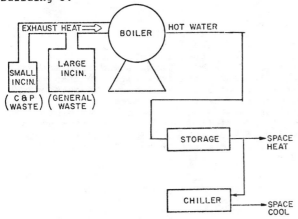

Figure 4 - Flow Diagram for Alternative C

Virtually all of the heating required by Building 3 can be provided by this system as well as 54% of the cooling requirements. An estimated $1.15 million is the installation cost. Operating costs are $81k per year and annual savings are $90k, providing a net operating savings of $9k the first year. This is equivalent to an average annual savings of $517k for 20 years.

ALTERNATIVE D:

DESCRIPTION: Incorporate steam turbine-generator behind boiler and ahead of storage tank to generate electricity.

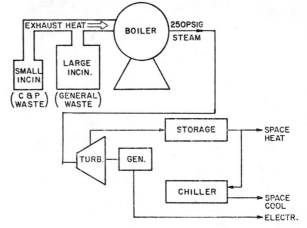

Figure 5 - Flow Diagram for Alternative D

The maximum amount of electricity that can be generated with this type of system and 10,000 lbs of paper waste per day is 300kW (net). In addition to this, the system still provides 100% of the building space heating requirements and 38% of its cooling needs. Installation cost is $1.3 million and the annual operating cost is $90k. Annual cost savings the first year is $144k, providing a net savings of $54k the first year and an average $957k over the next 20 years.

ENERGY FLOW COMPARISON:

An estimated 22 tons of general waste and four tons of classified and proprietary waste is generated each week. This waste has an average heat value of 8100 Btu/lb, 80% of which can be recaptured. This will produce 5 million kWh of energy per year. The energy released from combustion is shown in Figure 6 which is a summary of the energy flow of each alternative. As the energy flows through the systems, energy loss and useful energy are also depicted.

As each alternative was considered, it was found that as the system grew, the economic analysis became more favorable. On that basis, Alternative D was selected. It is the largest system that can be supported on the amount of combustible waste available at WDL. Under this alternative, the existing disposal company would continue to collect paper trash material in dumpsters. In lieu of transporting the waste to the city refuse dump, it would discharge the material into a recessed collection basin at the WDL incineration plant located adjacent to the Building 3 Boiler Room. From there, it is conveyed to the incineration load table where it is fed into the incinerator combustion chamber. Classified and proprietary waste would be removed from secure areas by security personnel and brought to a smaller incinerator also at the WDL incineration plant where it would be incinerated under security supervision. The exhaust heat from both incinerators is passed through a boiler that generates high pressure steam. It is used to drive a turbine-genera-

ENERGY FLOW DIAGRAM

Figure 6.

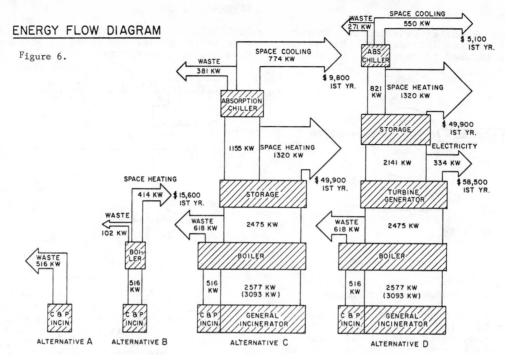

tor that produces approximately 300 kW of electricity for sale to the local utility. Hot water from the turbine is circulated through a heat exchanger and an absorption chiller. There, hot and chilled water are generated for space heating and cooling.

As energy and waste disposal costs increase, it may become feasible to expand the incineration system and offer a service to other local companies facing similar disposal difficulties. The system is presently sized to process WDL waste during a 40 hour work week. However, extra work shifts can be added to process more waste without requiring additional capital outlay. Also, incineration of hazardous materials such as solvents and PCB's are being considered. The incinerator burns hot enough to break down these materials into more harmless constituents which can be handled by air pollution control devices and their heat value would be recovered for practical use.

FINANCIAL CONSIDERATIONS

There are three possible methods of financing the incineration system. The first is, of course, direct capital appropriation. Since Ford Aerospace is a subsidiary of Ford Motor Company which has been losing money, many of the tax advantages generally available for such an appropriation cannot be applied here. The Time Adjusted Rate of Return of the incineration purchased under capital appropriation is estimated to be 27% per year.

Partial financing is available from the State of California at low interest rates through bond issues from sources such as the California Pollution Control Financing Authority and the State Energy Task Force. This offers several advantages over capital expenditure such as conservation of capital, which is crucial for Ford at this time, and incremental outlays. The estimated Time Adjusted Return is 40% under this method.

The final consideration is third party leasing. Under this plan, the third party can obtain low cost money (possibly through the state) and Ford can trade deferred Investment Tax Credits and provide a depreciation tax shield. In trade, Ford would receive a low implicit interest rate and incremental outlays would be obtained from real cost savings from the system. There are various forms that third party leasing can take, and WDL is considering several of them. Based on some preliminary quotes, the Rate of Return is expected to be about 35% using this method.

CONCLUSION

Waste incineration is one of those few projects in which everybody benefits. In one system, WDL has found a solution to effective destruction of classified and proprietary material, general waste disposal and high energy costs. Investors are interested in the economic attractiveness of a system that provides good revenue for the investment. Utilities benefit from the decreased demand of gas and electricity, and the community has an extended life of the municipal landfill area and possibly an alternative disposal method.

128

Chapter 20

MECHANICAL COGENERATION IN A FOOD PLANT/WAREHOUSE

J. R. Haviland

Ralphs Grocery Company is one of the leading supermarket chains in Southern California. Ralphs operates over one hundred supermarkets, along with office, manufacturing and warehouse facilities. Ralphs was a family owned and operated business until 1968 when it became a division of Federated Department Stores. Federated and Ralphs are very energy conscious, and have been very supportive of all projects designed to increase energy efficiency.

Ralphs' Central Facility Complex consists of two buildings housing manufacturing operations and refrigerated warehouse space. This is the central storage and distribution point for all frozen food, dairy, deli and meat products for all stores. The complex has 360,000 square feet, of which 300,000 is refrigerated. All utilities are provided from a central plant, including refrigeration, compressed air and steam. Each building has a separate hot water system utilizing steam from the central plant. The refrigeration system is a two stage ammonia system utilizing screw compressors.

Because of the critical nature of the plant and its refrigeration system, excess capacity was always required. By 1980, load growth had reduced excess capacity on the second stage of the system to 10%. Since this stage was served by four large compressors, we were exposing ourselves to considerable risk if we should lose one compressor during peak load periods. Thus the purchase of an additional compressor with an electric motor drive was included in the 1981 Capital Budget. During a conversation with a gas turbine manufacturer near the end of 1980, this installation was mentioned and the manufacturer suggested we investigate the feasibility using a gas turbine to drive the compressor with a waste heat recovery system to provide steam. As a result of this conversation, we approached the compressor supplier about their experience with such systems. They had some experience with natural gas internal combustion engines from the "Total Energy" days, and suggested this approach. Thus we had three alternatives to consider: electric motor, gas turbine and gas engine.

The electric motor drive was the initial choice and had several advantages. We already had four identical units including a spare motor. The first cost was considerably lower. The system could be on line much sooner because the spare motor could be utilized pending delivery of the motor, the longest lead time item. The motor is 700 horsepower, 3600RPM, 4160 volt, so it is not an off-the-shelf item.

The gas turbine we were considering was the Garrett 831 distributed by Onan. This is the smallest gas turbine available for stationary applications. Its power output was ideally suited to the load requirements of the compressor. In addition, a heat recovery boiler would be utilized to provide 100 psig steam from the exhaust heat. One potential problem was the starting torque

required by the screw compressor, which might mean using an auxiliary electric starter. With the compressor operating at full load, the system would provide 3900 pounds per hour of steam. We were fortunate to be in an area where the gas utility could provide us with 150 psig gas so that the amount of compression required to provide the 220 psig gas which the turbine required was not a major concern.

The compressor supplier had an arrangement with a local Caterpillar dealer, so the gas engine we considered was the 16 cylinder G399. The engine would provide the required horsepower operating at 1000 RPM. This would require the use of a speed increaser since the compressor was rated at 3600 RPM. We did not want to use a larger compressor which would not match the existing units. Heat recovery would be accomplished in two stages because of the requirement for 100 psig steam. The engine jacket cooling water would be piped through a heat exchanger to heat water for plant use. A load balance condensor is piped in series with the heat exchanger to ensure that the required heat rejection is provided. The exhaust passes through a heat recovery muffler which provides the steam.

The economic benefits of the two alternatives to the electric motor were compared using the Internal Rate of Return (IRR) method. This analysis was compounded by uncertainty in the amount of heat which could be utilized and expected maintenance costs. To account for these, the IRR was calculated for three cases: best, expected and worst for each system. The parameters for each are as follows:

	BEST	EXPECTED	WORST
NATURAL GAS ENGINE			
Initial Cost ($)	275,000	275,000	275,000
Investment Tax Credit ($)	27,500	27,500	27,500
Energy Tax Credit ($)	3,000	3,000	3,000
Fuel Cost ($/MMBTU)	3.585	3.585	3.985 (1)
Fuel Use (MMBTU/HR)	5.8	5.8	5.8
Displaced Electricity ($)	243,900	243,900	243,900
Heat Value ($)	99,100	43,000(2)	43,000(2)
Maintenance Costs (Annual) ($)	2,000	3,000	5,000
Major Overhaul Schedule	Yr 10	Yrs 6,12	Yrs 4,8,12
Minor Overhaul Schedule	Yrs 5,15	Yrs 3,9, 15	Yrs 2,6,10 & 14
IRR	50.2%	37.7%	30.1%

	BEST	EXPECTED	WORST
GAS TURBINE			
Initial Cost	$400,000	$400,000	$400,000
Investment Tax Credit	$ 40,000	$ 40,000	$ 40,000
Energy Tax Credit	$ 8,000	$ 8,000	$ 8,000
Fuel Cost ($/MMBTU)	3.585	3.585	3.985 (1)
Fuel Use (MMBTU/HR)	8.4	8.4	8.4
Displaced Electricity ($)	243,900	243,900	243,900
Steam Value ($)	169,000	126,800 (2)	126,800 (2)
Maintenance Costs Annual ($)	500	500	500
Major Overhaul Schedule	Yr 10	Yrs 7,14	Yrs 4,8,12
IRR	38.2%	30.2%	23.9%

(1) This assumes that the system would not qualify for a special reduced rate for cogeneration systems.

(2) This assumes only 75% of steam could be utilized.

The analysis was completed over a fifteen year period. Inflation factors of 15% for utility costs and 7% for maintenance costs were used. Based on this analysis, we elected to proceed with the Natural Gas Engine. The expected return and the worst case return are better for this system than for the gas turbine. The need for an additional large piece of equipment, the waste heat boiler and the relatively low heat requirements on a continuous basis adversely affected the economics of the gas turbine system.

SYSTEM DESCRIPTION

The final system consists of the following components. The mechanical load is a Stahl Screw Compressor on an ammonia refrigeration system. This has a power requirement of 625 bhp at our operating conditions. This machine is designed to operate at 3600RPM. A Caterpillar G399 engine was chosen as the power source. This unit provides 730 bhp at 1000RPM, turbocharged and aftercooled. A speed increaser is used to drive the compressor at 3600RPM. This was done so that the same compressor could be used as we already had in the facility.

Two levels of heat recovery are required. The engine jacket cooling water is used to heat domestic hot water through a heat exchanger. A load balance condensor is piped in series with this heat exchanger. This unit has four fans which are cycled based on return water temperature to the engine. This ensures that the proper amount of heat rejection occurs at all times. The engine exhaust goes through a Maxim heat recovery muffler. This unit provides 100 psig steam for process use in the facility. This is not an ideal configuration, but was necessary to match existing conditions.

The expected heat balance and operating efficiency are as follows:

INPUT: 5660 CFH X 1030 BTU/CF = 5.83 MM BTUH

OUTPUT: 625 HP X 2545 BTUH/HP = 1.59 MM BTUH
1000 lb/HR Steam X 1018 BTU/lb = 1.02 MM BTUH
100 GPM Hot Water @ 30°F Rise = 1.50 MM BTUH
4.11 MM BTUH

OVERALL EFFICIENCY: $\frac{4.11}{5.83}$ X 100% = 70.5%

HEAT RECOVERY: $\frac{2.52}{5.83}$ X 100% = 43.2%

MECHANICAL EFFICIENCY: $\frac{1.59}{5.83}$ X 100% = 27.3%

Thus the system should operate more than twice as efficiently as a standard thermal power plant. Even if only one third of the hot water capability is utilized, as expected, the efficiency is still greater than 50%.

REGULATORY AGENCY INTERFACE

California Governor Jerry Brown has jumped on the cogeneration bandwagon. As a believer in alternative energy strategies to the exclusion of traditional sources, he sees this as a way to increase energy system efficiency and thus reduce dependence on imported oil. He has set up a special cogeneration desk in the Office of Planning and Research and directed all State agencies to expedite permitting and otherwise take all possible steps to encourage cogeneration projects. In addition, he has also committed the State to provide 400 MW of cogenerated power at State facilities by 1985.

The California Public Utilities Commission (CPUC) took several steps which were designed to encourage cogeneration. First it required the electric utilities to provide standy power at reasonable rates and to buy power at full avoided costs. Second, it required the gas utilities to provide financial assistance for feasibility studies, engineering design and construction, when warranted. Third, it also authorized the gas utilities to provide gas to cogenerators at the same cost as electric utilities pay, if that is less than the regular rate for the facility. Unfortunately all these incentives apply only to cogeneration systems which generate electricity. Even on appeal, the CPUC specifically denied the benefits for mechanical cogeneration systems. This was not a major roadblock for our project, but could hinder some similar projects.

The South Coast Air Quality Management District (SCAQMD) is responsible for emissions from stationary sources in the Los Angeles area. Issuance of construction permits is covered by their Rule 13, New Source Review. A portion of this rule provides easier requirements for cogeneration projects, specifically those which generate electricity. Permit issuance is almost automatic provided that the project uses Best Available Control Technology and that the company take steps to provide "offsets" or emission reductions at its own facilities. Otherwise it would be necessary to offset all increased emissions above set limits by working with other installations in the immediate vicinity. The SCAQMD did accept that mechanical cogeneration should receive the same treatment as electrical cogeneration which made the necessary permit much easier to obtain.

Several discussions were held with SCAQMD engineers concerning the definition of Best Available Control Technology for the engine. Catalytic convertors have been installed on similar engines, but they have only proved useful on rich burning engines while the Caterpillar is a lean burning engine. The catalytic convertor people had indicated to SCAQMD that this should not make a difference, but the engine manufacturer disagreed. Caterpillar provided some data which suggested that the engine could only operate in a very narrow range with the catalytic convertor. In addition, there were no operating units other than a test installation in San Diego at Sea World. SCAQMD accepted that no proven control technology existed for these units, but did require that space be provided to add a convertor at a later date when (if) the technology did develop.

OPERATING RESULTS

It is difficult to assess the operating results because of minor problems that have occurred with the installation. Some of these involved the compressor but several also involved the gas engine and its controls and the heat recovery piping. They were minor and did not suggest any difficulty with the concept. When the unit was operating, we did achieve the expected results. The fans on the load balance condensor operate rarely during the day. We can shut down a second boiler which was idling to provide surge capacity. We fully expect the unit to be completely operational in the near future, and have every expectation that we will receive the expected return on investment.

Chapter 21

COGENERATION ENERGY SYSTEMS ASSESSMENT

C. R. Cummings, R. T. Sperberg

INTRODUCTION AND BACKGROUND

The origins of natural gas-fueled cogeneration, as addressed herein, are based upon the "total energy" concept pursued by the natural gas industry from the early 1960's to early 1970's. This total energy concept evolved from earlier onsite power generation concepts, with the effective recovery and utilization of "waste" or reject heat being the key to the successful application of the concept. For the most part, these systems were "isolated" from the electric utility grid, thus requiring redundancy and/or over-sizing. Also, since these total energy systems followed the user's electric load, typically the available thermal energy was either partially "wasted" (when excess thermal energy was generated) or inadequate (requiring supplementary boiler capacity when the system could not meet the building's peak thermal requirements); electric load following also meant that the prime mover was not always operating at its most efficient load level. Improved prime movers, the development of effective and efficient heat recovery equipment, and advances in air conditioning equipment (absorption, in particular) compensated for these limitations and combined to yield both practical and economical total energy packages.

Between 1972 and 1976, however, natural gas prices increased in many parts of the country by as much as a factor of five; in this same period, electricity prices increased by only a factor of two to three. Moreover, the harsh realities of the long-term implications of the oil embargo finally became understood and this realization initiated a national policy directed toward lesser dependence upon oil and gas. The net effect of these factors was to diminish the economic attractiveness of the gas-fueled total energy concept.

More recent events have dramatically altered the prospects for onsite thermal and electric (or mechanical) energy generation, or cogeneration. In particular, the Public Utilities Regulatory Policy Act (PURPA), one of five bills contained in the National Energy Act of 1978, provides significant incentives for cogeneration. The key points are summarized below:

o   Qualified cogenerators are exempt from burdensome state and federal regulations applicable to utilities.

o   Electric utilities are required to connect qualified cogenerators to the utility grid.

o   Electric utilities are required to provide standby or back-up electric power to qualified cogenerators under non-discriminatory rates and policies.

o   Electric utilities are required to sell or buy power from qualified cogenerators at fair and reasonable prices.

These policies represent a completely different environment for onsite generation than that encountered in the total energy effort of the 1960's. Many other federal and state initiatives provide various incentives to cogeneration, however, the PURPA-related factors noted above are the primary reasons for the renewed interest in cogeneration.

A primary mission of the Gas Research Institute (GRI) is to promote the efficient utilization of natural gas by developing higher efficiency natural gas equipment and technology, thereby lowering consumer energy costs and decreasing energy consumption. Cogeneration systems which produce either electricity or mechanical work and thermal energy have demonstrated the potential for yielding significant energy savings in the residential, commercial, and industrial sectors. The energy and cost savings potential of gas-fueled cogeneration, plus the incentives afforded by the Public Utilities Regulatory Policy Act (PURPA), led GRI to conduct an assessment of the technology, economics, market, and institutional factors relevant to gas-fueled cogeneration. A team comprising Science Applications, Inc. (prime contractor) and Insights West, Cogeneration Development Corp., and Mechanical Technology Inc. (subcontractors) was selected to perform this assessment. The specific objectives of this project included:

1.   Develop an overview of gas-fueled cogeneration technologies, applications, and markets.

2.   Based on this overview, recommend appropriate technology development, systems analysis or developmental installations that would maximize the benefits of gas-fueled cogeneration to the customer and the utility.

The scope of this study was limited to the consideration of gas-fueled cogeneration system sizes between 100 kW and 10 MW and to applications in the commercial market sector (including multifamily dwellings). The use of natural gas in industrial cogeneration applications is relatively well proven. High utilization of recovered heat, relatively constant operation, and the relatively low cost of large cogeneration systems combine to provide attractive economics for industrial-sized systems and applications. The proven economic feasibility of these larger systems made it unnecessary to address the industrial market in this initial assessment.

SUMMARY OF RESULTS

Commercial cogeneration applications entail appreciably more risk due to uncertain thermal loads, varied hours of operation, and higher costs of small systems. Generally, the smaller the cogeneration system, the higher the system costs (on a "$/kW" basis); this result was expected. In addition, no proven gas-fueled prime movers with demonstrable, long-term continuous-duty track records were found below approximately 100 kW. Although data exist on applications and technology below 100 kW, this initial phase of the assessment effort focused on commercial applications and technologies greater than 100 kW. Assessment of the smaller cogeneration systems and the more complex market below 100 kW is suggested for future study.

Significant potential for cogeneration systems in the commercial sector was found to exist. In addition, research and development (R&D) efforts can have a beneficial impact on the market potential for cogeneration, particularly in the smaller size ranges (less than 1 MW). This study has also shown that:

1. Prime mover technology exists and is well proven above about 100 kW.

2. Although favorable in most areas of the country, the economics for cogeneration are site- and utility-specific.

3. Significant institutional constraints are evident; however, none appear to be overwhelming. The majority of the constraints are expected to be overcome or ameliorated as more experience with the regulatory process is acquired.

4. The impacts of cogeneration on the gas consumer and utilities appear favorable.

COGENERATION TECHNOLOGY

For systems larger than approximately 100 kW, the basic prime mover technologies considered in this project (reciprocating engines and combustion turbines) are well proven. At least 18 domestic manufacturers offer gas-fueled prime mover/generator sets, ranging in size from approximately 100 kW to 20 MW. None of these was originally or specifically designed for cogeneration applications; most are conversions of engines originally designed for mobile, agricultural, aviation, or standby power generation applications. Because the operation of these prime movers in a cogeneration mode is often less rigorous than the mode for which they were designed, their performance record in cogeneration applications has been very good.

Currently, cogeneration systems typically are designed specifically for individual sites, with only limited standardization of packages. If systems could be pre-engineered and pre-packaged to meet a wide sector of the cogeneration market, and could be produced in significant quantities (i.e., thousands per year), total system installed costs might be expected to be reduced by 25-50%. This development would greatly expand the cogeneration market. Manufacturers or system packagers are unlikely to initiate this effort by themselves because of the large costs and large risks involved. To ameliorate these risks and costs, a multi-phase project is recommended to lay the foundation for the development and manufacture of pre-packaged and pre-engineered systems.

It was also found that a number of other component and system improvements could potentially improve the implementation and utilization of gas-fueled cogeneration systems. Some of these are summarized in Table 1 and are characterized as a function of cogeneration system size. Although no detailed quantitative assessment of the effects of these improvements was made, the accomplishment of some or all of these technology improvements would enhance the outlook for gas-fueled cogeneration.

TABLE 1. RELATIVE VALUE OF TECHNOLOGY IMPROVEMENTS

	TECHNICAL CONSTRAINT	COGENERATION SYSTEM SIZE			
		<100 kW	100-1,000 kW	1,000 kW-10 MW	>10 MW
PRIME MOVER	1. Develop Improved Prime Mover	Very Valuable	Valuable	No Need (exists)	No Need (exists)
	2. Dual-Fuel Capability		Valuable	No Need (exists)	No Need (exists)
	3. Higher Allowable Jacket Temperatures (Reciprocating Engines)		Valuable	Valuable	
SYSTEMS DEVELOPMENT	1. Pre-Engineered Packaged Systems	Very Valuable	Valuable	Valuable	Not Practical
	2. Steam Systems Integration (Combined Cycle)	Not Practical	Valuable	Valuable	Valuable
COMPONENT DEVELOPMENT	1. Grid-Excited Generator	Very Valuable	Valuable	Valuable (at low end)	Not Practical
	2. Low-Temperature Heat Recovery (Exhaust)	Valuable	Valuable	Valuable	Minor Value
	3. Thermal Storage (Daily)	Very Valuable	Valuable	Valuable	Minor Value

Potential improvement in cogeneration system performance is possible via advanced prime movers, such as closed- and open-cycle Brayton engines, organic Rankine systems, and Stirling engines. Because these technologies are being developed primarily for other applications, no specific recommendations are suggested until technical feasibility is demonstrated. The one exception is the onsite fuel cell being developed primarily for cogeneration applications. Because of the fuel cell's unique and highly attractive performance characteristics, continued support for this program is recommended.

For systems smaller than approximately 100 kW, no technology has been proven in long-term, continuous-duty applications. Many standby generators are offered in this size range, but few are rated for continuous duty. The costs of cogeneration systems (capital, operating, and maintenance) tend to increase (on a per $/kW basis) as system size decreases. In addition, although potentially very large, the market in this size range is more complex and uncertain. However, many companies are involved in various aspects of technology development of prime movers and systems smaller than 100 kW for heat pump and cogeneration applications. Most of these efforts are focused on modified automobile engines, one notable exception being the onsite fuel cell. All of these developments are in the pre-commercial stage of hardware development, primarily focused on reducing fixed and O&M costs. R&D could provide a valuable contribution to these efforts, and therefore a project to characterize cogeneration market applications less than 100 kW and survey the technology status in this size range is recommended.

COGENERATION ECONOMICS

Cogeneration Applications

In general, if the energy outputs (electric and thermal) can be utilized, cogeneration systems provide the best economic return when they are operated at the highest possible capacity factor. Therefore, the most attractive applications tend to be those with constant thermal loads and significant electricity requirements. When a cogeneration system is connected to the grid, however, sizing is more flexible. The building's electric demand becomes a secondary consideration for sizing and operating the cogeneration system, which leads to two conclusions:

1. Cogeneration systems generally should be sized and operated under conditions providing maximum utilization of the energy produced, (i.e., typically sized for the thermal baseload).

2. Increasing the need for thermal baseload uses in buildings (e.g., thermally-driven cooling) will increase the size of the potential commercial cogeneration market -- in both market size and in types of applications.

The first conclusion can be illustrated by examining the sizing and operating options for a "typical" hospital application. If the load profile shown in Figure 1 is examined, both thermal and electric baseloads exist (on an annual basis). Caution should be used here: the data do not demonstrate a true "baseload," since no daily load variations are included. Although this point is minimized in the hospital application because of its 24-hour operation, it may be significant in other commercial applications. If the cogeneration system is sized to meet the thermal baseload of the hospital, all thermal energy would be utilized. In this example, however, significant electricity in excess of the building

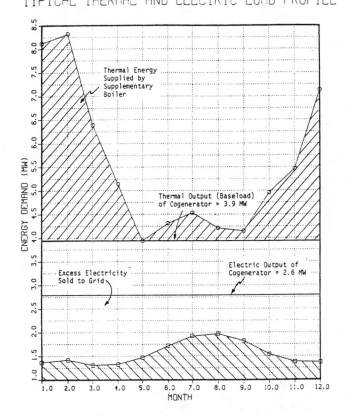

FIGURE 1.   OUTPUT OF COGENERATOR ASSUMING IT IS SIZED TO MEET THERMAL BASELOAD

demand would be produced. If this electricity can be sold to the electric grid at a price greater than the cost to produce it, then operating the system in the mode where electricity is exported is advantageous:

If, however, the price for electricity sold to the grid is less than the cost to produce that electricity, reducing the size of the cogeneration system to increase the utilization of energy produced from the system would be more economically attractive. In the extreme, a system sized for the smaller of the thermal baseload or the electric baseload (in this case, the electric baseload) would provide energy that is always used within the building. If all the electricity is used within the building, the buy-back price for electricity sold to the electric grid is not relevant; only the price of the displaced electricity is important.

These operating and sizing concepts are in contrast to those employed in the "total energy" projects. Due to their isolation from the electric grid, "total energy" systems were sized to meet the electric peak of the building, including additional oversizing and redundancy for equipment outage, and were operated to follow the electric demand of the building. The economic potential of these systems, therefore, was tied to the electric load factor of the building, which is generally below 50% for commercial applications, and to the requirement for operation at times when there was electric demand but no use for the thermal energy produced (and the inverse case). In many cases, these practices resulted in non-economic installations.

In general, connection to the electric grid is important, if not critical, to the economic success of a cogeneration installation. However, the buy-back price for electricity sold to the grid is important only to the size of the installed system and therefore the size of the cogeneration market. The buy-back price is not a key factor in the existence of a significant market.

This result leads to the second conclusion that increasing the need for thermal baseload uses in buildings will increase the potential market for cogeneration systems. One primary example is the use of thermally driven cooling devices, especially those using low temperature recovered heat. As the hospital example shows, the lowest thermal requirements occur during the summer months. Increasing this load would increase the potential size of the economically attractive cogeneration system.

As time-of-day pricing for electricity reflects the increasing differential value between on-peak and off-peak electricity, another potentially attractive mode of operation may become feasible. Operation of the cogeneration system only during on-peak hours with daily thermal storage to meet off-peak thermal needs may become attractive under these conditions. This concept has been examined analytically by Dr. Clark Bullard in a separate effort sponsored by GRI.

Economic Feasibility

Many factors affect the economic feasibility of a cogeneration system at any particular site, including cogeneration system costs and operating parameters, building parameters, and utility-related parameters, such as energy rates, standby charges, electric utility reliability, buy-back rates, capacity credits, and grid connection costs. As expected, study results indicate that economic feasibility is site-specific, as well as utility specific, and varies over a wide range, even within traditional geographic regions.

A key determinant for assessing the economic feasibility of gas-fueled cogeneration was found to be the differential between the value of electricity and the cost of gas. In a survey of the commercial gas and electric prices for 76 cities, 64 exhibited a price differential considered economically feasible for gas-fueled cogeneration. As mentioned previously, the buy-back rate is not critical to economic feasibility (for cases where export of electricity is not necessary), but this rate is important to the size of the cogeneration market. Electric utility standby rates are critical and can result in a major barrier to the economic viability of cogeneration. A data base establishing the high reliability and availability of cogeneration systems, as individual systems or as groups of systems, could reduce standby rates. An effort to evaluate statistically the actual track record of reliability for natural gas-fueled cogeneration systems therefore is recommended.

R&D can impact the economics of cogeneration by reducing the cost of cogeneration systems on a "$/kW" basis through improved component and system performance. Also, potential cost reductions can be achieved through the optimization and significant production of pre-packaged, pre-engineered systems. Further analysis of key specific utilities in the context of specific sites is also recommended to further characterize R&D opportunities and to refine and prioritize other important parameters relevant to economic feasibility.

Institutional Constraints

Even if cogeneration is economically attractive in many applications, market uncertainties and institutional factors can severely inhibit the viability and growth of a cogeneration market in the commercial sector. Cogeneration is the subject of a large number of regulatory, legal, and institutional issues, many of which are currently in a state of change. Some of these factors are summarized briefly below:

o Implementation of PURPA - Many utilities remain in the early stages of implementing provisions for mandatory grid connection, for non-discriminatory rates for buying and selling electricity to cogenerators, for capacity credits, and for standby rates.

o Fuel Use Act - Although not as important to the commercial sector because of the size considerations and possible cogeneration exemptions, the Fuel Use Act creates market uncertainties because of its clear intent to reduce the use of gas and oil in fuel burning installations.

o Natural Gas Policy Act - The incremental pricing provision again creates market uncertainties, even though "qualified" cogeneration can be exempted.

o Federal Tax Issues - Some tax incentives apply to cogeneration, although most of these exclude gas-fueled cogeneration. However, new tax rules on accelerated depreciation of capital equipment provide strong incentives for cogeneration investments.

o Conflicts in Federal Policy and Regulation - Inconsistencies exist within the previously identified legislation which again create market uncertainties. In many areas, one law conflicts directly with another.

o Electric Utility "Resistance" - Uncertainties in cogenerator/utility contracts, in rates and in attitudes create market uncertainties that discourage cogeneration investments in certain utility territories.

o Lack of Cogeneration Constituency - Presently, no overall infrastructure addressing the cogeneration market exists. Such an infrastructure would be composed of equipment manufacturers, system packagers, engineers, system installers, parts and service organizations, financing entities, insurance specialists, and others. The gas industry has not taken an active role in cogeneration since the early 1970's.

o Environmental Regulations - Uncertainties in environmental restrictions and permitting costs provide barriers to cogeneration, especially in smaller sizes, where these costs can be a significant part of the total project costs.

o Perceptions of Future Energy Price and Supply - Rapid fluctuations in price and supply projections of natural gas and electricity in the past several years have created uncertainty in the projection of economic feasibility of cogeneration. Guaranteed fuel supply is also an issue.

Although these institutional factors and market uncertainties cannot be directly addressed through R&D, they serve to increase the perceived risk of cogeneration investments, particularly in the commercial sector. If R&D projects could reduce the cost of cogeneration systems, an increased incentive for cogeneration investment would be provided, which would compensate for the perceived risks.

Market Size

Although the "technically feasible" market potential and the "economically feasible" market potential for gas-fueled cogeneration in the commercial sector are quite large, the projected market penetration is less, but still substantial. Based upon a simplified analysis, gas-fueled cogeneration can be expected to result in an increase in gas demand in the commercial sector by 10-20%. Gas demand for cogeneration in this sector is projected to be in the range of .5 to 1.2 TCF per year by the year 2000 (approximately .3 to .5 TCF by 1990), resulting in 5,000 to 13,000 MW of installed electric capacity. This level of capacity is approximately the same as that projected for the industrial cogeneration market by Resource Planning Associates in a recent study completed for the Department of Energy.

Impacts on Gas Consumer and Utilities

The net impact of the accelerated growth of gas-fueled cogeneration appears to be quite favorable for the gas consumer, the utilities, and the country as a whole. Because the fuel utilization efficiency is improved, oil consumption and overall environmental pollutants are likely reduced, resulting in lower and/or more stable energy costs to the consumer.

The specific impacts on the electric utility industry strongly depend on each utility's unique circumstances; however, on the whole, the impacts appear favorable. Gas-fueled cogeneration offers the potential for improving capacity reserve margins in a relatively short time frame (without commensurate utility capital investment) and the possibility of improved seasonal load factors.

The net impact on the gas utility industry appears to be quite positive, with a strong potential for increased gas sales, leveling of seasonal load factors, and new and expanded business opportunities.

RECOMMENDATIONS

The results, conclusions, and impacts cited in this report have led to the recommendation to pursue three areas of R&D (in order of priority):

1. Pre-packaged, pre-engineered systems.

2. Analyses of:

   o the less than 100 kW market and technology

   o the potential for cogeneration at specific sites in specific utility service territories

   o cogeneration system reliability

   o characterization of the light industrial market

3. Technology development.

Each of these recommended R&D activities is discussed in the following paragraphs.

Pre-packaged, Pre-engineered Systems

Pre-engineering and pre-packaging standard cogeneration systems could reduce total installed system costs by 25-50%. In addition to lowering fixed costs, O&M costs potentially could be reduced because of resulting improvements in design, matching of components, and reliability of operation. R&D efforts should be directed toward initiating feasibility and engineering studies of pre-packaged cogeneration systems, which will lead to the development and availability of a spectrum of systems for application in the commercial market. The initial phase of this effort would not involve hardware development, but would be aimed at defining standard system packages and developing engineering cost estimates based upon different market penetration scenarios.

This effort, which might involve parallel cost-shared contracts with multiple contractors, would include the following tasks:

a. Define common market characteristics (applications and utility interfaces) which would provide a basis for a standard package.

b. Define a cogeneration system with the characteristics defined above, including: prime mover, heat recovery equipment, control system, generator, grid interface equipment, and system peripherals (e.g., absorption chillers, thermal storage, etc.). Output would be a detailed system specification.

c. Provide a detailed cost estimate of the packaged system under early entry and mature market conditions (i.e., less than 50 units per year and greater than 1,000 units per year). Estimate the O&M cost for the system. Estimate the nonrecurring costs associated with production quantities of the system.

d. Define the technology, development, and/or field test requirements for initiating commercialization of the pre-packaged concept.

Specific Areas for Analysis

Because of the limited scope of the first phase of this effort, the following specific areas noted above have been identified for further analysis.

o less than 100 kW market

o potential for cogeneration at specific sites in specific utility service territories

o cogeneration system reliability

o characterization of the light industrial market

The Less than 100 kW Market. The complexity of this market segment precluded its inclusion in this project. No proven technology was found to exist for continuous duty cogeneration applications less than about 100 kW. Since the potential market in that size range is quite large and quite complex, it is recommended that GRI undertake R&D aimed at better characterization of the applications in this market segment, assessment of the state-of-the-art of cogenerators under 100 kW, and an overall determination of the technical and economic feasibility of pursuing cogeneration for applications less than 100 kW. Specific tasks would include the following:

a) Characterize the less than 100 kW market in terms of building type, geographical region, and energy consumption characteristics.

b) Define generic characteristics for gas-fueled cogeneration systems in various size ranges (less than 100 kW) including maximum allowable costs (including capital and O&M) which would allow penetration of this market sector.

c) Assess current and planned domestic and foreign technology efforts in prime movers and cogeneration systems to determine the technology status. The output should provide prioritized recommendations for R&D opportunities.

d) Define the specific advances in technology, which must occur and the cost/benefit ratio of the technology development required. Evaluate the overall feasibility of penetrating the less than 100 kW cogeneration market.

Potential for Cogeneration at Specific Sites in Specific Utility Service Territories. One clear result of this study is that the economic attractiveness of cogeneration is highly variable from site to site and from utility region to utility region. Due to limitations in project scope, no comprehensive attempt was made to address the relationship between site or utility-specific characteristics and the economic feasibility of cogeneration, and the resulting impact on R&D priorities. It is therefore recommended that analyses be conducted to determine the impact of the following site-specific and utility-specific factors on economic feasibility and overall market potential:

o Technical characteristics including: grid connection requirements; electric efficiency of prime mover and overall efficiency of system; O&M costs; thermal to electric ratio of building and cogeneration system; effects of thermally driven cooling on load profiles.

o Economic characteristics including: standby rates; buy-back rates; capacity credits and requirements; gas and electricity rates; rate structure.

o Institutional characteristics including: utility/cogenerator contracts; insurance requirements; permitting requirements/cost; third-party contracts; PURPA implementation status; electric utility characteristics.

Cogeneration System Reliability. The expanded market acceptance of cogeneration is highly dependent upon the demonstrated reliability and performance of installed systems. This record of reliability and performance will also be a key factor in the calculation of standby rates charged by electric utilities. It is therefore recommended that a research effort be initiated to develop a reliability data base of cogeneration systems installed throughout the U.S. When established, this data base would provide an accurate, statistical record of the performance of cogeneration systems under a spectrum of operating conditions, applications, equipment types, etc.

Characterization of the Light Industrial Market. Because of the limited scope of the present effort, only commercial cogeneration applications greater than 100 kW and less than 10 MW were considered. DOE studies of the industrial sector have been limited to consideration of the five largest energy-consuming industries and have focused primarily on coal as the fuel of choice for backing out oil consumption. Therefore, a gap remains regarding the characterization of the light industrial market (i.e., industrial applications less than 10 MW). It is recommended that applications in this market be characterized in terms of load profiles, thermal requirements, and numbers of applications to determine if significant potential exists for cogeneration in this part of the industrial sector.

Component and System Technology

A number of technical constraints to gas-fueled cogeneration were identified. These constraints are noted below (in no particular order):

o water jacket temperature limitation (reciprocating engines);

o development of grid-excited generators specifically for cogeneration applications;

o dew-point limiations on exhaust heat exchangers;

o thermal storage systems optimized for and integrated with cogeneration systems;

o lack of dual fuel capability (i.e., oil and natural gas) for smaller systems (less than 1 MW);

o noise attenuation;

o emission reduction/control;

o gas compressors for input fuel;

o supplementary firing of exhaust;

o steam injection in turbines/reciprocating engines;

o compact heat exchangers for packaging;

o air-to-air heat exchangers; and

o measurement techniques for hot gases, liquids.

A framework for analyzing the potential benefits of specific component performance or cost improvements should be developed. The framework should include an overall system model for identifying tradeoffs and effects on the overall system performance and cost. This model should be valuable in evaluating and prioritizing specific R&D opportunities in component/system technologies for cogeneration.

Having developed this analytic tool for estimating the value of potential advancements in cogeneration-related technology, GRI could proceed with specific activities directed toward component and system hardware improvement.

Chapter 22

EVALUATION OF INDUSTRY/UTILITY COOPERATIVE EFFORTS IN COGENERATION

D. R. Limaye

ABSTRACT

The changing economic and institutional aspects of energy in the U.S. have made cogeneration an attractive option for industry and utilities. The benefits of cogeneration include efficient energy utilization, conservation of non-renewable resources, reliable electricity supply and stabilization of energy costs. However, the implementation of cogeneration is constrained by uncertainties in regulation and reluctance of industry to get into the power generation business. In this environment, cooperative efforts among industry, utilities and possibly third party investors, may be the most attractive approach to implement cogeneration.

This paper describes an approach to estimate the costs and benefits of cogeneration to industry and utilities (and appropriate third parties) and explores alternative institutional arrangements to implement cogeneration. Specific computer models for detailed evaluation of the impacts of cogeneration are described and applications to specific case studies are illustrated.

INTRODUCTION

According to the General Accounting Office, U.S. industry and electric utilities use nearly half the primary energy consumed, and the waste heat from power generation and process energy use amounts to over seven million barrels per day oil equivalent (1). Cogeneration can offer a method to reduce the amount of waste heat by simultaneously producing electricity and useful thermal energy from a common primary energy source. Because of its potential for efficient use of energy, cogeneration is receiving increasing attention in the U.S.

The concept of cogeneration is not new. Industrial generation of electricity has been practiced for a long time. In the early 1900's, most industrial plants generated their own electricity and approximately half of this was using cogeneration (2). On-site generation/cogeneration was more reliable and less expensive than utility generated power. However, in the 1920's and 1930's, the regulation of electric utilities, first by state agencies and then by the Federal government, resulted in elimination of unproductive competition, and consolidation and extension of utility service areas. Coupled with the availability of inexpensive fuels for power generation and technological

advances in central station utility generation and transmission of electricity, industrial generation/cogeneration became economically less attractive. From the 1920's to the mid-1970's, there was a generally declining trend in the proportion of electricity cogenerated in industry (3). Other factors contributing to this declining trend included the following:

- Industry was hesitant to invest in generation because of the possibility of Federal and state regulation as a utility, and the related reporting requirements.

- Utilities offered very low prices for excess power sold by an industry to a utility.

- Utilities charged high prices for standby or supplemental power needed by the cogenerator.

As a result, industrial generation declined from 18% of total electric generation in 1941 to about 4% in 1977 (4).

THE CHANGING ENERGY SITUATION

In the last decade, the energy situation in the United States has undergone a significant transition. The nation has faced increasing prices and decreasing availability of conventional energy sources, energy supply disruptions, environmental constraints to the utilization of coal, and high capital costs for expanding the energy delivery system. Efficient utilization of our energy resources has become a very high priority and cogeneration has become economically attractive. At the same time, Federal legislation has attempted to remove some of the institutional barriers to cogeneration and small scale power production. Moreover, the problems faced by electric utilities have resulted in increased interest, on their part, in industrial cogeneration.

Industry facing rapidly escalating energy costs is searching for alternative methods to obtain its future energy requirements. Cogeneration offers the potential for increased efficiency of energy use, less uncertainty in energy costs and more reliable supply of energy. Moreover, the recent regulatory changes (discussed below) provide industry an opportunity to obtain significant economic benefits from cogeneration.

Many electric utilities are facing financial problems of unprecedented magnitude. New generating capacity committed in the 1968-1974 time frame, when demand forecasts were growing at an annual rate of 7-10 percent, has been mostly deferred or cancelled. Few large projects have been completed. The basic problems faced by the utilities include high costs of new

---

*This paper is adapted from the author's paper titled "Cogeneration in the U.S.: Problems and Prospects" presented at the American Power Conference, Chicago, April 1982.

capacity, high interest rates, escalating fuel costs, environmental/siting constraints, increased customer resistance to rate increases and regulatory lag.

These problems, coupled with slower load growth, have led to lower revenues than forecast, while the capital requirements for new capacity have continued to escalate rapidly. These utilities, looking ahead to the late 1980's, see their best prospects in completing plants now almost completed, and to some extent, discouraging increases in load growth with the expectation that a two percent annual growth rate will be manageable, allowing time for their economic situation to stabilize before having to undertake another new plant. As part of this basic approach it would be advantageous to flatten the system load curve and to reduce or eliminate use of expensive peaking generation requiring use of high cost fuels in relatively inefficient power plants. Cogeneration could contribute significantly in this approach. In addition, utilities may be able to raise capital through innovative financing schemes such as joint ventures or third party arrangements to build new capacity for cogeneration.

The significant changes in the economic and institutional aspects of power generation, which occurred in the 1970's and are expected to continue in the 1980's, have created a trend towards increased interest in and acceptance of industrial cogeneration by utilities. These changes have led utilities not only to consider industrial cogeneration in their planning for future capacity needs, but have also resulted in the growing recognition of cogeneration systems as a utility business opportunity. Cogeneration ventures, owned and operated by a utility, can be highly complementary to traditional utility operations and possibly offer a potential for higher profits than the traditional utility business. Utilities are therefore increasingly interested in examining opportunities for participation in industrial cogeneration projects (6).

COGENERATION - OLD GAME, NEW RULES

A number of significant changes have occurred in the last few years relative to the institutional and regulatory aspects of cogeneration. The National Energy Act (NEA) of 1978 contains a number of important provisions which attempt to remove institutional barriers to cogeneration. The most important provisions are in the Public Utility Regulatory Policies Act (PURPA), which provides the following for facilities that "qualify" by meeting certain operating and efficiency requirements (7):

- Utilities must purchase any and all power that the qualifying facility (QF) wants to sell.

- The rate offered by the utility for such power purchase should be based on the "avoided cost" of the utility.

- The rates charged by a utility to a QF for standby/backup power must be non-discriminatory.

- The QF is exempted from utility regulation under the Federal Power Act, the Public Utility Holding Company Act and state regulations related to rates and financial reporting.

A qualifying facility must not be more than 50% owned by an electric utility.

In addition to PURPA, three other parts of the 1978 NEA also provide incentives for cogeneration. The Powerplant and Industrial Fuel Use Act (FUA) allows cogenerators to be exempted from prohibitions on the use of oil and natural gas. The Natural Gas Policy Act (NGPA) provides an exemption from incremental pricing of natural gas to cogenerators. The Energy Tax Act (ETA) provides a 10% investment tax credit for certain property which may be used with cogeneration systems. Also, additional incentives were provided in subsequent legislation passed by the 96th Congress (8).

THE NEED FOR INDUSTRY/UTILITY COOPERATION

While the current regulatory environment provides significant incentives to industry to invest in cogeneration projects, it appears that there is a need for cooperative efforts among industry and utilities for financing and implementing cogeneration. The reasons for considering such cooperative efforts are:

- Cogeneration is likely to be more capital intensive than a conventional energy system, and industry may have other uses for capital which are more attractive.

- Industry is hesitant to make major cogeneration investments because of many uncertainties relative to PURPA implementation.

- Industry may not have the skilled staff needed to operate and maintain a power generation system.

- Industry may not consider power generation a natural extension of its primary business, even when such generation is economically attractive.

- Utilities are generally willing to accept a lower rate of return than industry.

- Industrial plant managers may be hesitant to face the problems related to the handling, storage and use of coal (the preferred fuel for cogeneration) and the associated environmental requirements.

- Utilities can offer the necessary expertise in the construction, operation and maintenance of cogeneration systems.

Many utilities are currently actively seeking cooperative ventures with industry. Thus, industries interested in cogeneration may find the local utility a willing and cooperative partner.

Options For Industry/Utility Cooperation

A number of options exist for cooperative efforts among industry and utilities to implement cogeneration (10). These include:

- Sole utility ownership of the cogeneration plant with sale of thermal energy by the utility to industry

- Joint venture between industry and utility (with utility owning 50% or less to qualify the cogeneration facility for PURPA benefits).

- Third party ownership with contracts for thermal energy and electricity sales to industry and utility respectively.

- Partial ownership with the utility owning the power generation equipment and industry owning the remaining plant

- Sole industry ownership but operating control (dispatch) by utility.

The main theoretical justification for a multi-party approach is to share the risk of a project. This reduces the total risk to any one participant, while commensurately reducing the possible returns. In addition, a joint venture arrangement should reduce the "moral risk" of a project where two or more participants must cooperate: if all participants have a stake in the operation, they will all have an incentive to do their part. This is particularly appropriate in the case of cogeneration, where cooperation between the industrial user(s) of the thermal energy and the utility purchaser of the electricity is essential.

EVALUATION OF COGENERATION - THE EPRI PROJECT

While the concept of Cogeneration is not new, recent studies sponsored by the U.S. Department of Energy and other public agencies have pointed out the lack of adequate and reliable information to objectively evaluate the potential for cogeneration, and the costs, benefits and impacts of cogeneration implementation. In 1977, the Electric Power Research Institute (EPRI), the research arm of the U.S. electric utility industry, sponsored a Dual Energy Use Systems (DEUS) Workshop to initiate a dialogue among utilities, industrial energy users, equipment manufactureres and government agencies relative to the development of a consistent and detailed resource base of information that would allow identification and evaluation of attractive DEUS options, and definition of the needed research, development and demonstration programs (9). DEUS options include cogeneration, district heating and other systems for simultaneous production of two forms of energy from a common source. The major findings of the DEUS Workshop were:

- An accurate evaluation of the potential for and the benefits of DEUS requires an evaluation of site-specific factors.

- Utility perspectives must be properly represented in DEUS evaluation.

- Cooperative efforts among industry and utilities are likely to lead to greater total benefits from DEUS.

- A systematic evaluation of grid-integrated DEUS systems is needed.

In early 1979, EPRI initiated a project to address the techno-economic and institutional aspects of DEUS (10). The principal objectives of this project are to:

- Develop a methodology to assess DEUS options, giving explicit attention to utility perspectives and impacts

- Identify promising DEUS applications

- Identify and assess utility options for participation in DEUS

- Identify research, development and demonstration needs and priorities.

The principal focus of the project is on industrial cogeneration. An overview of the project structure is shown in Figure 1. The project consists of three major sets of activities. The first step in the project was to conduct a set of surveys and case studies, covering both industrial cogeneration and district heating, to develop an understanding of the techno-economic and institutional factors affecting the success of DEUS. The second set of activities, classified as systems analysis, encompasses the development of a methodology and a set of computerized analytical tools for screening and evaluation of cogeneration, and for assessment of impacts of cogeneration on utilities. The final set of activities is a series of conceptual designs of promising cogeneration applications.

Methodology for Evaluation Costs and Benefits of Cogeneration

The methodology for cogeneration evaluation consists of two steps (11). In the first step, the aggregate benefits, costs and impacts of cogeneration are calculated, taking into account the total impacts on the utility, industry and society. This calculation is based on the value of electric and thermal energy used, the costs of producing these outputs, and the related social and environmental considerations. Institutional and regulatory considerations such as standby and buy-back rates (PURPA rates), tax credits, alternative arrangements for ownership and operation, etc., do not affect the overall benefits of cogeneration from the systems viewpoint, but do determine how the benefits, costs and impacts are shared by the various affected parties. Such institutional and regulatory factors are therefore considered in the second step under each type of arrangement for ownership or operation. These considerations influence the negotiated position of each party relative to the cogeneration venture.

An overview of the methodology is shown in Figure 1. In the first step, information regarding the characteristics of cogeneration technologies, the energy needs for the application, and local utility data, is used to determine the size of the cogeneration system under alternative sizing options. Calculations are then made for the performance of the cogeneration system and its capital, operating and maintenance costs. The performance calculations provide information regarding the amount of thermal and electric energy generated by the cogeneration system under different operating strategies. The value of the power generated is then calculated based on data on the utility's generation mix and expansion plan. Similarly, the value of thermal energy produced is calculated based on the alternative costs of thermal energy generation for the industry. An economic analysis is then performed, taking into account the value of the thermal and electric outputs relative to the capital and O&M costs under each sizing and operating option. The economic data are then compared to the conventional energy generation systems to determine the aggregate costs, benefits and impacts of the cogeneration option. By performing these sets of calculations for different cogeneration technologies and different sizing and performance options, the most attractive options can be identified.

Figure 1 also shows an overview of the second step, the detailed analysis of the cogeneration options. For each option considered to be an attractive option, an analysis of the institutional and regulatory constraints is performed. Based on this analysis, the alternative organizational and financial

Figure 1

METHODOLOGY FOR EVALUATION OF COGENERATION

Step 1. Identification of Attractive Options.

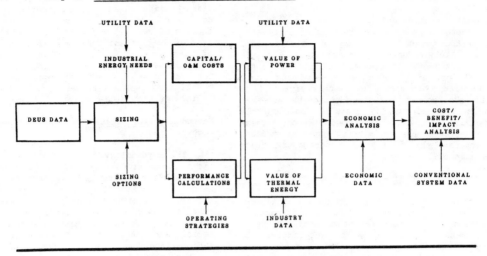

Step 2. Detailed Analysis of Options.

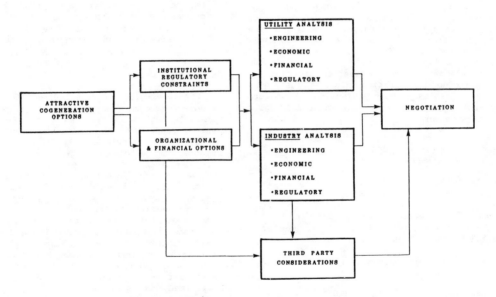

options are identified. For each of these options, an analysis of the impacts on the utility and industry is then performed. Where appropriate, if third party considerations are important, the analysis includes the impact on such third parties. In this step, a detailed evaluation of the economic, financial and regulatory aspects is performed from the point of view of the utility and industry to provide information regarding the alternative methods of allocating the benefits of cogeneration. It is hoped that this analysis will provide all concerned parties with adequate information to enter into a meaningful negotiation process which will lead to the implementation of the most attractive cogeneration systems.

Computer Evaluation of Dual Energy Use Systems (DEUS)

In order to perform the sizing and performance calculations, and to screen and evaluate the costs and benefits of cogeneration options relative to a conventional system, an analytical model called DEUS – Computer Evaluation of Dual Energy Use Systems, has been developed (12). This model accomplishes step 1 of the evaluation methodology. The model can evaluate up to twelve systems, (including a no-cogeneration base case) taking into account industrial requirements for heat and power, fuel types, utility rate schedules (including industrial and PURPA rates), economic data, operational ground rules, and various ownership types.

142

In many industrial processes, the actual process thermal and power demands vary with time-of-day and/or seasonally. To be compatible with anticipated PURPA rate schedules, the program has the capability to represent 36 time periods per year. For example, the 36 time periods might be used to cover four seasons, three types of days per week, and three time periods per day (on-peak, near-peak, and off-peak). The program has the capability to evaluate cogeneration configurations incorporating up to four fuel streams, with each fueling a given type energy conversion system (ECS).

COPE - Cogeneration Options Evaluation

A computer model called COPE - Cogeneration Options Evaluation, has been developed to calculate after tax cash flows to the utility, industry and, where appropriate, third parties (13). COPE can handle all practical ownership and financial arrangements and account for tax credits, depreciation and other relevant financial and economic parameters, taking into account the most recent legislation and regulations. COPE is designed to provide information to all potential participants in a cogeneration venture so as to identify mutually beneficial institutional arrangements.

The magnitude and distribution of after-tax costs and benefits of a cogeneration system are significantly influenced by its ownership structure (utility, industry, third party), operating mode (thermal dispatch versus utility economic dispatch) and the electricity sales arrangement (simultaneous buy-sell, buy-shortage/sell-excess). COPE is designed to evaluate alternative combinations of ownership, operating modes and sales arrangements.

In the past, a common assumption was that a cogeneration system is owned entirely either by an industry or a utility. With the increased interest in cogeneration, a number of innovative arrangements are being considered. For example, joint ventures among industry, utility and third parties may offer benefits to all the participants. One arrangement to form a joint venture is to create a separate corporation for the sole purpose of owning and operating the cogeneration project. In this arrangement, the cogeneration project would be taxed as a corporation.

The partnership arrangement can also be used to form joint ventures. Partnerships do not pay a Federal tax on earnings comparable to the corporate earnings tax; however, each partner pays Federal tax on his share of earnings from the partnership. Also, partnerships enjoy considerable flexibility in the apportionment of tax and depreciation benefits as well as profits (or losses) among partners. It is possible, therefore, to design partnership arrangements so as to attract private (or "third party") investors by offering them substantial tax-related benefits. At the same time, third parties, having no site-specific thermal or electric requirements, are unlikely to insist on specific operating modes. Thus, partnerships between utilities, industries and "third parties" could often be mutually beneficial.

COPE is being designed to analyze any one of the following ownership arrangements. The utility can be either an investor-owned or a tax-exempt utility.

- 100% Ownership

    - 100% Utility Ownership

    - 100% Industry Ownership

    - 100% Third Party Ownership (or Separate Corporation).

- Joint Ventures

    - Partnership - Utility/Industry

    - Partnership - Utility/Third Party

    - Partnership - Industry/Third Party

    - Partnership - Utility/Industry/Third Party.

- Leasing Arrangements

    - Lessor/Lessee - Third Party/Utility

    - Lessor/Lessee - Third Party/Industry.

Illustrative Examples

Illustrative results of the application of these models for the economic evaluation of cogeneration in a pulp mill are shown in Table 1. The results indicate that a 59 MW cogeneration system offers a 25% rate of return on incremental investment over a no-cogeneration case. A 100 MW cogeneration system offers a 15.6% rate of return. The revenues from electricity sales in the 100 MW case are comparable to income from pulp sales (14). Figure 2 shows the rate of return versus size for the pulp mill application.

Illustrative results for cogeneration in enhanced oil recovery are shown in Table 2.

CONCLUDING REMARKS

The changing economics of cogeneration have led to a great deal of activity and interest on the part of both industry and utilities to evaluate and implement cogeneration projects. A large number of applications have been filed with FERC for certification as a qualifying facility (15). An analysis of these shows that the greatest activity is in California and the Southwest.

A recent assessment of cogeneration potential for the U.S. Department of Energy estimated a total of over 42,000 MW (16). While these estimates may appear to be high (they were developed using a 7% ROI criterion), they nevertheless indicate the likelihood that substantial activity related to cogeneration is likely to be undertaken in the U.S., particularly if the uncertainties in PURPA are resolved in a manner favorable to cogeneration.

Because of the efficiency of cogeneration systems, they offer economic benefits to both the industry and the utility. Cogeneration results in conservation of our energy resources and lower environmental impacts. Cooperative efforts between industry and utilities can lead to the implementation of the optimum cogeneration systems, providing benefits to industry, utilities and society.

ACKNOWLEDGEMENT

This paper is based upon the author's current research as part of a project sponsored by the Electric Power Research Institute titled "Evaluation of Dual Energy Use Systems (DEUS)," Project RP1276.

Table 1

ILLUSTRATION OF COGENERATION ECONOMICS

PULP MILL EXAMPLE

	NO GENERATION	ASSUMED THERMAL MATCH	MAXIMUM COGENERATION
Gross MW Output	0	59.1	100
Total Installed Cost of Power Plant (Million $)	88	146	202
Cost Chargeable to Power Plant Generation (Million $)	---	58	114
Annual Operating & Maintenance Costs (Million $)	2.98	6.09	7.01
Annual Fuel Costs (Million $)	7.29	17.48	33.10
Annual Cost of Purchased Electricity (Million $)	13.04	13.91	15.02
Annual Electric Revenues (Million $)	0	37.71	61.07
Projected Return on Investment (%)	0	25.4%	15.6%

Figure 2

PROJECTED RETURN ON INVESTMENT VS MEGAWATT SIZE

1985 CONCEPTUAL DESIGN, 1000 TONS/DAY BLEACHED KRAFT PULP MILL

WEST COAST, U.S.A.

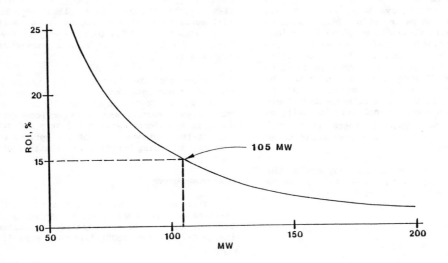

144

Table 2

ILLUSTRATION OF COGENERATION ECONOMICS

ENHANCED OIL RECOVERY

	GAS TURBINE WITH FIRED HEAT RECOVERY BOILER (RMR SYSTEM 1-B)	GAS TURBINE WITH TWO BOILERS PLUS A STEAM TURBINE (RMR SYSTEM 2-E)
Gross Electric Output - MW	3.7	4.3
Gross Steam Output - Million Btu/Hr	51.0	97.3
Firing Rate - Million Btu/Hr	74.7	142.5
Total Installed Cost* - Million $	3.9	4.3
First Year O & M Costs* - Million $	0.1	0.1
First Year Fuel Cost* - Million $	3.2	6.2
Purchased Electricity - MWH	0	0
First Year Electricity Revenues* - Million $	2.6	3.0
First Year Steam Revenues* - Million $	3.3	6.3
Internal Rate of Return - Percent	51.5%	54.2%
Simple Payback - Years	2.04	1.95

*Costs are expressed in 1984 dollars.

REFERENCES

1. U.S. General Accounting Office, Industrial Cogeneration - What It Is, How It Works, Its Potential, EMD-80-7, April 29, 1980.

2. Fred J. Sissine, "Energy Conservation: Prospects for Cogeneration Technology," Issue Brief No. IB81006, Library of Congress, Washington, D.C., February 1982.

3. Frederick H. Pickel, Cogeneration in the U.S.: An Economic and Technical Analysis, M.I.T. Energy Laboratory Report, MIT-EL-78-039, Boston, Massachusetts 1978.

4. Synergic Resources Corporation, Cogeneration Data Base, Draft Report submitted to Electric Power Research Institute, September 1981.

5. Synergic Resources Corporation, Industrial Cogeneration Case Studies, EPRI EM-1531, Electric Power Research Institute, September 1980.

6. Thomas C. Hough and Dilip R. Limaye, Utility Participation in DEUS Projects: Regulatory and Financial Aspects, Final Report, submitted to Electric Power Research Institute, December 1981.

7. For further details, see Final Rules Implementing Sections 201 and 210 of PURPA, 45 Federal Register, 12214 and 17949.

8. The legislation providing additional incentives to cogeneration includes the Crude Oil Windfall Profits Tax Act, the Energy Security Act and the Housing and Community Development Act of 1980.

9. Dilip R. Limaye, "Utility Perspectives on Industrial Cogeneration Systems", paper presented at the Annual Meeting of the Association of the Pulp and Paper Industry, Atlanta, October 1981.

10. Synergic Resources Corporation, Evaluation of Dual Energy Use Systems: Volume I, Executive Summary, Draft Report, to EPRI, May 1982.

11. Dilip R. Limaye, <u>Methodology for Evaluation of Cogeneration Projects</u>, paper presented at the National Fuel Cell Seminar, Norfolk, Virginia, June 1981.

12. General Electric Company, <u>Computer Evaluation of Dual Energy Use Systems, Volume 1: Program Description Manual</u>, Draft Final Report, November 1981.

13. B. Venkateshwara, <u>Technical Documentation of COPE-Cogeneration Options Evaluation</u>, unpublished draft report, Synergic Resources Corporation, Bala Cynwyd, PA, April 1982.

14. KPFF Engineers, <u>Pulp Mill Design for Maximum Cogeneration</u>, Draft Final Report to EPRI, October 1981.

15. "FERC Applications for Qualifying Facility Status", <u>Cogeneration World</u>, March/April 1982.

16. "Industrial Cogeneration Potential (1980-2000)", <u>Cogeneration World</u>, March/April 1982.

Chapter 23

SECO-DOW CORNING'S WOOD FILLED INDUSTRIAL COGENERATION PROJECT

H. Rolka, P. E. Sworden

ABSTRACT

In 1979 Dow Corning Corporation decided to use wood as a primary fuel for its new steam and electric cogeneration (SECO) power plant. Cogeneration is the production of both steam and electricity in a single operation. This was prompted by the high cost of fossil fuel and an abundant supply of wood in mid-Michigan.

The SECO project underscores Dow Corning's commitment to conservation and alternative energy sources at a time when reducing U. S. dependence on foreign oil is a national priority. This power plant will have multi-fuel capability, but wood in the form of chips will be the primary fuel. Approximately half of the wood will come from forest land within a 75-mile radius of Midland, Michigan. The remaining wood will come from urban and rural wood residue.

To help accomplish this, Dow Corning has assembled a natural resources team of three foresters, two wildlife biologists, and a field ecologist to plan and carry out sound forest and wildlife management programs.

SECO is a comprehensive, thoroughly planned multi-benefit energy project. It will bolster the mid-Michigan economy by adding jobs and money. Also, the natural resource management will enhance the value of forests and improve wildlife habitat.

INTRODUCTION

A decision in 1979 by Dow Corning Corporation to consider wood as a primary fuel for its new steam and electric cogeneration power plant at Midland, Michigan was prompted by aging boilers, the high cost of fossil fuels, and an abundant supply of wood in mid-Michigan.

Dow Corning is a Fortune 500 company founded in the forties as a joint venture by Corning Glass Works and The Dow Chemical Company. We are the technological pioneer of a unique material called silicone. Dow Corning manufacturers more than 1100 silicone products that are used in virtually every industry.

Currently our largest manufacturing facility, the Midland Plant, is working to make these products with greater energy efficiency through a project called SECO - steam and electric cogeneration. Operating the Midland site chemical complex requires great amounts of energy. Steam used in our processes and for heating buildings is currently produced by antiquated oil- and gas-fired boilers. Our electricity is purchased.

SECO DESCRIPTION

Costs of oil, natural gas and electricity have increased dramatically in recent years. After studying the inherent energy costs of our present system, we decided cogeneration is more practical and cost efficient as a means of supplying power for the Midland Plant.

The illustration shows how cogeneration at Dow Corning will work.

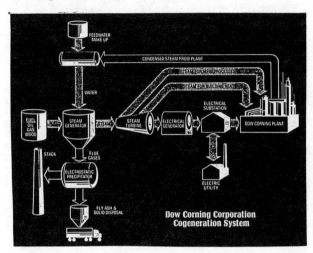

Heat produced by combustible fuel -- it could be wood, gas, oil -- is used to generate steam in a boiler (275,000 pounds/hour). Steam at 1250 pounds per square inch (psi) and 900 degrees (F) turns a turbine, which drives a generator that produces electricity. Steam extracted from the turbine at 165 psi for process steam and at 30 psi for heating is piped throughout the plant in two separate systems. Maximum power production capability of the new system is 22.4 megawatts. The electrostatic precipitator separates exhaust gas and particles. The gases are exhausted through a 200 foot high stack. The ashes are collected for disposal and have some value as a low grade fertilizer.

BUILDING SCHEDULE

In February 1980, an Air Emissions Permit was approved by Michigan's Department of Natural Resources -- an effort that was two years in the making and a legal requirement for getting this project underway. Construction started the next month. The following is a summary of activities:

1980:
  June   - Concrete poured for
               turbine/generator pedestal.
  July   - Plant foundation completed.
  August   - Began erection of structural
               steel.
  September - First piece of upper support
               steel for boiler is set.
  October   - Structural steel "topped
               off" with the American flag.
  December   - Poured 99 cubic yards con-
               crete at 80 foot level for
               the forced draft and induced
               draft fan floor and foun-
               dations.

1981:
  January   - Steam generating tubes be-
               tween upper and lower boiler
               drums are installed.
  February   - Nine water well panels
               erected for boilers.
             - 35 ton Dresser crane
               installed in generator room.
  March   - 10 foot diameter steel
               exhaust stack installed. Its
               top is 200 feet above grade.
  April   - Structural steel is com-
               pleted. Air ducts and
               heater for boiler installed.
  May   - 70 ton turbine and its
               generator on the foundation.
  June   - Wood handling system
               erected.
  July   - Alignment and welding of 480
               collector plates and install-
               ation of 2880 high voltage
               wires.
             - Dow Corning silicone roof
               installed in high bay areas.
             - Installation of exterior
               siding on building.
  August   - Erection of cooling towers
               completed.
             - Mechanical and piping con-
               tract awarded.
  September - Boiler is completed and
               hydro test is taken to
               verify system integrity.
  October   - Power transformers installed
  November   - Completed traveling grate
               erection in boiler.
  December   - Start erection of ash hand-
               ling system.

1982:
  January   - Installation complete of
               boiler and electrostatic
               precipitator.
  February   - Material handling equipment
               installed.
  4th Quarter- Start up and on line
               operation.

There are many safety considerations being
designed into the plant to minimize hazards.
These include fire protection of wood pile
and conveyor galleries, special safety re-
requirements for personnel working in conveyor tunnels, emergency exists, and much
more, including a hazardous area classification for electrical equipment to prevent
wood dust explosions.

GENERAL SOURCE INFORMATION

As mentioned earlier, the SECO power plant
will have multi-fuel capabilities. It is designed to burn wood, natural gas, oil, or with
modification, coal. Due to the high cost of
fossil fuels, wood in the form of chips, will
be the primary fuel for our SECO power plant.
Dow Corning has opted for wood fuel for
several reasons.

There is an abundant supply in central Michigan, and transportation to the Midland Plant
poses few problems. Wood is relatively clean-
burning and a renewable energy resource.

According to our estimates, Dow Corning will
require up to 180,000 tons of dry wood per
year. The actual tonnage required will depend
on plant needs and final arrangements with the
local utility for purchase of our excess electricity.

About 50 percent of the wood fuel to be used
at the SECO power plant will come from non-
traditional sources. These include such projects as removal of diseased and dying trees
in municipalities and pruning of branches
around power lines. Wood fuel will also be
derived from clearance of land for agricultural uses, reforestation and new construction. Residue from sawmills, including bark,
end pieces, and sawdust, is a third source.
Broken, unserviceable pallets and wood scraps
now being discarded by industries can be made
into wood chips for use as boiler fuel. Some
of the wood from these sources is currently
being buried in sanitary landfills as waste.

The other 50 percent of the wood will come
directly from forest land within a 75-mile
radius of Midland. The majority of this
forested land, 69 percent, is privately owned.
The Michigan Department of Natural Resources
controls 21 percent, with the U. S. Forest
Service accounting for the remaining 10 percent. The total forested land of all three
owner groups is just over 4,000,000 acres.

Dow Corning will harvest up to 4,000 acres,
or about .1 percent of the four million
forest acres annually.

One of the most attractive aspects of the SECO
Project is that the cogeneration power plant
can burn wood materials now going to waste.
A Michigan Public Service Commission report
estimates that wood now going to waste within
a 75-mile radius of Midland could supply Dow
Corning's energy needs for more than 50 years.
Large tracts of forest land -- both public
and private -- have high rates of decay. In
a Michigan DNR Forest Management Division
Summary Report (January 11, 1982, "Apparent
Annual Timber Surpluses for the Northern
Two-Thirds of Michigan" by R. Bertsch), it
notes that 27.6 percent of the annual timber
surplus is in dead wood. That accounts for
2.64 million cords of timber annually.

HARVEST OPTIONS

On forested acreage, depending on the type and
condition of timber stands and landowners'
management objectives, there are two basic
harvest management techniques: "selective
thinning" and "clearcutting."

Selective Cutting

148

Selective thinning involves removal of designated trees, such as diseased, rotted and otherwise inferior timber, in order to improve the growth of new and remaining trees. In the past, many of the commercially undesirable trees thinned from a forest were left to clutter the ground and inhibit new growth. Because Dow Corning will be using all harvested wood for fuel, there will be no waste of such poor quality wood.

Clearcutting

Timber Benefits: Although selective thinning has its advantages, certain species (aspen-birch type) of Michigan's forests will benefit most from clearcutting -- removing all trees from a specified area. Under the Dow Corning plan, clearcutting will be limited to between 10 and 50-acre harvests. While harvesting, the advantages of clearcutting far outweigh the temporary loss of appearance. These advantages include benefits for wildlife and reforestation.

Wildlife Benefits: Demonstrations have shown that removal of trees by clearcutting reduce fire potential, and opens the forest to unrestricted sunlight which encourages rapid growth of young, vigorous trees. Young fruit-bearing plants also grow profusely under the direct stimulus of the sun. This young vegetation is an important food source for wildlife, especially during our winter months when snow hides food on the ground. For example, deer and the ruffed grouse need mature and young forests to meet all their habitat requirements. Rabbits prefer the dense underbrush of young forests wheras squirrels prefer mature trees for food and shelter. The real key to a plentiful wildlife population is to provide a balance of their habitat needs.

HARVEST OPERATIONS

The harvesting will be accomplished by three independent operators under contract to Dow Corning. Their modern machinery contributes to the efficiency of today's harvesting practices. A feller-buncher hydraulically clips trees cleanly at ground level and then stacks them neatly in bunches. The wood is picked up by the grapple-skidder, which transports the entire bunch to the chipper which reduces them to wood chips. The chips are then blown into a van for transport to Dow Corning. Transport vehicles will be routed on major highways. We anticipate about 50 trucks per day, five days per week.

Unloading is quick and simple by using the two 65-ton hydraulic truck dumps at the SECO site. Wood chips will be transferred from the storage pile to the boiler as needed. Our fuel requirements will range from 390 dry ton per day to 500 dry ton per day depending on steam demand and sales of electricity.

DOW CORNING'S NATURAL RESOURCES TEAM

Private Land Management

Dow Corning has developed a wood fuel purchasing specialist and assembled a natural resources team for SECO. That natural resources team is made up of three professional foresters, two wildlife biologists, and a field ecologist.

One of the most important tasks for our forest management team is to work closely with landowners to accommodate their goals while developing balanced forest and wildlife management plans. The process beings with a careful inventory of the timber, soils and wildlife of an owner's property. After input from and approval by the owner, this forest and wildlife management plan becomes part of the formal contract between the owner and Dow Corning. We will closely monitor all phases of the harvesting program during the contract period.

Conservation of our valuable forests is a top consideration with Dow Corning. Overall, the objectives of the natural resources management team are to ensure optimal regeneration of trees, to enhance wildlife habitat and to practice proper soil conservation while respecting the priorities of the landowner.

We have already entered into over 70 land management agreements with private owners and have contracts covering all our forest-derived fuel needs for 1982 and 1983. We also have about 40 percent of our 1984 needs under contract.

Dow Corning's scientific forest management practices can go a long way toward minimizing the waste resulting from poor management or no management at all. It is designed to assist nature in achieving maximum productivity.

U.S.F.S. - Mio Burn

One unique example of minimizing waste is Dow Corning's harvesting efforts near the small northern Michigan community of Mio. In 1980, the Mio area fell victim to one of the most devastating fires in Michigan history. A prescribed fire to burn and reforest 200 acres of forest land was planned by the U. S. Forest Service to improve the habitat for the Kirkland Warbler, an endangered songbird. The Kirkland Warbler needs the young jack pine timber for nesting which is the result of fires burning in mature jack pine.

Weather conditions were not ideal and the fire took off -- burning 25,000 acres and approximately 40 homes before it was under control. Today there stands charred dead trees as a somber reminder of the nightmare.

In an unique agreement initiated by the U. S. Forest Service, Dow Corning will remove dead wood from 867 acres of the burn site to use as fuel in the SECO power plant. This program is advantageous for everyone. The dead wood will be removed so the forest service can replant with a more fire resistant timber type. The residents will have their area reforested in a way that will better protect their investments and Dow Corning has a chance to find out if it is feasible to economically salvage this residue.

Wood Residue

As mentioned earlier, 50 percent of our wood fuel will come from the non-traditional sources, such as the Mio burn. We have five multi-year contracts with sawmills that make up a large part of the amount in this category. They will be delivering sawdust and chipped slabs, bark, and end pieces.

We are also developing a source of wood residue which now ends up in local landfills. This, over a longer time frame, could easily become a major source of wood fuel.

With all these non-traditional sources added together with the traditional forest-derived sources, we have all our fuel needs met at this time through the end of 1983.

SECO PILE MANAGEMENT

To insure these daily requirements, we will construct a wood storage pile. The pile will be 30 to 35 feet deep and cover a four-acre area. A concrete surface will allow for clean fuel storage as well as full recovery of all run off. This is particularly important if the run-off needs some type of treatment before it can be transferred off our site.

Another large unanswered question concerns what the moisture pattern and percentages will be throughout our large pile. To answer this and a few more unknowns, a small (100' x 30') test pile was constructed in August 1981. Within the pile, there are 16 heat sensors which are monitored periodically. Chip and run-off water samples are collected for analysis each month.

Although the data is not complete, we do see some definite trends. The pile reached its hottest temperature (70°C) within the first three weeks. The outside layer of chips absorb a large amount of the rain that falls. With this, their average moisture content went from 41 percent moisture when the pile was constructed to a high of 75 percent moisture after a very wet period. However, the inside of the pile went the other direction; from a high of 41 percent moisture to a low of 32 percent moisture after about 60 days of storage. We have seen a trend of the chips and run-off becoming more acidic as the storage time increases.

When the pile is moved to the SECO site in late spring 1982, we will be able to tell the true pattern of moisture in the pile. With this information, the actual pile management standard operating procedures can be developed. Proper pile management will improve boiler efficiency.

SUMMARY

That, essentially, is what the SECO Project is all about. You can see that it is a comprehensive, thoroughly planned multi-benefit energy project. SECO will also help bolster the mid-Michigan economy by adding an estimated 60-70 jobs at the power plant, in harvesting and hauling. In addition, over three million dollars yearly in Dow Corning energy costs will be returned to the Michigan economy. As an added benefit, our natural resource management will enhance the value of our forests and improve wildlife habitat.

The Dow Corning Midland Plant is uniquely suited to this project because of the size of our operation and location, near a ready supply of wood fuel.

We believe our SECO Project is the first of its kind in the chemical industry. We are proud to be a leader in reintroducing this old, proven technology and applying it to today's critical energy problems.

SECTION 6
EQUIPMENT AND SYSTEM
APPLICATIONS

Chapter 24

$500,000/YEAR — ENERGY MANAGEMENT SUCCESSES IN THE CONVENIENCE STORE AND SNACK FOOD BUSINESS

W. E. Johnston, T. C. Kraemer

As in other types of endeavors, the success of an Energy Program is dependent on the effort and intelligence invested in the project. The practical application of logic and technical skills to any problem is the proper first step in developing profitable solutions.

So it is with the Convenience Store and Snack Food industry. The basic questions to be addressed are:

1. How much Energy is being used?

2. Where is the Energy being used, and in what quantities?

3. What does this Energy cost at each area?

4. Is its use necessary?

5. What can be done to eliminate or reduce the Energy usage?

6. Is there an adequate information system available?

Only by getting to the roots of each of these questions can Energy usage be defined, understood, and modified to reduce the basic dollar cost.

The General Host Corporation is the parent company of:

```
450 "Little General" convenience stores) Retail
135 "Hot Sam's" pretzel stores         ) Food
450 "Hickory Farms" cheese stores      ) Group

  2 "Van de Kamps" frozen food plants
  5 "Milk Specialties" animal food plants
  2 "American Salt Company" salt mines
      and processing
  1 "Cudahy" ham processing plant
  1 "Cudahy" sausage plant
```

These companies together use over $15,000,000 per year in Energy. Little General accounts for over $5,000,000/year; Hickory Farms uses $2,000,000/year; and Hot Sam's uses approximately $500,000/year.

We will address basically the "Little General" and "Hot Sam's" programs.

Auditing

A basic Energy Audit was performed on a number of Little General stores. The historical Energy usage of the store was reconstructed in terms of KWH usage, peak KWD, total dollar cost, and cost per KWH; including all extras -- taxes and surcharges.

Following is an example of an audit for the Hillsborough Street Little General Store, Tampa, Florida:

Period	KWH/Month	Average Daily KWH	Peak KWD	Dollars Per KWH	Total Month's Cost
BASE PERIOD					
January 1981	13,177	412	25	$.049	$ 643
February 1981	13,770	430	25	.048	665
March 1981	13,558	466	25	.048	657
April 1981	15,587	537	28	.048	748
May 1981	15,726	542	30	.057	891
June 1981	18,101	566	33	.056	1013
July 1981	16,817	561	32	.058	973
August 1981	16,344	564	32	.058	951
September 1981	16,481	515	30	.057	946
October 1981	15,116	521	29	.052	790
November 1981	13,891	479	27	.053	730
December 1981	13,840	419	26	.052	723
January 1982	13,923	409	25	.054	749
IMPLEMENTATION OF ENERGY PROGRAM -- TEST PERIOD					
February 1982	11,954	398	26	.056	622
March 1982	10,859	374	25	.057	673
April 1982	10,956	378	25	.063	749

The second step was to individually audit each
user with the store and identify its cost components. Below is the result of that audit:

Item	KWH Per Month	KWH Dollar Cost Per Month	KWD (Actual)	KWD Dollar Cost Per Month (@$5.50/KW)	Total Dollar Cost Per Month
A. CONSTANT ELECTRICAL DRAW					
Lights, Row 1	435	$ 18.83	.81	$ 4.46	$ 23.29
Lights, Row 2	435	18.83	.81	4.46	23.29
Lights, Row 3	498	21.55	.69	3.80	25.35
Lights, Row 4	744	32.20	1.04	5.72	37.92
Cooler Lights	288	12.46	.40	2.20	14.66
Storage Room Lights	141	6.10	.20	1.10.	7.20
B. TIME OF DAY OPERATION					
Yard Lights (Night)	243	10.52	1.61	-0-	10.52
Gas Lights (Night)	177	7.66	1.18	-0-	7.66
Outside Lights (Night)	84	3.63	.70	-0-	3.63
Lighting Contactor (Day)	45	1.94	.08	.44	2.38
Sign Lights (Night)	91	3.94	.77	-0-	3.94
C. OFF-ON OPERATIONS					
AC #2 (August)	3198	138.44	6.05	33.28	171.82
AC #1 (August)	3276	141.81	4.95	27.23	169.04
Cooler Compressor	2625	113.63	3.85	21.17	134.80
Frozen Food	720	31.16	1.75	9.63	40.79
Koolee Machine	942	40.78	1.73	9.51	50.29
Ice Maker	123	5.32	1.61	8.86	14.18
Ice Cream	741	32.07	1.50	8.25	40.32
Hot Water Heater	123	5.32	1.44	7.92	13.24
Gas Pumps	42	1.82	1.38	7.59	9.41
Ice Storage	744	32.20	1.15	6.33	39.53
Cooler Door Heaters	828	35.84	1.15	6.33	42.17
Beer Cooler	204	8.83	.81	4.46	13.29
Soda Fountain	462	19.92	.39	2.14	22.13
Double Jet Spray	297	12.85	.41	2.26	15.11
Single Jet Spray	135	5.84	.25	1.37	7.21
Coffee/Hot Dogs	54	2.33	.08	.44	2.77
Sandwich Cooler	207	8.96	.29	1.59	10.55
Low Current Koolee	42	1.81	.12	.66	2.47
	17,946 KWH/ Month	$766.88/ Month	37.20 KW	$181.08	$958.96/ Month

Comparing the $958.96 total audit estimated dollar cost and the KWH estimate against the 1981 August
actuals, we find:

	Actual	Estimated	% Error
KWH	16,344	17,946	9
Dollar Cost	$951.00	$958.96	1

From this, we establish that the audit is valid, and can be depended upon as reference material for
future recommendations.

Analysis

As the two major users of electricity are the Air Conditioning units, the question arises as to the causes and effects of all the other heat generators within the stores, and by the outside ambient weather conditions. Assuming that we are committed to maintaining a 75°F interior temperature, the additional heat loads will create additional work for the air conditioners. Graphically, this can be represented as follows:

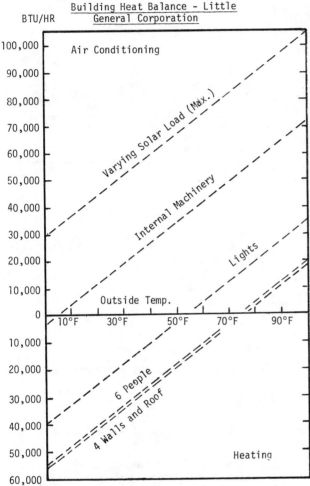

Building Heat Balance - Little General Corporation

It can be seen that the interior of a convenience store is a heat generator, and theoretically should not require comfort heating down to approximately 10°F. Practically, this is not true, as such conditions as the opening and closing of doors, etc., has not been taken into account. On the other hand, it does show why the store requires 6-7 tons of air conditioning on a 95°F day.

The first dotted line in the above graph represents the amount of BTU's per hour (heating or cooling) necessary to maintain a 75°F inside temperature due to heat loss in the walls and windows of the building as the outside ambient temperature varies. The second dotted line represents the heat added by the normal continuing lighting loads. The third dotted line represents the heat added to the building by the estimated body heat of six people. The third dotted line represents the heat added by

the normal continuing lighting loads. The fourth dotted line represents the heat added by all the store machinery loads such as ice storage, frozen foods, soft drink machines, etc. The fifth line represents the maximum amount of heat transmission into the building by solar loads. Naturally, this will vary from day to night and will also be affected by cloud cover.

From the above, it is seen that not only are we paying for the basic Energy used in a store but also for the air conditioning energy to remove the heat generated by lights and equipment. Consequently, for every unit of energy used in the store, we also must use approximately 1.25 times that amount in air conditioning to remove the generated heat. Now that we understand the dynamics of store energy, let's look at the basic causes.

Individual Users

The next step was to determine those areas where changes could effect energy savings and test the hypotheses under operating conditions. Five stores in the Tampa, Florida, area, were selected as test facilities and a number of energy modifications were effected. These were:

1. Installation of a four-channel microprocessor, on-off times which did the following:

   a. Controlled air conditioner #1
   b. Controlled air conditioner #2
   c. Controlled the front row of store lights to be off during the daylight hours and on after dark.
   d. Controlled the beer and soft drink coolers to be off during closed hours.

2. Installed voltage reducers on all fluorescent fixtures.
3. Replaced all non-energy saving bulbs with energy saving bulbs.
4. Installed 400 watt high pressure sodium night lights in place of using 2000 to 3500 KW of fluorescent store lights that were being used for internal night security.
5. Humidity controls were installed on the internal heaters of walk-in cooler doors and frames which cycled the heaters off during periods of low humidity.
6. All refrigeration was checked and set for correct operating temperatures.
7. Set up in-store testing of roof mist cooling to reduce air conditioning loads (Fan Jet Corporation).
8. Set up in-store testing of water assisted refrigeration on both the air conditioning units and the walk-in coolers (W.A.R. Company, Tampa*). Eight units were installed on three stores.

   *W.A.R. = Water Assisted Refrigeration

Results

The bases for energy use over the base and test periods were selected as the daily average number of total KWH used per store. Below are six graphs representing the 1981 base period and the 1982 test period for each of the five test stores and one special store utilizing only the W.A.R. system. Store No. 5 was a new facility in 1981 so base data was not available.

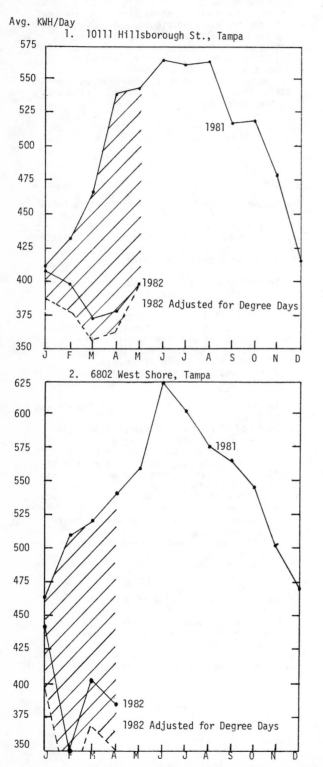

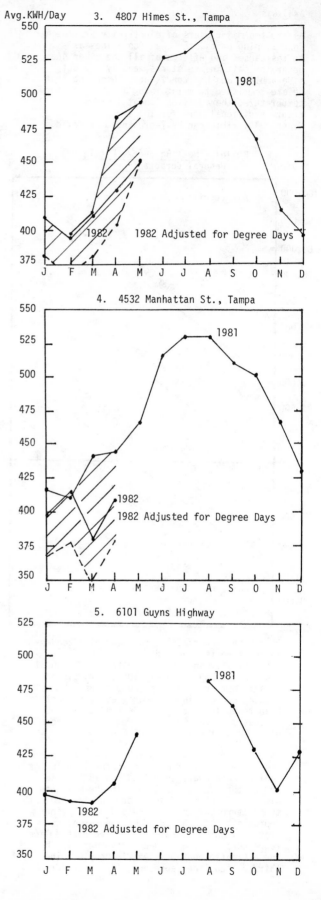

SAVINGS PER YEAR PER STORE			Dollars Saved
1. a. Control AC Unit #1	(estimated)		$ 50
b. Control AC Unit #2	(estimated)		100
c. Front Row of Lights			
5.5 fixtures X 170 watts x 12 hours/day x 365 days/ x .062 $/KWH		=	254
(effective in about 60% of the stores)			
Estimated air conditioning savings @ 50%		=	127
d. Soft drink and beer coolers off at night		=	36
Air conditioning savings		=	18
2. Voltage Reduction on Fluorescent lights			
20 fixtures x 22 watts x 21 hrs/day x 365 days x .062 $/KWH		=	209
All stores air conditioning savings		=	104
3. Replacement to non-energy bulbs with watt misers			
4000 bulbs x $7.70/yr ÷ 450 stores		=	68
Air conditioning savings		=	34
4. 400 W high pressure sodium night lights			
(2000 watts-400 watts) x 6 hrs./day x 365 days x .062 $/KWH		=	217
Air conditioning savings		=	100
(Effective in about 35% of stores)			
5. Install humidity sensors on walk-in coolers			
1 KW x 40% of time x 24 hrs x 365 days x .062		=	217
Air conditioning savings		=	100
Effective in 30% of stores			
TOTAL POTENTIAL YEARLY SAVINGS			$1,634
6. Roof mist cooling (savings potential unknown)			
7. W.A.R. System (savings potential unknown)			

In order to project these numbers into corporate-wide savings, they are factored over the estimated number of stores where the concept is applicable.

	Savings		Store %		# of Stores		Corporate Yearly Savings
1. a. AC Unit #1	$ 50	x	10%	x	450	=	$ 2,250
b. AC Unit #2	100	x	20	x	450		9,000
c. Front row of lights	254	x	60	x	450	=	68,580
Air conditioning*	127	x	60	x	450	=	34,290
d. Coolers off at night	36	x	70	x	450	=	11,340
Air conditioning*	18	x	70	x	450	=	5,670
2. Voltage reduction on lights	209	x	100	x	450	=	94,050
Air conditioning*	104	x	100	x	450	=	46,800
3. Replacement of non-energy bulbs							
4000 bulbs x $7.70 ea						=	30,800
Air conditioning savings*						=	15,400
4. High pressure sodium night lights	217	x	40	x	450	=	39,060
Air conditioning*	100	x	40	x	450	=	18,000
5. Humidity Sensors	217	x	30	x	450	=	29,295
Air conditioning*	100	x	30	x	450	=	13,500
TOTAL POTENTIAL CORPORATE YEARLY SAVINGS							$418,035

*Air conditioning has been factored by 50% as it is most difficult to prove.

In comparing the actual test data (see KWH charts in beginning of this paper) it can be seen that five test stores are producing KWH savings that exceed the above calculations by significant amounts, at the writing of this paper. By September, 1982, when this paper is presented, the data changes will be included.

Hot Sam's Pretzels

Pretzels are baked in a semi-enclosed rotary tray
oven at up to 400°F, taking]7 minutes to cycle
through the oven. This is done in full view of
the customer, for good marketing impressions. The
interior of the oven is heated by five quartz
radiant heaters across the top and five Calrod
heaters across the bottom. The oven draws a total
of 15 KW, when baking and 9 KW when idling.

The five quartz heaters also light the oven interior
so that the customers can observe the oven operation
when pretzels are being baked, and when they are not.
This is a major marketing and point of sales aid.
However, when pretzels are not being baked, the
oven is producing 9 KW of heat needlessly.

To modify this waste, a separate switch was added
which activates only one quartz heater (in the top of
the oven) which provides sufficient light for the
customer to see into the oven. A set of stainless
steel and glass doors were added to retain heat
within the oven.

Prototype models of these changes have produced
results showing $900 per year per store can be
achieved.

158

Chapter 25

DOVER TEXTILES - A CASE HISTORY ON RETROFITTING FACTORIES WITH A BOILER SYSTEM FUELED ON COAL, WOOD AND WASTE

R. D. Pincelli

ABSTRACT

The relentless price escalation of natural gas due to phased natural gas deregulation and the impending $5.00/barrel Federal levy on imported oil means that even with the recent OPEC price declines and temporary oil glut the composite costs of natural gas and fuel oil for firing the typical Eastern United States industrial boiler has held steady or increased slightly in the last 18 months. Coal continues to be, in the Eastern United States, approximately 50% the cost of gas/oil per million BTU's.

Several projects are now well documented with respect to conversion to coal, wood and waste fired boiler systems. These projects include Dover Textiles, Shelby, North Carolina, and Continental Grain, Cameron, South Carolina. Projects under construction include Ti-Caro's - Piedmont Processing, Belmont, North Carolina; Cross Cotton, Marion, North Carolina, a second Dover project at Cherryville, North Carolina; (2) Burlington conversions to coal at Erwin and Cordova as well as a non-textile industrial plant, Anchor Continental, Columbia, South Carolina, burning coal, wood and processed tape waste.

This case history of Dover Textile, Shelby, North Carolina will document a payback of less than 2½ years and a return on investment of 35%; benefits of North Carolina and Federal investment tax credits; EPA considerations, which in this case required no additional capital investment; fuel supply; material handling; ash removal; and other design considerations.

DOVER TEXTILES

Dover Textiles of Shelby, North Carolina, a major producer of finished goods in the textile industry, with dye house and manufacturing plants approximately one-half mile apart, made the decision in July of 1980 to proceed with a boiler plant located between the manufacturing facilities and positioned to serve both plants. This system would be required to burn local wood waste, as well as coal, and backup fuels of gas and oil. Mr. Dave Roberts, Energy Manager at Dover, led the corporation into the strategy of eliminating vulnerability to oil cutoff and in addition, maintain their competitiveness by the ability to generate steam at approximately 50% of the former cost.

Dover Textiles has three plants, J & C Dyeing, Dover Mill Company and Ora Mill Company located approximately 2,000 feet apart. The project calls for J & C Dyeing and Dover Mill to be tied into a central solid fuel boiler facility near the J & C Dyeing plant, which will require approximately 25,000 - 30,000 lbs/hr. of steam with the remainder of the 42,000 lbs/hr capacity being available for Dover and ultimately for Ora Mill when that tie-in is accomplished at a later date under a Phase Two Construction program.

The former boilers that supplied all three plants are firetube, Scotch Marine, gas/oil only boilers. The three plants had seen massive increases in the cost of their oil and gas over the last five years, as has all southeastern industry. Current cost of No. 6 oil is $.70 per gallon and natural gas is approaching $5.00 per MCF. Coal, under long-term contract being purchased at the equivalent of $2.05 natural gas and $.31 per gallon No. 6 oil on a delivered basis. Wood waste is available at approximately $1.30 per million BTU. After allowing additional maintenance, electrical load for intermittent conveyors, as well as some additional labor, the payback period for Dover is 2.1 years. Substantial tax credits from North Carolina and the Federal Government of 27½% plus were obtained. These tax credits are 10% investment and 10% energy tax credits from the Federal Government and 15% investment from the North Carolina Department of Revenue. One half the North Carolina investment tax credit was off-set by Federal taxation, therefore; 7½% net North Carolina investment tax credit was used. The ROI on this project exceeds 35%. Obviously, each passing year the project becomes more economically attractive to Dover as the costs of gas and oil continue to outpace the cost of coal and wood by 10-15% per year. After tax credits, Dover's total cash requirement was approximately $1,000,000.

Dover Textiles made the decision to install a multi-fuel boiler system capable of burning coal, wood waste (both dry and green), selected textile and plant waste, as well as gas, oil, and propane/air mixtures after observing units already in operation for Standard Knitting Mills of Knoxville, Tennessee, Division of Stanwood Corporation in Charlotte; Spring Mills unit at Calhoun, Georgia; Mount Vernon Mills of Greenville, South Carolina, the Tallassee Mills Division in Tallassee, Alabama, as well as Steel Heddle Corporation at Greenville, South Carolina currently burning coal, wood and oil.

The savings from this project at the current cost of gas and oil is nearly $500,000.00 per year. The Dover solid fuel boiler system consists of two T-600P CNB multi-fuel boilers. The design pressure is 150 PSI saturated and the operating pressure is approximately 130 PSI. These boilers generate, on a combined basis, 34,500 lbs/hr of steam when fired on coal, wood waste, or other solid fuels. When "over-fired" with gas, oil, or propane, the two units are capable of generating a combined steam output of 42,000 lbs/hr. The units are firetube with water walls. Efficiency is guaranteed at 78½% on coal without economizers or blow-down heat recovery.

Turn-key delivery at current industry activity is approximately 12 months. Complete EPA compliance was guaranteed by the design and build contractor, ESI,

Chattanooga, Tennessee, as well as operational satisfaction of the material handling system for all types of fuel specified. Site preparation, soil analysis, concrete buildings, material handling equipment, ash removal equipment, water side auxiliaries, as well as boilers, piping, wiring, instrumentation and controls were all supplied thru the sole-source contractor eliminating substantial involvement, time and expense on Dover personnel, as well as eliminating problems between equipment suppliers and installing contractors at start-up time.

Relative to financing of this project, Dover elected internal financing. However, several "shared-savings" programs were offered to Dover. One program would have essentially worked as follows:

The investor group would have paid for the entire installation at Dover.

The investor group would retain all investment tax credit benefits and depreciation tax benefits.

Dover and the investor group would share equally in the fuel savings as documented by historic gas/oil use and the actual coal, wood, and waste use.

The investor group would take 50% of these savings for a period of between 7 and 10 years. At the end of this time the entire project would be turned over to Dover at no cost, meaning essentially that Dover could have had, without capital investment, 50% of the fuel-savings for the first 7-10 years and then obtained a multi-fuel boiler system at no cost.

Obviously, these financial programs are structured for the benefits of both parties and in-depth tax and other ramifications must be studied. We also participate in programs offered by Manufacturers'-Hanover Bank and others in which lease-purchase options are employed to maintained positive cash flow throughout the life of an investment. If further details on either of these financial options are of interest, please contact the author at E.S.I.'s Chattanooga office.

Environmental considerations on the Dover project were:

1. Particulate Emissions

2. SO_2 Emissions

3. Opacity.

Applications were submitted to the North Carolina environmental authorities and permits have been issued to construct and operate without clean-up equipment. This is possible, primarily due to the combination of the underfed ram stoker and the conservative furnace design of the boiler, which is inherently capable of very low particulate emission rates. To date, over 80 of these multi-fuel boilers have been installed without clean-up equipment. Each site must be evaluated on its own individual characteristics, including emission off-sets from prior sulphur content of the oil, ambient air quality conditions, BTU input, etc. Again, a sole-source contractor taking responsibility for total construction, as well as EPA compliance, has considerable advantage to the client.

Material handling at the Dover site is kept simple. A large bunker is filled by front-end loader with coal or wood. The bunker, which has a live bottom unloader to a primary "u-trough" feed screw, automatically maintains the desired fuel level in each stoker hopper. The Dover system is designed for expansion to a third

multi-fuel unit, which will boost capacity to over 50,000 lbs. on solid fuel and over 60,000 lbs. when over-firing with gas, oil or propane. This over-fire burner capability is not required for EPA compliance. It's purpose is two-fold:

1. To allow rapid and automatic fluctuations in steam demand to be taken with liquid and gas fuels which have a quicker response characteristic than do solid fuels.

2. Should a problem develop in the solid fuel feed system, the liquid/gas fuels can automatically supply needed steam for plant operations during repair of solid fuel systems.

The Dover system will utilize solid fuels, i.e. coal, wood, and selected plant waste to maintain base load conditions anticipated to be about 30,000 lbs/hr. The liquid and gas fuels, i.e. natural gas and fuel oil, will be used to make up any sudden swings in load on an automatic, quick response basis. The burners are not continually operating, but only come on a decay in pressure. This feature makes the gas/oil fuel consumption minimal. The fuels which are currently being contracted for by Dover include 50% moisture content green wood in the form of sawdust and mill residue, as well as wood from local furniture sources; coal, via railroad delivery to an existing railroad facility on the Dover property. Textile waste is also being considered as a fuel. Dover Textiles anticipates being able to eliminate waste haul-away costs, land-fill charges, and potential EPA water contamination problems by burning this material in the multi-fuel boiler system. More development work is required, for the processing and introduction of this fuel source to the boiler system.

Ash removal is semi-automatic with the operator manually dumping the ash grates, raking ash to a drag chain ash removal system, which then automatically deposits this material in a truck supplied by a local concrete block manufacturer. This block manufacturer uses the ash as a filler material for his product. The county and state highway department is interested in utilization of the ash as a road anti-ice media during the winter.

Labor input on the Dover project is approximately 4 hours per shift from one boiler operator. This involves dumping ash one to two times per shift, monitoring all systems, water softener regeneration, logging of critical temperature, pressures, etc., routine maintenance and stoker bed maintenance depending on the coking tendencies of the coal. Skill level of the operator would be that of a typical $5.00-$6.00 per hour textile worker with normal aptitude and job interest.

Space requirements for the entire Dover project including boiler room, fuel storage bunker, ash removal facilities, etc, is approximately 100 x 200 ft. outside dimensions. Fuel storage will include approximately 2-3 weeks of storage on-site of combined wood and coal.

Additional reasons that Dover Textiles chose E.S.I. of Chattanooga, Tennessee to do the turn-key design and build construction on this central solid fuel boiler facility included numerous Carolina jobs completed on the same basis and in the same approximate size range. These jobs included Monarch Furniture Company, High Point, North Carolina, a 21,000 lbs/hr unit burning wood and coal; and Steel Heddle in Greenville, South Carolina, a 12,000 lbs/hr unit burning wood, coal, gas and oil; and Continental

Grain, Cameron, South Carolina a 42,000 lbs/hr unit.

An interesting observation in the simultaneous mixing and burning of wood waste with coal is that at several current projects, this combined burning has proven that low grade, inexpensive types of coal can be fired in the boiler. The technical reasons include the fact that the wood has "high volatility" and burns out prior to carbon burn-out in the coal, thus leaving a very porous bed in which oxygen from the underfire air fan can readily finalize combustion of the carbon remaining in the coal. Again, many of these technical advantages are available from a turn-key contractor experience with this size and type of installation and fuel utilization that would not be available from other groups.

Our company, in conjunction with Dover, is also exploring the possibilities of mechanical cogeneration. This would involve utilization of steam turbines to replace large continuous running electrical motors, such as those found on air compressors, chillers, etc. In this case, 125 PSI steam would be introduced to the turbine, 15 PSI discharge steam would be available from the turbine. This 15 PSI steam could then be used for plant heat, process water heating, absorption air conditioning, etc., with most of the heating benefit still available. The payback on this type mechanical cogeneration is approximately 12-18 months.

An economizer, which can be lowered directly down the existing stack will be considered as a Phase 2 add-on project when the economics dictate. Design and construction consideration is given to both very adequate soot blowing and pressure drops on the water and flue-gas side to prevent any build-ups or corrosion from sulphur or vanadium. In addition, "cold-end" corrosion will also be specifically designed out of the system.

The cost of the Dover system including over 2,000 ft of 6 inch and 500 ft of 8 inch steam tie-ins between J & C Dye House and Dover Mills was less than $1,500,000. Dover saw a 50% reduction in boiler operating costs and this reduction includes the increased labor, preventive maintenance and electrical requirement imposed by the multi-fuel boiler system.

These systems make excellent technical and economic sense for intermediate size plants requiring steam and currently burning gas and oil on a 4,000 hour per year basis or more. The attractiveness of the investment increases each time OPEC, natural gas deregulation, curtailments, and other supply and price problems impact on the gas/oil fuel costs. Not only are the economics outstanding, but the competitive edge offered by reducing gas/oil costs and eliminating potential gas or oil cut offs means that Dover will not get caught with their "plants down".

Chapter 26

COOLING TOWERS -
THE OVERLOOKED ENERGY CONSERVATION
MACHINE

R. Burger

We are all aware that cooling towers are the stepchildren of the chemical process plant, electric power-generating station, and refrigeration system.

While our engineers are pretty well convinced of the importance of their sophisticated equipment, and rightly so, they take the cooling towers and cold water returning from them for granted.

Design conditions are specified for the particular requirements before a cooling tower is purchased. After it becomes operational and the cold water temperature or volume becomes inadequate, they look to solutions other than the obvious. While all cooling towers are purchased to function at 100% of capability in accordance with the required design conditions, in actual on-stream employment, the level of operation many times is lower, downwards to as much as 50% due to a variety or reasons:

1. The present service needed is now greater than the original requirements which the tower was purchased for.

2. "Slippage" due to usage and perhaps deficient maintenance has reduced the performance of the tower over years of operation.

3. The installation could have been originally undersized due to the lower bidder syndrome (1).

4. New plant expansion needs additional water volume and possibly colder temperatures off the tower.

OPTIMIZE PERFORMANCE BY RETROFIT

This article does not inquire into specifications and testing of performance of new towers, but will investigate the answer to the question: "We have this piece of equipment which is a part of the production or refrigeration chain. It is not giving us a sufficient level of cold water. What can we do to remedy this situation?"

In order to upgrade a cooling tower to produce higher levels of colder water which will conserve energy with it's utilization, let us briefly review the elements involved and compare the conventional construction with modern day technology.

There are three major elements of the cooling tower we will investigate, whether it be counterflow or crossflow, which consists of:-

1- Air handling
2- Water distribution
3- Heat transfer surfaces

The upgrading of these elements will be investigated in order of ascending costs and then case histories will be presented to document the rapid return of investment (ROI) by optimizing the performance of the cooling tower.

AIR HANDLING

a. The least expensive fix to improve air handling and increase the volume is by pitching blades to a maximum angle consistant with the motor plate amperage requirements.

b. Installation of velocity regain (VR) fan cylinders will generate approximately 7% more air due to relieving the exit pressure which the fan works against. This brings into effect Bernoulli's principle which states that increasing the velocity of a fluid will reduce it's pressure.

c. In conjunction with the VR stack, air flow can be enhanced if a right angle fan deck and fan cylinder are removed and an eased inlet is provided for.

d. Providing a larger motor and/or fan should be considered to increase the air flow to obtain higher levels of performance.

WATER DISTRIBUTION

a. Crossflow water distribution patterns are set by the flow requirements and orifices (holes) located in the hot water distribution basin which are fixed in position. The conventional orifice drops the solid column of water on a lattice work of wood which splashes the water about the top of the fill. A higher level of performance can be obtained by removing the splash deck and replacing it with efficient cellular redistribution decks or target orifice nozzles. Both of these procedures will

FIG. 1   SPLASH BARS WHERE WATER DROPLET BOUNCES FROM SLAT TO
SLAT COOLING EXTERIOR OF DROP ONLY.

provide a more uniform distribution at
the top of the tower thereby utilizing
the entire height for cooling rather than
a portion of it for water breakup resulting in a net lower temperature.

b. Water troughs or enclosed flumes in
counterflow towers should be changed to
a low-pressure spray system.

c. Existing spray systems can be greatly improved by installing noncorroding PVC
piping in conjunction with nonclogging,
noncorroding square spray ABS plastic
nozzles.

HEAT TRANSFER SURFACES

a. The most dramatic improvement in the performance of the tower can be obtained by
installing, on a retrofit basis, cellular
fill, available from a number of manufacturers.

b. Whether it be counterflow or crossflow
configurations, change out to high-efficiency, dense-film fill also known as
cellular fill, can result in improvements as much as 10 to 15°F (5.5 to 8.4°
C), colder water which could equal upwards to 50% or more increase in capacity of the existing tower.

c. The main factor in this extraordinary
improvement is that in utilizing conventional splash bars, the water must bou-
nce from slat to slat on it's way from
top to bottom of the crossflow or counterflow tower and the exterior of the
droplet is cooled, Fig. 1. If a high enough vertical travel is engineered, the
water will be cooled to design conditions. Bear in mind, however, the more
wood that is put into the tower, the
higher the static pressure loss and
larger horsepower motor and fan diameters are required.

FIG. 2   CELLULAR FILL PRINCIPAL STRETCHES WATER INTO THIN FILM
THEREBY COOLING MORE RAPIDLY AND COLDER

CELLULAR FILL PRINCIPLE. Cellular fill works on the principle that the droplet of water is stretched into a thin film as the water proceeds vertically downward through the cells, thereby permitting the available crossflow or counterflow air to cool the entire droplet more rapidly, Figure 2.

Even though the static pressure per cube of cellular fill is much higher than the static pressure per cube of wood fill, considerably less depth of cellular fill is required to produce the same or greater cooling results, thereby generating less total pressure drop which will provide a greater available volume of usable cooling air.

    d. While the fill upgrading changeout from wood splash bars to cellular could be the most expensive retrofit procedure, it produces the highest level of improvement in performance which will provide for a rapid ROI.

ROLE OF COLDER WATER

The role of the cooling tower is to remove waste heat in the chemical or petrochemical reaction. In a condensation process, it is obvious that colder condensing water temperatures will increase unit production at a lower unit cost. The degree of elevation of the process temperature above ambient conditions is the sum of the tower's approach of the wet bulb temperature, the cooling range (which equals the temperature rise in the heat exchanger), and the terminal difference in the exchanger. A reduction in operating temperature, always desirable for economic reasons, may be obtained by increasing the capability of the cooling tower's performance. (2)

The importance of colder water for the compression of gasses is evident in that all compressors have one thing in common: a major portion (80%) of the energy is converted to heat. This rejected heat must be continuously removed at the same rate it is generated or the compressor will overheat and shut down. Reducing the operating temperature of the compressor will proportionately reduce the energy input to the equipment, the less energy is required to produce the same degree of work at lower costs (3).

In a refrigeration system, whether it be for process cooling or room temperature reduction (comfort cooling), colder water requires less energy to operate this system. Enthalpy charts indicate that for every degree of colder water returning from the cooling tower within the operating range of the compressors, 3 1/3% less energy will be required by the compressor to produce the same cooling results (4). Thus a cooling system comsuming $350,000.00 of electric or steam energy to power the turbines can reduce this utility cost by approximately $46,666.00 per year for a 4°F (2.2°C) colder water from the cooling tower.

REFINERY EXPANSION

This case history clearly indicates the importance in value of the utilization of colder water from a well-engineered and rebuilt cooling tower functioning at optimum levels of performance.

Design conditions of the two cell counterflow tower were to originally cool 6,500 GPM (2,460 L/min) from entering the tower at 108°F (42.2°C) leaving at 90°F (32.2°C) during an ambient wet bulb temperature of 78°F (25.6°C). The manufacturer's data and associated fan curves indicated that two 18 ft (5.5m) diameter air foil aluminum fans utilizing 40 HP (30 Kw) drivers will produce 300,000 ACFM of air (8,500 M^3/min) per fan. This was designed to provide a delta T of 18°F (10°C) with a 12°F (6.6°C) approach to the wet bulb.

During the field inspection of this unit, the cooling tower Engineers interviewed operating personnel who indicated at most, the tower was producing 70 to 75% capability at full heat load.

Pullman-Kellogg was awarded a contract to double the capacity of the Alkylation Unit and determined quite early in the engineering phase of the program that the cooling tower was indeed a weak link in the process. To yield the projected quanity of increased product, 10,500 GPM (3,974 L/min) of water, approximately 60% greater capability of the existing 6,500 GPM (2,460 LMP) rate must be circulated through the system, which initially indicated erection of a new tower.

However, new tower delivery, scheduling, plot plan constraint problems and budgetary requirements dictated that if the existing tower could be rebuilt to accommodate the newer capacity, considerable savings in time, real estate, and dollars would accrue over the alternative installation of new OEM equipment.

In order to provide sufficient cooling water for the doubled output of the new projected requirements, the old 6,500 GPM (2,460 LPM) tower would have to be upgraded to 10,500 GPM (3.974 LPM) of water from entering at 108°F (42.2°C) leaving at 90°F (32.2°C) at a 78°F (25.6°C) wet bulb.

It is also evident that if the additional gallonage was not added to the tower after retrofit, the rebuilt facility could now cool the 6,500 GPM (2,460 LPM) from 108°F (42.2°C) to 84°F (28.9°C) at 78°F (25.6°C) wet bulb indicating a new capability of providing a 24°F (13.3°C) delta T with a 6°F (3.3°C) approach to the wet bulb.

A further discussion at management level indicated that a new requirement of doubling the plant's capacity again is being considered which would now require cooling 15,500 GPM (5,866 LPM) of cooling water. Engineering calculations (5) indicate that with additional rebuilding, the existing tower will have the capability of cooling this increased requirement.

While spending considerably more money, it would be cost-effetive when the new cooling water rate could be used to produce approximately 1,000 additional bbl (159 M^3/d) of lead free, high-octane gasoline. The savings in plot area and new construction would more than offset the labor and materials involved in excavating the underground piping to change from 12 to 14 ft (3.7 to 4.3 m) diameter together with installing new 60-HP (45-kW) motors and 20 ft (6.1 m) diameter fan blades required

NOTE CLOGGED INOPERATIVE
NOZZLE

FIG. 3   INEFFICIENT WATER DISTRIBUTION THROUGHS AND LOW HEAT TRANSFER
WOOD SLATS.

to generate sufficient cooling air.

This clearly illustrates the viability of
modern retrofit techniques which can upgrade
this original design from cooling 6,500 GPM
(2,460 LPM) to a new level of 15,500 GPM
(5,866 LPM).

The upgrading and modern retrofit consisted of
changing the water distribution from old-
fashion water troughs and inefficient wood
splash bar decking fill, Figure 3, energy-
absorbing, high-static pressure wood drift el-
iminator blades, to high-efficiency PVC cellular fill, Figure 4, and PVC non-corroding
water distribution system utilizing large
orifice [1¼ in. (3.175 cm) diameter] ABS non-
clogging square spray pattern nozzles.

The installation of the high thermal transfer,
cellular-fill surfaces provided over a 50%
increase in capability, new water distribution
patterns were extremely uniform, and the new
cellular drift eliminators functioned at lower
static pressure levels.

CHEMICAL PLANT CAPACITY INCREASE

A study was performed by the operating Engineering Department of a monohydrate soda ash
plant facility. Soda ash is a white crystalline material that forms a mildly basic solution when dissolved in water. It is used primarily in the soap industry, where it is employed to saponify oils for soap and as a
water softener in detergents. Soda ash also is
a major ingredient in glass manufacturing

FIG. 4   HIGHLY EFFICINET NON-CORRODING NON-CLOGGING SQUARE SPRAY
PATTERN NOZZLES.

where it serves as a fluxing agent.

The product, trona ore, is refined from deposits located between 800 ft (243.8 m) and 1,500 ft (457.2 m) below ground level. The ore that is mined is about 90% pure sodium sesquicarbonate (a double salt of sodium carbonate and sodium bicarbonate). Due to the increases in price of both energy and raw materials, the synthetic process for making soda ash is becoming too expensive to operate. As this condition has worsened, the importance of the natural soda ash industry has increased dramatically.

The process most used for converting trona into refined soda ash is the monohydrate process. In the mono-process, the trona ore is brought into the plant, crushed and sent to a calciner. In the calciner, sodium bicarbonate, which is very corrosive to mild steel, is converted to sodium carbonate by driving off CO_2 and water under high temperatures.

The calcined ore is then dissolved and the resulting liquor, which is about 30% by weight sodium carbonate, goes through clarification and filtration steps to remove any insoluable material. Filtered liquor is pumped to an evaporator where water is driven off to make a supersaturated solution which precipitates out sodium carbonate monhydrate crystals. A slurry of the crystals is sent to a cyclone where it is concentrated and then dewatered in a centrifuge. The final step in the process is the drying out of the product in a fluid bed dryer to give anhydrous soda ash.

The evaporators are multiple-effect evaporators, which allows the system to be operated with boiler-generaged steam fed in only at the first-effect evaporator. Steam for the second effect comes from evaporated water in the first effect, and steam for the third effect comes from evaporated water in the second effect.

This highly efficient usage of steam is made possible because the second-and third-effect evaporators are run at pressures lower than atmospheric. The reduced pressures come primarily from the rapid condensation of third-effect vapors by cold water in a barometric condenser.

Studies have conclusively shown that the lower the cold water temperature, the more vacuum can be drawn on the evaporators, and ultimately more product can be put out on the production silo feed belt. Because of the tests and the studies that have been conducted by plant personnel, a much greater interest is being taken in the operation of the plant cooling towers. By lowering the temperature of the water returning to the barometic condenser during the warm summer months, a serious bottleneck to plant production can be removed.

Data collected during the spring and summer of 1980, verifies that the relationship between plant production and cold water temperature returning from the cooling tower is linear.

Records indicate that the plant cooling towers

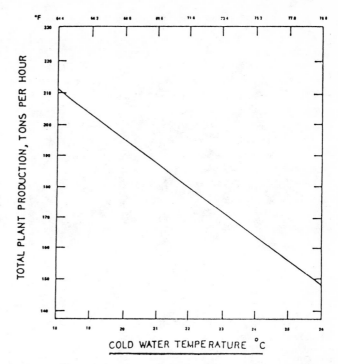

FIG.5. MONO PLANT PRODUCTION AS A FUNCTION OF COOLING WATER TEMPERATURE

produced 80°F (2,617°C) water during the summer months when the ambient wet bulb temperature was around 58°F (14.4°C). The potential increase in production that could be made if water were available at 70°F (21.1°C) would be worth approximately $1,000,000.00 additional for the summer months alone, with all other operating conditions and energy input remaining the same. Figure 5.

During the rest of the year, this increase in cooling capacity could support an estimated 20% expansion in total plant production. These significant increases in production could be engineered and purchased for a capital expenditure of around $700,000.00, chargeable to the maintenance budget for rebuilding the cooling towers, without any increase in manpower or energy requirements. (6)

It is the development of these statistics that have caused management, even in this time of tight money, to take a long, hard look at what can be done to the plant cooling towers to reclaim this currently lost production potential and reduce imput process energy.

The existing cooling tower complex is 20 years old, Figure 6, but well constructed of Clearheart Pressure Treated California Redwood. However, the state of the art then dictated enclosed water troughs for distribution which have become considerably clogged up and malfunctioning, wood slat splash bar fill, and heavy, two-pass drift eliminators. The water splash pattern in the cold water collecting basin indicated broken areas of fill and clogged water distribution orifices, Figure 7.

FIG. 6   OLDER BUT SUBSTANTIALLY BUILT CALIF-
ORNIA REDWOOD TOWERS PROVIDE A GREAT
POTENTIAL FOR THERMAL UPGRADING

Original design conditions which were never
met, were to cool 13,500 GPM (5,110 LPM) of
circulating water entering the tower at 122°F
(50°C) leaving at 65°F (18.3°C) during an
ambient wet bulb temperature of 58°F (14.4°C).
This 57°F (13.9°C) range and 7°F (3.8°C)
approach to the wet bulb was to be accomplished by one tower having an 18 ft (5.5 m) diameter fan and 60 HP (45-kW) driver and the
second tower with a 24 ft (7.3 m) diameter fan
utilizing a 124 HP (93-kW) driver.

Records maintained by operating personnel
indicate that the best average approach to the
wet bulb was 17°F (9.4°C) or a 10°F (5.5°C)

deficiency from design. Retrofit of the cooling tower is required to achieve a 10°F (5.5°
C) colder water discharge and can be accomplished by engineering and installing noncorroding PVC piping, ABS nonclogging wide-orifi-
ce square spray nozzles and high-density,
high-efficiency cellular fill.

It is obvious that the colder the water to
the barometric condensers, the more product
can be produced making the ROI a rapid, cost-
effective payback within a few months.

IN CONCLUSION

The cooling towers are hidden bonanzas for
energy conservation and dollar savings when
properly engineered and maintained. In many
cases, the limiting factor of production is
the quality and quanity of cold water coming
off the cooling tower. The savings accrued
in energy conservation and additional product manufactured can be an important factor
on the operator's company's profit and loss
sheet.

Energy management analysis is a very important
consideration in today's escalating climate of
costs of energy. It is advisable to consider
a thorough engineering inspection and evaluation of the entire plant to leave no stone
unturned in the search to reduce energy consumption (7). The cooling tower plays the major
role on waste heat removal and should be given
a thorough engineering inspection and evaluation by a specialist in this field. This
can be performed at nominal cost and a formal
report submitted with recommendations, budget
costs, and evaluation of the thermal, structural and mechanical condition of the equipment.

FIG. 7   UNIFORM WATER SPLASH PROBLEM INDICATES A WELL OPERATING HIGHLY
EFFICIENT COOLING TOWER

168

This feasibility study will assist in determining the extent of efficiency improvement available with costs and projected savings.

It can be stated that practically all cooling towers can be upgraded to perform at higher levels of efficiency which can provide a rapid, cost-effective payback. However, while all cooling tower systems might not provide such a dramatic cost payback as these case histories, the return of a Customer's investment in upgrading his cooling tower can be a suprising factor of operation and should not be neglected.

LITERATURE CITED

1. "Energy Considerations in Cooling Tower Application," The Marley Cooling Tower Co. (April 22, 1976)

2. "ASHRAE Equipment Handbook," Chapter 21, p. 11 (1975)

3. Stoven, K., "Energy Evaluation in Compressed Air Systems," Plant Engineering (June, 1979)

4. "The Pressure Enthalpy Diagram: It's Construction, Use and Value," Allied Chemical Corp.

5. Burger, R., Chapter 8, "Cooling Tower Technology" (1979).

6. McKee, G., "Results of Mono I and Mono II Cooling Tower Performance Evaluation Tests," Report to Management, Industrial Chemical Division.

7. Mathur, S., "Energy Today," Hydrocarbon Processing (July, 1980)

Chapter 27

A JOINT ENERGY CONSERVATION EFFORT BETWEEN A UTILITY AND A MANUFACTURING PLANT

P. Payne

In 1969 the American Cast Iron Pipe Company, Birmingham, Alabama, purchased the Waterworks Valve & Hydrant Division of the Darling Valve Company in Williamsport, Pennsylvania. The purchase included a small operation located in Beaumont, Texas, later to become known as American Valve & Hydrant Manufacturing Company (AVAH). After that acquisition the plant was expanded and renovated, and today it is a modern manufacturing facility representing $10 million in investment and 257 total employment.

Products manufactured at this Beaumont plant also were broadened. To the line of gate and swing check valves in sizes from 2" through 12" there was introduced in 1980 the "CRS" gate valve -- compression resilient seated -- in 4" through 12" sizes, for water and sewage services. Fire hydrant production at AVAH now ranges from 4" through 6", and it is at this plant that the largest known trench-depth hydrant was built for construction of the Alaskan pipeline.

AVAH markets its products locally as well as throughout the world. The waterworks valves and hydrants are primarily used in municipal water systems, and the industrial lines are used in industrial plants for fire protection. Many of the refineries and petrochemical plants in Texas and Louisiana use the American hydrant exclusively.

Electrical service to the AVAH plant is by way of an overhead 13,200 volt line. Gulf States Utilities (GSU) meters the plant at the 13,200 volt level, thus AVAH is responsible for voltage transformations, internal plant distribution, and all equipment protection schemes within the plant. The metering equipment used to determine AVAH's monthly electric bill consists of a KWHr meter, a KVARH meter, a mag-tape recording unit and the associated relays, potential transformers and current transformers used in operating the meters. During 1980 and 1981, AVAH's average KW demand was near 1350, with peaks of 1470 KW and 1512 KW noted.

During 1980, the management at AVAH became increasingly concerned about the rising cost of energy consumption within their plant. AVAH officials sought advice and direction from GSU and state agencies involved in industrial energy conservation efforts, and a walk-through survey of the plant was performed by an energy engineer.

Although much useful data was obtained during the course of the walk-through, and thereafter by billing histories obtained from the plant's energy suppliers, there were concerns about the limitations presented by having only one metered point to analyze AVAH's electrical consumption. Efforts to resolve these concerns resulted in the development of a joint energy conservation research project involving the utility and the plant. The extent to which GSU agreed to participate in the program was as follows:

1. GSU would furnish all potential transformers, current transformers, and the necessary load monitoring devices for three test locations designated by the engineer.

2. AVAH would take care of the installation expenses.

3. The utility's industrial representative and the plant's engineer would provide overview for the project, and data obtained would be analyzed by GSU's computer.

There are several types of load monitoring devices available on the market today. This made the choice of metering equipment a question of some pertinence to achieving the desired results. After much deliberation it was agreed upon that GSU would furnish mag-tape metering units at each of the three in-plant locations.

FIGURE 1. GSU'S SUB-METERING PANEL

This decision provided two distinct advantages over other load monitoring equipment:

1. It allowed GSU to obtain data which was compatible with the utility's existing system for recording industrial customer usage data, and

2. It accommodated the production of valuable computer-generated information using GSU's in-house equipment capabilities for data processing. The computer could read, analyze and produce printouts in various formats from data on the mag-tape recordings each month.

This latter advantage was most beneficial to the project. The computer could select from the myriad of daily load information data those readings for meaningful peaks at each metered location.

The location of the three survey meters was also a significant part of the project. The functional distribution of power within the AVAH plant -- among the major manufacturing operations -- had been estimated by the energy engineer's initial survey, and was relied upon to give the approximate percentage breakdown of electrical consumption. The engineer's original pie chart is shown below as Figure 1. This preliminary analysis was sourced from the equipment inventory and operating data gathered during the walk-through survey, using the plant's monthly billing records for KWH usage and costs.

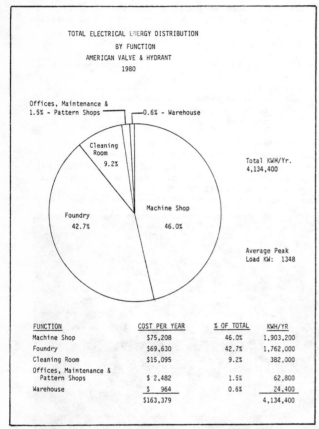

FUNCTION	COST PER YEAR	% OF TOTAL	KWH/YR
Machine Shop	$75,208	46.0%	1,903,200
Foundry	$69,630	42.7%	1,762,000
Cleaning Room	$15,095	9.2%	382,000
Offices, Maintenance & Pattern Shops	$ 2,482	1.5%	62,800
Warehouse	$ 964	0.6%	24,400
	$163,379		4,134,400

FIGURE 1.   TOTAL ELECTRICAL ENERGY DISTRIBUTION, BY FUNCTION, ESTIMATED PRIOR TO SUB-METERING.

It was determined from this information that a survey meter would be located at the machine shop, another at the foundry, and a third on the cleaning room.

This was a primary objective -- solid quantification of usage -- and should be an aid to management at AVAH who now knows how much money is spent each month on power to the various plant operations. Accuracies in calculating the payback for various energy conservation measures are prequisites for implementation decisions.

The three survey meters were installed in September 1981 and data has been obtained since October 1981. All participants feel that the information which has already been obtained and the information which will be obtained in the future benefits both the utility and plant

management. Information already extracted from the project is listed below.

1. Power Factor Correction -- The power factor at AVAH has been averaging around 77% during the past two years. The management has considered the merits of bringing this factor up, at least to a point exceeding the utility's penalty level, which on GSU's General Service rate schedule is specified to be 80%. The plant, in other words, would have to maintain power factor averages above 80% in order to avoid being penalized for poor power factor on their monthly electric bill. Although power factor correction is not a difficult task to achieve, the optimum placement of capacitors to improve power factor can be difficult to determine. By analyzing the data from the three test locations at AVAH, it has been determined that the average power factor measured at the meter located on the machine shop building has been around 65%, while the average power factor at the other locations has consistently averaged above the 80% level. Therefore, if AVAH would like to initiate corrective measures to bring their power factor above 80%, they could install the appropriate size capacitors on the machine shop loads.

2. Functional Distribution of Power -- After reviewing the information from the computer-generated contributor's report, a pie chart could for the first time be constructed which proportioned the actual incremental electrical consumption within the plant. See FIGURE 2, which defines the foundry as using almost twice as much power as the machine shop operation.

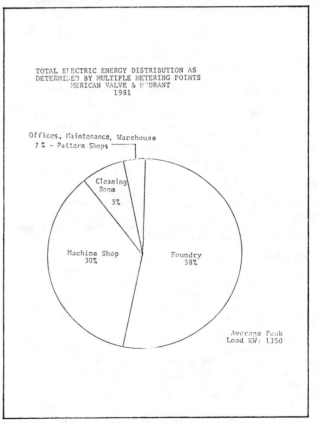

FIGURE 2.   TOTAL ELECTRICAL ENERGY DISTRIBUTION, AS DETERMINED BY MULTIPLE METERING POINTS.

The computer-generated data used to construct the pie chart of actual distributionwas sourced from readings from the three survey meters. Called "Contributor Report", this data is exemplified in FIGURE 3.

Another report entitled "Peak Summary" provides a tabulation of the six (6) highest peaks occurring each month at each location. See FIGURE 4, which is a report of peaks in location 1209S (the Machine Shop).

CONTRIBUTOR REPORT ON TOTALIZED ID GENE AMER. DARLING VALVE 02/04/82
SELECTED 30 MINUTE INTERVALS

#1 MACHINE SHOP — 1209S
#2 FOUNDRY — 1210S
#3 CLEANING BLDG. — 1211S

DATE TIME (CONTR. ID)	FILE	CONTRIBUTOR CHN	ORIG VAL	B VALUE ST KW	PCT	C VALUE ST KVAR	PCT
02/01/82 10:30 GENE				1287.14	100.0	863.21	100.0
1209S	B5	B	371.84	371.84	28.9	.00	.0
		C	421.76	.00	.0	421.76	48.9
1210S	B6	B	859.95	859.95	66.8	.00	.0
		C	393.30	.00	.0	393.30	45.6
1211S	B7	B	55.35	55.35	4.3	.00	.0
		C	48.15	.00	.0	48.15	5.6
				1287.14	100.0	863.21	100.0
02/01/82 10:00 GENE				1247.44	100.0	843.11	100.0
1209S	B5	B	365.44	365.44	29.3	.00	.0
		C	402.56	.00	.0	402.56	47.7
1210S	B6	B	811.80	811.80	65.1	.00	.0
		C	381.60	.00	.0	381.60	45.3
1211S	B7	B	70.20	70.20	5.6	.00	.0
		C	58.95	.00	.0	58.95	7.0
				1247.44	100.0	843.11	100.0
02/11/82 09:00 GENE				1226.55	100.0	798.03	100.0
1209S	B5	B	364.80	364.80	29.7	.00	.0
		C	391.68	.00	.0	391.68	49.1
1210S	B6	B	753.30	753.30	61.4	.00	.0
		C	307.80	.00	.0	307.80	38.6
1211S	B7	B	108.45	108.45	8.8	.00	.0
		C	98.55	.00	.0	98.55	12.3
				1226.55	100.0	798.03	100.0
02/11/82 08:30 GENE				1221.98	100.0	818.76	100.0
1209S	B5	B	366.08	366.08	30.0	.00	.0
		C	399.36	.00	.0	399.36	48.8
1210S	B6	B	748.35	748.35	61.2	.00	.0
		C	314.10	.00	.0	314.10	38.4
1211S	B7	B	107.55	107.55	8.8	.00	.0
		C	105.30	.00	.0	105.30	12.9
				1221.98	100.0	818.76	100.0
02/12/82 10:00 GENE				1214.77	100.0	738.43	100.0

FIGURE 3. COMPUTER-GENERATED REPORT PROVIDING BREAK-DOWN OF PLANT POWER USAGE EACH LOCATION

PEAKS SUMMARY ID: 1209S 891101405.1 03/02/82
AMER. DARLING VALVE HOLLYWOOD, BMT., TX.

FROM 01/29/82 09:49 TO 03/01/82 14:00 PRE-SUMM 30 MIN.

6 HIGHEST KW	KW	KVAR	KVA	PF	85% KVA
02/05/82 10:00	392.	414.	570.	.6872	484.
02/04/82 13:30	391.	396.	557.	.7025	473.
02/05/82 13:30	391.	376.	543.	.7205	461.
02/04/82 08:00	390.	397.	557.	.7013	473.
02/05/82 08:30	388.	360.	529.	.7332	450.
02/05/82 07:30	387.	348.	520.	.7442	442.

6 HIGHEST KVAR	KW	KVAR	KVA	PF	85% KVA
02/23/82 11:00	375.	438.	576.	.6506	490.
02/23/82 10:00	367.	435.	569.	.6449	483.
02/22/82 11:30	370.	433.	570.	.6493	484.
01/29/82 12:00	383.	426.	573.	.6681	487.
02/22/82 11:00	362.	426.	559.	.6481	475.
02/18/82 15:00	353.	425.	553.	.6393	470.

6 HIGHEST KVA	KW	KVAR	KVA	PF	85% KVA
02/23/82 11:00	375.	438.	576.	.6506	490.
01/29/82 12:00	383.	426.	573.	.6681	487.
02/05/82 10:00	392.	414.	570.	.6872	484.
02/22/82 11:30	370.	433.	570.	.6493	484.
02/23/82 10:00	367.	435.	569.	.6449	483.
02/01/82 10:30	372.	422.	562.	.6613	478.

	KWH	KVARH	KVAH	AVG PF	85% KVAH
TOTAL USAGE	96080.	107505.	144183..	.6664	122556.

FIGURE 4. SUMMARY OF SIX HIGHEST PEAKS OCCURRING IN ONE MONTH (MACHINE SHOP METER)

Records of the plant's daily KW peaks had allowed the energy engineer to include in the initial survey report some graphics which intrigued plant management. One graph of monthly peaks revealed a July peak considerably higher that months adjacent. A second graph showed that peak to have occurred on only one day of that month.

However, records were unavailable to determine the time of day for the peak, and the engineer's survey data could at best provide a guestimate of the peak occurrence. Computer-generated graphics were able to define this with accuracy, displaying the recorded KW and KVAR peaks in 30-minute intervals. FIGURE 5.

```
LOAD GRAPH                            ID: 12093   891101405.1    PRE-SUMM:  30 MIN.              02/05/82 00:01              02/04/82
AMER. DARLING VALVE       HOLLYWOOD, BMT., TX.                                            TO  02/05/82 24:00            PAGE    1

                                                                         PEAKS:  K =    391.7   R =    414.1
  DATE   TIME      KW      KVAR      KVA    PF                            0%         20%       40%       60%       80%       PEAK

02/05/82 00:30    21.12     5.76    21.89   .965                         :R W                 .         .         .         :
         1:00    10.88      .00    10.88  1.000                         RW                    .         .         .         :
         1:30    10.88      .00    10.88  1.000                         RW                    .         .         .         :
         2:00    10.88      .00    10.88  1.000                         RW                    .         .         .         :
         2:30    10.24      .00    10.24  1.000                         RW                    .         .         .         :
         3:00    10.88      .00    10.88  1.000                         RW                    .         .         .         :
         3:30    10.88      .00    10.88  1.000                         RW                    .         .         .         :
         4:00    10.24      .00    10.24  1.000                         RW                    .         .         .         :
         4:30    10.88      .00    10.88  1.000                         RW                    .         .         .         :
         5:00    10.88      .00    10.88  1.000                         RW                    .         .         .         :
         5:30    10.24      .00    10.24  1.000                         RW                    .         .         .         :
         6:00     8.32      .00     8.32  1.000                         RW                    .         .         .         :
         6:30    60.16    31.36    67.84   .887                         :     R   W           .         .         .         :
         7:00   304.00   160.00   343.53   .885                         :                R.             W.                 :
         7:30   387.20   347.52   520.20   .744                         :                          .       R       W       :
         8:00   366.08   357.12   511.42   .716                         :                          .       R    W          :
         8:30   387.84   359.68   528.95   .733                         :                          .       R    W          :
         9:00   384.64   369.28   533.21   .721                         :                          .       R    W          :
         9:30   361.60   331.52   490.57   .737                         :                          .    R   W              :
        10:00   391.68   414.08   569.90   .687                         :                          .               S       :
        10:30   383.36   392.96   548.98   .698                         :                          .           R  W        :
        11:00   382.72   396.00   551.29   .694                         :                          .             RW        :
        11:30   369.28   394.24   540.18   .684                         :                          .            WR          :
        12:00   387.20   387.20   547.58   .707                         :                          .           R  W        :
        12:30   328.96   221.44   396.55   .830                         :                    R          .  W              :
        13:00   365.44   384.64   530.56   .689                         :                          .            RW         :
        13:30   391.04   376.32   542.71   .721                         :                          .           R  W        :
        14:00   366.72   374.40   524.00   .700                         :                          .           R  W        :
        14:30   364.16   384.64   529.68   .688                         :                          .             S         :
        15:00   360.96   385.28   527.95   .684                         :                          .            WR         :
        15:30   298.24   348.16   458.43   .651                         :                          .       W . R           :
        16:00   161.28   199.68   256.68   .628                         :                    .W   R                        :
        16:30   165.76   236.80   289.05   .573                         :                    .W        R.                  :
        17:00   184.96   251.52   312.21   .592                         :                    . W        R                  :
        17:30   176.64   244.48   301.62   .586                         :                    . W        R.                 :
        18:00   170.24   239.36   293.73   .580                         :                    . W         R.                :
        18:30   160.00   227.84   278.41   .575                         :                   W.  R                          :
        19:00   149.76   192.00   243.50   .615                         :                   W. R                           :
        19:30   165.12   238.72   290.26   .569                         :                    .W        R.                  :
        20:00   158.72   231.68   280.83   .565                         :                    . W        R.                 :
        20:30   125.44   146.56   192.91   .650                         :                 W R                              :
        21:00   153.60   221.44   269.50   .570                         :                   W        R.                    :
        21:30   160.00   247.04   294.33   .544                         :                   W              R               :
        22:00   164.48   248.96   298.39   .551                         :                   .W            R                :
        22:30   158.08   241.28   288.45   .548                         :                   W           R.                 :
        23:00   158.72   248.96   295.25   .538                         :                   W          R.                  :
        23:30   155.52   241.92   287.60   .541                         :                   W          R.                  :
        24:00   102.40   149.76   181.42   .564                         :            W    R                                :

DAILY HIGH    391.68   414.08   569.90   .000
DAILY AVERAGE 196.84   213.13   292.81   .742       W = KW                    R = KVAR               S = BOTH
DAILY LOW       8.32      .00     8.32   .538
DAILY TOTAL  4724.16  5115.20
```

FIGURE 5.   THIRTY-MINUTE READINGS OF LOAD DATA AND GRAPHIC PRESENTATION OF POWER TO MACHINE SHOP

Other mutual benefits to be gained from this joint effort between GSU and AVAH are as follows:

1. Determine Peak Shaving Techniques -- GSU feels that after properly analyzing the available graphs, AVAH will be able to determine when simultaneous peaks at various parts of the plant are most likely to occur. Therefore, by properly scheduling the start-up and shut-down sequences of their various process-es, AVAH should be able to successfully reduce the maximum KW of demand reached during any given month. This would in turn allow reduction in the demand portion of their bill.

2. Establish a Base Period -- Once this research effort is complete, the data obtained will serve as a base period for studies in pursuit of energy conservation opportunities.

3. Valuable Load Data -- The data obtained from this study allows GSU to establish a reference source as to the magnitude of load and the type of load cycling that a similar industrial customer might impose on GSU facilities.

4. Productivity Measure -- An indirect application of this study is the measurement of productivity, i.e., units of product per KWH of usage, by function within the plant. This type of analysis could prove to be of value in determining which process is more economical, for cost comparison studies, and for production scheduling.

The feeling on the part of GSU representatives is that the utility should play an active role in assisting and simplfying the search for energy conservation opportun-ities by their customers. Joint energy conservation efforts such as this one can overcome roadblocks that deter and squelch ECO implementation.

SECTION 7
LOAD MANAGEMENT

Chapter 28

DEMAND MANAGEMENT -
AN ALTERNATIVE TO NEW ELECTRICAL
CAPACITY

M. M. Yarosh, D. E. Stein

Introduction

In few other sectors of the U.S. energy industry has the revolution in costs and prices manifested itself as profoundly as in the electric utility industry. This is true in part because of the impressive track record with which the industry entered the decade of the 1970s, including a history over the previous six decades of almost uninterrupted reductions in energy costs. Between 1906 and 1970 the average revenue per residential kwh used decreased from 11.2 to 2.1 cents.[1] This despite the concurrent rise in the Consumer Price Index by approximately a factor of four.

During the majority of this century, utilities participated in a variety of programs designed to encourage expanded electrical use. Such expanded use in turn increased the need for new facilities. Generating unit sizes increased dramatically and the economies of scale contributed to reducing generating costs. New generating capacity had to be committed for construction to meet ever increasing demand. In Florida, between 1960 and 1972 electrical generating capacity increased at the incredible rate of 11 percent annually.[2] Utilities devoted much attention to both stimulating demand and meeting that demand by increasing generating capability. Relatively little attention was given to the management of electrical demand. Rather supply management was the essential tool used to meet such demand. Since new generation provided reduced busbar costs, there was little argument with supply side management. During this period, regulatory bodies were periodically called upon to approve requests for rate reductions as reduced generating costs from newer, larger, more efficient facilities increased the rates of return beyond recommended limits.

Over the past dozen years profound changes affecting the utility industry have occurred. The long term downward trends in electrical prices have been sharply reversed and costs and prices for electrical energy have erased much of the gains of the previous six decades. Rapidly escalating construction costs, high interest rates, long construction lead times and the rapid rise in fuel costs have all combined to drive energy prices to increase at astonishing rates.

Normal mitigating factors to price increases such as economies of scale and improvements in thermodynamic efficiencies had already achieved most of their effects before the 1970s.

*This is an extended abstract of the paper to be presented at The 5th World Energy Engineering Congress - Atlanta, GA, September 15-17, 1982.

These profound effects on utility costs are being reflected by frequent rate hike requests by utilities. Because the marginal generating costs from new capacity are now often significantly higher than average generating costs, approval of new facilities to meet increasing demand almost guarantees that increased rates will be required. Regulatory bodies must therefore consider alternative options which may be more cost effective for meeting such demand and the existence of viable cost effective alternatives to new generating capacity must be a key issue in future regulatory decisions.

Load or demand management (LM) appears to be a viable option which can permit at least a deferral of a portion of otherwise required new capacity and therefore a slowdown in the rate of rise in the cost of electrical energy. Much of a utility's generating capacity is idle much of the time; the effect of LM is to reduce the amount of capacity which, while seldom needed, must nonetheless be built. This option, to be effective, must be aggressively marketed by utilities in much the same way as earlier utility efforts to market electrical sales. Such aggressive commitment has not yet been generally displayed. In a survey of 100 companies early in 1981, only six had filed LM rates. Only half of the companies were even experimenting with LM, yet only 16 had determined that LM was not economical for their system.[3] In addition to the utilities, regulatory bodies can and should encourage such efforts. In Florida load management has received little encouragement from its regulatory body.

Load Management Techniques

Load management has been defined as "influencing the level and state of the demand for electrical energy so that the demand conforms to individual present supply situations and long run objectives and constraints".[4] Such influence can occur as a result of actions initiated by the utility (direct load control), or actions by the customer to alter the demand in such a way that his demand is reduced during peak load periods. Typical of direct load control methods is interruption of residential service through the shutdown or cycling of specific loads such as water heaters, air conditioners or other equipment. Direct load control also includes the exercise of interruptible or curtailable loads in commercial and industrial service in accordance with previous customer-utility agreements, and more widespread actions such as voltage reductions or general load shedding.[5]

†The views expressed herein are those of the authors, and do not necessarily reflect the views of the Florida Solar Energy Center.

Indirect control through actions exercised by the customer can also influence electrical demand. Customer actions can be encouraged by positive or negative incentives such as Time of Use (TOU) rates, or high demand charges which can be placed on the customer. Unlike direct control, indirect control carries a somewhat higher degree of uncertainty as to its effects and such effects may vary significantly between the short run concerns and longer range objectives. That a TOU rate will effect a reduction in air conditioning loads during peak demand hours on an extremely hot day is uncertain. This is an example of where the effectiveness of such rate structures as a load management tool is unclear. (A much stronger justification for application of TOU rates probably lies with the improved approach it presents to a cost based rate structure.) Realistic TOU rates, however, may encourage the development and availability of a wide range of thermal storage technologies enabling customers to make use of TOU rates.

Some load control devices are not controlled directly by the utility or the customer but nevertheless serve to limit the maximum demand which can be placed on the utility grid. Typical of such equipment are residential interlocks which may defer water heating loads to times when air conditioners or heat pumps are not operating or defer other loads to off peak times. For customers in the commercial sector that have a variety of loads, customer control of demand can be instituted and controlled by microcomputers programmed to limit customer loads below set point values in a desired manner and sequence. Pacific Gas & Electric, for example, has instituted a Group Load Curtailment Program in San Francisco and other urban areas in which the group of customers determines the specific loads to be curtailed once the utility has requested a curtailment. Failure to meet agreed upon curtailment carries stiff penalties. This type of management has the advantage of achieving utility demand reduction requirements, but leaving detailed decisions of load management to the customer.

Utilities in their previous forecasting of growth in electrical demand, preferred to err on the side of conservatism. They preferred excess capacity rather than be faced with an inability to meet peak loads. Excess capacity often meant newer base load plants were contructed which had lower marginal generating costs than average costs of existing plants. In times of rapid electrical growth, such growth in demand soon reduced any excess capacity. The profound changes in both costs and rate of growth in demand have dramatically changed this picture. There are now serious penalties associated with excess capacity and strong incentives for seeking ways to avoid building new capacity. Today, when capital is scarce and expensive, and when environmental degradation has economic effects, it can be argued that surplus electrical generating capacity may hurt the economy, particularly if less costly alternatives exist. Many cancellations of planned power generating facilities in the past few years are reflections of the decline in demand and failure of such demand to meet previous predictions. Recent statements resulting from studies by energy consultants suggest, however, that utilities are continuing to forecast anticipated demands above demand forecasts performed by nonutilities. [6]

This overforecasting may reflect the traditional fear of consequences of underestimating actual demand. Load management, however, can be seen as a much less expensive way of hedging one's bet. The costs for load management control devices are but a fraction of the cost of equivalent new generating capacity per Kw of demand. Like power plants, load management provides an assurance that rotating brownouts can be avoided, but its installation purchases such assurance more cost-effectively.[7] Such installations provide a cushion which, if necessary, can be invoked to reduce demand.

Utility Load Management Programs and Costs

Nationwide, utilities are beginning to embrace direct control as well as other load management practices.[8] By the latter half of 1980 utilities had over 500,000 radio control points installed, and implementation is growing so rapidly that today the total of all installed direct control points may be approaching one million.[9] Although radio control systems are by far the most common type of direct control, ripple control, power line carrier systems, and various hybrid systems are also extensively used. Radio control systems represent perhaps the simplest, most mature technology and among the lowest in cost per control point.

Many individual utilities have in excess of 5,000 control points; fewer have in excess of 50,000 points. Detroit Edison which controls approximately 200,000 water heaters in Michigan operates perhaps the largest and one of the oldest (1968) residential direct load control systems.[10] Data from the many load management programs already operating suggest installed cost ranges for per point direct load control vary from approximately $80 to $200 when the control points number in the thousands.[7] Radio control costs tend toward the lower range while bidirectional power line carrier systems require somewhat higher costs.

The cost per Kw for demand management must be compared with the costs of providing a Kw of new capacity. Estimated costs of providing new base load generating capacity vary widely. In Florida recent coal-fired plants have come in around $800 per Kw, but coal plants committed today are expected to come in at costs of $1200 - $1500 per Kw of capacity. Capital cost escalation rates of 20% per year are not unheard of. Peaking units (gas turbines with combined cycles) were recently purchased at around $500 per Kw but these costs too are expected to rise significantly.

The cost of new generating capacity to meet peak demands must be compared with the range of costs to provide direct control load management. Such comparisons conducted by many utilities have resulted in utilities offering consumers cash incentives for joining in direct control load management programs.

Evaluating the cost-effectiveness of load control is a complex matter. According to conventional practice, new generating capacity is selected not only on the basis of reserve margin but also to preserve reliability and minimize cost of electrical generation. The change in reliability and operating cost resulting from a capacity addition depends not only upon the size and forced outage rate of the proposed unit, but also on the characteristics of the existing system mix. Similarly, load control also improves system reliability and reduces costs to a degree which depends not only on its own operating characteristics, but also that of the existing system. Moreover, the actual cost per Kw of controlled capacity is, of course, a function of the fraction of controllable load which happens to be "on" when called by the utility. This varies according to weather conditions, the success of marketing efforts, and other factors. Thus, the cost-effectiveness of load control can vary considerably among utilities. But although it may not be attractive to a few, it appears to offer advantages to most. Simplified calculations are often adequate to suggest the desirability of future implementation.

The Florida Experience

Over the past three decades Florida has been among the fastest growing of the states. In 1950 Florida's population was 2,771,000 which by 1980 had grown to 9,739,000.[11] During this period Florida's utility generating capacity increased from approximately 1000 MW in 1950 to over 28,000 MW in 1980.[12, 13] The predominant fuel base selected for new generating plants from 1950 through the embargo of 1973 was natural gas or petroleum so that by 1973 these fuels comprised about 75 percent of the installed capacity.[13]

Between 1960 and 1973 the generation rate grew at approximately eleven percent per year.[2] Between 1973 and 1975 there was essentially no growth, and since 1975 the growth rate dropped to an average annual rate of approximately five percent.

These significant changes in growth rate obviously present enormous problems to Florida utilities in their forecasting efforts. Previous construction plans for new generating capacity were based on significantly higher growth rates, and there was and is, uncertainty concerning the credibility of the more recent lower growth in demand. Perhaps partly as a result of this uncertainty and partly because of commitments to construction already made, new plant construction continued, and as committed plants became available reserve margins in Florida increased significantly.

In the past two years, many have argued that with the projected continual increase in petroleum prices, construction and operation of new coal-fired generating stations would be cheaper than the operating cost alone of existing oil-fired plants. New capacity has been recently justified on the basis of economics rather than reliability. A 1981 report by the Florida utilities asserted that reserve margin levels would range between 50-80% if all coal capacity found to be economical were added.[14] However, that conclusion is very sensitive to a number of highly uncertain assumptions: cost of capital, relative fuel prices, electricity sales, and capital cost escalation. Thus the degree to which new generation will be built on this basis is far from clear.

What is more clear, however, is that billions of investment dollars are at stake, and that justifying new capacity for economic oil displacement may be a major blow to the attractiveness of both energy conservation and direct load control. With excess capacity, the financial incentive to pursue these options is reduced; further power plant construction will lessen the attractiveness of direct load control even more. Thus the belief among some utilities and regulators that LM is not a useful tool may be a self-fulfilling prophecy.

Margins before maintenance and excluding interruptible loads are projected to increase in Florida from 37 percent now to a high of 53 percent in 1984 for summer margins and from 32 to 44 percent for winter margins during the 1980s.[15] In 1980 for the U.S. electric utility industry the margin of reserve based on non-coincident peak load was 30.8 percent.[16]

With the rapid rise in the cost of new capacity and the significant drop in growth rates of electrical demand, the costs of excess capacity becomes much more significant. The incentives for improving forecasting techniques are increased and the incentives for initiating programs which increase the utility flexibility for meeting demand have become significant.

An analysis and a series of calculations were performed to estimate the cost to Florida consumers of excess generating capacity assuming that capacity is added only to maintain adequate reliability. To be conservative, only capital carrying charges were included. Typical values were taken for discount rate, escalation, fixed charge, capital cost, reserve margin as well as other parameters. A sensitivity analysis was performed to determine the impact on results from changes in the various assumptions. Within reasonable assumptions it appears that if presently planned capacity additions are completed, the costs to Florida consumers of excess capacity in this decade may exceed four billion dollars or between seven and nine dollars per month for the average residential ratepayer. This is about 12-15% of the average utility bill.

In many cases demand management can defer the need for new generating capacity. Florida Power Corporation, the state's second largest utility with over a fourth of the residential customers, embarked on a systemwide residential and commercial load management program within the past year. The residential portion, which started January 1, aims for direct radio control of 644,000 water heating, space conditioning, and pool pump loads by the year 2000. The company projects that this will save ratepayers over $470 million by the year, compared to the alternative of 518 MW of new coal-plant construction.[17]

Customer response to Florida Power's program (with monthly bill reductions) has been overwhelming. While the utility has originally set an end-of-year target of 6,000 installations, more than 7,000 control switches were in place by May.[18]

It is worth noting, however, that this success reflects customers seeking out the program more than the other way around. In general, in Florida, demand projections do not take into account the considerable potential of demand management devices.

This potential contribution can be roughly estimated. At the end of 1980, Florida had a generating capability of 28,000 MW. In 1980, of the 4.1 million residential electrical customers in Florida [11], approximately eighty percent use electrical resistance heating to provide hot water.[19] This alone represents an estimated diversified load at time of peak demand of approximately 2000 MW. Such load can be easily managed.

Central air conditioning in the residential sector in Florida in 1980 served approximately 2.4 million households and by 1990 this figure may reach 2.75 million households.[20] Although the diversified load from AC units is not precisely defined it is estimated that by 1990 air conditioning will represent a diversified load in Florida of at least 2700 MW and perhaps substantially more. Because of the heavy air conditioning load, summertime utility peaks last for many hours. Management of residential AC loads requires staggered application of on/off cycles during periods when peak generating capacities are being approached. All told, it is likely that by 1990 the diversified demand from sources subject to load management (water heating, air conditioning, pool pumps, irrigation, curtailable commercial and industrial loads) will easily exceed 12,000 MW and may be significantly higher. While it is unreasonable to assume that all of this load can be brought within a load management program, it is not unreasonable to assume that by 1990, through an aggressive load management program a significant portion of this load could be managed.

An analysis was conducted of the potential savings which could be achieved by 1990 through an aggressive program of load management in Florida. Results suggest that significant savings can be achieved. Moreover, the state's official oil reduction goal of 25% in this decade may be feasible without a large increase in generating capacity. Such results echo similar efforts in California, another rapidly growing state with a substantial fraction of oil-fired capacity.[21]

References

1. Historical Statistics of the Electric Utility Industry - Through 1970 EEI Publication 73-74, Edison Electric Institute.

2. "Energy in Florida", a Publication of the Florida Energy Committee, March 1, 1974.

3. "The Future of the Electric Utility Industry", Leonard S. Hyman, V.P., Securities Research Div., Merrill Lynch, Pierce, Fenner, and Smith, Nov. 1981.

4. Utilities Action Plan, Dr. Douglas C. Bower: Proceedings of the Conference on "The Challenge of Load Management", June 11-12, 1975, Washington, D.C.

5. Impact of Load Management - A Regulators Perspective, Richard D. Cundaly, Proceedings of the Conference on "The Challenge of Load Management", June 11-12, 1975, Washington, D.C.

6. Energy Ventures Analysis Inc. Studies, Reported by Energy Users Report, May 20, 1982.

7. Evaluation of Load Management Systems & Devices, EPRI EM-1423 TPS 78-807, Final Report, June 1980.

8. Utilities Embrace Direct Load Control, Electrical World, February 1982.

9. Survey of Utility Load Management Projects, (Third Revised Report), ORNL/Sub 80/13644/1, July 1981.

10. Ten Years of Operating Experience with a Remote Controlled Water Heater Load Management System at Detroit Edison, B.F. Hastings, IEEE Summer Meeting, July 15-20, 1979.

11. 1981 Florida Statistical Abstracts.

12. Statistical Abstract of the United States, 1952 (Federal Power Commission Data).

13. Statistics of the Florida Utility Industry 1980, published Dec. '81, the Florida Public Service Commission, p. 26.

14. "Peninsular Florida Economic Oil Displacement Study", Florida Electric Power Coordinating Group, March 18, 1981.

15. 1981 Ten-year Plan, State of Florida, Florida Electric Power Coordinating Group, July ,1981.

16. Statistical Year Book of the Electric Utility Industry, 1980, Edison Electric Institute.

17. "Petition of Florida Power Corporation for Approval of Residential Load Management Rate Schedule", before the Florida Public Service Commission, November 3, 1981; "Residential Load Management Cost-Benefit Analysis and Explanation", Florida Power Corporation, December 3, 1981.

18. Jim Stitt, program director, Florida Power Corporation, personal communication of June 4, 1982.

19. Based on data from Florida Power & Light report for their service area extrapolated to state.

20. Information based on "Residential Air Conditioning Load Study", Florida Power & Light, Energy Management Research Department.

21. "Load Management Status Report" to California Public Utility Commission from Energy Resources Conservation and Development Commission, July 31, 1981.

Chapter 29

UTILITIES AND INDUSTRY: PARTNERS IN POWER?

H. J. Edwards, Jr.

Historically, the electric utility industry has best been characterized by the concept of "supply-planning": total reliance on central station service and construction of new generating capacity to meet any level of demand. Today, of course, the utility world is best characterized in terms of "demand-planning": the introduction of conservation measures, load control, and time-differentiated pricing to adapt demand as well as supply. In addition, unconventional generating technologies have become more important. Their importance has arisen because of a simple physical fact: conventional steam generation is believed to be running out of scale economies since heat rates are nearly at the minimum level possible. Thermodynamic factors now raise a barrier to further reduction in heat rates, and consequently to further scale economies. Therefore, it is natural to see intense interest in alternative energy supply, and just as natural to see interest in demand modification.

But there is more to it than that, because the problems the industry faces are too imposing. Roles are changing, and the roles involved are not just those of the planner, the forecaster, and the ratemaker. In fact, as a result of increased interest in cogeneration, it seems that even the roles of producer and consumer are no longer clear.

NEEDED: FINANCING OF ALTERNATIVE ENERGY
AND
ALTERNATIVE FINANCING OF ENERGY

In this new climate, there may be a need for industrial customers to abandon their traditional role as only a consumer of electricity. Their new role may involve partnership, or even a reversal of the old role into a new one where the industrial becomes a power producer with the utility being the consumer of excess capacity.

Innovation is essential if utilities are to see to the power needs of the next twenty years and beyond. For some utilities, the question of financial feasibility for new or existing generating technologies is not a question of alternative ways to develop power, but alternatives to doing without power.

A RATE DESIGN CHALLENGE: STAY FINANCIALLY WHOLE, BUT DON'T GIVE ANYBODY AN INCREASE

Those who manage and regulate electric utilities are in trouble. Everybody knows that, and everyone thinks they know why. But those who rely on electric utilities for industrial power are in even more trouble, and no one has bothered to inform them. Part of the reason is the old story that the electric utility industry is caught in a capital squeeze. As

it happens, that isn't exactly true. That is, the capital squeeze is not a cause, but an effect. Actually the utility industry is caught in a revenue squeeze, because it must sell its service at half-price. That's the squeeze, because investors demand higher prices for their capital; prices which utilities cannot always meet. In short, rates are not covering costs, and the shortfall is coming out of the investor's pocket. Unfortunately for the utility, the investor has better things to do with his capital.

As a result of this squeeze, many utilities have reached the conclusion that the power business is not a very good business to be in. Some would like to get out of it, for understandable reasons. The reasons are as obvious as they are incredible. Imagine a business where prices are based on costs incurred in 1972; where return on equity is pegged to a level beneath the cost of double-A bonds; where customers of the business believe that they have a God-given right to prices that existed when Spiro Agnew was Vice-President; and where political careers rely on preserving that belief. Who would enter such a business?

THE NEW ECONOMICS OF ELECTRICITY

Some utilities are beginning to evolve a rational response to that question: thinking in terms of getting out of an unrewarding business, just as the economics textbooks say they should. That does not mean nationwide blackouts, but it does mean that those who rely on access to firm power and energy for industrial purposes may feel the first pinch, in the form of unavailability of power. They are already feeling it in the price of power, and they will feel it even more in the future in order to ease the pain of escalation on others. When that expedient is gone, they will face uncertain availability (or perhaps it should be termed certain unavailability), because utilities may not be able to provide it to them. Because of this, a day may be coming when both utilities and industrials will have to enter partnerships for the production of power. Utilities and industrials face continued troubles and the rational response would be to solve their problems together.

Why? A better question is to ask, why are we in trouble? In one way, the answer is, very simply, that utilities are regulated. That does not mean that regulators are incompetent, for they are not. It does not mean that regulation has beaten utility managers into incompetents either, for they are not. It means that utilities and regulators are captives: captives of the law. Not the law in general; regulatory or administrative law in particular. Regulatory law has a life of its own, and it can create a very uncomfortable life for those it covers.

Two examples will make the point. As a matter of fifty years' worth of case law, nearly all utilities are constrained to rate levels based on net investment, per books. Now, the average book life of a thermal powerplant is about thirty years. As an industry, the utility business is (or used to be) about one-third depreciated, which makes the average net investment ten years old. That means, effectively, that rates are based on costs which represent 1972, not 1982.

It also means that depreciation charges are about 3 percent per year. That in turn means that the old accountant's adage, "Depreciation Generates Cash Flow," doesn't mean much. This situation is more like running a fleet of taxis on the cash flow generated by depreciating 1972 Chevys over ten years, when their economic life is three years.

The second example stems from the first. To ease the pain of financing new generation at 1982 costs, utilities have struggled to get units still under construction included in current rate levels. Some of the benefits of this treatment are well known. Among the obvious virtues is the fact that pressures of external financing are eased and coverages are improved. A less obvious virtue is the fact that rate levels ease upward more gradually. This means that our ratepayer, who is still paying 1972 costs, is not wrenched violently into the eighties on the day the unit goes commercial. He eases into the eighties at a lower ultimate level of cost. And if the new unit happens to be a large coal-fired unit reducing the utility's use of expensive oil, it would seem rational to adopt this practice. Since everyone benefits, everyone has a stake in making this happen, or so it seems.

It doesn't work that way, of course. One understandable reason is that a large coal-fired unit might cost $900 per kilowatt to complete in 1982, versus an average or "1972" cost of about $250-300. Because these numbers are so different, it is natural for the following bit of folklore to emerge in utility rate cases:

> "Present ratepayers should not pay the cost of serving future ratepayers."

This means, in effect, that existing customers are believed to be entitled to hang onto their privileged position, and new customers should be made to pay for the $900 power plant. Guess which kind of "new" ratepayer it is who adds load to an electric system in large, easily-identifiable lumps?

This is not really case law, because it is not yet widely accepted. But it does spell trouble for industrial users of electricity. Many industrial firms, facing obsolescent processes and adverse labor markets, need to relocate or modernize in order to remain competitive. There was a time when the friendly local utility would jump through hoops to encourage new industry to take power: not any more. Today, some utilities are questioning their franchise to electrify everything in sight. Some utilities are discouraging or even turning away large prospective industrial customers. At 1982 prices, the utility system cannot absorb the high-priced increment of new capacity and average it in with 1972 costs. This is true not only because of long lead times, cash flow problems, chronic dilution, and poor quality of earnings; it's true because of a "social reality". That reality is as follows: a utility manager simply cannot go to Joe Ratepayer and tell him that his electric bill must increase by 30 percent because Consolidated Metals just built a plant nearby. That

statement doesn't make any sense; it's irrational. But, so is the way utility rate levels are determined, because they're based on 1972 costs.

Unfortunately case law works from history, and ratemaking must work within it. For example, a large industrial customer who seeks to get new service today from the Bonneville Power Authority must pay for his power based on the costs of expensive new generating resources. He gets no access to the lower average cost of the BPA system, but buys from the top of the stack, while others buy from the lower-cost resources at the bottom. That Federal law may come to be the new ratemaking standard for other utilities as well.

THE RATE-CASE GAME

In many ways, rate cases have become an elaborate game of "Stick the Industrial". Many different philosophies are introduced in this game, but most of them are based on the fact that new generation costs more than old generation. The game boils down to finding out who pays what cost first. As a matter of political reality, some regulators must play this game. So must utilities, at times, because the first task of the utility is to obtain enough overall revenues to survive, even if survival means holding down residential rates at the expense of industrial customers.

But this is not enough, because sooner or later the other ratepayers will have to pay "true" costs for their power. Some utilities seem to have lost faith in the ability or the inclination of the regulator to recognize that fact. They have concluded, sometimes without knowing it, that a day of reckoning is on the way. On that day, no matter how strong the case for an increase, the regulator will not even grant subsistence in rate relief - because Joe Ratepayer can't afford it.

In anticipation of that day, utilities will buy, and are buying, more time. The way to do that is easy: stake out a position that no new capacity will be built. Obviously, if customers ever have to be turned away, the first will be the large industrial customers. That is already happening in some utilities, as well as steady price increases, and perhaps someday utilities will be forcing interruptible service on all industrial loads, existing as well as new.

In addition, there may be another development. It is well known that a recent issue in utility rate cases is that of "excess capacity". The question revolves around whether the ratepayer should pay for new capacity which turns out to be more than is immediately necessary. The merits of the issue are not important to this discussion. What matters are two facts: first, as a result of this issue, utility managers are going to be slower to construct new capacity because of forecasting risk; second, forecast error with regard to industrial sales will now draw increased attention. A likely outcome will be that new industrial customers will face increasingly stiff take-or-pay contracts, possibly even assessments in the form of a Contribution in Aid of Construction in order to recognize construction risk as well as escalation.

THE UTILITY MANAGEMENT GAME: THEM VS. US
OBJECT: FIND OUT WHO "US" IS, AND WORK TOGETHER

How does all of this connect with either conventional or innovative generating technology? In this way. The electric utility industry may no longer be able to see to all of the power needs of the country; the regulatory burden is too great. Conventional scale economies may still be there, but they have been drowned out by inflation and environmental law.

If utilities do get to the point where the franchise obligation can't be met, it may or may not "get dark out". But there should be no doubt on one point: it won't get dark on Joe Ratepayer until after it gets dark on others.

To reverse this, two breakthroughs are needed. One is technological: the need to move past the constraint of the Carnot cycle by diversifying the energy inputs and outputs. The second is financial: the need to move past conventional balance sheet funding by diversifying the ownership.

Mr. Drexel D. Journey, former General Counsel of the Federal Power Commission, put it another way:

"Capitalization and cash flow demands are now limiting the sizes of generating units and related transmission facilities that can be built without joint ownership or divided interests. Absent a partner or partners, some systems are facing 'no go' decisions."

"Under Title I of PURPA, utility managements find governmental agencies busily engaged in regulatory programs which are designed to restructure utility markets without changing the common law and statutory obligations to serve all who may seek electric services.

For those utility systems which are subject to rate and service controls under regulatory-type statutes, there is a second major regulatory impact. Here, managements find themselves operating under a stifling burden of historical legal precedent which, when closely examined, will be found to be economically—biased against innovative technological efforts and major large scale energy projects."

In other words, Mr. Journey believes utilities cannot provide the necessary breakthroughs on their own. Case law is risk-averse; regulators who will not even fund construction of known technology will certainly not fund R&D at ratepayers' expense. Utilities cannot do it on their own; they have the burden of worrying about that ten years' worth of inflation still waiting in the pipeline.

That is why the title of this discussion is "Partners in Power". There is little else left to try but to solve the problems by forming financial alliances between utilities and industry. This will not be easy. For example, regulation limits utilities to unrealistically-low levels of ROE, while industrial hurdle rates are often far higher. Utilities are hemmed in by outmoded depreciation rules and tax treatment, while industrials have far more flexibility. Utilities sometimes risk loss of investment tax credits due to low operating income and low tax liability, while industrials often have far more flexibility in ITC treatment.

This all means that the economics of self-generation (as well as cogeneration) may be turning around. Compared to uncertain availability of power, even at new-resource costs, industrial investment in power supply will become more economic as time goes on. In particular, joint ventures between the utility and an industrial firm, or among several industrial firms, show promise.

We are now at the point where questions can be raised that were outlandish only a few years ago. For example, what if an industrial firm constructed a generating unit which provided not only its own power but also leased a portion of the unit to a utility? What if third-party arrangements could be constructed among a utility and various members of an "industrial energy park"? What if the generating technology were a qualified cogeneration facility? Some of the problems might be:

-- The required hurdle rate of an industrial enterprise may be too high for feasibility.

-- Industrial firms are not now in the power business and probably don't want to be.

-- Utilities don't like the prospect of losing industrial load.

-- Utility financial officers don't like unusual financing arrangements.

-- Utility investors don't like the risk feature associated with off-balance-sheet transactions.

-- Regulators don't like to lose control.

These are imposing problems, but the possible alternatives to trying such projects could be far worse:

-- Industrials will pay more for reliable power than they should - if they can get it.

-- Plant modernization will be held back due to difficulties in getting power on a timely basis.

-- Some utilities will lose both load and economic benefits as industry relocates to areas where power is available in the near-term.

-- Scale economies will decline as more and smaller units become the only feasible construction choice.

-- Residential rates will face additional upward pressure as the industrial cushion shrinks.

There are wide varieties of partnership, operating, and financing options available for such projects. There are also many obstacles (e.g., ROE, risk, etc.) to overcome. But many factors can play a part in overcoming these obstacles. Among them is the fact that today's true capital costs for utilities can often exceed those of industrials. In addition, if a take-or-pay lease by a utility is involved, it helps reduce the overall risk of the project which should reduce the return requirement. The key is to recognize that it will be a case-by-case, not a blanket solution.

In order for unique partnerships to occur, several things must happen. Utilities must break the habit of equating growth in rate base to growth in earnings per share. They must also remember that there have always been two reasons for promoting industrial development: load factor improvement and jobs. The industry may be nearing a point where high load factor from a large and growing industrial sales base is of dubious value, but there is nothing dubious about jobs. They pay electric bills.

Industrial firms must keep in mind that utilities plan for the future based on a very central assumption: that the annual revenues needed to support the plan will materialize. Today that assumption is a shaky one; when it does not hold, the plan selected may not be a least-cost plan because there is no longer the assurance that the "right" plant, once built, can be paid for. As a result, alternate plans are emerging. Among them is the option to sit tight and avoid expansion, since there is reason to wonder whether the existing customer can afford expansion.

THE REGULATORY GAME: ASSURE FAIR RATES, BUT SURVIVE

Utilities should keep in mind that every rate request is a plea for endorsement by the regulator. There can come a time when even the most enlightened regulator can no longer "go to bat" for the beleagered utility because the price is too high, in more ways than one.

Everyone should recognize that regulators face the hardest task of all. No matter what utilities or industrials do, regulators operate in a world that is hostile to future considerations, because future ratepayers seem to have no standing in a rate case. Regulators carry the burdens of excess legal baggage and a constituency interested only in the here-and-now. These burdens force their gaze backward and compel them to discount the interests of future ratepayers.

Utilities, industrials, and even regulators must begin to enter into a dialogue to find ways to develop power resources together. It will require innovation in financing as well as in technology to do this, and many traditional barriers exist for all parties. If these barriers cannot be surmounted, utilities will face not only a declining industrial sales base and lower efficiencies, but obstacles to economic development of their service areas as well. Industrials will face increasing uncertainty in obtaining adequate and reliable sources of power, and regulators will face increasing political woes.

Without ample access to power at reasonable rates, plant modernization and economic recovery may be seriously hampered. Industrials, who today face worldwide competitive pressures, must be more vigilant than ever in finding ways to keep the power bill under control. Those ways may include "getting into the power business" in order to avoid "getting the business".

Chapter 30

IMPLEMENTATION OF CUSTOMER DIRECTED LOAD MANAGEMENT PROGRAMS IN A LARGE INVESTOR OWNED ELECTRIC UTILITY

R. P. Thompson

ABSTRACT

In recent years, rising electrical energy costs and declining operating reserve margins have caused many utilities to closely examine their corporate commitments to supply their customer electrical energy requirements. Load management programs that are directed towards reducing the usage of electrical energy during the utility's peak generating period are becoming an important part of these utilities' operating plans. Pacific Gas and Electric Company has undertaken an aggressive program to carry out customer directed load management programs in its service territory as a cost-effective alternative to building new generating capacity. This paper will provide an overview of the various load management programs currently being implemented by this large investor-owned combined gas and electric utility.

INTRODUCTION

Until the last decade, electric utility economics were relatively uncomplicated. The load growth of the Pacific Gas and Electric Company (PGandE), as with utilities generally, resulted in economies of scale that led to decreases in rates. The task of pricing electricity and promoting or encouraging its use was, accordingly, very straightforward. The objectives were simple:

Set rates to collect adequate revenues to cover costs, including capital costs.

Design rate structures to distribute charges for electric service equitably among classes of customers. (This was performed in accordance with the different costs associated with serving the different classes.)

Encourage consumption of electricity so the utility could take advantage of the economies of scale and building new power facilities.

Encourage use of electricity at off-peak times in order to spread fixed costs of plant and equipment over as many units of electricity as possible, plus further lowering costs and enhancing earnings, and ultimately leading us to lower rates.

In 1970, this half-century long period of growth in the electric utility industry drew to a close. This was the year in which inflation caused capital costs to rise to the extent that average costs per kilowatt hour of additional generation increased, rather than declined. By 1973, fuel costs were accelerated as well. This was not a problem unique to PGandE. Rather, it was a nationwide development that the electric utility business was no longer a declining cost industry.

The traditional view of pricing and promoting energy used to reduce average costs was no longer applicable. Utilities and regulators were seeking new theories to govern how rates should be designed in the face of rising average costs of providing electricity. In late 1974, the National Association of Regulatory Commissioners (NARUC) called for an industry wide study of the technology and costs of shifting some electric utility usage from peak to off-peak periods. The same year the California Public Utilities Commission, responding to a California legislative resolution, opened Case No. 9804 to investigate the feasibility of time-of-use pricing. These two events were the genesis of load management for PGandE.

Load management is defined as the economic reduction of electric energy consumption during PGandE's peak generating periods. Load management differs from conservation in that the former shifts usage from peak times while the later is concerned with reducing usage over the entire 24 hour load profile. The peak period energy to which load management programs are directed is graphically shown in Figure 1.

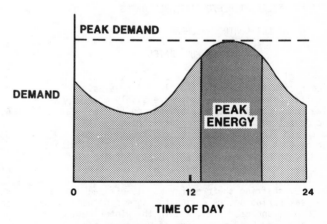

FIGURE 1. EXAMPLE OF TYPICAL UTILITY SYSTEM LOAD PROFILE WHERE LOAD MANAGEMENT PROGRAMS ARE DIRECTED TOWARD REDUCING ENERGY DURING PEAK ENERGY LOAD PERIOD.

The objectives of load management are to:

Reduce the need for additional generation, transmission, and distribution capacity;

Conserve scarce resources through the encouragement of more efficient use of existing resources;

Provide adequate revenues to support company operations, and;

Ensure equitable levels of service to customers consistent with the prices they pay.

Customers are provided with a choice of service options that correspond to various price incentives. Energy usage decisions can then be made on the customer's side of the meter based on their own individual situations. Load management thereby, provides a measure of product differentiation reflecting the diverse costs incurred by the company in day and year.

PGandE has been involved in load management programs since 1973, and it is an important element in PGandE's electrical resource plan. The PGandE electrical resource plan for 1990 currently includes over 630 megawatts of electrical peak demand reduction from load management programs. This capacity reduction will help defer or eliminate the need for an amount of additional electrical generation capacity. The 1990 capacity reduction is equivalent to the electrical capacity needs of approximately 130,000 homes.

The implementation of load management programs requires the interaction of many company departments and the integration of generation resources and power system operations. In the 1980-81 period, PGandE spent about $38,000,000 on load management. In the 1982-83 period, expenditures are expected to exceed $88,000,000 with the most aggressive and far reaching load management programs of any utility in the United States.

Load Management programs will be carried out in all customer sectors: residential, commercial/industrial, and agricultural. The programs are divided into three categories:

    o   TIME-OF-USE RATES

    o   INTERRUPTIBLE/CURTAILABLE RATES

    o   DIRECT CONTROL PROGRAMS

TIME-OF-USE RATES

On any given day, the average cost of generating electricity rises as the peak load increases. This results because units with progressively more expensive running costs must be brought on line to meet the peak load. Also, with today's high capital costs and long lead times for construction, new capacity is added at higher cost than past additions, leading to higher average unit costs for the electricity provided to customers. With TOU rates, customers can avoid higher costs during the peak hours by shifting their energy use to off-peak hours. The peak period occurs during the summer between 12:30 p.m. and 6:30 p.m., and in the winter between 4:30 p.m. and 8:30 p.m. Time-of-Use rates are currently provided to agricultural, commercial and industrial customers, and in the later part of 1982, a Time-of-Use program for residential customers is scheduled for initial testing. These Time-of-Use rate programs can be separated into three categories: Mandatory, Experimental, and Voluntary.

Mandatory TOU Programs: Mandatory Time-of-Use rates for large customers were ordered by the California Public Utilities Commission on March 16, 1976. Subsequently, PGandE began a gradual implementation of Time-of-Use rates with the introduction of the A-23 rate schedule for commercial and industrial

customers whose monthly maximum demands exceeded 4,000 kW. The next rate schedule implemented was Schedule No. A-22, applicable to customers with demands between 1,000 and 4,000 kW. The most recent TOU rate, Schedule No. A-21, is applicable to customers with demands between 500 and 1,000 kW.

Figure 2 illustrates the load shifting reports of TOU rates for the A-23 customers in the PGandE area. Recent studies indicate that energy usage during the peak period has been reduced by approximately 2 percent.

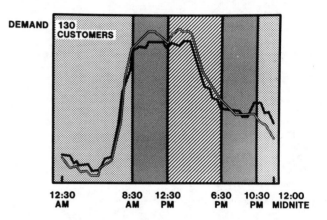

FIGURE 2.   LOAD SHIFTING DUE TO TOU RATES FOR A-23 (4,000 kW DEMAND AND GREATER) COMMERCIAL/ INDUSTRIAL CUSTOMERS.

Experimental TOU Programs: Experimental programs enable PGandE to ascertain the responsiveness of customers to a particular Time-of-Use rate.

Presently, there are two experimental Time-of-Use programs in existence. One involves over 500 industrial and commercial customers with demands under 500 kW. The major objective of this experiment is to determine the feasibility of applying Time-of-Use rates to medium and small commercial or industrial customers. The experiment will also provide PGandE with data concerning innovative metering technology. The second experimental program involves approximately 6,000 customers and is being conducted to determine the implications of offering an agricultural Time-of-Use rate.

Voluntary TOU Programs: Voluntary Time-of-Use programs give customers the option of switching from their regular rate schedule to a time-of-use rate schedule. Currently, a time-of-use rate is offered to agricultural customers with demands over 35 kW per month.

A TOU rate schedule will be made available to residential customers on a voluntary basis in 1983, and commercial and industrial customers with demands under 500 kW will have the option of being served on time-of-use Schedule No. A-21.

Cooperative Electricity Management Program: The Cooperative Electricity Management (CEMP) promotes load management at the community level. This joint PGandE/community program offers financial incentives to the participants to help reduce peak energy consumption. Media coverage, seminars, community events, and other means are used to tap into the "community spirit" so as to achieve the reduction. PGandE will be operating the CEMP Program in nine communities during 1982, and monitoring energy consumption in several other new communities. The monitored communities are expected to be added to the program during 1983.

INTERRUPTIBLE/CURTAILABLE PROGRAMS

There are several options for commercial or industrial customers who allow PGandE to interrupt their service for limited periods of time. These customers receive compensation for their willingness to give up service when asked to by PGandE.

Interruptible Service: Large industrial customers have the option to be served under an interruptible rate. Electrical service can be automatically interrupted with a digital under-frequency relay if an emergency situation arises. Full interruption excludes plant lighting, fire protection loads, and other safety and security end-uses. Customers are given substantial rate discounts in exchange for exposing their power requirements to immediate shut-down.

Curtailable Service: Commercial or industrial cus-tomers on rate schedule A-23 may sign up for one of three curtailable service options. The customer must have at least 500 kW of curtailable load. The advance notification times vary with each curtailable service option.

Under the terms of Special Condition No. 10, customers retain control over the specific loads to be curtailed when reductions are requested by PGandE. Curtailment is accomplished by reducing the use of equipment in a manner conducive to a safe and economic plant opera-tion. Customers designate a firm level of service, and curtailment of the exposed load will require prior notification by the utility. Should the cus-tomer elect to disregard PGandE's curtailment request, a penalty charge is added to the monthly bill.

The load reductions from an A-23 Special Condition No. 10 customer during the 1982 Summer is shown in Figure 3. Although customers may request any level of curtailment, the average curtailment for customers currently participating in the program is approxi-mately 5 MW.

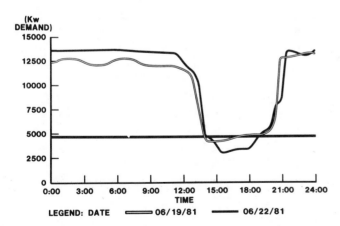

FIGURE 3. LOAD REDUCTION ACHIEVED FROM CURTAILABLE CUSTOMER ON A-23 SPECIAL CONDITION 10 RATE.

Customers with standby power generating equipment capable of generating 500 kW or more, may sign up for the Auxiliary Power Sources Program. This program allows PGandE to shift some of the customer's load to the customer's own power source, relieving the load on the PGandE system.

Another curtailable option enables qualifying commercial and industrial customers to participate in the Group Load Curtailment Program (GLC). This program involves small groups of customers who agree jointly to shed a certain amount of combined load when it is requested by PGandE. Upon notification, the members of the group curtail loads independently to reduce their combined electric demand. If one member cannot meet a request, the others can jointly make up the deficit. The group is not asked to curtail its load over 90 hours per year nor more than 6 hours in any one day. The program is only operated on weekdays between 12:30 p.m. and 6:30 p.m. To qualify, each group member must have a demand of at least 750 kW per month with the ability to shed 200 kW of load on a peak day. Experience with operation of the initial GLC program during the 1981 summer provided demand reduction of approximately 5 MW as shown in Figure 4. This curtailment level is expected to increase as additional members join the groups and experience is gained in system operations.

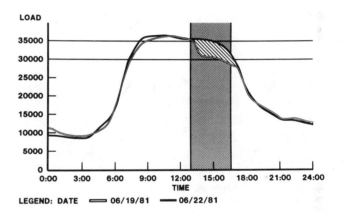

FIGURE 4. COMPARISON OF DAILY LOAD PROFILES FOR CURTAILMENT AND NON-CURTAILMENT DAYS FOR GROUP LOAD CURTAILMENT CUSTOMERS.

Presently, PGandE has Group Load Curtailment Programs in Oakland and San Francisco. Plans are being made to develop additional groups in these two cities as well as in San Jose, Sunnyvale, and parts of the East Bay.

DIRECT CONTROL PROGRAMS

Direct control programs involve the installation of equipment by PGandE on the customer's premises. The equipment includes a receiver as well as control hardware that enables PGandE to control the use of air conditioners and water heaters. Transmission of the signal to the customer premises is by radio, power line carrier, or ripple signals.

The Residential Air Conditioning Program: The residential air conditioning program involves a con-cept referred to as cycling. When a customer signs up for this program, PGandE installs (at no cost to the customer), a radio-controlled switch on the customer's air conditioner. This enables PGandE, when it is necessary because of capacity shortages, to switch the air conditioner on and off. The air conditioner will only be cycled on weekdays between noon and 9:00 p.m. and for no more than 20 minutes in any half-hour. It is not anticipated that PGandE will cycle the air conditioner for more than 20 days between May and September. A representative diversified load impact of the air conditioners after the strategy is terminated.

The program is open to residential customers with central air conditioning who reside in Fresno, Davis, Yuba City, Bakersfield, Walnut Creek or Concord. Customers who choose to participate in the program receive monthly reductions in their electric bills during the months of May through September. The reduction ranges from $2 to $6, depending on the climate in which the customer resides.

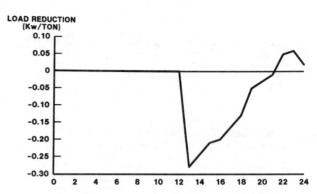

FIGURE 6. TYPICAL LOAD REDUCTION FROM OPERATION OF COMMERCIAL DIRECT CONTROL AIR CONDITIONING PROGRAM.

SUMMARY

PGandE's load management programs have brought about changes to the traditional types of electrical service. As new technologies become available and more information is gathered, the present programs will be refined and new programs will be implemented. Peak load impact will increase, operating costs will be further reduced, and system reliability will be enhanced. The continued success of the programs depends on PGandE and its customers working together. With the customers becoming ever more aware of PGandE's operations, the relationship between the two will strengthen and broaden.

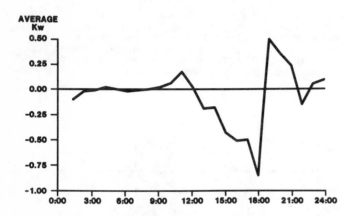

FIGURE 5. DIVERSIFIED LOAD IMPACT FOR RESIDENTIAL DIRECT CONTROL AIR CONDITIONING PROGRAM IN FRESNO AREA WITH A 15-MINUTE CYCLING STRATEGY AT A MAXIMUM AMBIENT TEMPERATURE OF 107.5°F.

The Water Heater Program: The water heater cycling program is open to residential customers with electric water heaters who reside in Fresno or Red Bluff. Customers who sign up for the program receive a monthly reduction of $1 to $2 in their electric bills all year around. Water heaters will only be cycled on weekdays, never on weekends or holidays. They are only cycled between noon and 10:30 p.m. and are not turned off for more than 6 hours in any one day.

Commercial Air Conditioning Load Deferral: This program selectively concentrates the benefits of control on days of peak electrical use. Because of the sensitivity of many commercial establishments to temperature, a method of load deferral which prevents the internal temperature of the building from rising excessively was designed. Three types of this control are now being used: 1) increases the thermostat setting gradually until the temperature reaches a maximum of 82°F, 2) changes the on-off pattern of the air conditioner in a predetermined manner, and 3) "remembers" the on-off pattern of the air conditioner during the precycling period and maintains that pattern throughout the day. Participating customers receive a rebate of $1.50 per ton of air conditioning available for cycling. A typical load reduction which is representative of the demand reduction realized from cycling of commercial air conditioners, is shown in Figure 6.

Chapter 31

LOAD MANAGEMENT: CUSTOMER OPTIONS

J. P. Stovall, J. M. McIntyre

ABSTRACT

Load management offers the residential customer an opportunity to reduce energy costs by changing their pattern of electrical energy usage in a way to assist the electric utility to more effectively and efficiently provide electric service. Two customer load management options will be discussed in this paper. The first option is customer energy storage for space heating and water heating. The benefits or costs of such equipment will depend upon the penetration of installed units, their load characteristic and hours of operation, and the load characteristics of the particular utility. The second option is a residential demand limiter which is being demonstrated by Omaha Public Power District. The demand limiter is set by the customer to the desired level of demand which is related to the expected monthly bill and then controlled by the utility to limit demand when most beneficial to the utility system.

INTRODUCTION

Historically, electric utilities have supplied electric energy on demand from the customer. This demand varies with the time of day and is greatest during daytime, or peak demand periods, and lowest during nighttime, or off-peak periods. This pattern of electric energy demand is principally caused by the greater daytime activities in commerce and industry and the greater requirement for space conditioning by residential, commercial, industrial, and municipal customers during the day. This fluctuating pattern of energy use places a heavy burden on the electric utility's energy delivery system because electricity cannot be stored conveniently and must be produced in response to demand. Utilities supply electric energy known as base generation with more efficient oil-fired, coal-fired, and nuclear generating plants. To meet additional demand during peak periods, most electric utilities supply electric energy from peak generation using hydroelectric plants and inefficient gas- or oil-fired units. In the past, utilities have easily financed and built new generating plants to meet the growing peak demand for electric energy; however, the increasing expense of oil and gas, stricter utility industry regulations, and less favorable financial conditions have changed the electric utility industry's philosophy about meeting the customers' electric demand. As a result of all these factors, utilities are now motivated to look at means to tailor the electric energy demand of consumers to match energy supply, in order to delay financing and building new base generating facilities and to reduce the need for peak generation.

Concept of Load Management

Load management is a method to alter or reshape the electric utility load as a function of time. The purpose of load management is to reduce peak demands to level the daily or annual electric demand. The definition of load management given by the U.S. Department of Energy is as follows [1]:

> Load management is a systems concept of altering the real or apparent pattern of electricity use in order to (1) improve system efficiency, (2) shift fuel dependency from limited to more abundant energy resources, (3) reduce reserve requirements, and (4) improve reliability to essential load.

An important point of this definition is that load management is a systems concept. The entire utility system, including generation, transmission, and distribution systems is affected.

The concept of load management has been further divided into use-management and supply-management according to the location at which the action is taken. Supply management alternatives involve the use of concepts such as central station storage and expanded interconnections to affect the bulk supply. Use-management alternatives apply to the load side of the meter in which the customer responds to an incentive offered by the utility.

The use-management alternatives include direct control of customer loads, indirect control of customer loads, and customer energy storage. Direct control of customer loads has been implemented by utilities in the cycling of air conditioners, water heaters, and other appliances. Indirect control of loads has been experimented with and implemented by various forms of electricity pricing and rate structures to encourage customers to change usage patterns. Emerging customer energy storage technology offers the potential of altering electric energy usage patterns by allowing customers to save energy for use at a later time.

In this paper, two use-management concepts will be explained and representative projects demonstrating the concepts will be discussed. The first use-management concept is customer energy storage for space conditioning and water heating in residential homes. The second concept is local control of residential customer loads using a demand limiter.

CUSTOMER THERMAL ENERGY STORAGE

Customer energy storage has the characteristic of changing energy use patterns and may be classified

*Research sponsored by the Division of Electric Energy Systems, U.S. Department of Energy under contract W-7405-eng-26, with the Union Carbide Corporation.

according to the type of storage medium. In this paper, the discussion will consider thermal energy storage (TES) in which electric energy is converted to heat energy, stored and then released. For space heating and water heating, TES is an ideal application and can offer the same quality of service to the customer. TES can allow the utility to deliver electric energy to the customer at a time of day most advantageous to the utility and allow the customer to use that energy during the times desired. More specifically, space heating and water heating are controlled by the customer setting a thermostat to the desired level of comfort. With conventional appliances, electric energy usage is controlled by thermostats during all times of day and usually coincident with the daily system peak. However, with thermal energy storage, the utility may deliver energy which is stored as heat during off-peak periods, usually nighttime. Then, during on-peak periods of high system demand, the stored energy is released under the control of the thermostat. The result is that the same quality of service is delivered by the conventional appliance but electrical demand is shifted to a time most advantageous to the utility.

Demonstration Program

In an effort to determine the costs and benefits of customer TES, the U.S. Department of Energy, Division of Electric Energy Systems embarked upon a demonstration program of load management using customer side thermal energy storage. Eight utility companies were selected and are participating on a cost-shared basis. There are five heat storage and five cool storage projects. Each utility's project consists of the installation of 30 to 50 TES appliances in residential test homes in their service territory. An equal number of control homes with conventional appliances are included and both test and control homes have been instrumented to collect load research data. In the program, TES equipment was installed beginning in late 1978 and data was collected during 1980 and 1981. A more complete description of the demonstration program can be found in Ref. 2. Public Service Electric and Gas Company is one of the utilities participating in the program which focused on space heating and water heating.[3,4]

Public Service Electric and Gas Company

Public Service Electric and Gas Company (PSE&G) is a combination electric and gas utility serving approximately 75% of New Jersey's population located in a 1,400 square mile, highly industrialized, densely populated corridor between Philadelphia and New York City. As of December 1980, PSE&G was serving about 1.6 million electric customers. PSE&G is essentially a summer peaking company with a relatively high saturation of summer weather-sensitive load and a very low saturation of electric heating. The 1980 PSE&G system peak hourly integrated load of 7,159 MW on July 27, 1980. The total net system output for 1980 was on the order of 32.7×10^6 MWh, and the corresponding 1980 system annual load factor (ratio of average load to peak load) was about 52 percent.

Despite the fact that PSE&G is a summer peaking utility, there are possible benefits to encouraging the installation of TES heating systems. Although PSE&G currently has a very low saturation of electric heat, present forecasts indicate that residential electric systems will become increasingly more attractive for home heating. TES heating and water heating yield specific benefits derived from the use of these devices to replace conventional uncontrolled systems.

Description of PSE&G Demonstration

The demonstration consisted of retrofitting 30 single-

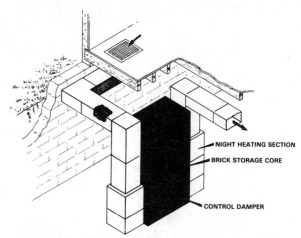

FIGURE 1 - DIAGRAM OF TPI FURNACE IN TEST HOMES

family homes previously utilizing conventional electric furnaces and water heaters with a central ceramic brick TES furnaces and a TES water heaters equipped with load leveling controls. The storage units are controlled by a bi-directional communications system which utilizes voice grade telephone lines to transmit both commands and status requests from PSE&G's General Office to the remote field locations. In the test homes, both the TES furnace and the water heater have customer-initiated overrides to protect against any possible communications and/or equipment failures.

A group of 30 single-family homes using conventional electric furnaces and water heaters serves as the control group. All the energy use in these test and control homes is being metered in detail using magnetic tape recorders. Data collection began in December of 1979 on the units then installed. All installations were completed by late spring 1980. Figure 1 illustrates the installation in a test home.

Central Ceramic Brick Storage TES Furnace: Central ceramic brick TES systems are designed to heat an entire residential home or multiple-living area from one unit. To date, the systems are of the forced-warm-air variety, making them very compatible with central heating. The ceramic bricks, which are housed in a heavily insulated metal cabinet, are heated to temperatures of 649 to 760°C (1200 to 1400°F) by electric heating elements distributed through-out the bricks. Air is circulated as in a conventional forced-air furnace, with dampers to divert a portion of the air over the heated bricks. This heated air is mixed with unheated air to provide a discharge temperature between 52 and 60°C (125 and 140°F).

The selected furnaces were manufactured by Tennessee Plastics, Incorporated (TPI) which produces a central ceramic brick TES furnace under license with Creda, Ltd. of Great Britain. The unit operates in series with a backup electric furnace. The 30-kW TPI TES furnace used in the test is capable of storing approximately 200 kWh and is designed to operate for a 16-hour off-period each day by charging for a 8-hour period. The unit as shown in Figure 2 consists of four major sections: a night heater or auxiliary furnace, a storage core, dampers, and controls. The night heater, a 15-kW electric furnace, is sized to provide heat to the entire residential home while the storage section is charging. In the storage section, the ceramic bricks are fitted together in layers, with electric heating elements in each layer. The bricks are fitted together to form air passages for directly heating air. The storage section is surrounded with several layers of insulation to maintain storage temperatures and reduce heat loss from the surface or skin of the unit.

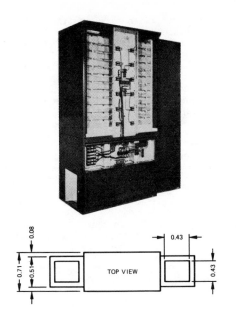

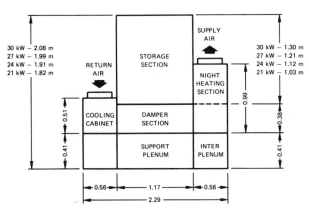

DIMENSIONS ARE IN METERS

FIGURE 2 - TPI CENTRAL CERAMIC BRICK TES FURNACE

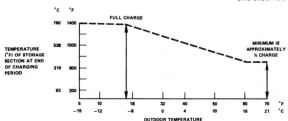

FIGURE 3 - TPI LOAD-LEVELING CONTROL

the 21-kW unit has a storage capacity of 140 kWh, the 24-kW unit has a capacity of 160 kWh, and the 27-kW unit has a capacity of 180 kWh.

In the unit selected, storage capacity is designed to be 200 kWh at a full charge temperature requiring a maximum charge period of eight hours. The standard unit has three single-phase 240-V supply circuits; the top two circuits each supply four 2,625 W elements and the bottom circuit contains three elements. Total rated demand is 28.875 kW. In PSE&G's demonstration, the uppermost element pair was disconnected, because the ten-hour charge period utilized was two hours longer than that for which the unit was designed. This reduced the maximum demand of the storage section in PSE&G's unit to 23.625 kW.

The night heating section contains the open coil resistance-type heating elements used for nighttime heating of the test homes and for stand-by on-peak heating. The standard night heating section contains three elements, each rated at 5 kW. In PSE&G's demonstration a fourth element was added in the form of a duct heater located outside of the unit, but wired to the control circuit of the night heating section. This brought the size of this section above the then estimated 17 kW heat loss of the home, and therefore added a safety factor.

TES Water Heater: Conventional electric water heating is accomplished using a storage tank with an upper and lower element which are usually interlocked to prevent simultaneous operation. As hot water is drawn from the tank and replaced with incoming cold water, thermostats located on the tank will activate the electric elements to reheat the water in the tank to operating temperature.

The storage concept that has been applied to water heating uses a larger tank and allows the electric elements to operate only during fixed daily off-peak periods to heat water. During these off-peak periods the electric elements operate identically to a conventional unit to heat the incoming replacement cold water to operating temperature and to heat the cold water already in the tank. During the on-peak period the electric elements are kept off. In a storage water heater, the cold replacement water is piped into the bottom of the tank. As hot water is drawn from the top, this cooler water rises. Typically, a storage water heater can supply 80-90% of its rated capacity at temperatures suitable for domestic needs.

The TES water heater selected was a 454-liter (120-gallon) A. O. Smith Conservationist with load-leveling controls. The water heater has two elements; a single-staged 4.5 kW element located in the top portion of the tank, and a two-staged element in the lower portion of the tank. The two-staged element is composed of a 1 kW and a 2 kW element located adjacent to one another. The unit has four thermostats, spaced from the top to bottom of the tank. The elements are enabled or allowed to come on only during the off-peak period. During this period, the four thermostats give

A hydraulic limiter in the storage section is connected in series with the storage unit's safety contactor and the solid state charge control to prevent overheating of the units. The damper section controls (1) air movement through the heat storage section, and (2) the volume of bypass air. The control section, or logic panel, which is not shown in the schematic, contains all the sensing and control devices for the unit.

During charging of the unit, the charge level of the storage section varies with outdoor temperature (Figure 3). The control section senses outdoor temperature and storage section temperature; thus, only those electric heating elements necessary to bring the storage section to a predetermined temperature and to level the storage charging over the on-period are energized. When the house thermostat calls for heat, the dampers are set so that circulated house air only flows through the night heater. During on-peak periods, the storage section heating elements and night heater are deenergized; then, when the house thermostat calls for heat, the dampers route a portion of return house air through the storage section.

At present, TPI units are available in four sizes ranging from 21 to 30 kW. In addition to the 30-kW unit, priority control to the upper 4.5 kW element. The lower elements are controlled by the lower thermostats and may operate simultaneously. The maximum demand of the water heater may not exceed 4.5 kW.

There is no automatic operation during the on-peak period. In the test homes, an override device or guest-cycle timer has been provided which can be energized for cycles up to 1-1/2 hours each through manual operation of this timer. In this study, the homes selected did not activate their guest switches during the day used in the analysis.

Control Homes: The control homes utilize a four element, 22 kW, 240-V single-phase electric warm air furnace to provide space heating. This furnace operates by engaging all elements whenever the day/night thermostat calls for heat. Based on an 18.5 kW heat loss, the conventional furnace is sized 1.2 times the heat loss. Water heating is accomplished with an 303-liter (80-gallon) water heater having 4.5 kW interlocked elements.

Base Appliances: Both test and control homes selected for study contain an electric central air conditioning unit, an electric range, a dishwasher, and other electric appliances normally found in a total-electric home.

Data

In analyzing impacts of space heating systems on the utility system, the critical design condition occurs during a peak day. While the load duration does affect the life of a distribution transformer, the sizing criteria used by most utilities generally is based upon peak conditions, including load duration. During the PSE&G demonstration, the day December 25, 1980, provided an opportunity for observing peak conditions for several reasons. First, the homes involved are located near a southern New Jersey weather station which recorded an average 24 hour dry-bulb temperature of -13.7°C (7.3°F) and an average 24 hour wind-speed of 4.56 meters per second (10.2 miles per hour). These conditions were quite close to design conditions for southern New Jersey. And second, the occupancy levels were high (homes selected were occupied during the entire day). Figure 4 illustrates the half hour integrated demand for the TES and conventional homes on December 25, 1980.

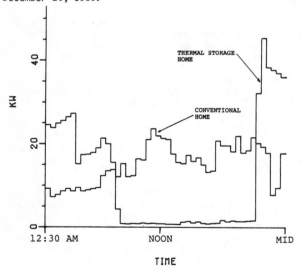

FIGURE 4 - INTEGRATED DEMAND FOR TES AND CONVENTIONAL HOME FOR DECEMBER 25, 1980

Distribution System Analysis

For the secondary and primary distribution system analysis, two case studies were completed. Case Study I analyzed the TPI unit as tested with a storage section and night heating section. Case Study II analyzed a proposed TES system redesigned to lower power demand by eliminating the night heating section and operating solely off the storage section. Based upon the data collected from these homes, designs for the service entrance, secondary service, and distribution transformer are discussed in the following paragraphs.

Service Entrance: All electrical work in the State of New Jersey is governed by the State's Uniform Construction Code and refers to the most recent edition of the National Electrical Code (NEC). Electrical service entrance sizing is referred to in Article 220 - Branch-Circuit and Feeder Calculations. Using the Optional Calculation for Single-Family Dwellings, as stated in Section 220-30 of the 1981 Edition of the Code, the following calculation would result for the conventional total electric home used in this study.

TABLE 1 - OPTIONAL CALCULATION METHOD
NATIONAL ELECTRICAL CODE

Connected Loads	Watts (W)
General Lighting Load	4,500
Two 20-amp Appliance Circuits	3,000
Laundry Circuit	1,500
Range	12,000
Water Heater	4,500
Dishwasher	1,200
Space Heating	22,000
Clothes Dryer	5,000
TOTAL	53,700

Using NEC Diversity Rules	
First 10 kW @ 100% =	10,000
Electric Space Heating @ 65%	14,300
Remainder @ 40%	8,680
	32,980

Serving these loads at 240-V would require 137-A service entrance, which means a 200-A service would adequately supply the home.

In the TES homes, the space heating load was to be a maximum of 43.625 kW. This made the required size of the service entrance equal to 196 A. Applying a 20-25% safety margin, necessitated an above 200-A service entrance. Using this safety margin implicitly changes the diversity of 65% used for conventional space heating to 85-100% diversity for storage heating. This additional margin was necessary since codes do not yet account for TES systems. Because of the difficulty in obtaining an odd-size 250-A service, a 300-A service entrance was selected.

Secondary Service: In New Jersey, all new residential developments comprised of three or more homes must have electricity supplied by buried underground distribution (BUD) systems and the case studies in this paper will follow PSE&G BUD standards. Homes with resistance heating (baseboard or central furnace) or heat pumps, normally require 200-A rated customer service entrance equipment and would be provided a 1/0 aluminum service cable. Homes with TES furnaces require customer service entrance equipment in excess of 200-A rating and would be provided a 4/0 aluminum service cable.

Distribution Transformer: Most utilities base their distribution transformer loadings standards on the American National Standards Institute (ANSI) Appendix C57.91 (1974) Guide for Loading Mineral Oil - Immersed Overhead - Type Distribution Transformers with 55°C or 65°C Average Winding Rise. This loading guide is conservative in the sense that the expected loss of insulation life for a single recommended overload will not be greater than the amount stated in the tables. PSE&G selects distribution transformer sizing (kVA) based on an initial loading between 60% and 120% of transformer nameplate rating, thus allowing sufficient time to elapse before load growth would require replacement due to overload.

Case Study I

Based upon the preceding design criteria, incremental utility costs for typical BUD layouts were calculated. Table 2 shows the incremental utility costs for a 50% saturation of TPI customers per transformer. Table 3 shows the costs for a 100% saturation per transformer. Table 4 shows the size of distribution transformer selected for both penetrations. Based on experience relating to the numbers of customers connected per transformer, an average incremental cost of approximately $173 per TPI customer can be expected.

TABLE 2 - INCREMENTAL UTILITY COST FOR 50% TPI TES
CUSTOMERS PER DISTRIBUTION TRANSFORMER

Customers Per Transformer	Distribution Transformer	Service Cable	Total Cost Per Transformer	Average Cost Per Customer
2	$215	$ 47	$262	$262
4	455	126	581	291
6	0	221	221	74
8	*			

TABLE 3 - INCREMENTAL UTILITY COST FOR 100% TPI TES
CUSTOMER PER DISTRIBUTION TRANSFORMER

Customers Per Transformer	Distribution Transformer	Service Cable	Total Cost Per Transformer	Average Cost Per Customer
2	$215	$ 94	$309	$155
4	455	252	707	177
6	*			

TABLE 4 - TRANSFORMER NAMEPLATE RATING FOR
TPI TES CUSTOMERS

Customers Per Transformer	Transformer Nameplate Rating (kVA)		
	Conventional Electric Heating	50% TPI Customers	100% TPI Customers
2	50	75	75
4	100	167	167
6	167	167	+

* Not possible with one distribution transformer with initial design criteria.
+ Exceeds initial design criteria for a single distribution transformer installation.

Proposed Redesigned TES System

As a prelude to further field testing, an algorithm was devised which would use the actual heating data from the test to investigate the use of the TPI system, without the auxiliary furnace. That is, there is no night heating section, or duct heaters incorporated in the device. During off-peak or charging periods, the unit charges or restores pile temperature and coincidentially "bleeds" off energy to the home interior. During the on-peak periods, the design provides house heating requirements, as does the TPI system described earlier.

In this analysis, it was assumed that the two disconnected elements would be reconnected, bring the pile demand up to 28.9 kW. In addition, the night heating section and duct heater would be disabled. Figure 5 depicts the revised total simulated home load for the redesigned TES system. The reader will note the marked load leveling from the previous load shape on the storage home. Most importantly, this system would reduce peak distribution impacts by approximately 13 kW of 28% of the total home load.

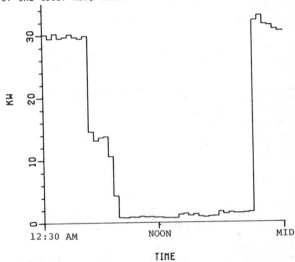

FIGURE 5 - SIMULATED INTEGRATED DEMAND OF
PROPOSED REDESIGNED TES HOME

Redesigned Service Entrance: Substituting a 28.9 kW TES heating load in the service entrance size calculation would yield a total expected house demand, for sizing purposes, of 37,465 W or 156 A at 240 V. This would easily be supplied with a conventional 200-A home service effecting considerable savings in the customer's installation costs. A summary of the service entry design for Case Study I and II is shown in Table 5.

TABLE 5 - SUMMARY OF SERVICE ENTRANCE SPECIFICATIONS

	Conventional Home	TES Home	
		Case I	Case II
Heating Load (W)	22,000	43,625	28,900
Connected Load (W)	53,700	75,350	60,600
Rated Load (W)	32,980	47,040	37,465
Rated Current (A)	137	196	156
Service Entrance (A)	200	300	200
Secondary Cable	1/0	4/0	1/0

Case Study II: Based upon the redesign TES unit and the preceding design criteria, incremental utility costs for typical BUD layouts were calculated. Table 6 shows the incremental utility costs for a 50% saturation of redesigned TES customers per transformer. Table 7 shows the costs for a 100% saturation per transformer. Table 8 shows the size of distribution transformer selected for both penetrations. Based on experience relating to the numbers of customers connected per transformer, an average incremental cost of approximately $61 per redesigned TES customer, can be expected.

TABLE 6 - INCREMENTAL UTILITY COST FOR 50% REDESIGNED TES CUSTOMERS PER DISTRIBUTION TRANSFORMER

Customers Per Transformer	Distribution Transformer	Service Cable	Total Cost Per Transformer	Average Cost Per Customer
2	$215	$ 0	$215	$215
4	455	0	455	218
6	0	0	0	0
8	*			

TABLE 7 - INCREMENTAL UTILITY COST FOR 100% REDESIGNED TES CUSTOMER PER DISTRIBUTION TRANSFORMER

Customers Per Transformer	Distribution Transformer	Service Cable	Total Cost Per Transformer	Average Cost Per Customer
2	$215	$ 0	$215	$108
4	455	0	455	114
6	*			

TABLE 8 - TRANSFORMER NAMEPLATE RATING FOR REDESIGNED TES CUSTOMERS

Customers Per Transformer	Transformer Nameplate Rating (kVA)		
	Conventional Electric Heating	50% TES Customers	100% TES Customers
2	50	50	50
4	100	100	100
6	167	167	167
8	167	167	*

* Not possible with one distribution transformer with initial design criteria

+ Exceeds initial design criteria for a single distribution transformer installation.

Residential Primary Circuits and Substations

In Case Study II, an analysis of the impact on primary circuits and substations was made. Approximately 50% of the distribution circuits on the PSE&G system are predominantly residential in nature and are in non-metropolitan or outlying areas. The circuits serving this area have a sizeable number of "all electric" homes. These areas have the greatest probability of housing development growth and have the highest probability of TES heating growth. Of the estimated 121,000 residential electric heating customers on PSE&G's system in the year 2000, over 90% (109,000) will be on residential distribution circuits.

Assuming the 109,000 customers would be evenly divided between resistance heating and heat pumps, the number of primary circuits and substations required to supply the load was determined. The coincident loads of these customers on the circuits result in a winter daytime peak which is 22% greater than the winter evening peak. This is due in part to the fact that there is no utility load control on the resistance heating or heat pump systems considered in this paper.

Calculations on the effects of the addition of TES heating were based on the premises that the total number of "electric heating" type residential customers will remain at 109,000 (i.e., TES customers will be replacements for some resistance heating and heat pump customers used in the base calculation). Also, the designed electrical demand on a 10,000 kVA, 13-kV primary circuit will be averaged at 8,500 kW due to the addition of new circuits required annually to serve a growing distribution load.

On a system basis, PSE&G's defined off-peak rate period is 9:00 a.m. to 7:00 a.m. EST. The coincident peaks of the resistance heating and heat pump customers approximate 82% of their day peak at 9:00 a.m. Because TES heating peaks as it comes on, the overlapping of TES load with other electric heating load in the 9:00 p.m. to 10:00 p.m. period creates a new distribution system peak. This either limits the number of TES customers which may be accommodated and/or requires additional facilities to serve the same number of customers.

Based on PSE&G's forecast of 50% TES saturation in the year 2000, an analysis of the impact on primary circuits and substations is illustrated on Table 9. This analysis reveals that for the 9:00 p.m. to 7:00 a.m. off-peak period this saturation would require an additional 8 substations and 61 13-kV primary circuits.

Resistance heating and heat pump loads drop off significantly around 11:00 p.m. By delaying the start of the thermal storage charging cycle (off-peak period) until 11:00 p.m., a much better daily circuit load factor can be realized. Additional comparisons were made under these conditions. The results of these calculations are shown in Table 9. Only one additional substation and nine 13-kV primary circuits would be required under this condition, and the 50% reduction in primary circuit day peak occurs at a time when the system is experiencing its winter day peak which results in a reduction of the system winter peak.

A comparison was also made of retaining an 8,500 kW day peak and coincidentally increasing TES customers so that the night peak was also 8,500 kW, using both of the above off-peak periods. This resulted in reduction of substation and primary circuit facilities required. This increased the number of customers on the circuit, but necessitated a reduction in the saturation of TES customers. The results indicate that the ideal "mix" of customers would be 10% TES saturation for those with the off-peak period starting at 9:00 p.m. and 27% for those with the off-peak period starting at 11:00 p.m. The results are also shown in Table 9.

An additional comparison was made maintaining the number of customers on each circuit at the same level as in the base case but allowing the circuit to peak at night. By varying the combination of TES customers, the same facilities are required as in the base case, but, as is shown in Table 9, maximum TES saturation would be 21% for those starting the off-peak period at 9:00 p.m. and 47% for those starting the off-peak period at 11:00 p.m. Also by using a 11:00 p.m. to 9:00 a.m. storage cycle, the primary circuit day peak would be only 55 percent at the time most advantageous to the system.

TABLE 9 – EFFECT OF REDESIGNED TES SYSTEMS ON NEW RESIDENTIAL PRIMARY CIRCUITS AND SUBSTATIONS

	Base Case	50% TES Saturation		Equal Day/Night Demand		Night Peak of 100% Customers Per Ciruit	
		Storage Cycle		Storage Cycle		Storage Cycle	
		9:00 p.m.- 7:00 a.m.	11:00 p.m.- 9:00 a.m.	9:00 p.m.- 7:00 a.m.	11:00 p.m.- 9:00 a.m.	9:00 p.m.- 7:00 a.m.	11:00 p.m.- 9:00 a.m.
Heating Saturation (%)							
Resistance	50	25	25	45	36.5	39.5	26.5
Heat Pump	50	25	25	45	36.5	39.5	26.5
Thermal Storage	0	50	50	10	27	21	47
Requirements							
Number of 13kV Circuits	248	309	257	224	184	248	248
Number of Substations	31	39	32	28	23	31	31
Day/Night Peak Ratio (%)	122	42	50	100	100	80	55
Incremental Utility Costs							
Total (1980 x $1,000,000)	Base	37.1	5.05	(14.25)	(38.00)	0	0
Cost Per TES Customer	Base	681	93	(1307)	(1291)	0	0

This study of primary circuits and substations concludes that there can be electric thermal storage heating saturations of from 10 to 47%, depending on the mode of operation and number of customers per circuit with no increase in primary circuit and substation costs. The number of circuits and substations could be reduced, thereby saving costs from the base case values.

DEMAND LIMITING DEVICES IN RESIDENTIAL LOAD MANAGEMENT

The Omaha Public Power District (OPPD) is a municipal utility involved in the generation, transmission, and distribution of electric power. In 1980, OPPD served 212,000 customers distributed over 5,000 squares miles. The majority of the load falls in residential and medium commercial classes. Peak loads were 1,304 MW in the summer and 925 MW during the winter. Total customer energy sales of $5,144 \times 10^6$ MWh resulted in a load factor of 46% in 1980. Annual load growth is predicted to be between 3.1 to 3.5% until 1990. The customer mix and contribution to peak demand is shown in Table 10.

TABLE 10 - LOAD DATA BY CUSTOMER CLASS

Customer Class	Fraction of Total Load	
	Winter	Summer
Residential	35.5%	54.1%
Greater Than 5000 KW Demand	6.6%	5.9%
Other Commercial and Industrial	57.9%	40.0%

While all-electric homes represent a small fraction of the total population, their peak demand coincides with the system winter peak. However, other homes which do not use electricity for heating are heavy users of power in the cooling season, approximately 300 MW at the system summer peak. For these reasons, OPPD is participating in a two-phase project to reduce air conditioner use at summer peak times, and to implement demand-limiting devices year-round in all-electric residences. The results of this project will be evaluated to determine if wider use of these methods will enable the utility to defer construction of new generating units needed to meet peak loads. The air conditioner control portion of this project will be described in a separate report.

Test Concept

There are 50 residences involved in the demand limiting experiment. Local demand limiting devices have been installed in each of these homes as shown in Figure 6 to keep peak power use within bounds set by the consumer. Each device measures peak power use and limits that use to a level selected by the customer by cutting off power to selected electric loads. Associated data acquisition equipment reads and records energy consumption for billing use. The utility can exert direct control over the local devices to lower the demand setting when desired by means of the consumer's telephone line. A customer override is provided to negate utility control signals. The controlled appliances include heat pump, baseboard, or resistance space heating, central air conditioning units, hot water heaters, and clothes dryer resistance heat coils. The customers are billed on an experimental demand rate; no further incentives are given. The project began in the spring of 1980, data collection was initiated in January 1981 and data analysis will continue through November 1982.

Customer Selection: Customers were recruited from a group that met the following requirements:

o All-electric residential service, including central air conditioning; heat pump, baseboard or forced air resistance heating; electric water heating; and electric clothes dryer.

o Single family dwelling.

o One year prior service to provide a base for group classification.

o Home located near the selected telephone switching office; this was necessary to allow mass addressing of customers via the telephone equipment at that office.

The residential all-electric customers were stratified by energy using during the winter peak month as shown in Table 11. The sample size was calculated for a 90% confidence level, and 5% sampling error about the mean. A previous experiment had indicated an expected move-out rate of 13% and an equipment failure rate of 30%. Using these figures, 39 customer points were required; however, 10 points were considered the minimum count for validity within any one stratum. Thus 50 controlled points were included in the final project.

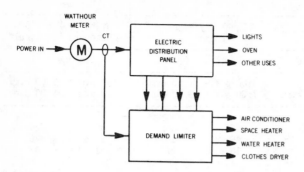

FIGURE 6 - DIAGRAM OF A RESIDENTIAL DEMAND LIMITER
TO CONTROL SELECTED ELECTRIC LOADS

TABLE 11 - CUSTOMER STRATIFICATION BY ENERGY USE

Stratum	kWh/Peak Month	Sampling Meters
1	0 - 3000	10
2	3001 - 5500	15
3	5501 - 8500	15
4	Greater than 8500	10

Customers were selected at random from within the strata, and the final sample included 19 heat pump systems and 31 central resistance heat systems. A similar uncontrolled group was chosen from the OPPD load survey population for comparison to this group.

No incentive was offered to the consumers, except the opportunity to reduce their cost of electric service by altering usage patterns to take advantage of the demand rate. This rate consisted of a customer minimum charge, a demand charge, and an energy charge; in contrast, the standard residential charge is based on a declining block rate. A high load factor (defined as average load/peak load) is encouraged by demand rates. The rates used are detailed in Table 12.

TABLE 12 - OPPD RESIDENTIAL ELECTRIC RATES

Experimental Demand Rate

Minimum Charge	$3.95 /month	
Override Charge	$0.50 /override	
Energy Charge	$0.013/kWh	

Demand Charge	Summer	Winter
0 - 2 kW	$7.50/kW	$5.65/kW
Greater than 2 kW	$6.50/kW	$3.50/kW

Standard Residential Declining Block Rate

Minimum Charge	$3.95 /month	

Energy Charge	Summer	Winter
0 - 300 kWh	$0.0396/kWh	$0.0396/kWh
301 - 1200 kWh	$0.0396/kWh	$0.0285/kWh
Greater than 1200 kWh	$0.0331/kWh	$0.0239/kWh

Summer: June - September
Winter: October - May

Customer Education: A continuous program of customer education was undertaken to explain to participants why peak demand reduction was beneficial to the utility and the customer, and how a demand limiting device could best be used to reduce peak demand and costs. As part of the process of signing up randomly selected customers, OPPD personnel visited each selected customer home for one to two hours for a detailed explanation of the purpose of the project and demonstration of the equipment involved. They also described the standard residential rate and the experimental demand rate. When the equipment was installed, owner's manuals provided by the manufacturer were given to customers. For the next few months, follow-up letters were mailed containing further information. A survey of customer demographics and attitudes was conducted at the start of the project, and feedback was encouraged during operation by means of preaddressed postcards and surveys periodically included in monthly bills.

Customers on the program were supplied with a self-adhesive sticker to place near the home telephone. This sticker contained the program name and a special telephone number for service. A log of all calls or complaints was kept by OPPD service personnel.

Communication System: During a previous project in Omaha involving telephone communication for remote meter reading, the customer's existing telephone connection was used for communication between the residence and the control hardware. This method proved effective and was selected for the communication medium.

Demand Limiter Control

A Dencor residential demand limit controller was installed in each home. This device calculates 15-minute peak demand from kWh readings over the period. It uses current sensors on the 240-V input line to determine the load by multiplication with a set voltage value; the voltage can be manually reset by utility personnel when adjustment is necessary. Space conditioning units are controlled by means of the 24-V thermostat wiring, and other loads are controlled by opening contacts in series with the load. When power usage is found to exceed the set point, this controller opens contacts to deenergize selected loads in time-delayed steps until power use fails within the desired range. The shedding priority is determined at installation of the device. Loads are restored in reverse order from the shedding sequence, so that the last shed is the first restored. If all controlled loads are disconnected and power use exceeds the set point, a buzzer sounds to alert the consumer that a new peak will be set unless the uncontrolled loads are manually reduced. A custom expansion circuit board was designed and built by the manufacturer to allow direct control of this unit by OPPD upon command via telephone lines. Figure 7 shows the demand limiter containing electronic logic circuits and appliance relays, an the current transformers used at the 240-V service to the house.

The companion Dencor set station shown at the lower left was used by the homeowner to set the desired peak power level, and to override utility control signals during any 15-minute period. Pulse initiators were fitted to each electric watthour meter to allow digitizing the meter output for input to the D511 transponder. The Darcom D511 load survey transponder stored in volatile memory 15-minute energy use for up to five days, until the central memory could poll and clear each unit. A battery backup helped prevent loss of data during short power outages. The Darcom Automatic Meter Reading (AMR) transponder stored the number of times that the override switch was used to negate direct utility control.

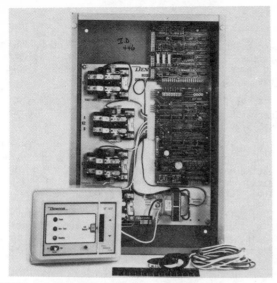

FIGURE 7 - DENCOR RESIDENTIAL DEMAND LIMITER, DEMAND
SET STATION, AND CURRENT TRANSFORMERS

An Energy Controller reduces the customer set point for the 15-minute demand period in progress by means of frequency-coded signals over the telephone line. This controller was located at the telephone switching office to enable easy access to the telephone system mass address frame. All data from D511 and AMR transponder memories was read and recorded on magnetic tape cassettes by the Data Acquisition Controller. This controller is also located at the telephone switching office; two modems provide communication between OPPD offices and the two controllers.

The D511 was read twice weekly to obtain energy use. Data collected at the 50 homes included total 15-minute kWh use at each home, the status of the set station at each home, and a count of the number of times the override switch was used. There were also two magnetic tape recorders in the neighborhood which collected ambient temperatures.

The Energy Controller automatically sends out load control signals using a preprogrammed schedule determined by the utility. The frequency-coded signal sent over the customer's existing phone line causes a contact closure, reducing the peak kW set point for a 15-minute interval. An override switch allows the customer to reset the original peak for the remainder of the time period. Three levels of demand reduction are possible. The amount of load relief will depend on the amount of connected load at the time the control signals are sent.

The control strategy was varied during the course of this project. An initial period of adjustment was allowed from December 1980 to March 1981 to enable the consumers to become familiar with the device and its operation. In March 1981, a control scheme was begun which allowed control from 9:00 a.m. to 3:00 p.m. on Monday, Wednesday, and Friday of each week. This was changed in the fall of 1981 to a weather-dependent schedule in which control would be exercised on days when the forecasted minimum temperature was below 15°F. The times of operation were reset at 7:00 a.m. to 1:00 p.m. and 3:00 p.m. to 10:00 p.m. The days were varied randomly to prevent the customers from recognizing and reacting to a pattern. This schedule was planned to include the winter peak load periods in the morning and evening while allowing some energy catch-up at midday.

During normal operation, personnel from OPPD visited the telephone switching office twice weekly to change tape cassettes in the data acquisition unit and to take the data collected to OPPD offices for storage and analysis. Occasional power outages or tape jams have necessitated more frequent trips to reset the control equipment or individual D511 units.

Several modifications have been necessary during the course of the project. A time delay was designed between successive load sheds in the demand limiter, as well as a time delay in each 15-minute period before any load can be shed. As a result, the peak demand can exceed the desired value even during correct operation of the unit. Many central space conditioning units and thermostats also incorporate a time delay, typically around five minutes. When these two delays are added, it can be impossible to adequately control the load. OPPD has attempted to bypass this problem by splitting the heating elements of the furnaces involved into two or three stages with different priorities. Heat pumps must perform a defrost cycle periodically in the heating season and control could interfere with this, causing inefficient operation or damage. The heat pumps involved were equipped with a temperature bypass which allowed free operation when ambient temperatures fell below 5°C (40°F).

The power use at the customer's home was calculated by multiplication of the actual current reading and a fixed voltage. If voltage was not within the range designed in the demand limiting device, the resulting power reading is in error. A calibration adjustment was necessary at many of the homes to allow for voltage values higher or lower than the norm. The dial setting on the unit used by the customer to select the peak power level was designed to ±10% tolerance, but some were found to exceed this range. For billing purposes, peak demand (kW) was measured by the demand limiting device, and total energy use (kWh) was determined by the modified residential watthour meter.

Test Results

Preliminary analyses show that the demand limiting device does help to reduce peak demand both for individual customers and for the composite group. However, frequent direct control by the utility to reduce the demand further was counterproductive, as it caused the customers to reset their peak demand level at a higher value. The net effect when this occurred was that the demand limiting devices were set at too high a level, and had little effect on the individual loads.

Many of the customers made significant changes in their lifestyle, and experienced colder or warmer households than usual. The winter of 1980-81 was unusually mild and offered little opportunity for savings. By July 1981, many customers had lost money on the demand rate when compared to the standard rate and were dissatisfied with results of their efforts. In order to keep the customers from defecting prior to the end of the project, OPPD made the decision to reimburse the customers at that time for any difference which the customers had paid from January through June. It was emphasized that this was a one-time adjustment.

The demand rate can be a crucial element in the success or failure of a project. The consumers were willing to make substantial changes in their patterns of use if they could gain financially from their efforts. A few were unable or unwilling to modify their usage, and demanded the equipment be removed after a short time. The largest consumers of energy show the greatest reductions in demand and have higher load factors than others in the controlled population.

Homes using the demand controller used less energy than a similar uncontrolled group. More extensive analysis will be necessary to quantify this result, but a reduction in comfort levels in the home translates into less electrical consumption. Data analysis is continuing, and more information will be available late in 1982.

After one year on the program, some of the customers were still confused and unsure of the proper actions to take when the buzzer sounded. Customer education cannot stop with the head of the household. It should include all residents in the home, particularly those who are at home for the majority of the time.

Demand limiting equipment shows promise of offering the electric consumer some control in the load management process.

SUMMARY

The two customer options for load management discussed here represent different technologies, but they are alike in several respects. First, both methods require customer recognition of the time-varying nature of the electric system load curve and its implications. Second, both depend on an encouraging a rate struture for their success. Finally, these two methods require the customer to become an active part of the electric utility system. In return, the consumer can be assured of a choice in the future control and use of electric energy and contribute to controlling its cost.

REFERENCES

1. Department of Energy, Division of Electric Energy Systems, "Program Plan for Research, Development and Demonstration of Load Management on the Electric Power System," DOE/ET-0004 (1978).

2. D. T. Rizy, "Utility Controlled Customer-Side Thermal Energy Storage Tests: Heat Storage," ORNL-5796 (1982).

3. C. W. Gellings, J. R. Redmon, J. P. Stovall, T. W. Reddoch, "Electric System Impacts of Storage Heating and Storage Water Heating - Part I of II (Background Information and Demonstration Description),"IEEE Paper No. 81 TD 632-9.

4. C. W. Gellings, J. R. Redmon, J. P. Stovall, T. W. Reddoch, "Electric System Impacts of Storage Heating and Storage Water Heating - Part II of II (Primary and Secondary Distribution System Analysis)", IEEE Paper No. 81 TD 633-7.

SECTION 8
FLUIDIZED BED COMBUSTION

Chapter 32

FLUIDIZED BED BOILER: COMMERCIAL RESULTS AND PRACTICAL GROUND RULES

D. Ormston

Introduction:

Fluidized bed combustion (FBC) has been promoted as being the means by which to:

- Cleanly and efficiently burn conventional fuels such as coal, oil and gas and
- low-cost fuels and wastes such as wood wastes, petro-chemical oily wastes, rubber scrap, etc.,
- and to do so without excessive feedstock preparation such as crushing, hogging and grinding
- in low-cost capital equipment not needing stokers, precipitators, scrubbers, chaingrates, and so on,
- and be simple to operate, with low maintenance

These objectives should be the yardsticks by which to measure FBC plant design and performance.

This paper analyses conventional (or "vertical") FBC plants and defines four weaknesses inherent within their design. As the result of these weaknesses, and the range of remedial measures necessary to overcome them, the plants built so far have, in the main, departed from the original objectives as set out above. The plant equipment and operations have become far more complex, and the FBC's advantage of using low-cost alternatives have been whittled away by extra preparation and running costs.

The paper then sets out the origins, development, and commercial installation of a simple, proven and thoroughly effective FBC process which overcomes all of the weaknesses apparent in conventional FBCs. This FBC process eliminates the need for excessive fuel preparation, it simplifies FBC operation, and obviates downstream recapture/recirculating chambers. 10 plants have been built in Britain, Japan and Sweden. The most advanced of these is described. It has run continuously since October 1980, producing 6,000-10,000 PPH steam.

Finally, the paper sets out practical non-theoretical ground rules by which to design and operate FBC plants that will work reliably and efficiently.

This paper is in four sections:

Section 1: Conventional (Or "Vertical") Fluidized Combustion.

Section 2: Non-Conventional "2nd Generation" Circulating FBC

Section 3: Commercial Circulating FBC Water-tube Boiler Results

Section 4: Practical Rules To Design And Operate Commercial FBC Plants

SECTION 1.0 - CONVENTIONAL OR "VERTICAL" FLUIDIZED COMBUSTION DESIGNS

1.1 Aircraft With It's Wings On Upside Down

Most of the FBC plants built so far can be compared to building an aircraft with its wings on upside down. The aircraft will not fly. It can be made to fly by adding wings upfront and at the tail: by adding vertical thrust: by adding more power. Such changes would demand more complex controls, more capital expenditure and more operating costs. All of these repercussions start with the one design flaw - "the wings are on upside down".

Similarly, conventional (or vertical) FBC designs have three physical characteristics which lead to four inherent flaws in the same way as the aircraft has if its wings are on upside down. From these four basic flaws flow some sixty additional secondary problems, which do much to make conventional FBC plants far more complex, and more difficult to operate than is necessary.

This section of the paper now diagnoses these weaknesses, traces the resultant effects, and then shows how the symptoms rather than the causes have been treated.

1.2 Conventional/Vertical FBC Design Characteristics

The conventional/vertical fluidized bed design is shown in Figure 1. It has three main characteristics:

1.2.1 a horizontal distributor plate
1.2.2 vertical, uniform, fluidizing air flow
1.2.3 vertical walls

Whenever these three design characteristics are present, four serious weaknesses follow:

1.3 Four Inherent Weaknesses

The conventional/vertical FBC has four inherent weaknesses, shown in Figure 2.

1.3.1 Residence time of fuel in the bed is short
1.3.2 Lateral mixing is poorer than optimum
1.3.3 Elutriation of fine particles of carbon/sorbent/bed is high
1.3.4 Ash/heavy incombustibles are difficult to remove.

Flowing from these four (wings on upside down) faults, are a range of symptoms, which engineers have counteracted in various ways:

1.4 Short In-bed Residence Time: Repercussions, And Treatment Of Symptoms

1.4.1 Main repercussions: - Below optimum carbon burnout
  - Below optimum reduction of toxic products of combustion.

CONVENTIONAL FLUIDIZED BED

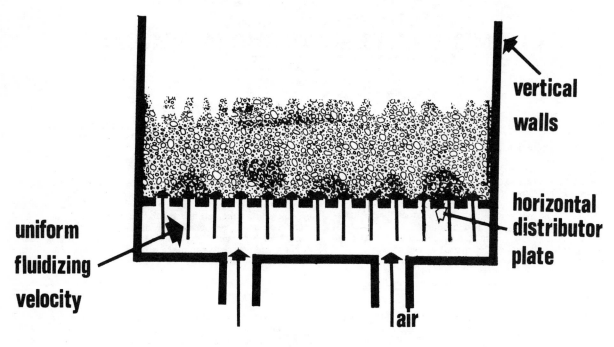

vertical walls

horizontal distributor plate

uniform fluidizing velocity

air

FIG.1

FOUR INHERENT WEAKNESSES - CONVENTIONAL FBC

1. short in-bed residence time

t

2. poor lateral mixing

3. high elutriation

4. difficult to clear heavy incombustibles

FIG. 2

1.4.2 Consequences:
- Lowered thermal efficiency
- Lowered combustion efficiency
- Incomplete burnout of toxic gases
- Lowered thermal output
- More feedstock preparation
- More equipment to re-capture and re-circulate unburned carbon
- More sorbent to kill SOx
- Higher excess air
- More secondary air to oxidize CO to CO_2

1.4.3 The treatment of the symptoms (i.e., not treatment of the causes).

- Deepen bed (increase fan size)
- Re-circulate the carbon (and other solids)
- Inject super-prepared feedstock from below the bed
- Lower bed velocity (and heat flux/ft^2)
- Increase freeboard and accept freeboard combustion
- Add more carbon than optimum
- Add more sorbent than optimum
- Avoid feeding light solids and liquid fuels
- Add a second bed in the freeboard
- Add freeboard burners to achieve burnout

1.4.4 All of the remedies in 1.4.3 flow from a failure to achieve full and proper combustion within the fluidized bed itself. So, the poorer the fluidized bed's performance, the greater the emphasis on remedying the symptoms.

The better the fluidized bed's performance, the fewer will be the additional remedies.

1.5 Poor Horizontal Mixing:

1.5.1 Main repercussion:

- Bed fails to thoroughly mix fuel, sorbent, bed material, air, products of combustion and heat

1.5.2 Consequences:

- Parts of bed become starved of fuel (inadequate temperature)
- Other parts clinker the fuel
- Temperature variations cause control problems
- Heavy solids sink to bottom and locally block the distributor
- Poor catalytic action in parts of bed
- Higher than optimum levels of air/sorbent
- Difficulty of start up from cold

1.5.3 The treatment of the symptoms:

- Spread the fuel with a "spreader stoker"*
- Or with a multiplicity of feed points and splitters
- Introduce "bubble caps" to prevent blockage of distributor, and a dead zone of bed material
- Introduce partial-bed start up procedures
- Start up on shallow bed, and deepen bed gradually as temperature rises
- Accept a higher than optimum Ca/S ratio
- Accept a higher than optimum level of excess air
- Restrict fuels to those which readily burn

1.5.3 The Treatment Of The Symptoms: (Continued)

- Avoid smoky and smelly fuels

*This increases elutriation of unburned carbon, and causes clinkers on bed surface.

1.5.4 All of the "remedies" in 1.5.3 flow from a failure to achieve full and thorough lateral/horizontal mixing within the fluidized bed itself.

1.6 High Elutriation
Repercussions And Treatment Of Symptoms

1.6.1 Main repercussions:
- Solid materials, bearing heat (and unburned carbon) are carried out of the in-bed combustion zone.

1.6.2 Consequences:
- Loss of sensible heat in the solids
- Loss of unburned carbon-lowered combustion efficiency
- Combustion takes place in freeboard (or not at all)
- Blockage of downstream equipment - firetube boilers, baghouses,
- Instruments fail
- Fires and downstream explosions

1.6.3 The Treatment Of The Symptoms:
- Build FBC with very high freeboards
- Add a re-capture and recirculation chamber
- Add a carbon burnout cell
- Lower fluidizing velocity
- Add more dust collection equipment
- Have outage time to clear blockages in boiler firetubes
- Add heat exchanger to recover sensible heat in elutriated solids
- Add a second bed to the freeboard
- Increase bed material particle size
- Avoid feeding light particles/light oils
- Add a cyclone in the furnace zone

1.6.4 All of the "remedies" in 1.6.3 flow from a failure to "kill" high elutriation at its source.

1.7 Difficult Ash Removal:
Repercussions And Treatment Of Symptoms

1.7.1 Main repercussion:
- Heavy incombustibles block distributor and defluidize parts of bed.

1.7.2 Consequences:
- Loss of full fluidization
- Less air distribution throughout bed
- Bubbling and tracking
- Cold spots/hot spots
- Reduced thermal output
- Reduced mixing
- Downtime

1.7.3 The Treatment Of The Symptoms
- Introduce ash slots (and hope ash finds them)
- Introduce "bubble caps"
- Have a defluidized layer at bottom of bed
- Crush/shred/hog fuels to eliminate ash/metals
- "Blow off" ash
- Introduce wier above bed level to syphon off ash

1.7.4 The "remedies" in 1.7.3 flow from a failure to design the fluidized bed to self-clear itself of heavy incombustibles.

SECTION 2.0
NON-CONVENTIONAL "2ND GENERATION" FBC

2.1 Origins

During 1968-1970, a group of engineers was trying to develop an FBC to burn garbage. They found that paper, and similar light materials, flew off the bed, whilst metals and heavy incombustibles sank to the bottom of the bed to block the distributor.

To overcome these weaknesses, they:

2.1.1 Bent one of the chamber walls inwardly over the bed to contain light particles

2.1.2 Sloped the distrubutor plate to give an airslide effect

This had an immediate and remarkable effect. The bed "circulated" – see Figure 3. The paper was burned to a greater extent <u>within</u> the bed, and heavy incombustibles gathered at the bottom of the sloping distributor.

Extensive pilot plant burning trials showed that such wastes as metal cans came out in the ash, bright and clean, while rubber from tires burned off without smoke or smell, leaving the tire reinforcing metal in the ash - bright and clean. The list of materials burnt is given later in this paper.

ORIGINAL "2nd GENERATION" CIRCULATING FBC (1968)

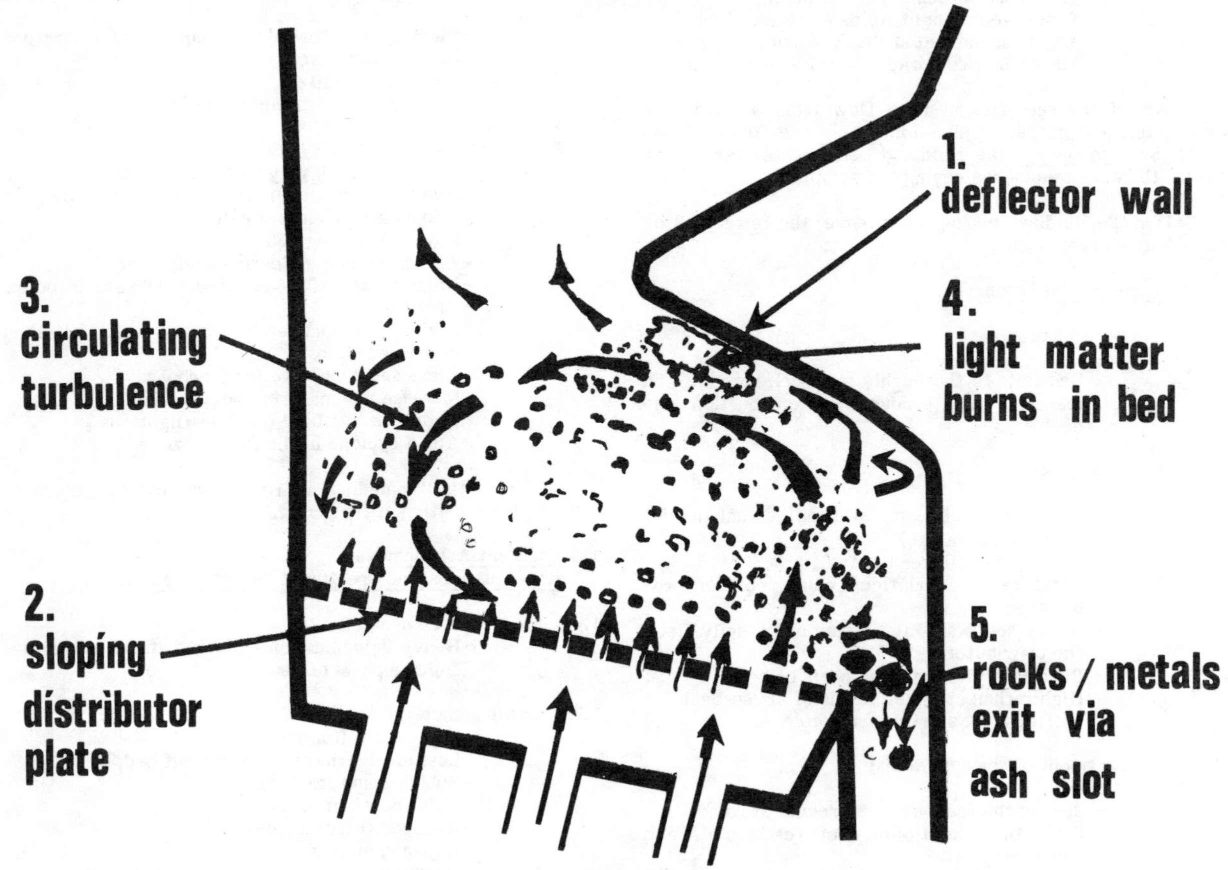

1. deflector wall

3. circulating turbulence

4. light matter burns in bed

2. sloping distributor plate

5. rocks / metals exit via ash slot

FIG.3

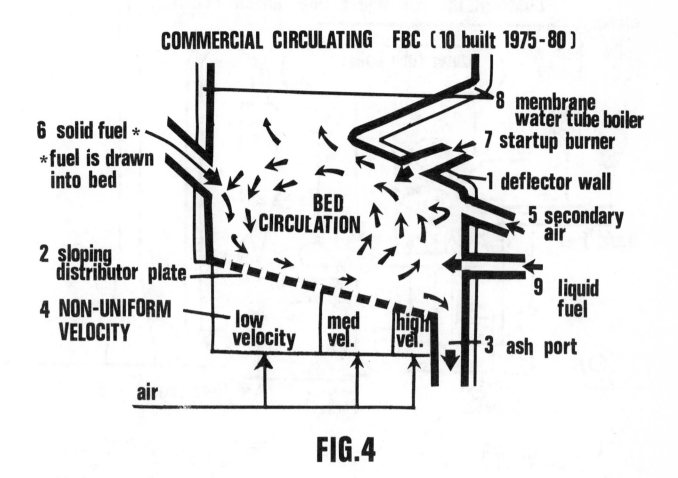

COMMERCIAL CIRCULATING FBC (10 built 1975-80)

6 solid fuel *

*fuel is drawn into bed

8 membrane water tube boiler

7 startup burner

1 deflector wall

BED CIRCULATION

5 secondary air

2 sloping distributor plate

4 NON-UNIFORM VELOCITY

low velocity

med vel.

high vel.

9 liquid fuel

3 ash port

air

FIG.4

2.2 "Circulating" FBC - 2nd Generation Design

Several refinements followed, until the "circulating" FBC had the following characteristics. See Figure 4.

2.2.1   - Deflector wall
2.2.2   - Sloping distributor plate
2.2.3   - Ash port at lower end of distributor plate
2.2.4   - Non-uniform fluidizing velocities (low velocity, medium and high)
2.2.5   - Secondary air over bed
2.2.6   - In-bed fuel injection (solids)
2.2.7   - Over bed start up burner
2.2.8   - Membrane wall boiler
2.2.9   - Liquid fuel - no atomizer required.

2.3 Advantages Of "Circulating" FBC

2.3.1   Increased in-bed residence time
2.3.2   Tremendous lateral turbulence
2.3.3   Very low elutriation
2.3.4   Easy removal of ash/solids/metals

2.4   These improvements overcome the flows in the conventional bed, and lead to:

2.4.1   No need for elaborate fuel distribution, spreader stok-

ers, multiple fuel entry points, underbed feeds (the bed distributes the fuel)
2.4.2   No need for re-capture/recirculation of carbon/carbon burnup cells (combustion in-bed is 99%)
2.4.3   No need for excessively high freeboard/second bed in freeboard (the bed "kills" eclutriation)
2.4.4   No need for baghouse (except for light ash fuels e.g. wood)
2.4.5   No need for in-bed tubes (deflector wall is tube bank)
2.4.6   No need for "bubble caps" (bed self-cleans)
2.4.7   No need for downtime to clear firetube blockages (low elutriation/properly designed boiler eliminates this)
2.4.8   High thermal efficiency (78-83%) in one cohesive unit
2.4.9   High output/ft^2 of grate area
2.4.10 Tremendous range of fuel and wastes can be burned: Tolerance to solids, liquids, gases, fine particles, with full combustion, and tolerance of ash, metals, without undue fuel preparation.
2.4.11 Elimination of temperature variations across the bed - not hot spots -no cold spots
2.4.12 No clinkering
2.4.13 Easy, automatic ash/incombustibles removal
2.4.14 Lowered Ca/S ratio to reduce SOx
2.4.15 Simple start up procedure
2.5.16 Easy to control
2.5.17 Relatively lower overall capital and operating costs

COMMERCIAL FBC WATERTUBE BOILER PLANT

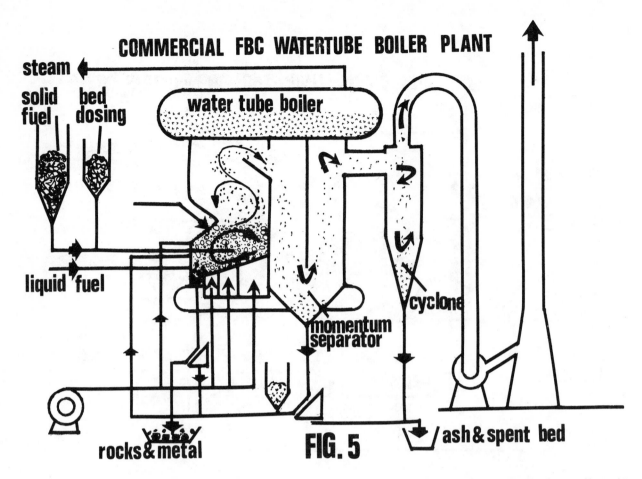

steam

solid fuel

bed dosing

water tube boiler

liquid fuel

cyclone

momentum separator

rocks & metal

FIG. 5

ash & spent bed

SECTION 3
COMMERCIAL "CIRCULATING" FBC
WATER TUBE BOILER PLANT

3.1 This plant raises up to 10,000 PPH of steam from the combustion of solid, liquid or gaseous fuel or wastes, fed singly or in combination.

The plant is shown diagrammatically in Figure 5.

3.2 The boiler is a membrane wall water tube boiler, specifically designed to suit the 6' x 3' circulating fluidized bed. It is crucial that the boiler be designed to suit the FBC, otherwise there will be shortfalls in performance and reliability.

3.3 This particular application burns tarry sludge containing 37% sulfuric acid, and also burns "light ends" waste oils. These liquids are pumped into the bed through 3/8" bore pipes. No atomizing nozzles or tips are required - the bed does the atomizing.

3.4 Bed temperatures are 750°-800°C, uniform to $\pm2^\circ$C. Freeboard temperature is no more than 20°C above in-bed temperature.

3.5 Residence time in-bed is not less than 0.5 seconds through the shortest gas path, but is significantly more for heavy tars and all solids. Due to this, combustion efficiency is 99%. There is no smoke when burning "smoky" fuels such as light ends, or tire rubber.

3.6 In-bed turbulence is tremendous, surf-like violent and thorough. When viewed through a side port, bed particles are seen to be travelling HORIZONTALLY.

3.7 Elutriation is minimized by this dampening effect within the bed. In addition the gas path is forced to overturn several times which assists drop out of particles. As the result the amount of net bed material loss is only 2.2 lbs per hour. This is only a fraction of the figures experienced in conventional FBCs, and reported by others.

3.8 There are no steam tubes IN the bed. Steam tubes are OVER the bed within the membrane deflector wall.

3.9 Ash removal is simple and automatic with this circulating design. The action of the bed "sweeps" rocks, clinker and metals down to the ashport.

3.10 Feedstock preparation needs to be only sufficient to provide reliable feed. Coal can be 1" to 2" size. Wood can be crudely hogged. Rubber tire waste can be 2" to 4" with metal left in. Hydrocarbons can be crude liquids. The circulating FBC does the atomizing, and breaking-up.

3.11 Emission cleanup is by multi-cell cyclone for this application. A cyclone will often be all that is necessary. This is because the bed itself minimizes elutriation, followed by the several momentum separator effects.

3.12 Commercial performance. The entire facility is engineering to British/ASME codes and NFPA safety standards. The client, required the plant to pass stiff standards of performance, reliability and emissions.

This was Britain's first truly commercial FBC watertube boiler plant.

3.13 Modular design: The boiler can be manufactured in sizes of 10,000 PPH, 20,000 PPH, 30,000 and 40,000 PPH, without changing proven factors, and without introducing scale up problems.

3.14 Bed output. The heat output is 770,000 BTUs per sq. ft. of bed area.

3.15 Hours or operation and past experience. Ten plants have been built, 3 of which were pilot plants. The longest running was commissioned in 1975 in Japan giving excellent results in an incinerator application. The British boiler plant was handed over in October 1980, and has run on a continuous basis ever since, without any abnormal downtime.

3.16 Energy savings. The use of the FBC to raise steam/hr. from petrochemical wastes, eliminates completely:

- Prime fuel oil to raise 6,000-10,000 PPH steam/hr.
- Costs of disposal of the waste tars
- Purchase of a new conventional boiler

The payback period was approximately one year to eighteen months.

3.17 Potential fuels. The following materials have been cleanly and efficiently burner in the circulating FBC, since its inception in 1973 at pilot plant status:

Note: all without support fuel. Figures given are "as received" BTU/LB.

Polystyrene	19,500	Carpet waste	8,500
Polyethylene	18,000	Acid tars	8,000
Fuel oil	18,000	Decorative Laminates	8,000
Fried Fats	16,000		
"Light ends"	14,000	Biomass/wood	8,000
Polyamides	14,000	Newspaper	8,000
Carbon Black	14,000	Meat waste	7,000
Poly-methyl-Acrylate	14,000	Brown paper	7,000
Tire rubber	14,000	Shoe leather	7,000
Polyester	13,000	Cardboard	7,000
Coke	13,000	Rotten wood	6,000
Coal(s)	12,000	Garbage	4,500
Wax cartons	11,000		
Phenolic resins	10,000	Garden refuse	3,700
P.V.C.	9,000		

3.18 So What?

When full and complete combustion takes place IN the fluidized bed itself, with low elutriation and proper ash removal:

When the associated boiler is designed to suit the FBC: then,

Fluidized combustion fulfills its promise as the most effective, reliable and versatile means of burning more or less any combustible matter, without resorting to elaborate fuel preparation or double beds, recirculation, carbon burnup, boiler cleanout, and so on.

4.0 PRACTICAL RULES FOR FBC DESIGN AND OPERATION

4.1 Slope distributor

- This assists ash/rocks/metals to fall down to ash port

4.2 Ash port

- Width of bed, not a 6" diameter hole

4.3 Deflector wall over bed

- Increases residence time
- "Kills" particle kinetic energy
- Forces lateral movement

4.4 Smooth distributor

- Avoid bubble caps
- Wire mesh covering (S.S.)
- Drilled distributor plate (S.S.)
- Stainless steel for non-water cooled parts
- Make in sections

4.5 Non-uniform fluidization

- Promote surf-like turbulence
- Get "tumble-drier" effect
- Breaks up laminar flow paths

4.6 Secondary Air

- Directly over bed, assisting circulation
- Promotes lateral movement
- Enhances in-bed burnout
- Be able to vary amount

4.7 In-bed fuel input

- Reduces elutriation
- Facilitates in-bed combustion
- Avoid nozzles, spreader stokers

4.8 Overbed burner (start up)

- Before fuel is introduced
- Plays into bed

4.9 Boiler: to suit bed

- Design boiler to suit bed
- Membrane wall water tube best
- No in-bed tubes if designed properly

4.10 Boiler: modules

- Make each bed 10,000 PPH steam
- 10:1 turndown for 40,000 PPH boiler
- 770,000 BTU/ft^2 distribution area

4.11 Boiler: Freeboard/furnace

- Membrance walls (4)
- 6' to 7' high above deflector wall
- Expansion zone to reduce elutriation

4.12 Boiler: convection bank

- Reverse gas path several times
- Vertical flows only (no horizontal paths)
- Momentum separator to reduce elutriation

4.13 Boiler: sootblowers

Avoid
- Provide sootblowers
- Horizontal surfaces in gas path
- Bubble caps in distributor
- Very large areas of bed in one unit
- Divided responsibility for success
- Ash slots in bed or small ash holes
- Underbed feed of pulverized coal
- Spreader stokers over the bed
- Excessive fuel crushing/handling
- Too much heat transfer surface in bed
- Double beds
- Recirculating systems
- Carbon burn up cells

4.14 Boiler: operating furnace pressure

- Inside freeboard: -1/2" to -1" W.C.

4.15 Bed temperature

- 750° - 800°C

4.16 Temperature probes

- Provide many
- Temperature excursion indicates need for operator action

4.17 Air distribution

- Provide butterfly dampers to facilitate control of bed

4.18 Bed depth

- 3 feet expanded is a good depth

4.19 Bed depth make-up

- By measuring ΔP between air plenum and to just over bed

4.20 Ash classification & re-injection

- Via intermittent operation of vibratory classifier below ash port
- Oversize to "reject" bin (+1")
- Undersize to "reject-fines" bin (dust)
- Remainder pneumatically reinjected into bed
- Reinject into body of bed material, away from boiler tubes and metal parts
- Reinjection system activated from ΔP across bed (bed level)

4.21 Standby oil or gas

- Have source of standby fuel (oil or gas) available, with guns, in case of solid fuel blockage

4.22 Sightports

- Include several

4.23 FBC is a process plant

- Fluidized bed plant is a process plant rather than a boiler.
- All parts of process must work cohesively, e.g. fuel feed, flame safeguard, combustion modulation and control, emission control, switch to standby fuels, electrical/instrument interlocks...
- Make sure user fully appreciates difference
- Few "engineering" firms have handled entirely successful FBC installations.

4.24 Use of FBC boiler

- Use as "workhorse" boiler wherever possible, providing base load
- Use gas/oil boilers, if installed, for peaks of loading

4.25 Weekend shutdowns (if required)

- FBC can be restarted, with two hours to full steam, after weekend shutdown

4.26 Avoid

- Purely vertical fluidization
- Uniform fluidization
- Horizontal distributor plate
- Purely vertical chamber walls

- Horizontal steam tubes in-bed
- Doing the wrong thing more efficiently
- Being hypnotized with R&D formulae and too much theory.

4.27 Provide purge air

- To clear sightports
- To clear instrument probes

4.28 Consider providing

- Standby F.D. and I.D. fans
- Dump hopper for hot bed material in case bed needs maintenance
- Ample working space under bed plenums to dismantle plenums

4.29 Provide preheated air

- For start up
- For bed temperature maintenance in case of fuel interruption

4.30 Provide controls

Record:
- Bed temperatures
- Freeboard temperature
- Flue gas temperature at boiler exit and stack
- O_2
- Steam quantity and pressure

Monitor and control:
- Fuel quantities
- Feed water temperature
- Electric power used
- Supply air temperature and quantity
- Bed material used
- Elutriated material and temperature
- Bed level
- Differential pressure in boiler
- Steam drum levels

4.31 Provide emission measurement

- Build-in the required straight length of ducts to facilitate measurement
- There will be interest in CO, O_2, SOx, NOx, particulates, smoke density: flue gas volume and temperature

4.32 Damp boiler/furnace movements

- Violent circulating turbulence in the smaller FBC boilers will cause inertial movement in the boiler/support structure.

4.33 Scanners

- U.V. scanner (not infrared) for flame detection in over bed burner, because I.R. scanner may be misled by glow from bed.

CONCLUSION

Reliable, continuous and efficient performance has been achieved by "2nd Generation" circulating FBC plants. This commercially proven technology eliminates the weaknesses found in conventional (or vertical) FBC designs, and obviates the need for many of the now commonly accepted extra steps in fluidized combustion operations. The paper gives a thorough diagnosis of the problems in conventional FBCs, and gives a set of practical ground rules for the design and operation of successful commercial fluidized bed plants.

Chapter 33

COMBUSTION OF WASTE TIRES FOR ENERGY RECOVERY

J. N. McFee, R. F. Vance, R. P. Apa

ABSTRACT

A scrap tire fluidized bed combustion system has been demonstrated on a pilot plant scale. The pilot plant tire combustion system was operated using an inert fluidized bed medium for the combustion of scrap tire chunks. Wire removal was accomplished by using a continuous bed media withdrawal and reinjection system which separated the wire magnetically. Fiberglass melting was avoided by careful system temperature control. Combustion efficiency information led to conclusions on the appropriate fluidizing velocity.

Stack sampling methods were used to establish a particulate loading and a particulate size distribution. Sulfur oxide and nitrogen oxide levels were measured and reported. These data were used in designing the off-gas system components.

Based on the success of the pilot plant operation, a tire combustion system to produce 1100 Kg/hr of steam was designed and is currently under construction. The general flow scheme includes a tire chopper, a fluidized bed combustor with a magnetic wire separator, a cyclone, a heat recovery boiler, and a baghouse. The specified off-gas processing components will lead to a stack gas which meets applicable regulations.

INTRODUCTION

The problems associated with scrap tire disposal are only recently coming to light. According to the Society of Automotive Engineers, 200 million truck and automobile tires are discarded in the United States annually.[1] Approximately 42 million (21%) of these tires are reclaimed for retreading. The remaining 160 million tires are disposed of in a number of ways.

The tires disposed in above-ground storage pose a health problem as well as creating an aesthetic problem. As a health problem the tires present a breeding ground for insects and rodents. As an aesthetic problem the nonbiodegradable scrap tire piles are simply unsightly.

Landfill disposal is not satisfactory for whole tires as they tend to "float" to the surface of the landfill as time passes. Shredding or chopping the tires prior to landfill alleviates the standing water problem, but the tire shredding operation is very energy intensive and therefore costly.

Since conventional disposal methods create problems and tires have an appreciable energy content, incineration of the tires seems to be the obvious solution for disposal. To capitalize on existing fluidized bed combustion expertise, Energy Incorporated was contracted to develop the process and perform tire combustion tests for the National Standard Company and the U.S. Department of Energy. This paper describes that effort.

OBJECTIVES

There were two different sets of objectives of the program. The first was to develop and demonstrate a process on a pilot plant scale. The second involved design of a facility from the test results.

The objectives of the test program were limited, but because of the nature of the material, these simple objectives required solution of some imposing problems. The objectives were:

(1) To develop a continuous fluidized bed combustion process,

(2) To demonstrate an efficient combustion system, and

(3) To characterize the stack gas for subsequent off-gas system design.

Inherent in the first objective was the solution of problems related to tire wire from both the tire bead and steel belts and to fiberglass from fiberglass belted tires. The tire wire was a known problem as it could form matted "birds nests" in the fluidized bed. The mats would be difficult to remove and could lead to defluidization. The problem presented by fiberglass was the formation of clinkers resulting from softening of the fiberglass strands which not only stick together but also capture and agglomerate bed material.

The second objective, efficient combustion, was complicated by the fact that tires contain a substantial amount of carbon black which requires vigorous high temperature combustion conditions to burn. In addition, it was important that the rubber burn completely from the wire so that a good wire separation would be possible.

The final objective simply entailed sampling the stack gas for both chemical and particulate species.

The design objectives included the detailed design and construction of a system which would provide 1,100 Kg/hr of steam for process use. The off-gas cleanup system must provide a stack gas which meets air quality regulations at the selected demonstration site.

After consideration of the problems to be addressed, it was determined that combustion tests should be preceded by some noncombustion fluidization tests to assess the problems of the tire wire and evaluate possible solutions to those problems. For this purpose, a 50-cm-diameter plexiglass fluidized bed was constructed. This unit utilized a perforated plate as an air distributor.

The plexiglass model was operated with a fluidized bed containing chunks of tire and pieces of wire left from burning tire chunks. After operation, the bed media was carefully removed with a vacuum cleaner and the locations and nature of the tire chunks and wire were noted. The observations from these tests were:

(1) Wire tramp material sinks rapidly to the bottom of the fluid bed.

(2) With limited wire tramp material, the wire which settled to the bottom was mixed evenly with the bed material.

(3) High concentrations of wire tramp material resulted in "birds nesting" of the wire into tightly woven mats at the bottom of the fluidized bed, which could be removed only by "tearing" apart the mats in clumps.

After reviewing the results of this test program, it was determined that an air distribution manifold with a bottom letdown for bed media removal would be more appropriate. Therefore, the model was rebuilt using a manifold system which is shown schematically in Figure 1. The subsequent test results were as follows:

(1) Wire tramp material could be removed through a bottom letdown port, provided the tramp material concentration was less than that at which "birds nesting" began.

(2) With minimal tramp material, rubber tire chunks appeared to mix freely and randomly throughout the bed.

(3) High concentrations of wire tramp material showed a tendency to "capture" rubber tire chunks and entrap them in the birds nests. This may not have happened in a hot fluidized bed in which the rubber would be expected to burn.

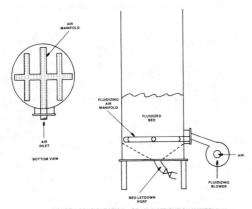

FIGURE 1 CONFIGURATION OF AIR DISTRIBUTION MANIFOLD IN PLEXIGLASS FLUIDIZED BED

Tire Analysis

Based on the success of the work in the cold model, it was judged that a fluidized bed combustion test was justified. In preparation for this test, an existing 45-cm-diameter fluidized bed combustion system was equipped with a manifold and bed letdown system similar to that used in the cold tests, and analytical chemistry tests were performed on tire samples. Samples of fiberglass belted, steel belted, nylon cord, polyester cord, and Kevlar belted tires were analyzed with the results given in Table 1. With this information which showed the tires to be a very high energy fuel, an experimental plan was developed to examine the effects of fluidizing velocity, bed temperature, and tire type on the combustion efficiency. The effects of wire concentration in the bed on birds nesting was a second parameter to be optimized.

TABLE 1

SCRAP TIRE ANALYSIS

Carbon	%	65 - 87
Hydrogen	%	5.0 - 7.4
Oxygen	%	1.7 - 5.4
Sulfur	%	0.9 - 1.5
Noncombustibles	%	2.5 - 25
Heating Value	Kcal/Kg	6,400 - 9,400

Equipment

The pilot equipment is shown schematically in Figure 2 and photographically in Figure 3.

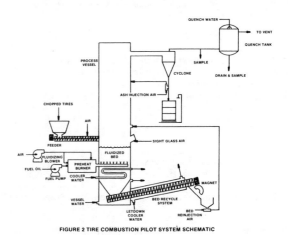

FIGURE 2 TIRE COMBUSTION PILOT SYSTEM SCHEMATIC

The process vessel was a hollow, refractory-lined right cylinder, 4.3 meters high with a 45-cm inside diameter. Fluidizing air was provided by a blower through a preheat burner (fired during warm-up operations only) and fluidizing manifold into a bed of alumina-silica bed media having the consistency of sand.

Figure 3 Tire Combustion Pilot Plant

Tires, chopped to a maximum dimension of about five cm, were fed from a hopper through an auger into the process vessel just above the fluidized bed.

Bed media was continually recycled for the purpose of removing the tramp wire which entered as bead and belting material in the tires. This recycle was accomplished by cooling the bed media below the fluidizing manifold and then withdrawing it from the bottom of the process vessel using an auger. Cooling was accomplished by passing water through tubes in the collapsed region of the bed below the fluidizing air manifold and through a jacket around the bed letdown hopper. The auger jacket was also water cooled. At the auger discharge the media passed beneath a magnet which removed the tramp wire. Periodically the magnet was manually cleared. The media was pneumatically reinjected into the process vessel slightly above the active fluidized region.

Combustion gases exiting the fluidized bed passed upward through the vessel overfire space, allowing the bulk of the entrained bed media to disengage and fall back. The gases were then ducted to the cyclone where most of the ash and abraded bed media was removed. During some of the tests, ash was withdrawn from the cyclone solids discharge line, by an air-operated eductor, and pneumatically reinjected into the process vessel to assess the effect of recycle on combustion efficiency. Otherwise, the ash and elutriating bed media captured by the cyclone were allowed to collect in a drum.

Samples were withdrawn from the cyclone off-gas exit duct for analyses. The EPA Method 5 was used for particulate sampling; a gas chromatograph was used for oxygen, carbon dioxide, and carbon monoxide analyses; and Drager tubes were used for sulfur dioxide and nitrogen oxide analyses.

After passing the sample port, the off-gas entered a spray quench tank which served to cool and scrub the gases. Fresh water was supplied to the quench tank nozzles, and the drain was directed to the sewer. The cooled, cleaned off-gas was vented to the facility stack.

Test Programs

A pilot system program consisting of six runs was conducted in the 45-cm-diameter fluidized bed burner. The first five were one-shift scoping tests, each using a unique tire type as feed. Feed materials for these tests were chopped nylon belted, polyester belted, Kevlar belted, steel belted, and fiberglass belted tires, respectively. From examination of the five scoping tests, operational parameters appropriate for incineration of a blend of all five types were established. An extended run applying a blend of tire types and using those parameters was then conducted.

Test Results

One of the principal parameters studied in the test program was the fluidized bed temperature. Common experience was that the most efficient combustion took place at the highest temperature consistent with other system limitations. In the case of scrap tires, it was found that above 800°C some melting of fiberglass cord strands was noticed. Therefore, 800°C was selected as the maximum operating temperature.

A second parameter which is important to efficient combustion is excess air. Because of the high energy content of the tires and because there was no energy removal from the bed, there was at all times considerable excess air.

Considerable attention was given to variation in the fluidizing velocity. Although there was considerable scatter in the combustion efficiency data due to the tire type variation in the experiments, there was a general trend which indicated that lower fluidizing velocities gave more efficient combustion. This was again consistent with general incineration experience, and therefore, the recommended fluidizing velocity was that giving minimum acceptable fluidizing properties. In this case a velocity of 1.8 m/sec was appropriate.

It was found that by maintaining a high bed recycle rate, the wire concentration stayed low enough so that birds nesting was not a problem. Since much of the fiberglass was captured in the cyclone, ash reinjection was halted to avoid fiberglass buildup in the system.

The outcome of the extended run which operated at these recommended conditions was that the system operated continuously with no problems from either the fiberglass or wire. In fact, the rubber was completely burned from the cord materials. The fiberglass strands were actually white in color. The cyclone ash contained some unburned carbon black resulting in an overall combustion efficiency of 95%.

Stack Gas Characterization

The stack gas analysis and particulate size distribution downstream of the cyclone are shown in Tables 2 and 3. It is noted here that the composition was determined prior to any cleanup operation and is presented to show that use of common off-gas cleanup systems can readily lead to acceptable emission levels. The sulfur dioxide level is equivalent to 1.4 pounds per million Btu's and will meet some air quality standards. The NO_x level was measured at 0.1 pound per million Btu's and is not generally regulated for small systems.

TABLE 2

OFF-GAS ANALYSIS

Temperature[1]	775°C
Composition[2], Vol. %	
Nitrogen	73.9
Oxygen	16.9
Carbon Dioxide	5.6
Carbon Monoxide[3]	0.0
Water Vapor	3.6
SO_x as Sulfur Dioxide, ppm (vol.)	350
NO_x as Nitrogen Dioxide, ppm (vol.)	35
Particulate[2], ppm (wt.)	786

(1) At process vessel exit
(2) At cyclone exit
(3) None detected by gas chromatograph

TABLE 3

STACK PARTICULATE SIZE DISTRIBUTION ANALYSIS

Micron Size	Wt. % in Size Range
12.7 -	1.0
10.1 - 12.7	2.0
8.00 - 10.1	4.0
6.35 - 8.00	6.0
5.04 - 6.35	8.0
4.00 - 5.04	11.
3.17 - 4.00	16.
2.52 - 3.17	16.
2.00 - 2.52	13.
1.59 - 2.00	11.
1.26 - 1.59	12.
1.00 - 1.26	0.0

DELIVERABLE SYSTEM

Based upon the successful pilot plant program, a deliverable system was designed to produce 1,100 kg/hr of 7.8 atm. steam from scrap tires. Figure 4 is a schematic representation of the system.

The feed to the system is 270 kg/hr of tires, which is approximately 30 tires per hour. These tires must be chopped to less than 10-cm chunks. The chunks are metered into the fluidized bed from a metering bin through a seal valve. The tires are burned in the combustion vessel at 775 to 800°C with a superficial gas velocity of 1.8 m/sec. Bed media is continuously drawn down through the fluidizing manifold over water-cooled cooling tubes which reduce the bed media temperature to a level compatible with a rotary magnetic wire separator. After separation of the wire, the cooled and cleaned bed media is conveyed back into the combustion vessel through a seal valve.

Off-gases pass from the vessel to a cyclone which removes the bulk of the elutriated ash, unburned carbon black, and fiberglass.

From the cyclone the gases pass to a heat recovery boiler which generates steam using water preheated by the bed media cooler. The boiler cools the gases to 220°C or lower before being ducted to the baghouse where final particulate removal is achieved. An induced draft fan is used to control the overfire space in the combustion vessel to slightly subatmospheric.

CONCLUSIONS

A fluidized bed scrap tire combustion system has been developed and demonstrated on a pilot scale. This system has demonstrated continuous operation and has overcome problems normally encountered from steel wire and fiberglass.

A commercial tire combustion system was designed from the pilot plant test results. The proposed system will incinerate 30 chopped tires per hour and provide 1,100 kg/hr of process steam.

REFERENCES

(1) E. L. Kay and J. R. Laman, "A Review of Scrap Tire Disposal Processes", Society of Automotive Engineers, Inc., Warrendale, PA (1979).

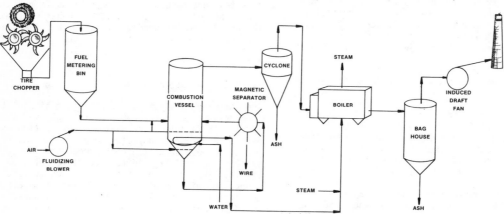

FIGURE 4 - COMMERCIAL TIRE INCINERATION FLOW SCHEMATIC

SECTION 9
INTERNATIONAL TECHNOLOGY
TRANSFER

Chapter 34

OPTIMAL DESIGN OF WATER DISTRIBUTION NETWORKS FOR DISTRICT HEATING

G. Caratti, F. Martelli, D. Miconi

ABSTRACT

A computer aided procedure directed at finding
out the minimum cost design of a branched network for district heating systems is presented.
A detailed description of the cost-function to
be minimized is given. The features of the numerical procedure are illustrated. The results
of an application are then reported and analysed.

INTRODUCTION

Since the rise in the price of fuel supplies
recovery of heat for district heating has taken
on increasing importance. Nowadays many European
cities are provided with vast underground piping
systems distributing heat to their inhabitants
and initiatives for the setting up of new plants
are coming into being.
Installation and management costs involved in
district heating systems are very high and a
priori investigations directed at ascertaining
investment feasibility are needed [1]. In order
to reduce these costs it is necessary to determine the optimal features of the network which
delivers the hot fluid from the thermal source
to a vast and scattered set of users.
The assessment of one of the feasible designs
for the network requires a great deal of work
if it is carried out with classical methods.
This fact and at the same time the vast amount
of possible designs to be selected make the use
of computer and optimization techniques imperative.
The aim of this paper is to illustrate the setting up of a computer program directed at finding out the minimum cost design for networks of
district heating systems.

OBJECTIVES

The outline of the system which is the object of
this study is reported in fig.1. It consists of
the following parts:
- a main station in which the heat from the source
  is transfered into the water of the network
  and where pumps ensure water circulation;
- a closed circuit network which delivers the
  heated water to users for heating and hot water supplies;
- a set of user sub-stations where heat is trans-

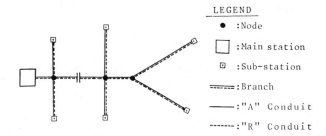

LEGEND
- ● :Node
- □ :Main station
- ⊡ :Sub-station
- ===== :Branch
- ——— :"A" Conduit
- ------- :"R" Conduit

FIG.1-EXAMPLE NETWORK

fered into the fluids that circulate inside
the buildings.
The location and heat demand of each sub-station
is estabilished, as is the site of the main station, the network layout, the types of valves
and fittings involved and the water temperature
at the main station outlet. The pipe, fitting
and valve sizes, the insulation wall thicknesses
and the features of the pumping system which minimize the total cost of system installation and
management are then to be determined.

COST EVALUATION

For the sake of clarity some definitions need
to be given in advance:
- a "conduit" is a pipeline in which flow neither
  divides nor combines;
- a "pipe-segment" is a portion of conduit consisting of a pipe having the same diameter
  throughout;
- a "branch" is the set of two parallel conduits:
  the advancement (A) conduit which conveys water toward one or more sub-stations and the
  return (R) conduit which conveys water back-
  wards to the source (•);
- a "node" is a point of convergence of two or
  more branches (the source shall be considered
  a node as well).
In the evaluation of the network cost-function
to be minimized only those items which depend
in some way upon the design variables are to be

(•) All quantities which refer to R conduits will
    be marked with (') and for the sake of simpli-
    city their definitions will be omitted.

taken into account. The costs which are related to the generic ith branch are reported here.

A) Cost of Piping

The costs of pipes, valves and fittings including trenching and installation are considered in this item. At first an index needs to be associated to each pipe-size of a chosen standardized series so that larger sizes will correspond to larger indices. The sub-sets of the indices allowed to the ith A and R conduits, J_i and J_i', may be singled out by introducing water velocity limitations:

$$J_i = \{r,..,j,..,s\} \text{ for which } \sqrt{\frac{4 M_i}{\pi \rho v_M}} \leqslant d_j \leqslant \sqrt{\frac{4 M_i}{\pi \rho v_m}} \quad ....(1)$$

$$J_i' = \{r',..,j,..,s'\} \text{ for which } \sqrt{\frac{4 M_i}{\pi \rho' v_M}} \leqslant d_j \leqslant \sqrt{\frac{4 M_i}{\pi \rho' v_m}} \quad ...(1')$$

in which d_j= the jth inside diameter; M_i=the mass flow rate in the ith branch; ρ=the average water density in all A conduits; v_m, v_M=the minimum and maximum allowed water velocity.

Henceforth the pipe-segment with the allowed jth diameter will be designated with the term "jth segment". Therefore the cost of piping may be expressed with the sum of the costs of the segments as:

$$Cp_i = \sum_{j \in J_i}(c_j \delta_i + \sum_{f \in F} \frac{A_{fj} n_{fi}}{L_i}) 1_{ij} + \sum_{j \in J_i'}(c_j \delta_i + \sum_{f \in F} \frac{A_{fj} n_{fi}'}{L_i}) 1_{ij}' \quad .....(2)$$

in which c_j=the cost per unit of length of the installed and buried pipe of the jth size; $\delta_i = 1$ if normal excavation is required in the ith branch (e.g. the trenching under paved road), $\delta_i \neq 1$ otherwise; F=the set of the indices which correspond to the types of valves and fittings involved in the network; A_{fj}=the cost of the installed valve or fitting of the fth type, jth size; n_{fi}=the number of valves or fittings of the fth type in the ith A conduit; L_i=the length of the ith A and R conduits; 1_{ij}=the length of the jth segment, ith A conduit.

B) Cost of Insulation

This is composed of a first term which is proportional to the volume of insulation and of a second one which is proportional to the surface of protecting casing. Therefore it may be written as (•):

$$Cw_i = \pi\{ \sum_{j \in J_i}[T(D_j+W_i)W_i + P(D_j+2W_i)]1_{ij} +$$
$$+ \sum_{j \in J_i'}[T(D_j+W_i')W_i' + P(D_j+2W_i')]1_{ij}'\} \quad ......(3)$$

in which T=the cost of the unit of volume of installed insulant; D_j=the outside diameter of the service pipe with jth size; W_i=the insulant wall thickness in the ith A conduit; P=the cost of the unit of surface of installed casing.

_____
(•)This expression refers to cased underground piping systems without air gap [2]. However, options for other kinds of insulation are enclosed in the program.

C) Cost of Heat Loss

The calculation of heat loss in the jth segment of the ith A conduit is carried out by expressing its thermal resistance, as (•):

$$r_{ij} = i_{ij} + g_{ij} = \frac{\ln(1+\frac{2W_i}{D_j})}{2\pi \, kw} + \frac{\ln\frac{4Y_i}{D_j}}{2\pi \, kg} \quad ....(4)$$

in which i_{ij}=the contribution of the insulant; g_{ij}=the contribution of the soil; kw=the average thermal conductivity of the insulant in all A conduits; kg=the average thermal conductivity of the soil; Y_i=the depth of burial of the pipe center line in the ith branch.

If c=the average water specific heat capacity in all A conduits, Ta=the enviromental temperature and Te_i=the inlet water temperature in the ith A conduit, the relative outlet temperature in the ith A conduit is given by:

$$To_i = Ta + (Te_i - Ta)\exp(-\frac{1}{M_i c} \sum_{j \in J_i} \frac{1_{ij}}{r_{ij}}) \quad ....(5)$$

Consequently the loss of thermal load in the ith A conduit may be expressed as:

$$M_i c(Te_i - To_i) = M_i c(Te_i - Ta)[1 - \exp(-\frac{1}{M_i c} \sum_{j \in J_i} \frac{1_{ij}}{r_{ij}})] \quad (6)$$

For the sake of brevity parallel considerations for the ith R conduit will be omitted. Finally the total discounted cost of the annual heat losses is given by:

$$Ch_i = Et \, Ot \, M_i\{c(Te_i - Ta)[1 - \exp(-\frac{1}{M_i c} \sum_{j \in J_i} \frac{1_{ij}}{r_{ij}})] +$$
$$+ c'(Te_i' - Ta)[1 - \exp(-\frac{1}{M_i c'} \sum_{j \in J_i'} \frac{1_{ij}'}{r_{ij}})]\} \sum_{r \in Ny}(t+1)^{-r} (7)$$

in which Et=the specific cost of thermal energy; Ot=the time of system operation in one year; Ny=the set of indices which correspond to the years in system life, t=the annual interest rate.

D) Cost of Energy for Pumping

The hydraulic power required for pumping water into the ith branch is given by:

$$P_i = gM_i\Delta h_i \quad ....(8)$$

in which g=the acceleration of gravity; Δh_i=the total head loss in the ith branch, which may be expresses as:

$$\Delta h_i = \sum_{j \in J_i}(a_{ij}1_{ij}) + \sum_{j \in J_i'}(a_{ij}'1_{ij}') =$$
$$= \frac{8M_i^2}{\pi^2 g}\{\frac{1}{\rho^2}[\sum_{j \in J_i}(\frac{f_{ij}}{d_j^5} + \sum_{f \in F}\frac{Cl_f n_{fi}}{L_i d_j^4})1_{ij}] + \quad ....(9)$$
$$+ \frac{1}{\rho'^2}[\sum_{j \in J_i'}(\frac{f_{ij}'}{d_j^5} + \sum_{f \in F}\frac{Cl_f n_{fi}'}{L_i d_j^4})1_{ij}']\}$$

in which a_{ij}=the total head loss per unit of length of the jth segment, ith conduit; Cl_f=the form loss coefficient of the valve or of the fitting of the fth type; f_{ij}=the pipe friction factor in the jth segment, ith A conduit, which derives from Colebrook equation:

216

$$\frac{1}{\sqrt{f_{ij}}} = -2\log_{10}(-\frac{2.51}{R_{ij}\sqrt{f_{ij}}} + \frac{\varepsilon}{3.715d_j}) \quad \ldots\ldots\ldots\ldots(10)$$

in which ε=the pipe roughness; R_{ij}=the Reynolds number which is given by:

$$R_{ij} = \frac{4M_i}{\pi d_j \eta} \quad \ldots\ldots\ldots\ldots\ldots\ldots\ldots(11)$$

in which η=the average water dynamic viscosity in all A conduits.

Therefore, the total discounted cost of the annual requirements of electric energy is given by:

$$Ce_i = \frac{g\, Ee\, M_i\, Ot\, \Delta h_i}{e} \sum_{reNy} (t+1)^{-r} \quad \ldots\ldots\ldots\ldots(12)$$

in which Ee=the total specific cost of electricity; e=the combined pump and motor efficiency.

E) Cost of Pumps and Motors

The cost of pumps and motors in the main station, C_S, may be approximated to:

$$C_S = Sf + Sv \sum_{i\varepsilon I} \frac{P_i}{e} \quad \ldots\ldots\ldots\ldots\ldots(13)$$

in which I=the set of indices which correspond to the branches; Sf,Sv=constants. Not considering the term Sf as it doesn't affect optimization one can write:

$$Cs_i = Sv \frac{g\, M_i\, \Delta h_i}{e} \quad \ldots\ldots\ldots\ldots\ldots(14)$$

Finally, the total cost is obtained by assembling expressions (2),(3),(7),(12) and (14):

$$C_N = \sum_{i\varepsilon I} (Cp_i + Cw_i + Ch_i + Ce_i + Cs_i) \quad \ldots\ldots\ldots(15)$$

OPTIMIZATION PROCEDURE

The problem has been tackled with the aid of both linear programming (LP) and dynamic programming (DP) optimization techniques. The outline of the proposed method is given in the flow chart of fig.2. To start the process initial values must be assigned to the water temperatures and to the insulant thicknesses. In this way the total cost of each branch depends only upon the head loss, Δh_i, and upon the lengths of the segments, l_{ij}, l'_{ij}. Furthermore this cost may be expressed as a function of the only head loss by solving the following parametrical problem:

$$f_i(\Delta h_i) = \min(\sum_{j\varepsilon J_i} c_{ij}l_{ij} + \sum_{j\varepsilon J'_i} c'_{ij}l'_{ij}) \quad \ldots\ldots(16)$$

subjected to:

$$\left. \begin{array}{l} \sum_{j\varepsilon J_i} l_{ij} = L_i \\[2mm] \sum_{j\varepsilon J'_i} l'_{ij} = L_i \\[2mm] \sum_{j\varepsilon J_i} a_{ij}l_{ij} + \sum_{j\varepsilon J'_i} a'_{ij}l'_{ij} = \Delta h_i \end{array} \right\} \quad \ldots\ldots(17)$$

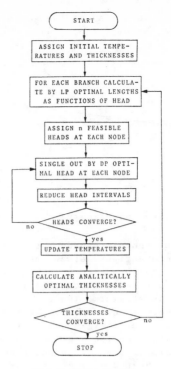

FIG.2-FLOW CHART

in which c_{ij}=the sum of the costs of piping, insulation and heat loss of the unit of length of the jth segment, ith A conduit, which may be deducted from expressions (2), (3) and (7). This problem has been worked out by means of LP algorithms [3]. Finally, the total cost of the generic ith branch, Ct_i, becomes:

$$Ct_i(\Delta h_i) = f_i(\Delta h_i) + Ce_i(\Delta h_i) + Cs_i(\Delta h_i) \quad \ldots(18)$$

In the next step the optimal heads at node, H, are found out by means of DP [4]-the "head" at a node must be regarded as the head difference between the ends of A and R conduits which coincide in the node (see fig.3).

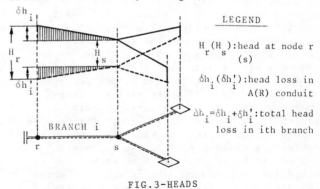

FIG.3-HEADS

LEGEND

$H_r(H_s)$:head at node r (s)

$\delta h_i(\delta h'_i)$:head loss in A(R) conduit

$\Delta h_i = \delta h_i + \delta h'_i$:total head loss in ith branch

A number of DP stages equal to the number of nodes, P, is to be considered. At first n feasible values of the head at each node ($H_{p1},..,H_{pn}$ for $p=1,..,P$) need to be assigned. Optimal heads at the nodes are then singled out over those assigned. An example which illustrates

the operations occurring at the generic qth
stage is given below (see fig.4).

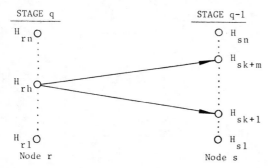

FIG.4-DYNAMIC PROGRAMMING STAGE EXAMPLE

In the branch connecting node r to node s only
m of the n combinations between the generic hth
head at r and the heads at s ($H_{rh}-H_{sk+j}$ for
$j=1,..,m$) do not cause water velocity to exceed
the assigned constraints. The cost of that portion of the network which lies downstream from
r is calculated for these m combinations. Only
the lowest value and the index of the corresponding combination are then recorded. This
operation is repeated n times in order to evaluate for each of the assigned heads at r (H_{r1},
..,H_{rn}) the minimum cost of the above-mentioned
portion. In the following stages the portion
taken into account becomes wider near the
source and the process continues until this one is
reached. The optimal head at the source corresponds to the lowest of the n minimum costs
met at the source; these costs refer, in fact,
to the whole network. Starting from this head,
the heads at the other nodes are singled out
descending downstream along an optimal path.
This path is pointed out, stage by stage, by
the above-mentioned indices that had previously been recorded. A new iteration operating with
other n values of the head at each node which are closer to each other takes place if the step (i.e. the
distance between two consecutive values) is
larger then the desiderable approximation. After reaching convergence the temperatures are
updated on the basis of the new values of the
diameters of the segments. The optimal thickness in each conduit is then calculated by determining the zero of a derivative assuming
temperatures and diameters as input data. If
non-admissible differences between the new and
the old values of the thickness in a conduit
are found a whole new cycle takes place.
It needs to be stressed that at each stage of
the LP and DP procedures only quantities which
are related to a few branches are processed.
This means that the memory requirements are
not directly affected by the dimension of the
network but, indirectly, by the recording and
updating of the current solution.

APPLICATION

A relatively small district heating system was
deliberately selected in order to report results
compactly and also in consideration of the
greater importance attached to methodology rather then to application.
The network which consists of 37 branches and
20 users is sketched out in fig.5.

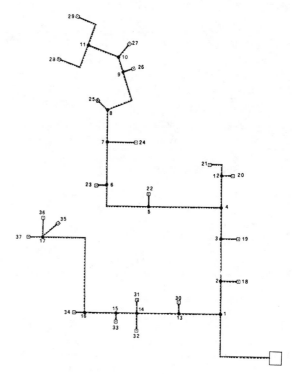

FIG.5-NETWORK

Each branch is specified by the index of its
first downstream node.
The input parameters are reported in tab.1;
both user requirements and optimal features
of the network are listed in tab.2. The parameters which are related to the heat demand
refer to average conditions and are calculated
on the basis of the annual curve of the heat
load required on the site.
The optimal costs of the network are reported
in tab.3.
As appears in tab.2, the optimal configuration
of the generic ith branch is the one which has
been sketched in fig.6

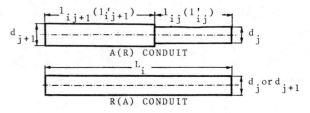

FIG.6-OPTIMAL CONFIGURATION OF iTH BRANCH

The relative difference between the current
value of the cost (C_N) at each iteration and
the final value ($\underline{C}_N$), through three cycles, is
shown in fig.7: as one can see the convergence

is satisfactory.

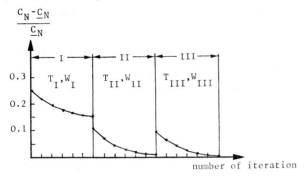

FIG.7-RELATIVE DIFFERENCE BETWEEN CURRENT
AND FINAL COSTS

CONCLUSIONS

The illustrated program enables the designer
to analyse quickly and inexpensively a vast
number of different solutions. In particular,
this is necessary for examining the effects
of uncertain parameters - such as the annual
interest rate and the user requirements - on the
optimal solution.

The program has been written in FORTRAN IV to
be used specifically on minicomputers: a network containing 200 branches may be analysed
in 10 minutes on a DIGITAL PDP 11/45 computer
within a storage capacity of 30 k-words; however, further investigations to reduce execution time and memory requirements are now being considered.

The next objective of this study is the extention of the program to analyse networks which
are fed by more main stations.

REFERENCES

[1] Silveri L., Il teleriscaldamento: da Brescia
una proposta su scala nazionale, Franco An-
geli Editore, Milano, 1980.

[2] British Standard Specifications for Thermally Insulated Underground Piping Systems,
BS 4508, Part 3, 1977.

[3] Bazaraa, M.S., Jarvis, J.J., Linear Programming and Network Flows, John Wiley &
Sons, New York, 1977.

[4] Bellman, R., Dreyfus, S.E., Applied Dynamic Programming, Princeton University Press,
Princeton, N.J., 1962.

[5] Filippini A., Miconi D., Spadoni G., Il
CRAL: un programma di calcolo automatico
per il dimensionamento di impianti di di-
stribuzione di liquidi (in Italian), Atti
dell'Istituto di Ingegneria Meccanica, U-
niversità di Firenze, Jan., 1980.

[6] Filippini A., Miconi D., Spadoni G., Otti-
mizzazione di un impianto di refrigerazio-
ne ad acqua in ciclo chiuso di uno stabi-
limento petrolchimico (in Italian), Atti

dell'Istituto di Ingegneria Meccanica, Univer-
sità di Firenze, Dec., 1980.

NOMENCLATURE

a_{ij} specific head loss in jth segment, ith A
conduit (m/m)

A advancement

A_{fj} cost of installed fth type of valve or
fitting, jth size (Lit)

c average water specific heat capacity in
A conduits (J/kg°C)

c_j specific cost of installed and buried pipe of jth size (Lit/m)

Ce_i cost of energy for pumping required in
ith branch (Lit)

Ch_i cost of heat loss in ith branch (Lit)

Cl_f form loss coefficient of valve or fitting
of fth type

Cp_i cost of piping in ith branch (Lit)

Cs_i cost of pumps and motors (Lit)

Cw_i cost of insulation in ith branch (Lit)

C_N network total cost (Lit)

C_S pump and motor total cost (Lit)

d_j jth inside diameter (m)

D_j jth outside diameter (m)

DP dynamic programming

e combined pump and motor efficiency

Ee electric energy specific cost (Lit/J)

Et thermal energy specific cost (Lit/J)

f_{ij} pipe friction factor in jth segment, ith A
conduit

F index-set of valve and fitting types

g gravity acceleration (m/s^2)

g_{ij} soil thermal resistance in jth segment,
ith branch (m°C/W)

H_r head at node r (m)

i_{ij} insulant thermal resistance in jth segment,
ith A conduit (m°C/W)

J_i index-subset of allowed sizes to ith A
conduit

kg average soil thermal conductivity (W/m°C)

kw average insulant thermal conductivity in
A conduits (W/m°C)

l_{ij} length of jth segment, ith A conduit (m)

L_i length of ith branch (m)

LP linear programming

M_i mass flow rate in ith branch (kg/s)

n_{fi} number of valves or fittings of fth type

	in ith A conduit	Te_i	inlet water temperature in ith A conduit (°C)
P	specific cost of installed casing (Lit/m²)	To_i	outlet water temperature in ith A conduit (°C)
P_i	hydraulic power required in ith branch (W)	v_m	minimum allowed water velocity (m/s)
r_{ij}	thermal resistance of jth segment, ith A conduit (m°C/W)	v_M	maximum allowed water velocity (m/s)
		W_i	insulant wall thickness in ith A conduit (m)
R	return	Y_i	depth of burial of pipe center line in ith A conduit (m)
R_{ij}	Reynolds number in jth segment, ith conduit		
Sf	constant (Lit)	Δh_i	total head loss in ith branch (m)
Sv	constant (Lit/W)	ε	pipe roughness (m)
t	annual interest rate	η	average water dynamic viscosity in A conduits (Ns/m²)
T	specific cost of installed insulant (Lit/m³)		
Ta	temperature of the air above ground (°C)	ρ	average water density in A conduits (kg/m³)
		[ ]'	quantity [ ] related to R conduit

```
************************
*  INPUT PARAMETERS  *
************************
```

```
        - PIPES SIZES -                                  - COMPONENT COSTS(KLIT) -
.........................................................................................
*REF* NOMINAL INS.DIAM OUT.DIAM SP.COST  ST.TEE  90 ELBOW  ANG.COM.  AXL.COM.  REDUCT.  GATE VALVE
 (J)  (INCHES)  (M)     (M)    (KLIT/M)   (1)      (2)       (3)       (4)       (5)       (6)
   1    0.75    0.021   0.027   67.870   60.800   52.800    70.000   135.000    54.800   110.000
   2    1.00    0.027   0.033   68.820   64.800   56.000    75.000   140.000    58.900   112.000
   3    1.25    0.035   0.042   74.480   68.700   62.000    80.000   150.000    65.300   115.000
   4    1.50    0.041   0.048   75.440   76.800   68.500   242.000   165.000    71.500   140.000
   5    2.00    0.052   0.060   83.640   88.600   75.000   248.000   177.000    78.500   208.000
   6    2.50    0.067   0.073   85.040  101.000   79.600   251.000   195.000    82.700   260.000
   7    3.00    0.083   0.089   95.720  108.100   87.700   291.000   240.000    89.800   280.000
   8    3.50    0.094   0.102  102.780  121.500   98.200   297.000   245.000    94.700   347.000
   9    4.00    0.109   0.114  104.160  126.500  102.500   329.000   261.000   105.600   364.000
  10    5.00    0.135   0.141  120.320  150.800  111.000   371.000   281.000   107.600   525.000
  11    6.00    0.162   0.168  138.550  156.900  140.000   384.500   393.000   124.900   538.000
  12    8.00    0.211   0.219  167.160  219.600  165.500   529.000   437.000   146.200   744.000
  13   10.00    0.265   0.273  201.700  288.500  221.000   650.000   567.000   165.600  1059.000
  14   12.00    0.315   0.324  236.000  367.300  274.300   779.000   739.000   197.400  1406.000
  15   14.00    0.346   0.356  249.070  470.000  300.500  1016.00    915.000   231.400  2050.000
.........................................................................................
FORM LOSS COEFFICIENTS:                    1.20-0.25-1.50  0.81     1.50      0.10      0.05      0.15

AIR TEMPERATURE(C): 10.00

VELOCITY FIELD(M/S)      MIN:0.50      MAX:3.00

DENSITY (KG/M**3)     ADV:948.    RET:986.

SPECIFIC HEAT (J/KG*C)     ADV:4237.    RET:4178.

DIN.VISCOS.(N*S/M**2)     ADV: 0.248E-03    RET: 0.511E-03

THERM.CONDUCT.(W/M*C)    INS.ADV: 0.290E-01   INS.RET: 0.267E-01     SOIL: 0.100E 01

DEPHT OF BURIAL(M):0.45     ROUGHNESS(M):0.500E-04

INSULANT COST(KLIT/M**3):939.17     CASING COST(KLIT/M**2): 52.94

ELECTR.ENERGY COST(LIT/KWH): 63.00      THERMAL ENERGY COST(KLIT/KCAL):0.1042E-04

PUMP SYST.COST FIXED(KLIT):0.1494E 04     VARIABLE(KLIT/KW): 0.5310E 03

ANNUAL INTEREST RATE:0.06     SYSTEM LIFE(YEARS):25
.........................................................................................
```

TAB.1-INPUT PARAMETERS

```
    ----------------------------------------
    |              *REF.EXAMPLE*            |
    |                                      |
    |         HEAT FLUX    NODE:I          |
    |    0------------>--------0           |
    |              BRANCH:I                |
    |                                      |
    ----------------------------------------
```

```
    *************************************
    * NETWORK PHISICAL CHARACTERISTICS *
    *************************************
```

-ADVANCEMENT-

REF (I)	FLOW RATE (KG/S)	DIAMETERS (M)		LENGTHS (M)		VELOCITY (M/S)		HEAD LOSS (M)	INSULANT (M)	TEMP (C)	TOT.HD LOSS (M)	HEAD (M)	HEAT LOAD (MCAL/H)
1	17.486	0.109	0.135	0.0	117.0	1.985	1.290	1.2151	0.031	114.968	2.5731	21.5386	3829.
2	9.257	0.094	0.109	1.6	14.4	1.417	1.051	0.1617	0.029	114.929	0.3655	21.1732	2020.
3	8.598	0.094	0.109	8.1	9.9	1.316	0.976	0.2142	0.029	114.913	0.3971	20.7760	1884.
4	8.080	0.094	0.109	14.0	0.0	1.237	0.917	0.2092	0.029	114.898	0.3401	20.4360	1770.
5	7.234	0.083	0.094	3.2	53.5	1.426	1.107	0.7193	0.029	114.861	1.6060	18.8299	1584.
6	5.938	0.067	0.083	0.0	54.0	1.796	1.171	0.8473	0.028	114.797	1.8204	17.0095	1300.
7	5.357	0.067	0.083	1.3	41.1	1.620	1.056	0.5782	0.028	114.736	1.3108	15.6987	1171.
8	3.936	0.067	0.067	0.0	30.0	1.190	1.190	0.6330	0.026	114.682	1.3040	14.3947	860.
9	1.615	0.041	0.052	0.0	40.0	1.294	0.787	0.5108	0.025	114.584	1.0796	13.3150	355.
10	1.331	0.041	0.052	0.0	22.0	1.067	0.649	0.1934	0.025	114.463	0.3982	12.9168	292.
11	0.635	0.035	0.041	16.6	15.4	0.693	0.509	0.3896	0.023	114.285	0.9261	11.9907	139.
12	0.846	0.021	0.027	8.7	6.3	2.587	1.601	4.3325	0.020	114.854	6.1567	14.2793	184.
13	8.229	0.083	0.083	0.0	37.1	1.622	1.622	1.1008	0.028	114.920	2.4867	19.0519	1796.
14	6.822	0.067	0.083	0.0	30.1	2.063	1.345	0.6190	0.028	114.887	1.2714	17.7806	1489.
15	4.858	0.067	0.067	0.0	22.0	1.469	1.469	0.6999	0.026	114.857	1.4380	16.3426	1061.
16	3.619	0.052	0.067	0.0	30.0	1.765	1.095	0.5378	0.026	114.815	1.0975	15.2451	790.
17	1.703	0.052	0.067	69.7	0.3	0.830	0.515	0.9850	0.025	114.667	2.0342	13.2109	372.
18	0.659	0.021	0.021	15.0	0.0	2.015	2.015	3.7754	0.019	114.876	21.1732	10.0000	143.
19	0.518	0.021	0.021	16.0	0.0	1.584	1.584	2.5140	0.019	114.842	20.7760	10.0000	113.
20	0.496	0.021	0.021	0.0	14.0	1.515	1.515	2.0169	0.019	114.759	14.2793	10.0000	108.
21	0.350	0.021	0.021	12.0	0.0	1.072	1.072	0.8801	0.019	114.747	14.2793	10.0000	76.
22	1.296	0.027	0.027	0.0	15.0	2.452	2.453	4.1011	0.021	114.804	18.8299	10.0000	282.
23	0.581	0.021	0.021	15.0	0.0	1.777	1.777	2.9526	0.019	114.712	17.0095	10.0000	126.
24	1.421	0.035	0.035	0.0	35.0	1.551	1.551	2.7519	0.022	114.646	15.6987	10.0000	309.
25	2.321	0.035	0.041	2.4	17.6	2.533	1.860	2.1246	0.023	114.635	14.3947	10.0000	504.
26	0.284	0.021	0.021	14.0	0.0	0.867	0.867	0.6820	0.019	114.412	13.3150	10.0000	62.
27	0.696	0.027	0.027	0.0	20.0	1.318	1.318	1.6185	0.021	114.350	12.9168	10.0000	151.
28	0.316	0.021	0.021	15.0	0.0	0.967	0.967	0.9022	0.019	114.061	11.9907	10.0000	69.
29	0.318	0.021	0.027	18.7	11.3	0.973	0.602	1.3385	0.020	113.959	11.9907	10.0000	69.
30	1.407	0.027	0.027	8.0	0.0	2.663	2.663	2.5722	0.021	114.891	19.0519	10.0000	306.
31	0.422	0.021	0.021	24.0	0.0	1.291	1.291	2.5290	0.019	114.756	17.7806	10.0000	92.
32	1.542	0.027	0.035	12.1	7.9	2.918	1.683	5.3981	0.021	114.841	17.7806	10.0000	335.
33	1.239	0.027	0.027	10.0	0.0	2.344	2.344	2.5011	0.021	114.824	16.3426	10.0000	269.
34	1.916	0.035	0.035	10.0	0.0	2.091	2.091	1.4131	0.022	114.775	15.2451	10.0000	416.
35	0.501	0.021	0.021	10.0	0.0	1.531	1.531	1.4703	0.019	114.505	13.2109	10.0000	109.
36	0.492	0.021	0.027	13.8	6.2	1.505	0.932	2.2199	0.020	114.461	13.2109	10.0000	107.
37	0.710	0.027	0.027	0.0	28.0	1.343	1.343	2.3516	0.021	114.458	13.2109	10.0000	154.

USER REQUIREMENTS

-RETURN-

REF (I)	FLOW RATE (KG/S)	DIAMETERS (M)		LENGTHS (M)		VELOCITY (M/S)		HEAD LOSS (M)	INSULANT (M)	TEMP (C)
1	17.486	0.135	0.135	0.0	117.0	1.240	1.240	1.1769	0.019	54.455
2	9.257	0.109	0.109	0.0	16.0	1.011	1.011	0.1416	0.019	54.378
3	8.598	0.109	0.109	0.0	18.0	0.939	0.939	0.1384	0.019	54.357
4	8.080	0.109	0.109	0.0	14.0	0.882	0.882	0.0956	0.019	54.345
5	7.234	0.094	0.094	0.0	56.8	1.064	1.064	0.6662	0.018	54.341
6	5.938	0.083	0.083	0.0	54.0	1.125	1.126	0.8239	0.018	54.293
7	5.357	0.083	0.083	0.0	42.4	1.015	1.015	0.5314	0.018	54.298
8	3.936	0.067	0.083	30.0	0.0	1.144	0.746	0.6154	0.017	54.257
9	1.615	0.052	0.052	0.0	40.0	0.757	0.757	0.5035	0.016	53.861
10	1.331	0.052	0.052	0.0	22.0	0.624	0.624	0.1925	0.016	53.867
11	0.635	0.035	0.035	0.0	32.0	0.666	0.666	0.5240	0.014	53.620
12	0.846	0.027	0.027	0.0	15.0	1.540	1.540	1.7121	0.013	54.598
13	8.229	0.083	0.094	28.4	8.7	1.560	1.211	0.9420	0.018	54.597
14	6.822	0.083	0.083	0.0	30.1	1.293	1.293	0.5992	0.018	54.563
15	4.858	0.067	0.083	21.9	0.1	1.413	0.921	0.6746	0.017	54.522
16	3.619	0.067	0.067	0.0	30.0	1.052	1.052	0.5244	0.017	54.460
17	1.703	0.052	0.052	0.0	70.0	0.798	0.798	0.9742	0.016	54.240
18	0.659	0.021	0.021	0.0	15.0	1.937	1.937	3.6156	0.012	54.803
19	0.518	0.021	0.021	0.0	16.0	1.523	1.523	2.4251	0.012	54.744
20	0.496	0.021	0.027	14.0	0.0	1.457	0.902	1.9484	0.012	54.669
21	0.350	0.021	0.021	0.0	12.0	1.030	1.030	0.8609	0.012	54.638
22	1.296	0.027	0.035	15.0	0.0	2.358	1.360	3.9064	0.013	54.764
23	0.581	0.021	0.021	0.0	15.0	1.709	1.709	2.8379	0.012	54.630
24	1.421	0.035	0.041	34.2	0.8	1.491	1.095	2.6198	0.014	54.550
25	2.321	0.041	0.041	0.0	20.0	1.788	1.788	1.7743	0.015	54.599
26	0.284	0.021	0.021	0.0	14.0	0.834	0.834	0.6728	0.012	54.255
27	0.696	0.027	0.035	11.9	8.1	1.267	0.731	1.0949	0.014	54.248
28	0.316	0.021	0.021	0.0	15.0	0.930	0.930	0.8861	0.012	53.912
29	0.318	0.027	0.027	0.0	30.0	0.579	0.579	0.5355	0.013	53.644
30	1.407	0.027	0.027	0.0	8.0	2.560	2.560	2.4453	0.013	54.871
31	0.422	0.021	0.021	0.0	24.0	1.241	1.241	2.4566	0.012	54.576
32	1.542	0.035	0.035	0.0	20.0	1.618	1.618	1.7756	0.014	54.792
33	1.239	0.027	0.027	0.0	10.0	2.254	2.254	2.3851	0.013	54.796
34	1.916	0.035	0.035	0.0	10.0	2.011	2.011	1.3513	0.014	54.755
35	0.501	0.021	0.021	0.0	10.0	1.472	1.472	1.4199	0.012	54.442
36	0.492	0.027	0.027	0.0	20.0	0.896	0.896	0.8109	0.013	54.324
37	0.710	0.027	0.035	2.1	25.9	1.292	0.745	0.6939	0.014	54.314

* TOTAL *   REQUIRED HEAD (M): 24.112
            REQUIRED POWER (KW): 6.894 (EFFICIENCY:0.60)
            REQUIRED HEAT LOAD (KW): 4456. = (MCAL/H): 3842.
            LOSS OF HEAT LOAD (KW):   58. = (MCAL/H):   50.

TAB.2-NETWORK OPTIMAL FEATURES AND USER REQUIREMENTS

221

- COSTS SUMMARY(KLIT) -

REF	PIPES		INSULATION		HEAT LOSS		FITTINGS,VALVES		ENERGY+PUMP.SYST.VAR.	TOTAL
(I)	ADV	RET	ADV	RET	ADV	RET	ADV	RET		
1	14077.	14077.	5781.	4567.	4743.	2797.	262.	262.	5581.	52148.
2	1664.	1667.	654.	519.	566.	333.	852.	752.	420.	7427.
3	1864.	1875.	712.	584.	622.	375.	479.	388.	424.	7321.
4	1439.	1458.	524.	454.	466.	291.	122.	127.	341.	5221.
5	5811.	5834.	2111.	1663.	1879.	1097.	512.	419.	1441.	20767.
6	5169.	5169.	1810.	1411.	1664.	962.	196.	196.	1341.	17918.
7	4044.	4059.	1415.	1108.	1302.	756.	487.	399.	871.	14441.
8	2551.	2551.	858.	664.	830.	477.	296.	296.	637.	9159.
9	3346.	3346.	983.	754.	1002.	568.	164.	164.	216.	10543.
10	1840.	1840.	541.	415.	551.	313.	89.	89.	66.	5742.
11	2398.	2383.	630.	452.	694.	376.	141.	69.	73.	7216.
12	1024.	1032.	217.	178.	270.	160.	119.	65.	646.	3711.
13	3551.	3612.	1244.	999.	1144.	677.	399.	495.	2538.	14659.
14	2881.	2881.	1009.	788.	928.	539.	108.	108.	1076.	10318.
15	1871.	1872.	629.	488.	609.	351.	101.	101.	866.	6888.
16	2551.	2551.	858.	665.	831.	478.	101.	101.	493.	8629.
17	5855.	5855.	1723.	1323.	1755.	1000.	164.	164.	430.	18268.
18	1018.	1018.	202.	150.	259.	145.	0.	0.	1731.	4523.
19	1086.	1086.	216.	160.	276.	154.	0.	0.	1335.	4314.
20	950.	950.	189.	140.	242.	135.	0.	0.	878.	3484.
21	814.	814.	162.	120.	207.	115.	53.	53.	621.	2960.
22	1032.	1032.	237.	178.	285.	160.	0.	0.	3027.	5951.
23	1018.	1018.	202.	150.	259.	144.	0.	0.	1227.	4018.
24	2607.	2608.	656.	499.	738.	418.	0.	65.	2767.	10357.
25	1506.	1509.	410.	317.	446.	255.	71.	0.	4143.	8657.
26	950.	950.	189.	140.	241.	134.	0.	0.	469.	3072.
27	1376.	1422.	316.	256.	379.	222.	0.	61.	1116.	5148.
28	1018.	1018.	202.	149.	258.	143.	53.	53.	471.	3364.
29	2047.	2065.	430.	353.	535.	315.	110.	56.	473.	6384.
30	551.	551.	126.	95.	152.	86.	0.	0.	3325.	4885.
31	1629.	1629.	324.	240.	414.	231.	0.	0.	931.	5398.
32	1421.	1490.	340.	285.	397.	239.	61.	0.	3401.	7633.
33	688.	688.	158.	119.	190.	107.	0.	0.	2511.	4461.
34	745.	745.	188.	142.	211.	120.	0.	0.	3623.	5773.
35	679.	679.	135.	100.	172.	96.	0.	0.	821.	2681.
36	1363.	1376.	284.	237.	355.	212.	56.	0.	806.	4690.
37	1927.	2074.	442.	393.	531.	330.	0.	65.	1163.	6924.
	173147.		48361.		41713.		9538.		52294.	325053.

FIXED COST FOR 2+1 PUMPS: 1494.
* TOTAL COST * >>>   326547.

TAB.3-NETWORK OPTIMAL COSTS

Chapter 35

DESIGN OF A HEAT PUMP AND STORAGE AIR CONDITIONING SYSTEM

L. D. Elms

SUMMARY

Experience with school libraries which have intensive use as central resource centres has shown that natural and mechanical ventilation is inadequate to maintain a pleasant learning environment. Studies showed that the nature of the heating and cooling loads was such that heat pumping and thermal storage principles, whilst providing desirable features of energy conservation would also reduce the maximum power requirement. The system described provides heating and cooling from a water chiller and thermal storage tank through an existing air handling unit. When required, morning heating is provided by circulating condenser water through a wide range coil whilst storing the chilled water in the tank for use in the afternoon during which cooling will be needed. For summer, cooling is provided by the combined operation of the water chiller and the water storage as necessary.

The functions of the equipment and controls is described for the various operating modes of the system. Estimates of energy consumption are presented which illustrates that the use of precious oil fuel for heating is not necessary and that the cost of the energy could be reduced whilst providing comfort and a satisfactory learning environment for the whole school year.

1.0 INTRODUCTION

The conditions in many existing secondary school Libraries have been found to be oppressive. This is due largely to the inadequacy of natural and mechanical ventilation systems to provide relief from the build-up of heat caused by the high occupancy and lighting loads.

In 1978 our Firm was appointed by the Australian Capital Territory Region of the Federal Government Department of Construction to design the cooling for six school Libraries. Three different types of systems were found to be suitable, each type to be installed in two schools. This was to allow for evaluation of their relative performance and operating characteristics.

The systems were:

(a) Direct expansion cooling equipment added to existing heating and ventilation units.

(b) Commercial air to air heat pump units to replace existing heating and ventilation units.

(c) Electric heat pumping system with chilled water storage connected to existing ventilation units.

This paper describes (c), the electric heat pumping system.

The systems involved design and control arrangements which, with the application of the thermal storage tank resulted in an unusual installation.

Cooling loads in the Libraries display characteristics which lend themselves to a heat storage system, these being:

(a) Heating demand in the winter mornings followed by an afternoon cooling requirement.

(b) Cooling demand in the summer lasting for a peak of only several hours.

Both of the Libraries had existing mechanical ventilation and heating systems. Heating was being provided by hot water heating coils connected to central oil fired boiler plant. Whilst the system had been designed with sufficient air handling capacity for future cooling they were operating on nearly 100% outside air which necessitated the addition of effective return air cycles.

The principle of the system is outlined in Figure 1.

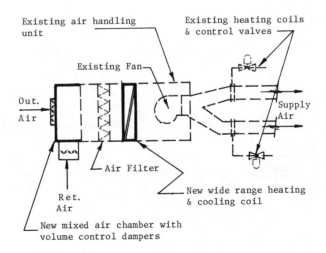

FIGURE NO.1

2.0 DESIGN PRINCIPLES

2.1 Criteria

The following is a summary of the design criteria used for the projects:

Ambient temperature : 35°C DB 21°C WB

Indoor temperature
- summer : 24°C DB, 55% RH (approx.)
- winter : 22°C DB

Lighting : 18 W/m² (average)

People : 100 (130 for short peaks)

Equipment : Photocopier, typewriters

Outside air : 7 l/s per person

Physical details were obtained from the actual buildings:

Windows : Single clear glazing

Roof : Steel deck with 75mm rockwool insulation

Exterior walls : Brickwork

Lighting : Recessed into ceiling

From the above details the instantaneous cooling load for each library was calculated to be 72 kW.

2.2 Type of System

The main components of the system selected comprise a small packaged chilled water set, insulated storage tank, closed circuit water cooler, wide range coil for both cooling and heating and three centrifugal pumps.

The concept of the design is, that during the cooling season, operation of the chilled water equipment in conjunction with the storage tank provides the cooling needed. During heating operation whilst using the condenser water for heating, the chilled water is stored in the tank for use during the later cooling cycle. Supplementary heating is available from the existing hot water heating coils, however preliminary calculations indicated that the supplementary heating would not be necessary. The system is illustrated schematically in Figure 2.

3.0 EQUIPMENT CAPACITIES

3.1 Water Storage and Chiller

The total cooling requirements for a design day were calculated on an hourly basis using CSIRO computer program STEP. The months of January, February and March were analysed using weather data tapes which allowed simulation of actual weather conditions.

From the total thus obtained was subtracted the cooling effect of the chiller during the operating hours. The difference between these two being the amount of cooling which would have to be provided from the store. The capacity of the chiller to cool the store also was checked to ensure that adequate cooling could be achieved during evening operation.

The volume of the water storage was calculated from the accepted temperature range and the total heat which had to be stored.

For this application the final selection was for a chiller of 25 kW capacity (approximately 1/3 of the instantaneous load) combined with a water storage of 14000 litres.

3.2 Indoor Coil

The selection of the indoor coil had to allow for the range of water temperatures during both the heating and cooling cycles. The temperature of the water during the heating cycle was relatively constant, however the temperature during cooling operation can vary as the storage tank temperature changes. The final selection was a 6 row coil to cater for all temperature ranges.

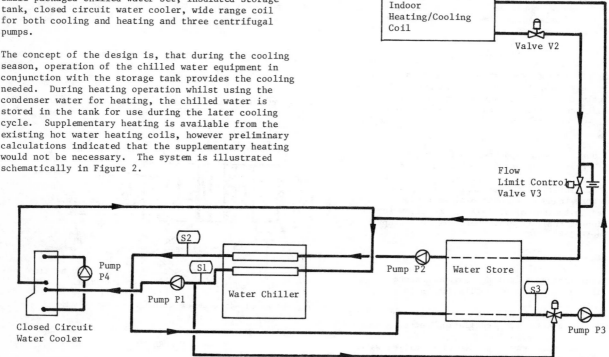

Note: For reasons of clarity hand valves, thermometer pockets, pressure gauge tappings, etc. have not been shown.

FIGURE NO. 2

4.0 OPERATING MODES

4.1 General

The system has five operating modes, each of which presents different system characteristics. A bypass control valve cannot be used on the cooling coil for when less cooling is required in the space than the chiller is providing, it is not possible to "store" the excess in the tank for later use. The control valve V2 (refer Figure 2) can therefore cause varying water flows through the system. Hence a separate pump P2 (refer Figure 2) maintains constant water flow through the chiller.

The five modes of operation are:

    Mode A : Cooling - coil flow less than chiller flow
    Mode B : Cooling - coil flow greater than chiller
                flow
    Mode C : Night time operation for storing chilled
                water
    Mode D : Heating - morning operation
    Mode E : Cooling - afternoon operation following
                morning heating

Each of these modes is now discussed in detail:

4.2 Mode A - Cooling
- Coil flow less than chiller flow

FIGURE NO. 3

Notes on Figure 3:

In this mode the space cooling requirement is less than the chiller capacity. Hence the excess cooling capacity of the chiller is available to lower the water temperature in the storage tank. Water to the indoor coil is drawn off from the bottom of the tank to ensure the coldest water possible, whilst the return water enters at the top of the tank from where the flow to the chiller is taken.

This ensures that the maximum cooling effect is obtained from the chiller whilst utilising the natural thermal stratification of the water in the tank. Condenser water is circulated through the closed circuit cooler by pump P1 in order to reject heat to the atmosphere.

Control of the chiller operation is by the leaving water thermostat S2. Should the temperature of the water drop below 3°C this thermostat switches off the chiller. When the temperature rises above 5°C the chiller will be started. Operation of pump P2 is regulated by the automatic control system which is described in Section 5.

4.3 Mode B - Cooling
- Coil flow greater than chiller flow

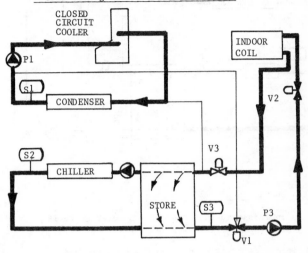

FIGURE NO. 4

Notes on Figure 4:

When operating in this mode the space cooling requirement is greater than the chiller capacity. Hence "cooling" is being removed from the store. Water draw off and return are in the same position as for Mode A. The sparge pipes are continuous pipes and all of the cooled water from the chiller will be drawn into the indoor coil together with some water from the store. This assists in maintaining the lowest possible temperature water to the indoor coil. Again the warm return water is conveyed direct to the chiller to ensure maximum cooling effect. Condenser water is again circulated through the closed circuit cooler by pump P1.

The chiller will be in continuous operation and when necessary will be controlled by thermostat S2 as previously.

4.4 Mode C : Cooling (Night time operation)
- For storing chilled water

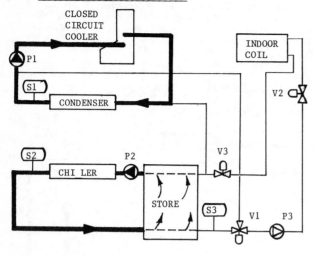

FIGURE NO. 5

Notes on Figure 5:

It is in this mode that the chilled water store is "charged" for the following days operation.

The indoor circulating pump P3 has been stopped as has the air circulating fan. The chiller continues to operate with the closed circuit water cooler. Control of the chiller remains with thermostat S2 which turns off the chiller when the water flow reaches 3°C. This indicates that the store is down to temperature. Once stopped a hold-in circuit is opened and the chiller cannot then start until the following morning.

4.5 Mode D - Heating
- Morning operation

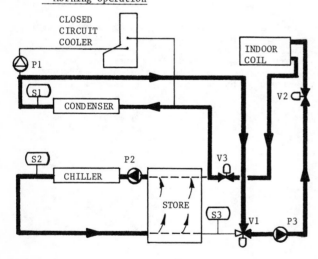

FIGURE NO. 6

Notes on Figure 6:

This operating mode applies when heating is required by the space.

Neither the condenser water pump P1 or closed circuit water cooler are operating and water is circulated through both the condenser and the indoor coil by pump P3. Thus the condenser heat is transferred to the space via the indoor coil. Butterfly valve V3 is closed and the water passes through the flow limiter. This ensures that the water quantity does not allow the maximum recommended velocity through the condenser to be exceeded. Temperature of the water circulated to the indoor coil is nominally 35°C. Operation of the chiller lowers the temperature of the water in the store.

The chiller is controlled by thermostat S1 which is located in the condenser leaving water piping. Should the temperature exceed 40°C the chiller will shut down and not re-start until the temperature reaches 30°C. In the unlikely event of the chilled water temperature becoming too low then thermostat S2 (in series with S1) will shut down the chiller.

4.6 Mode E - Afternoon Operation
- Follows morning heating

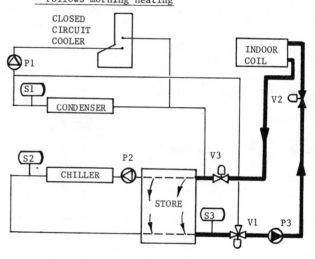

FIGURE NO. 7

Notes on Figure 7:

The load characteristics of the space generally demand cooling in the afternoon due to the lighting and occupancy load. This operating mode allows the cooled water to be drawn from the store to satisfy such a demand. The water chiller, closed circuit cooler, pumps P1 and P2 are not operating. Should more cooling be required than can be supplied from the store then thermostat S3 will start the chiller should the temperature flow to the indoor coil rise above 12.5°C.

5.0 AUTOMATIC CONTROLS

5.1 Equipment

As the plant is unattended all equipment is operated
automatically by a system of electronic controls.
These controls are shown schematically in Figure 8.

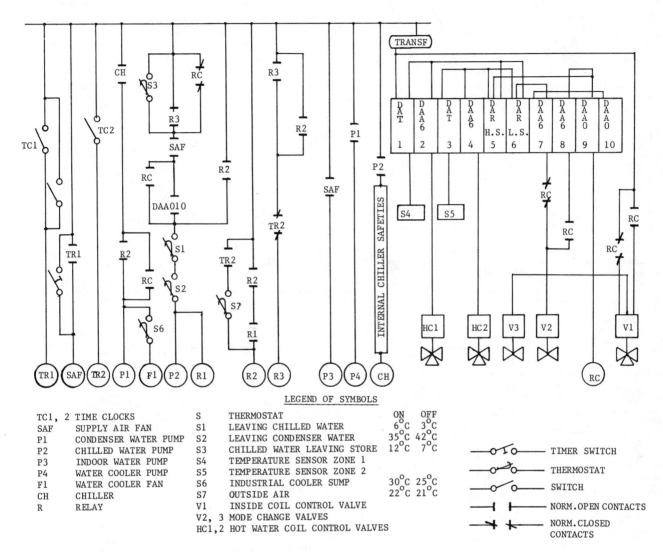

LEGEND OF SYMBOLS

			ON	OFF
TC1, 2	TIME CLOCKS	S	THERMOSTAT		
SAF	SUPPLY AIR FAN	S1	LEAVING CHILLED WATER	6°C	3°C
P1	CONDENSER WATER PUMP	S2	LEAVING CONDENSER WATER	35°C	42°C
P2	CHILLED WATER PUMP	S3	CHILLED WATER LEAVING STORE	12°C	7°C
P3	INDOOR WATER PUMP	S4	TEMPERATURE SENSOR ZONE 1		
P4	WATER COOLER PUMP	S5	TEMPERATURE SENSOR ZONE 2		
F1	WATER COOLER FAN	S6	INDUSTRIAL COOLER SUMP	30°C	25°C
CH	CHILLER	S7	OUTSIDE AIR	22°C	21°C
R	RELAY	V1	INSIDE COIL CONTROL VALVE		
		V2, 3	MODE CHANGE VALVES		
		HC1,2	HOT WATER COIL CONTROL VALVES		

— TIMER SWITCH
— THERMOSTAT
— SWITCH
— NORM.OPEN CONTACTS
— NORM.CLOSED CONTACTS

FIGURE NO. 8

Notes on Figure 8:

The main operating control is timeclock TC1 which
allows the plant to start through the respective interlocks. Space temperature requirements are then satisfied by operation of the equipment through the
respective thermostats. The control sequence is
illustrated in Figure 9.

Timeclock TC2 determines whether the equipment will
continue to operate after school hours to charge the
store for the following day. The timeclock contacts
close between 7.00 and 7.30 PM. During this period
should the outside air temperature be in excess of
24°C, then thermostat S7 will close which energises

lock-in relays R2 and R1, the chiller will then
continue to operate until switched off by the action
of thermostat S2. This breaks the hold in circuit and
the equipment cannot then re-start until timeclock TC1
allows it to do so on the following morning. Should
the outside air temperature be above 24°C relay R3 is
energised when TR2 is de-energised. This locks in
the chiller control circuit to ensure that the chiller
will operate from plant startup on the following
morning.

5.0 AUTOMATIC CONTROLS (Cont'd)

5.2 Temperature Control Sequence

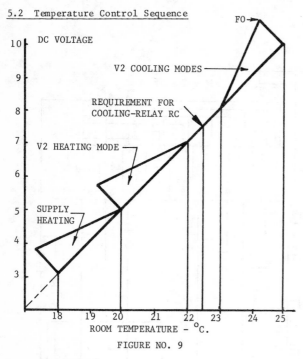

FIGURE NO. 9

Notes on Figure 9:

It can be seen that should the space temperature fall below 22°C then the indoor coil valve V2 will commence to open. At 21°C the chilled water set will be switched on and heat will be transferred to the space. Valve V2 will then modulate until the space load is satisfied. In the unlikely event of supplementary heating being required then the control valves will commence to open should the space temperature drop below 20°C. Later in the day as the space temperature rises the cooling relay (RC) is energised at 22.5°C. This relay changes the operation of the equipment to Mode E. At a space temperature of 23°C the indoor coil valve commences to open, admitting chilled water to the coil. The valve then modulates to maintain space temperature between 23 and 25°C.

It will be noticed that between 22 and 23°C no heating or cooling takes place.

6.0 CONSTRUCTION NOTES

6.1 Equipment

The existing air handling unit was located in the roof space and access to install the new coils and dampers was extremely limited. The remainder of the equipment was installed in the existing boiler rooms where adequate space was available.

6.2 Storage Tank

In order to provide water storage in the most economical manner the tank was designed to the same principle as an above ground swimming pool. This comprises a heavy galvanised steel tank of panel construction suitably reinforced with structural steel sections. Inside this tank is a 50mm thickness of urethane insulation to support the PVC liner. Details of construction are illustrated in Figure 10.

Water draw-off and return is through the sparge pipes located at the top and bottom of the tank. The pipe inlet and outlet are connected. To allow diffusion of water into and out of the tank a series of small holes were drilled in the sparge pipes.

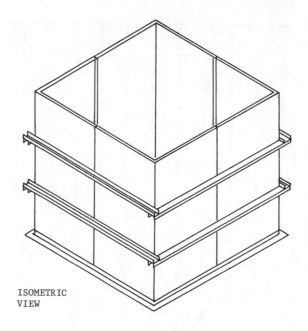

ISOMETRIC VIEW

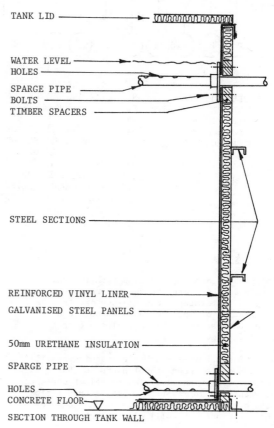

FIGURE NO. 10

7.0 ENERGY CONSUMPTION

7.1 General

The energy consumption and savings potential have been calculated and are detailed below. Due to operating difficulties during the winter season it was not possible to verify the savings. These difficulties have now been overcome and it is hoped that this verification will be carried out during the present winter season.

7.2 Existing Energy Consumption

7.2.1 Fuel Oil: The existing heating systems were surveyed and the rate of heating energy supply was found to be approximately 400 MJ/a.m^2. This was considered to be excessive and it was estimated that a more realistic target would be 270 MJ/a.m^2. This rate of 270 MJ/a.m^2 represents a quantity of approximately 4500 litres of oil per annum. Based on present costs the value of this oil would be some $1200.00 per annum.

7.2.2 Electricity: Electricity consumed by the existing systems was mainly for the supply air fans plus a minor component in such items as hot water circulating pumps, oil burners and controls. When surveyed this energy was found to be some 3000 kWh/a. Based on present costs the value of this electricity amounts to $120.00 per annum.

7.2.3 Total Cost: The total energy cost to provide heating and ventilation for the Libraries amounted therefore to some $1335.00 per annum. It should be remembered that conditions within the Libraries for a significant part of the year were considered to be unsatisfactory as no cooling was available to satisfy the cooling demand.

7.3 Estimated Energy Consumption

7.3.1 Fuel Oil: Based on analysis of the heating requirement profile it was considered that all of the heating required could be provided by the operation of the chilled water equipment in the heating mode. Hence no contribution would be required from the oil fired heating system.

7.3.2 Electricity: The increase of electricity for the supply air fan to overcome the additional resistance of the indoor coil is estimated at 2000 kWh/a, making the consumption 5000 kWh/a. Based on the operating pattern of the school it was calculated that the equipment would operate in the heating mode for 960 hrs/a and in the cooling mode for 900 hrs/a, a total of 1860 hrs/a. During both heating and cooling modes the operation of the various components will change. Accordingly the total installed electrical rating of 13 kW was discounted to 11 kW. Total electricity consumption to provide heating and cooling is therefore 25460 kWh/a which at current rates would cost $1020.00.

The reduction in energy cost to provide heating and "free" cooling is $315.00 per annum - which represents 24%.

7.4 Initial Cost

The initial cost of the system described was greater than the DX cooling equipment or air to air heat pump units. There are no energy savings connected with the DX cooling equipment as the existing oil fired heating system remained in use. The additional cost of the heat pump systems can be justified by the saving of both valuable liquid fuel reserves and money.

8.0 CONCLUSION

The characteristics of the cooling requirements of these school Libraries are similar to those in many other air conditioned spaces. It has been shown that the concept of small chiller capacity with storage can satisfactorily cope with the cooling requirements. One important design feature with this type of system is the elimination of costly and very often, troublesome, capacity controls on refrigeration equipment.

Comparison of the estimated energy costs show that electric heating with the added advantage of "free" cooling can be provided for less energy cost than oil fired heating with only mechanical ventilation whilst providing optimum learning conditions.

Where off-peak electricity tariffs are available this cost benefit would be even more significant. This type of system could also assist electricity supply authorities by lowering peak demands on generating plant.

A most important feature of these systems is the reduction in use of our valuable and limited reserves of oil and natural gas which can be conserved for more important tasks than providing comfort.

SECTION 10
ENERGY ACCOUNTING SYSTEMS

Chapter 36

ENERGY CONSERVATION ACCOUNTING PROGRAMS

J. N. Dougherty, D. Foley

ENERGY CONSERVATION ACCOUNTING PROGRAMS

As energy conservation programs have evolved over the past few years, it has become increasingly clear, that if these programs are to continue to be successful, a heavy emphasis must be placed on ACCOUNTABILITY. Therefore, energy managers must adopt a plan to measure not only a facility's performance, but more importantly, measure the economic merits of monies invested in energy conservation projects.

In the case of commercial properties, the two highest variable expenses are cleaning and energy. In industry, it's raw materials and energy. In both cases only energy costs are controllable.

In the past, energy costs have simply been passed through to the end user, whether it's the private consumer, or the building tenant. Today, though, the situation is more serious, especially in the case of commercial leased space. Since passed through energy costs can raise the total annual cost of the space significantly, prospective tenants are resorting to comparison shopping when looking for a place to establish their business. So, you can expect to find more tenants asking, what is being done to minimize the effects of rising energy costs on their pocketbooks.

Several issues will have to be examined before an investment and commitment is made to a particular energy accounting system. How much time will it save by analyzing energy use data and calculating facility performance. Can it document the financial benefits of conservation? Will the investment pay for itself? What are the available alternatives and which system is best for the particular situation? Most often overlooked is the question, "How can the report's information be used to minimize operating costs?

Without answers to these questions, you will not be able to select the optimum alternative and make the most of the energy conservation dollar. Also, top management's attention has to be gained to establish a plan which defines the intended objectives, establishes required policies, and sets definitive procedures which tie the whole process together.

Some of the more important features that an effective accounting program should include are:

Compare a building's current performance against its own historical performance.
. Document the financial benefits of energy conservation.
. Alert management of problems that require their attention.
. Require operating personnel to investigate and correct any problems.

Energy accounting systems which have either been developed in-house or purchased have not been designed to be simple to use by operating personnel. Typical problems that have existed with some of these systems are:

. An accouting system that is designed to be a statistical giant.
. An accounting system which requires excessive amounts of time to decipher the bottom line.
. And finally, a system of reports that does not satisfy both financial and engineering needs.

SYSTEM FOR MONITORING ENERGY CONSERVATION

A system designed to measure, analyze, and report energy conservation performance is an essential component of any effective energy conservation program. The following describes techniques for evaluating energy conservation performance of an energy cost center, a plant, a building, or an entire organization.

System Components

Any system designed to monitor energy conservation performance has certain elements. Three of these are:

. Establishing standards for measurement of performance.
. Comparing actual results against these standards.
. Analyzing deviations to take corrective action.

Below is a further description to emphasize their importance and value.

Establishing Standards: The first step in the controlling process is to establish standards. Standards are, by definition, criteria of performance; these selected points in an entire planning program give managers signals as to how things are going without their having to watch every step of the program. For many departments, verifiable energy conservation objectives established during the planning process can also serve as control standards. However, at both the

233

plant and corporate levels, many additional qualitative and quantitative standards may have to be set which, upon measurement, will give an indication of how the overall plans are working out.

Comparing Results Against Standards: The second step in the control process is to measure performance by comparing actual results against these standards. This evaluation must be reported to the people who can do something about it. Feedback on performance is a must if the entire energy conservation effort is to be effective. Ideally, it should be done with enough time in hand so that corrective action can be taken before it is too late.

Analyzing Deviations: The third step in the control process is to analyze deviations and then take corrective action. This involves finding out why we are off target and then seeing what, if anything, we can do about it. Usually, this is the responsibility of operating property managers.

Monitoring Techniques: A company wishing to monitor all facets of its energy conservation program will probably use a variety of techniques. Energy conservation accounting programs are relatively new in most companies. A brief description of the more commonly used techniques follows.

Monitoring BTU Consumption: The most commonly used technique is to monitor total BTU consumption for a facility. The BTU consumption of a facility for the current period is compared to that for the base period, and conservation progress is determined in terms of BTU or percentage of energy saved. For control purposes, the base year is generally chosen as the previous year. Before making this comparison, you must adjust base year data for changes in various operating conditions.

Monitoring Simple Energy Rate: Another commonly used method of moitoring energy conservation performance is to compare energy rate in the current period to the standard energy rate or the energy rate in the base period. Energy rate may be expressed as BTU (of units of fuel, or dollars) per unit of output or direct labor hours, etc. for productive energy. For nonproductive energy, the denominator might be square footage, cubic footage, or number of tenants. The energy rate method's main advantage is that it is simple and understandable.

Monitoring Adjusted Energy Rate: Possibly, you can adjust energy rates to reflect changes in product mix, production rate, degree-days, and other variables. If an adjusted energy rate can be determined, then it can be used to reflect changes that are a result of inefficient energy use.

Monitoring Energy Conservation Activities: If sufficient data are not avaiable for establishing a meaningful energy rate, you can consider using some of the other indicators to provide a measure of the overall energy conservation activities going on in the company. Some of these indicators are number of projects completed, and/or planned for in the future. Other measures might be, number of hours spent in energy conservation maintenance; or number of hours spent in motivation and training the company personnel.

You can require that each department submit a monthly or quarterly report, summarizing energy conservation activities by using a prescribed format. These reports indicate each department's degree of involvement and provide necessary feedback to top management.

Budgets: The most commonly used device for managerial control is the budget. Energy conservation performance can be evaluated through the use of many types of budgets. You can also get an idea of a plant's energy conservation efforts by looking at the capital budget allocated, used, and requested for energy conservation projects.

Using Charts: Energy consumption data for any process, equipment, or facility may be monitored through the use of charts. Consumption during the current period, when plotted on these charts, will indicate whether the conservation effort is proceeding as planned.

Cost Avoidance: This is a concept that must be incorporated into an effective accounting system, but is mot often overlooked. Consumption could decrease, and energy costs continue to rise, therefore, the cost avoidance method is the only way true economic merits of money spent can not be measured.

The following is a simple explanation of how cost avaoidance is determined. Even though you consumed less energy this year than last, costs still increased. However, the objective of computing cost avoidance is to determine what you WOULD have paid had you not reduced consumption. To do this you must re-calculate last year's consumption at current prices. The difference is your Cost Avoidance. Example: Last year in July, the consumption was KWH = 1,320,000 and KW (Demand) = 2832 for a total cost of $67,345. This year for the same month, consumption was KWH - 1,156,800 and KW (Demand = 2736 for a total cost of $94,602. If we were to re-calculate last year's consumption at today's costs the cost would have been $106,920. The difference or cost avoidance then is, $12,318.

Energy Budgets: Energy usage in facilities can be classified by several descriptions. The most common classification is to convert all fuels to a common energy equivalent such as BTU's. The total BTU's consumed by a facility divided by the gross square feet yields the energy usage in BTU per square feet, (called the Energy Budget). This is an invaluable measure of performance. Comparison of like facilities based on energy use, not cost, is more accurate because the prices of fuels are not equivalent in terms of energy supplied per unit cost. A facility could have a significant dollar difference between like facilities but consume equal amounts of energy. A specific example will clarify this point. Example: Oil costs $1.40 per gal and has an equivalent BTU value of 140,000. Electricity costs $.07 per KWH and has an equivalent BTU value of 3413. Obviously, it would cost $2.88 for electricity to generate the BTU equivalent of

one gallon of oil. Moreover, as energy prices
continue to climb, comparing total dollars
could hide the fact that energy use either
increased or decreased.

Features Of An Accounting System

To meet the need of owners and managers an
energy tracking and monitoring system should
at a minimum satisfy the following requirements:

  . Present the current monthly energy use
    and costs.
  . Compare current usage with its own
    history in terms of unit and percent
    change.
  . Report out-of-limits performance to
    initiate corrective action.

In summary, these essential ingredients are
required to minimize management's time to
determine the bottom line, provide both performance of energy and financial consequences
and finally alert management that follow-up
is necessary. In the case where there are a
large number of facilities or departments a
summary report should include the overall
performance of each facility on one report.

My presentation today will discuss commercially
available energy accounting systems, and more
specifically, will describe the features, advantages and benefits of each of the different
systems.

Chapter 37

LIFE CYCLE COSTING AND RISK ANALYSIS

J. M. Baker

ABSTRACT

A risk analysis method is presented which allows ranking of energy projects by their "expected" life cycle cost (LCC) values and the uncertainty surrounding these values. The inputs consist of a low, most-likely, and high value for each LCC cost element, such as construction costs and energy price projections. The model uses existing microcomputer software and provides powerful capabilities for "what if" projections, sensitivity and risk analyses, and management decision making.

INTRODUCTION

Total Life Cycle Cost (TLCC) analysis is important in energy management because the future recurring and non-recurring operational costs of many facilities are as important as initial acquisition costs. Operational costs over the life of the facility may range from 20 to 90 percent of the TLCC depending upon the type of facility. The combined effects of inflation and energy price escalation make future costs more important and also more uncertain. A method is needed which will incorporate uncertainty in the computation of traditional LCC analyses.

The Federal Energy Management Program (FEMP), which administers energy management in the Federal government, requires that LCC analysis be used to estimate:[1]

- Whether retrofitting an alternative building system to an existing Federal building is cost-effective and tends to minimize the life cycle cost of that building

- Relative cost-effectiveness of retrofit investments in buildings

- Whether an alternative building design for a new Federal building will minimize the life cycle cost of that building

- Payback time for solar demonstration projects

- Present value of net benefits or excess costs of a solar demonstration project compared to a substitute conventional non-solar alternative building system in an existing Federal building or in the design of a new Federal building

The LCC ranking indices to be used are savings-to-investment ratio (SIR) for retrofit projects; TLCC for new construction; net life cycle cost (NLCC), payback, and SIR for solar projects; and Btu/initial cost or Btu/discounted investment cost (E/C and E/INVEST) for secondary analysis.

As part of a task performed for the Assistant Secretary of Defense (Manpower, Reserve Affairs, and Logistics), a comprehensive LCC method was developed for use within the DoD at the installation, regional, and headquarters levels. The method satisfies the FEMP requirements, but also provides for sensitivity and risk analysis. The method increases the useful analysis that can be accomplished using an existing level of project cost information from the installation level. Table 1 summarizes the LCC analysis output.

TABLE 1. LCC ANALYSIS OUTPUTS

SIR (MEAN, VARIANCE)

E/C (MEAN, VARIANCE)

TLCC (MEAN, VARIANCE)

NLCC (MEAN, VARIANCE)

E/INVEST (MEAN, VARIANCE)

```
SIMPLE PAYBACK
DISCOUNTED PAYBACK
ANNUAL UNDISCOUNTED CASH FLOWS
ANNUAL DISCOUNTED CASH FLOWS
PRESENT VALUE OF:   ENERGY COSTS
                    MAINTENANCE COSTS
                    INVESTMENT COSTS
                    REPLACEMENT COSTS
MILCON SUBMITTAL PROJECTIONS
MBTU'S AVOIDED (DISCRETE)
SENSITIVITY ANALYSIS
DOE, DRI, EER, etc., ENERGY PROJECTIONS
SPW, UPW, UPW* TABLES
```

DATA INPUT AND ANALYSIS

To assure that inputs to the LCC analysis are as accurate as possible, the analyst should first construct a time line as shown in Figure 1. This helps to visualize the magnitude and frequency of all costs which occur throughout the life of the project. Ideally, installation personnel should provide estimates for design, construction (CWE), operation and maintenance (O&M), salvage, replacement, and energy costs. Other inputs such as energy escalation rates and discount rates should be provided by management staff personnel. The analysis would be performed simply and efficiently at a higher management level using a "user-friendly" microcomputer system. The proposed system is shown in Figure 2. Once a year, or whenever the latest energy and inflation cost projections are published, the stored values for fuel escalation rates and the projected GNP implicit deflator can easily be changed and all LCC analyses recomputed using the latest estimates. The installation personnel would use a simple input form which is transmitted to the central computer location where the information is entered into the computer system.

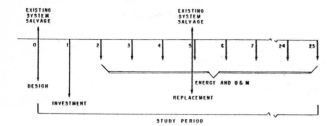

FIGURE 1.   LCC CASH FLOW TIMELINE

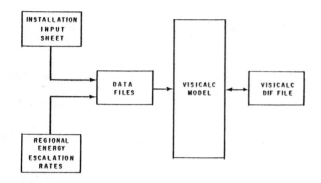

FIGURE 2.   LEDGER SHEET SYSTEM FLOW CHART

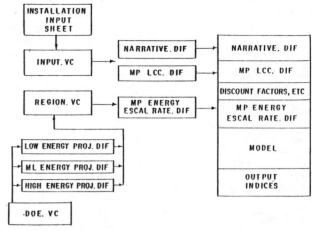

FIGURE 3.   SYSTEM FLOW CHART

The outputs from the program are a data base that is automatically updated by the computer program and hard copy reports. Thus, all data base management actions such as sorts, searches, statistics, graphics, and report generation are available to management decision makers. Additionally, all or part of the data base may be electronically transmitted to a home office to provide current information on projects. At the installation level, the analyst "sees" a simple system almost identical to the existing "discrete variable" manual computation system.

Two years ago, inexpensive "user-friendly" software packages (on either microcomputers or mainframes) were not available to create such a system. Developing a system in-house would have required extensive software development and dependence on a software programmer to maintain the programs and data bases. Now software for electronic ledger sheets, data base management systems, and management decision making systems are commercially available for $100 to $500. The linked series of ledger sheets used in the method are shown in Figure 3.

UNCERTAINTY ANALYSIS

The proposed LCC analysis method incorporates basic concepts of capital budgeting under uncertainty. The uncertainty in fuel and construction costs experienced in the last few years indicates the need for such an analysis. There are several established techniques for assessing the effects of uncertainty about future costs on capital budgeting decisions.[2]

The method uses a statistical approximation of Monte Carlo simulation techniques combined with elements of portfolio management techniques. The basic tenet of the method is that a decision maker given a choice between two energy projects of equal SIR would prefer the project with the least uncertainty of achieving the

computed SIR value. Another way of saying this is that he would prefer the project with the lowest percent variance.[3,4] Conversely, given two projects with the same percent variance, the decision maker would prefer the project with the higher computed SIR. Due to the laws of probability, if discrete values for all cost elements (investment, O&M, energy, salvage, escalation rates, discount rates) are used to calculate a discrete ranking index such as SIR, the probability of achieving that discrete SIR is remote because the SIR probability is the product of the probabilities of occurrence of each of the cost elements. The computation of an "expected" SIR allows the computation of a more realistic (and more nearly achievable) LCC index.

The method combines probability distributions of the input life cycle cost estimates and calculates the mean and variance of the output SIR indices. In effect, for each project the computer approximates (by means of equations derived from statistical theory) the performance of a Monte Carlo simulation using randomly selected values of each input cost element from its unique distribution curve of probable values.

As shown on the left side of Figure 4, each LCC input cost element must have a distribution curve associated with it. The analyst (either at the original preparation level or at a higher management level) must provide information to determine the distribution curves. To do this, the analyst should select a low, most-likely, and high value for each cost element. The model then develops a distribution curve by fitting a Beta distribution to the three values. The low and high estimates specify the lower and upper boundaries of a 90 percent confidence interval around the most-likely value. They are thus the practical bounds that may be realistically expected, not the absolute maximum and minimum estimates.

The middle section of Figure 4 indicates a Monte Carlo simulation of 1000 iterations of calculating "discrete" values of the ranking index for each project considered. The mean and variance of the resulting normal distribution curve is calculated for that project. The mean SIR value is then plotted versus the percentage variance. Each succeeding project is then simulated using its set of cost estimate distribution curves, and a mean and a percent variance are computed and plotted as before.

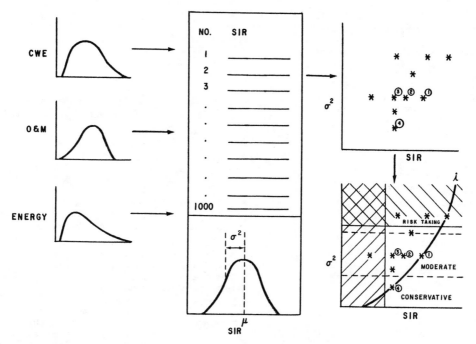

FIGURE 4.   CAPITAL BUDGETING UNDER UNCERTAINTY

After all projects have been plotted, the first graph shown on the right side of Figure 4 is obtained. Using the two criteria of preferring investments with high SIR and low percent variance mentioned earlier, our decision maker would prefer projects to the right and downward on the SIR versus percent variance graph in Figure 4. For example, Project 1 is preferred to Projects 2 or 3 because it has a higher SIR at the same variance. Likewise Project 4 is preferred to Project 3 since at the same SIR value it has lower variance, or a better probability of achieving the computed SIR. Consequently, projects to the right and downward are preferred. A "frontier" curve (i) may be drawn of the most favorable projects achievable.[5]

Finally, we can divide the graph into three distinct management decision making policy regions: conservative, moderate, and risk-taking. For example, a risk-taking management policy would invest in projects that had a possible high return (i.e. high SIR), accepting a larger risk (greater variance) in actually achieving that value. Conversely, a conservative policy would invest in projects with lower but more certain return (i.e. lower SIR). Management must decide which investment strategy is most suitable.

Screening criteria may be imposed: SIR must be greater than 1.0 or percent variance must be less than 20 percent, as shown in Figure 4 by vertical and horizontal lines. Our "feasible area" (unshaded in Figure 4) would include those projects meeting the screening tests and type of investment policy selected. Other LCC output measures (TLCC and E/C for example) may be simulated, plotted, and screened. The final "feasible area" of projects will be composed of those projects contained in the "feasible areas" of all screens.

The expected value of the index, after being simulated, is a more realistic estimate than the discrete value computed without simulation. Examples from

the literature suggest that a 20 percent discrete rate of return computed from discrete cost estimates would become about 7 percent after uncertainty analysis. This reflects the higher probability of achieving the 7 percent based on the uncertainty of the inputs. Project ranking by this "expected" index will be a more accurate ranking since it is based on higher probabilities of achievement.

MONTE CARLO APPROXIMATION EQUATIONS

By use of the Central Limit Theorem and logarithms, the results of the Monte Carlo simulation for the expected index value may be approximated accurately (within 10 percent) by the following equations.[6]

Individual Cost Elements

Expected Non-energy Cost:

$$\mu_i = n(0.15L + 0.7M + 0.15H)$$

Weighted Non-Energy Variance:

$$\sigma_i^2 = 0.09 \ n^2 (H-L)^2$$

Expected Energy Cost:

$$\mu_E = 1 + \left[\mu_{E_{k-1}}\right]\left[\frac{1 + e_k}{1 + r_k}\right]$$

Weighted Energy Variance:

$$\sigma_E^2 = \left[\sigma_{E_{k-1}}^2\right]\left[\frac{\sigma_E^2}{(1+r_k)^2} + \left(\frac{1 + e_k}{1 + r_k}\right)^2\right] + \frac{(\mu_E)^2 \cdot (\sigma_E^2)}{(1 + r_k)^2}$$

239

where n = total number of identical costs within each LCC cost element
L = low estimate for this cost element
M = most-likely estimate for this cost element
H = high estimate for this cost element
i = LCC cost element
N = total number of LCC cost elements
k = year

Once the weighted costs and variances have been calculated for each LCC cost element, then those may be used in the standard equations to calculate the expected TLCC, SIR, E/C, or other indices. The equations are:

Expected TLCC:

$$\sum_{i=1}^{N} (\mu_i)$$

Expected NLCC:

$$\sum_{i=1}^{N} (\mu_i)$$

Expected SIR:

$$\left(\frac{\Sigma(\mu_E) + \Sigma(\mu_M)}{\Sigma(\mu_I) + \Sigma(\mu_R)}\right)\left(\frac{1 + \sigma_I^2 + \sigma_R^2}{[\Sigma(\mu_I) + \Sigma(\mu_R)]^2}\right)$$

where all costs are mean costs or weighted variances of each LCC cost element:

E = energy costs
M = maintenance costs
I = investment costs
R = replacement costs

Expected E/C:

$$\frac{(\text{Energy in MBtu's})}{\Sigma(\mu_I)}$$

The weighted variances for these indices are:

Weighted σ_{TLCC}^2:

$$\sum_{i=1}^{N} (\sigma_i^2) + 2 \cdot \sum_{i<j}^{N}\Sigma \ \text{cov}(x_i, x_j)$$

where $\overset{N}{\underset{i<1}{\Sigma\Sigma}}$ means that the summation extends over all values of i and j, from 1 to N, for which i<j

where $\text{cov}(x_i, x_j) = \sigma_{x_j x_i}^2 = N \Sigma(x_i x_j) - \Sigma x_i \Sigma x_j$

Weighted σ_{NLCC}^2:

$$\sum_{i=1}^{N} (\sigma_i^2) + 2 \cdot \sum_{i<j}^{N}\Sigma \ \text{cov}(x_i, x_j)$$

Weighted σ_{SIR}^2:

$$\frac{1}{(\mu_I + \mu_R)^2}\left[\sigma_E^2 + \sigma_M^2 + \left(\frac{\mu_E + \mu_M}{\mu_I + \mu_R}\right)^2(\sigma_I^2 + \sigma_R^2)\right]$$

Weighted $\sigma_{E/C}^2$:

$$(\text{Mean Energy in MBtu's})^2 \cdot \sigma_I^2$$

The use of such approximation equations eliminates costly and time consuming Monte Carlo simulations. They are easily incorporated into the electronic ledger sheet method.

A possibility exists that improper estimates may be provided by the cost estimator for the low and high values of the cost elements. It is possible to increase the expected SIR and lower the percent variance by providing artificially narrow boundaries for each cost estimate input. This potential problem can be eliminated by providing guidelines for setting the low and high values. For example, if historical data on construction bids for similiar projects is available by construction region, the 90 percent confidence interval around the design estimate may be determined and applied as default values for the input analyst construction estimates. Other large cost elements such as O&M, replacement, and salvage may be determined accordingly. The estimated energy dollar avoidances may similiarly be bracketed using low, expected, and high scenario values for the differential escalation rates provided by EIA and other forecasting services such as EER or DRI.

SUMMARY

A more sophisticated economic analysis including sensitivity and risk analysis is needed for analysis of construction projects in which operational costs will be a substantial fraction of TLCC. Since projections of future energy costs are extremely uncertain, risk analysis and the associated decision making procedures like those presented in this paper are essential in obtaining reliable SIR and other LCC ranking indices. The computerized system developed in this paper is recommended as a tool for accomplishing this improved analysis. The method is consistent with the FEMP Program Rules for computation of LCC ranking indices. The incorporation of a better treatment of cost uncertainty and its interpretation for decision making are important additions to the FEMP guidance.

REFERENCES

1. Ruegg, R.T., "Life Cycle Cost Manual for the Federal Energy Management Program," National Bureau of Standards Handbook 135, U.S. Department of Commerce, Washington, D.C., December 1980.

2. Peterson, D. J. S., "Energy Efficiency of
    Energy Intensive Weapon Systems," Logistics
    Management Insitute, Washington, D.C. ML-111,
    (in press).

3. Hertz, D. B., "Risk Analysis in Capital Invest-
    ment," Harvard Business Review, January-February
    1964.

4. Hertz, D.B., "Investment Policies That Pay
    Off," Harvard Business Review, January-February
    1968.

5. Markowitz, H., "Portfolio Selection," The
    Journal of Finance, Vol. VII, No. 1, March 1952.

6. Dienemann, P.F., "Cost Uncertainty Analysis:
    Predicting the Reliability of Construction Cost
    Estimates," F. R. Harris, Inc. report to Royal
    Commission of Saudi Arabia, Contract No. SGC-02-
    1397, March 1979.

7. Meyer, P. L., Introductory Probability and
    Statistical Applications, Addison-Wesley Pub-
    lishing Co., Reading, Massachusetts, 1965.

SECTION 11
SOLAR PHOTOVOLTAIC
ELECTRIC POWER PLANTS

Chapter 38

A VERY DEEP PHOTOVOLTAIC-POWERED WATER WELL

R. E. Davis

Work Supported by the U.S. Department of Energy
Under Contract DE-AC08-76NV00020

ABSTRACT

This paper covers the activities associated with the design of a 13,000-watt photovoltaic system by an engineering firm not normally involved with photovoltaic equipment. The procurement and field modification of off-the-shelf modules is discussed.

BACKGROUND

The National Energy Conservation Act of 1978 included a provision for the use of Federal funding to install solar photovoltaic demonstration programs at Federal facilities. This program, under the direction of the Department of Energy (DOE), was called the Federal Photovoltaic Utilization Program (FPUP). The objectives of the FPUP were to reduce energy costs at Federal facilities through the development and use of photovoltaic power sources, and to provide a baseline demand for photovoltaic hardware. All Federal agencies were requested to prepare and submit a list of prospective application programs for study and possible funding.

The Nevada Operations Office (NV)/DOE and its contractors submitted a list of 246 potential application projects, from which 41 were approved and selected for funding. The Holmes & Narver, Inc. (H&N), proposal for a 13 kW photovoltaic system to power a continuously pumping hydrologic monitoring well was one of the 41 approved projects.

The project is located on the eastern slope of Yucca Flat, a wholly enclosed basin at the Nevada Test Site (NTS). The NTS is located approximately 65 miles northwest of Las Vegas, Nevada.

This project is under the management of a DOE Project Manager located at the Las Vegas office. H&N reports to this Manager through a Project Engineer located at the NTS.

APPROACH

Previous photovoltaic projects, either FPUP or other programs, have been accomplished within the photovoltaic industry. Projects were undertaken by a small group of acknowledged experts in the field of photovoltaic installation engineering or by photovoltaic manufacturers themselves. These installations were generally of the "turnkey" type, that is, complete design, procurement, and installation; in many cases, even the basic concept came from the company undertaking the project.

With the assignment of this project to H&N, the opportunity to seek answers to various questions dealing with the widespread application of photovoltaics was presented. While working for the DOE with the Solar Heating and Cooling Demonstration Program, the writer became construction industry (architects-engineers, constructors-workmen, and component suppliers) to deal with this new technology. To some extent those concerns also apply to photovoltaics. Could an engineering firm design a specified photovoltaic installation if that firm had no previous photovoltaic experience? Is the photovoltaic industry far enough along in development to provide component support (at reasonable costs) to a project not of their design? Could a construction company install such a system without the supervision of the photovoltaic manufacturer?

H&N will provide system design and component integration for this 13 kW, stand-alone water pumping system. Construction will be by Reynolds Electrical & Engineering Co., Inc. (REECo). For a number of years, H&N and REECo have supported the NTS as prime contractors in the area of general and specialized facility design, construction, and maintenance. Neither company has previous experience with photovoltaic systems.

THE PROJECT

To monitor the quality of the groundwater underlying the NTS, a series of wells are pumped continuously for periodic sampling. The well selected for this project is 2200 feet deep with a water level at 2000 feet. Due to the geological formations surrounding this well, it will only produce 1 3/4 gallons per minute at steady-state conditions. To pump this well in the manner of other NTS monitoring wells would require a 15-hp, 3-phase, 480-volt submersible pump.

An analysis of the energy required to operate the 15-hp pump 24 hours a day indicated that the photovoltaic array cost would exceed the total available funding. Further analysis led to the selection of a dc system, to eliminate inverter losses, and of a beam-balanced pump to reduce the power needs.

The final engineering design selected a 120-volt dc system powering a 3-hp dc motor which drives a Jensen Bros. oil field beam-balance pump.

By this stage of the design, it was apparent that the photovoltaic array was going to be very close to 13,000 peak watts in capacity. Final layout of the array field, design of the support foundations, and estimates of construction cost could not be completed until knowledge of the solar module that would be installed was obtained. To meet the requirements of both the Federal Procurement Regulations and FPUP, the array modules had to be procured by an open competition. Therefore, a Request for Proposals (RFP) format was used to effect that competition.

THE REQUEST FOR PROPOSALS

To prepare the RFP, the Jet Propulsion Laboratory (JPL) provided both the technical requirements by which FPUP module purchases should be evaluated and a potential vendor list comprised of some 35 companies and individuals [1].

H&N wrote the RFP and REECo (as the purchasing agent for NTS) issued it to those on the JPL list.

As the result of technical data submitted, and after an opportunity to supply additional data, six of the eight bidders were found to be acceptable. One bidder was not acceptable as only a third-party warranty was offered. Another bidder was not acceptable due to lack of technical data. The price proposals were then opened and the successful bidder was selected from among the six acceptable bidders. SOLAREX was declared to be the successful low bidder [2].

FINAL DESIGN

Fig. 1 is an artist's rendering of how this project will look upon completion. Note that the array field is composed of 8 rows, each containing 27 modules mounted on 7 individual structural elements. This structural arrangement was selected to be the one best able to withstand the seismic activity that occurs at the NTS.

This final configuration did present a problem with the individual J-box wiring of each photovoltaic module. Each of the acceptable RFP bidders based their price proposal upon supplying all or a major portion of the modules from existing inventory. The SOLAREX inventory modules were wired as low-voltage units (5 volts) with no provision for series wiring within the J-box. By the terms of the RFP, the module supplier was also to supply the module interconnecting wiring.

SOLAREX, with H&N concurrence, decided to make use of the new AMP SOLARLOC connector for this project. (This is the first installation of SOLAREX modules planned to use these connectors.) The decision to utilize the AMP connectors, coupled with the unusual structural arrangement and the string voltage of 135 volts, led to the need to modify the photovoltaic module J-box wiring. As the modules had by this time been delivered to NTS, it was decided to perform the modifications at NTS rather than face any possible freight damage associated with two additional transcontinental shipments. SOLAREX provided all of the necessary material for the modifications and H&N engineering technicians made the changes.

As a result of these modifications, the modules became "customized" to specific positions within each row, while row assignments were determined by each module's current-voltage relationship. As each module was modified, it was labeled, boxed, and stored to minimize field handling by the constructor.

The final impact of not using the "turnkey" approach was reflected in the need to present the constructor

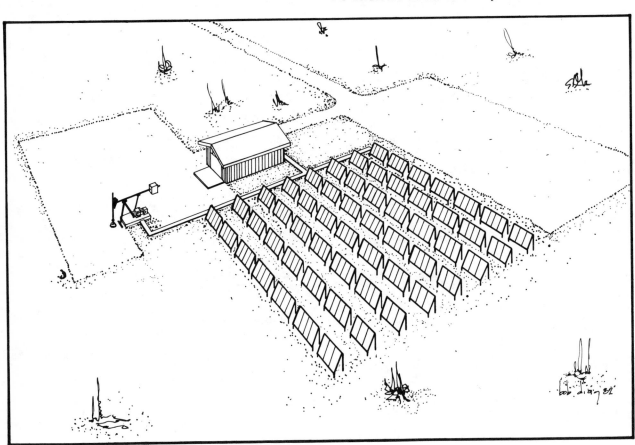

FIGURE 1. ARTIST'S RENDERING

with installation drawings reflecting physical arrangements rather than electrical logic. The electrical drawings normally prepared by SOLAREX reflect only the system wiring logic. The construction industry is accustomed to working from drawings that represent actual physical relationships and identify conduit sizes and wire counts.

CONCLUSION

No major engineering problems beyond the capabilities of H&N were revealed by this simple direct-current system. However, during the investigation of the alternating-current option, the availability of inverters and power conditioning equipment for this system did present a problem. Matching of voltages and current capabilities as well as potential sources presented the greatest difficulties.

The most frustrating aspect of the design effort was the need to stop the design team once they were started, to conduct the procurement for modules. Due to the wide variety of available modules, both in physical size and electrical characteristics, it is not possible to generate an installation drawing that would be applicable to more than one supplier. Therefore, prior to design completion and issuance for construction, the module to be installed must be determined.

One lesson to be learned from this project is that the full design team should not be activated until the specific module to be installed is determined. Initial effort, limited to system analysis and final system definition, should be restricted to a few key personnel. Once the module has been chosen, the full design effort can proceed in the same manner as other facility-type design efforts.

There are two options open for the module determination, each of which has serious drawbacks. The modules can be "sole-sourced" by the engineer or owner for procurement by the constructor. This option could lead to loss of economy, due to lack of competitive bids. It could also be detrimental to module manufacturers that do not have an active and successful sales organization even if they have a better product.

The second option is for the engineer or owner to procure the modules competitively and supply them to the constructor as "owner-furnished equipment." Two problems that can be encountered with this option relate to the responsibility for system operational checkout and the possible reluctance to bid on the project by constructors due to the loss of profit on a major part of the material while facing liability for damage to that material.

Photovoltaic modules should not be accepted from the supplier until an analysis has been completed of the need to modify the "off-the-shelf" modules for any reason. This analysis should be performed by the system designer and the module manufacturer working together. Should modifications be necessary, they should be made prior to shipment.

This project has been completed only through the design stage as of this writing. Construction is expected to begin in early May 1982.

With the exception of module mounting, no problems are expected during construction of this project. All structural and electrical work is common to the facilities building industry, or in the case of intermodule wiring, is done with prefabricated snap-in connectors.

The mounting of the modules is expected to be of concern from the standpoint that the installers will be tradesmen rather than technicians. The tradesmen will come from the local union hall and are not expected to have the delicate touch that module manufacturer installation technicians have developed over the years.

Thus far into this project, there appear to be no problems associated with photovoltaic system design that would prevent a facilities-oriented engineering firm from undertaking a photovoltaic project. Some minor additional costs could be experienced in the search for inverter equipment and in resolving the problem associated with module selection. Additionally, extra cost can be expected in preparing the electrical installation drawings. While photovoltaic systems of this size are rather straightforward electrically, they tend to be fairly complex in presentation.

The need to group photovoltaic modules in rows as a function of individual current-voltage characteristics is a task not normally encountered by a design engineering firm. Depending upon the arrangement between the design firm and the owner, this grouping effort can be accomplished either on the drawings of the installation package (perhaps by means of a table comparing serial number with field position) or in the field during installation.

For this 216-module, 8-row field, H&N devoted approximately 30 man-hours to current-voltage identification, grouping, and labeling of the modules to achieve the required electrical match of the modules.

REFERENCES

1. JPL Document 5260-5, "Solar Cell Module Design and Test Specification."

2. A 2000-Foot, Photovoltaic-Powered Water Well; Richard E. Davis, American Section of the International Solar Energy Society, 1982 Annual Meeting, June 1982.

SECTION 12
EVALUATING BOILER CONTROLS

Chapter 39

HOW TO EVALUATE AND SELECT A BOILER CONTROL SYSTEM

G. Trimm

INTRODUCTION

The recent profusion of new instrumentation and control system components for boiler control and optimization has made it increasingly difficult to make a proper system selection for a given application. A way to ensure project success is to select a control system based on its ability to achieve specific, well-reasoned and viable goals. Once these goals have been established on a factual basis, including technical and economic justifications, they can serve as a guideline for selection of the system strategy, hardware and software.

The process of control system selection for a multi-boiler, multi-fuel powerhouse will be explored in depth. Individual boiler and total powerhouse optimization will also be considered.

SYSTEM SELECTION

The procedure for control system selection involves four basic steps:

1. Evaluation of current operations.

2. Statement of desired benefits to be gained from system implementation.

3. Evaluation and determination of control strategy.

4. Selection of hardware and peripherals.

Evaluation of Current Operations

In order to measure the effects of any proposed changes in a control system or strategy, it is necessary to establish a factual baseline of operating conditions as a reference point. Such a task is best accomplished by conducting an "audit" of current operations. The audit should include both an energy and operating/maintenance evaluation geared toward yielding the following information:

1. Individual "as-found" boiler efficiencies over the complete load range.

2. Individual "tuned" boiler efficiencies over the load range at lowest possible excess air levels achievable within environmental constraints.

3. Manpower requirement history.

4. Individual boiler load history.

5. Total powerhouse load history.

6. Control system failure and maintenance history.

7. Electrical or steam power consumption records for all boiler auxilliaries, i.e., feedwater pumps, fans, sootblowing, heaters, etc.

Careful study of this data will illustrate potential areas for improvements in operations. Furthermore, it will serve as an aid in stating the desired benefits of the system and predicting the cost savings potential from the proposed changes.

Statement of Desired Benefits to be Gained from System Implementation

The first step in arriving at benefits to be gained from the system is to review the reasons that prompted the modification or upgrade under consideration. Some common factors are summarized below.

1. It is desirable to improve the operating efficiency of the boiler. (Optimization)

2. Downtime due to control system failure has become excessive.

3. Spare parts for the existing system are no longer available.

4. Altered fuel usage requires control changes.

5. Additional control and/or reporting requirements are necessitated by new environmental constraints.

6. It is desirable to centralize boiler control to a new location within the plant.

7. It is desirable to reduce manpower requirements.

8. It is desirable to improve overall system response to load swings.

An analysis of these factors coupled with consideration of information gathered in the evaluation of current operations should yield a statement of desired benefits to be gained from the new control system. It is imperative that all areas of the plant be involved in the formulation of such a statement. These benefits must now be justified as to the technical and economic advantages available to the plant.

Evaluation and Determination of Control Strategy

Selection of a control system strategy for a boiler and powerhouse centers primarily upon obtaining the required steam to meet process demands for the lowest possible cost. This involves three levels op optimization; operating individual boilers at their most efficient point, allocating the total load demand to the most efficient combination of boilers and fuels, and optimizing equipment, personnel, and maintenance costs.

Individual Boiler Combustion Optimization

The goal of boiler optimization is to minimize fuel usage in the furnace by operating near the maximum achievable thermodynamic efficiency. This is

251

accomplished through operation, as close as safety will permit, to stoichiometric conditions. This balance is most closely achieved by limiting the amount of excess air introduced into the combustion process. Losses represented by excess air are realized by heated air discharged to the stack without relinquishing its oxygen to combustion. (Refer to FIGURE 1).

In order to operate under low excess air conditions, reliable and accurate measurements of flue-gas constituents are required. The two most commonly applied measurements are Oxygen (O_2) and Carbon Monoxide (CO). The flue gases from a boiler furnace contain combustion products plus any air that may enter through cracks, open prots, etc. Complete combustion of the fuel at stoichiometric conditions will react all carbon to carbon dioxide (CO_2), and all hydrogen (H_2) to water (H_2O), leaving no free carbon, hydrogen or oxygen. In practice, however, this situation never exists. Products of incomplete combustion, mainly CO and O_2, will always be present in the stack. Measurement of these products can be used to control total air flow with the goal of maximum excess air reduction.

The decision to be made is which variable to use. Controlling air flow on O_2 provides only gross control of excess air. However, it may be the only feasible measurement in some applications. In cases where CO measurement is possible, control can be vastly improved. Reliable in-situ CO analyzers now available permit the measurement of minute quantities of CO in the flue-gas within the range of 0 to 1000 PPM. FIGURE 2 shows the relationship of excess O_2 to CO in the flue-gas. In the region of low excess air, a small reduction in O_2 results in a proportionately large increase in CO, thereby providing a high gain response control parameter that can successfully be used to control excess air to minimum levels. An added benefit is that CO is present only as a product of combustion and is therefore not affected by outside air infiltration as is the case with O_2. Flue-gas O_2 percentage is also affected by boiler load. Any system strategy employing O_2 control must account for this fact. A generalized relationship between %O_2 and %boiler load can be seen in FIGURE 3. Combining O_2 and CO control is the optimum course to follow. Numerous successful installations now control excess O_2 to as low as .3% - .5% using CO control. A common target on CO control is 0.5 to 1.5% O_2 over the full load range.

A suggested control strategy is illustrated by FIGURE 4. This configuration provides for the combustion of air flow to be controlled by the CO signal as long as the oxygen percentage is within preset limits. If the %O_2 exceeds these limits, control reverts to the O_2 signal. Override of both the CO and O_2 control signals is provided, in this case, by an opacity monitor which will increase air flow as necessary. This is required in order to ensure that local environmental regulation limits are not exceeded should CO or O_2 control cause intolerable opacity levels.

A decision on whether or not CO control is feasible for a particular boiler can be made based on the results of the energy audit. In this audit, using a CO and O_2 monitor, excess air can be reduced to determine at what level CO enters the control range (approximately 400 PPM). This level represents the excess air at which the boiler could be controlled on a specific fuel using CO as a measured control variable. For some boilers in poor mechanical condition, or which are burning fuels that do not combust cleanly, CO control is not feasible.

Sootblowing

Some fuels such as coal, oil, bark, black liquor, etc.,

over a period of time, generate soot deposits on boiler tubes. The result is a decrease in heat transfer capacity and a reduction in overall boiler efficiency. Systems are available that utilize microprocessor technology to automatically blow this soot away. The basis for this control can be a timed sequence, pressure drop calculation, or a heat transfer calculation to determine when, where, and how long to blow soot. A review of past history will show which alternative, if any, should be implemented. Those boilers requiring frequent sootblowing may benefit from such a system. Sootblowing control that initiates action only when needed results in significant steam savings over those that control on a set time basis only.

Load Allocation

The object of load allocation is to distribute the total plant energy demand on the powerhouse as optimally as possible. The concept ideally will reduce steam production costs to a minimum. Within a control system, this can be accomplished in several ways.

In the most simple case, bias of signals in the boiler master controller can selectively fire boilers known to be more efficient within a given range. In the most sophisticated systems, process inputs to a computer calculate the real-time efficiency of each boiler in the complex. These values are used to calculate the incremental steam cost for the next load change on a per boiler basis. Load increases are then assigned to the most cost-effective boiler. Load decreases are handled by the least cost-effective boiler as directed by the controlling computer.

Combination Fuel Optimization

In many installations, boilers burn a combination of fuels. For example, in a pulp & paper mill, "hog-fuel" boilers burn a base load of bark and other waste products. Oil or pulverized coal serve as the swing fuel for load deviations as combination boilers are not normally configured to vary base load parameters. A new control system should be capable of optimizing (maximizing) the use of these base load fuels in order to reduce steam production costs. The load swing capability, however, must take priority at all times. This means that waste fuel optimization will be dependent upon a particular plant's steam load patterns.

Equipment, Personnel, and Maintenance Costs

The wide variety of control system components now available can make hardware selection a complex matter. Again, the established goals and desired benefits must be used as a guideline in the equipment selection process. This direction, in conjunction with estimated return on investment calculations, provides the basis for equipment selection. The basic control system configurations available are generally classified as follows:

1. Traditional Analog Control - pneumatic or electronic.

2. Direct Digital Control - DDC.

3. Distributed Control - microprocessor based.

4. Distributed Control - computer based.

5. Distributed Control - autonomous microprocessor.

6. Computer Supervisory Control - with analog back-up.

The overall desire is to select a configuration that will meet the established goals at minimum cost. In the selection process, the following should be considered.

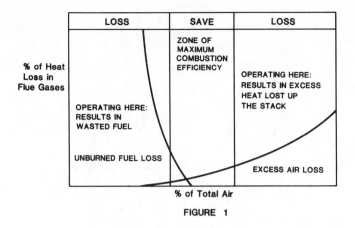

FIGURE 1

RELATIONSHIP BETWEEN CO AND PERCENT OXYGEN

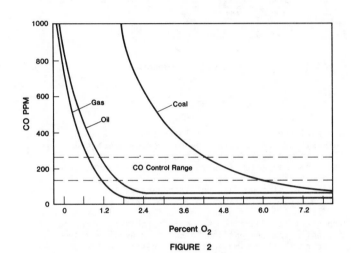

FIGURE 2

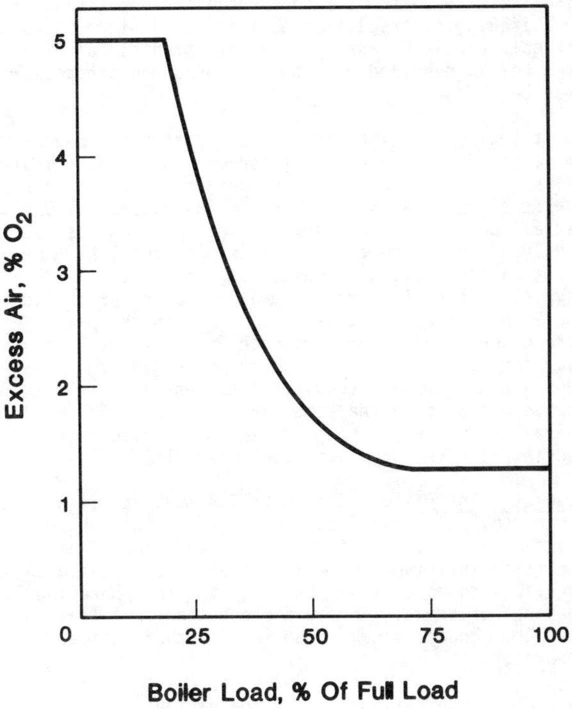

Excess Air, % O₂ vs Boiler Load, % Of Full Load

FIGURE 3

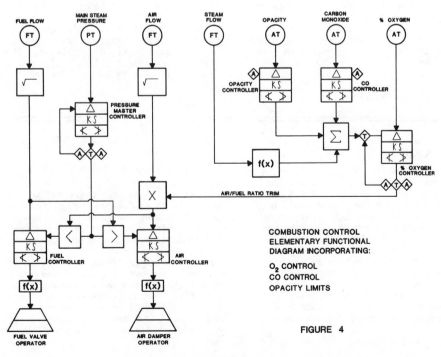

COMBUSTION CONTROL
ELEMENTARY FUNCTIONAL
DIAGRAM INCORPORATING:

O₂ CONTROL
CO CONTROL
OPACITY LIMITS

FIGURE 4

System Control Capability: The degree of complexity of the control strategy developed will be the most significant factor in making a first "cut" of the available systems. If a good deal of computation is required, e.g., real-time efficiency, a microprocessor or computer is necessary. If only traditional PID control is required, analog, DDC or a microprocessor may be used.

Distribution of Liability: The second "cut" may be made based on how liability for control is distributed within the system. Of utmost concern is the maintenance of boilers on-line despite instrument failure if at all possible. Each system configuration must be analyzed to ensure that redundancy is available for critical controls and monitored points. In analog systems, there is simply a manual loader serving as back-up to the main control system. In the microprocessor based scheme, however, suitable back-up support can take on numerous forms ranging from redundant electronics (equipment to continuously receive and process the same inputs and outputs in a parallel manner) to simple analog controllers in series with the processor controlled I/O.

Operator Interface: Consideration must be given to the number of operators required to handle the control systems as well as their training and background. Operator interface units vary from panel mounted controllers to color video CRT displays incorporating live process graphics. For multiple boiler installations, the CRT concept becomes the most cost-effective approach.

System Communications: With the various devices that comprise a state-of-the-art control system, i.e., CRT's, controller files, printers, computers, etc., it is important that inter-unit communication be secure and rapid enough to adequately control the system within established limits.

Installation Costs: Installation costs can vary significantly primarily due to the different methods of field wiring the inputs and outputs to the system. In facilities with numerous boilers controlled from a central location, the use of a data hi-way greatly reduces field installation costs. This is so due to the fact that with the data hi-way system, all required hard wiring can be locally clustered in an I/O module or local controller file. Information required for operator interface is normally transmitted to and from the control room via single or dual redundant co-axial cable. Panel costs can also be reduced by the use of CRT operator interface.

Conclusions: A typical powerhouse control system using

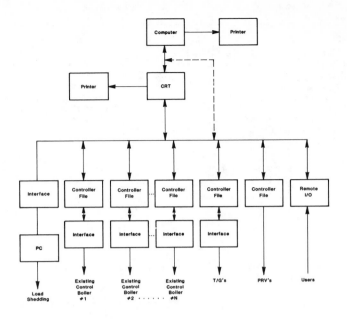

TYPICAL DCS FOR POWERHOUSE

FIGURE 5

a distributed approach is illustrated in FIGURE 5. This system incorporates several devices designed to place the control functions required in the simplest unit that can accomplish the task. This concept, known as hierarchial control, ensures maximum system reliability at minimum cost yet incorporates the optimization capabilities available. To arrive at the best configuration for a given application, it is extremely important to adhere to the goals established for the project. The project should be structured to achieve those goals at minimum cost. If this methodical approach is not followed, the results can be an excessively expensive system having under-utilized capabilities which yield little or no benefit to the plant.

In order to avoid this undesirable condition, the steps for control hard/software selection as detailed herein should be rigidly followed. If concentrated effort is put forth before choosing a system, the end result will be both technically and economically rewarding.

The Boiler Energy Audit

The single most effective way to cut energy costs in most industrial plants is to increase the operating efficiency of the power boilers and the overall efficiency of the powerhouse. With the recent profusion of new monitoring and control devices, it has become difficult to realistically evaluate what investments really make economic sense. The best approach is to do an energy audit of the boiler and powerhouse to establish a base line from which to measure any proposed changes to the system and also give an indication of just how far you can go in improving the system. This article discusses the purpose and methodology of conducting a boiler energy audit and offers guidelines for the interpretation and use of the results.

DEFINITION AND PURPOSE OF A "BOILER AUDIT"

The definition of the boiler audit is inherent in the term. It is an "accounting" of the energy flows into and out of a boiler to determine how efficiently the total available heat energy in the fuel is extracted and converted into useful energy in the form of steam.

The purpose of the audit is to determine the efficiency of the boiler over its load range and generally results in a graph of efficiency vs. load as depicted in FIGURE 1. Sounds simple enough. There are, however, several different connotations to the word "efficiency" and at least two generally accepted methods of determining the efficiency. These will be explored later.

As a goal in most energy audits, the engineer is attempting to determine how efficiently the boiler is operating under the existing control and operating strategy and also the efficiency at which the boiler can be operated, perhaps by modifying the control or operating strategy or by making mechanical changes in the boiler (e.g. chemical cleaning, sootblowing), or by adding a new heat recovery device. In this effort, the following terms become useful:

As Found Efficiency: That measured under current conditions of repair, control, and operation. This value is used as a base line for any subsequent efficiency improvements.

Tuned Efficiency: That measured after operating adjustments and minor repairs have been made.

Maximum Economically Achievable Efficiency: That which can be reached by economically justifiable improvements

Maximum Attainable Efficiency: That attainable regardless of cost considerations.

With these definitions in hand, the purpose of the boiler audit can now be stated as: "To determine the as found efficiency as a base line, tune the boiler while on-line to determine the tuned efficiency, and using the data obtained, evaluate the steps and costs necessary to reach the maximum economically achievable efficiency." The maximum attainable efficiency can be used as a guideline to determine how closely we can approach the theoretical.

METHODOLOGY

As was mentioned previously, there are two generally accepted methods of determining boiler efficiency. Both methods are, when taken back through their physical derivations, mathematically equivalent and would yield the same results if all measurements could be performed without error. It should be noted that these methods measure the "gross" efficiency and ignore heat inputs from boiler auxilliaries such as combustion air fans, pulverizers, stokers, etc. The measured efficiency can therefore be considered truly a measure of the boiler's "effectiveness" in extracting heat energy from the fuel.

Method 1 - Input/Output Method

$$\text{Efficiency (\%)} = \frac{\text{Sum of Heat Outputs}}{\text{Sum of Heat Inputs}} \times 100$$

This method requires the direct and accurate measurement of all energy flows (fuel, boiler feedwater, blowdown, steam, etc.). Due to the potential for significant errors in measurement, this method is not practical for most industrial boiler installations where precision instrumentation is not available. It can and should, however, be used as a check when practical.

Method 2 - Heat Loss Method

$$\text{Efficiency (\%)} = 100 - \frac{\text{Heat Losses}}{\text{Heat Input}} \times 100$$

This method requires only the measurement of flue gas oxygen, flue gas carbon monoxide, flue gas temperature, and combustion air (ambient) temperature and is much more accurate provided that the stack gas measurements are done properly and are not subject to air dilution by leakage or inaccuracies due to flue gas stratification. This method basically measures the amount of heat in the flue gas which is discarded to the environment without benefit to steam production. The O_2 con-

tent in the flue gas is a direct measure of the amount of air which passed through the furnace and was heated, but was not required for combustion and therefore represents wasted energy.

Performance Of The Test

The Boiler Audit should be performed by a qualified firm having the necessary equipment and an experience record in this specific field. Incorrectly performed audits by individuals who lack experience to do an on-the-spot evaluation of field data can lead to catastrophic results both in economic terms and safety considerations. Additionally, the firm selected to do the audit should have the engineering capability to properly evaluate the results both from an operating standpoint for boiler and personnel safety considerations and an economic standpoint. The potential for savings is great, but so is the potential for gross error in investing in improvements based on an incorrect evaluation of circumstances or an improperly performed audit. The bottom line is that there is a good portion of common sense, judgment, and experience that needs to be applied to any testing, data evaluation, or improvement recommendation.

The following general approach to conducting the audit should be followed.

The flue gas analysis instrument should be capable of measuring both Oxygen (O_2) and Carbon Monoxide (CO). The O_2 measurement will be used to calculate efficiency while the CO measurement will be used to indicate the approach to the most efficient boiler operation.

The first step is to determine the as found efficiency without any changes to the way the boiler is normally operated. Boiler load should be stablized at each firing rate for sufficient time to allow all variables to settle out. The next step is to find the tuned efficiency at the same load level by reducing the amount of excess air in the flue gas to a point where the CO meter reads approximately 400 ppm or just before smoking occurs and determining the efficiency at that point. This represents the most efficient operation possible given the mechanical condition of the boiler and its auxilliaries. Graphs of the as found and tuned efficiency vs. load can then be generated. (See FIGURE 1).

INTERPRETATION AND USE OF RESULTS

The first step in evaluating the results is to determine where the as found efficiency stands against the maximum economically attainable efficiency for the particular type of boiler and fuel. For a given fuel there are inherent restrictions on what level of efficiency can be achieved. These are summarized as follows:

1. For each type of fuel there is a certain minimum excess air required so that changes in ambient conditions, i.e., temperature and humidity, will not cause the boiler to enter a region of insufficient excess air levels or lead to smoking and exceeding of emissions standards. Other than economizer or air heater addition (if these are not already installed), reduction of excess air in the flue gas represents the single most significant potential for improvement of efficiency. The more closely a control system can safely approach this limit, the more efficiently the boiler will operate.

2. For each fuel type, a minimum amount of stack losses are inherent in the combustion process and cannot be avoided. These losses are composed of dry stack losses and moisture losses. Inherent

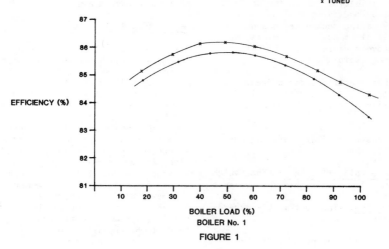

EFFICIENCY (%)

BOILER LOAD (%)
BOILER No. 1

FIGURE 1

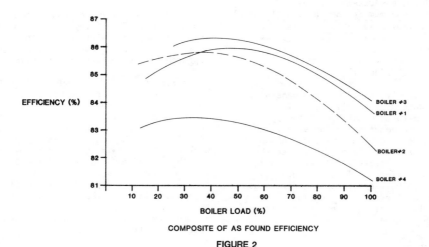

EFFICIENCY (%)

BOILER LOAD (%)

COMPOSITE OF AS FOUND EFFICIENCY

FIGURE 2

moisture losses are directly proportional to the amount of hydrogen in the fuel which is converted to water vapor in combustion.

3. The resultant flue gases from the combustion of each fuel type are potentially corrosive to the stack material if a temperature is reached in the stack at which acid is formed. Thus, there is a minimum stack temperature requirement to prevent cold end corrosion which limits the extent to which heat can be recovered from the flue gas.

Once these inherent limits are known, the maximum economically achievable efficiency can be determined.

The spread between this value and the as found value represents the total potential for improvements.

The next step is to economically evaluate possible changes in operations, controls, or equipment to reach the attainable level. First order economic studies should be made to determine First Cost, Annual Savings, Annual Operating Costs, and expected Return on Investment. Those that look promising should be given further careful consideration.

Experience has shown that savings potentials vary over a wide range for various options and general rules cannot be arbitrarily applied to specific installations. Typically, improvements in efficiency that can be identified in a boiler audit properly conducted can range from as little as 0.5% to as much as 10%. On an annual fuel bill of $15MM, for example, this could yield savings ranging from $90,000 to $1,750,000. A boiler energy audit is therefore the first step that should be taken in a concerted effort to improve total energy consumption. The individual boiler actual and potential performance curves should then be combined into a composite curve for the powerhouse which can be evaluated against the plant requirements. (An example is FIGURE 2). In all but the very rare cases, a boiler audit will pay for itself many times over. Even if major changes are not made, a composite, realistic look at powerhouse operations will show up inefficiencies that can be improved often without capital outlay.

SECTION 13
ENERGY MANAGEMENT
PRACTICES

Chapter 40

EXERGY MANAGEMENT:
THE DEFINITIVE THERMOECONOMIC
ASSESSMENT OF ENERGY

J. Soma, R. S. Krochmal, H. N. Morris

Abstract

Exergy management is based on the proposition that
the Laws of Thermodynamics and sound economic
principles represent the proper assessment of
energy. Unlike energy management which is endemically
a First Law practice, it embraces the Laws of
Thermodynamics in economic concert with capital,
resource and labor considerations. Exergy management's
goal is not energy conservation per se, but optimum
productivity through energy leveraging of an
activity's elements of productivity within extant
degrees of freedom and return-of-investment criteria.

There was a time when everyone believed that the Earth
was flat. To the unaided eye, in the absence of
mountains and valleys, it appeared to each person
that his local environment was indeed flat. Mapmakers
made their maps, and caravans and ships zigzagged all
over the flat Earth in the spirit of the economy of
the times.

The flatlanders of yore would be heartened to know
that they have a latter day counterpart, the energy
practicioners of today. Their flatland is the First
Law of Thermodynamics whereon all BTU's are equal.
Those BTU's cannot assume abovegrade or belowgrade
elevations or value. They are inexorably equal in the
most democratic sense.

Present day energy management is monoLawistic. It is
mired in the First Law and disinterested, if not un-
interested, in the other Laws. Though it had its
origins in the so-called Energy Crisis, it would be
hard pressed to explain why there was a crisis at all.
Surely, there are more BTU's in the free ambient air
than in all the costly oil under the sands of Arabia.
Could there be an intrinsic difference between BTU's?

For each electrical BTU a car battery discharges, a
BTU can be injected into the ambient air. Who ever
heard of the BTU's in the ambient air voluntarily
charging a drained car battery? The Second Law
doesn't admit of it.

A rigorous thermodynamic background is not a pre-
requisite for becoming an energy manager. Virtually
anyone can, from accountant through zoologist, by way
of dentist, engineer and lawyer, and oftentimes does.
When the First Law perception of the role is to count
BTU's much as one would count money, no wonder.

Students of thermodynamics are very comfortable with
the First Law, but not so, with the Second Law, which
may be articulated as: "The entropy of the universe
tends towards a maximum". Though rigorous and un-
deniable, this statement of it conveys nothing of a
practical nature. A more incisive statement of it
would be in terms of entropy's complement, exergy.
Thermodynamically, "The exergy (available energy) of
the universe tends towards a minimum". This identifies
a depletable resource in need of management.

In the context of the electrical BTU and the ambient
air BTU, the electrical one is pure exergy, in fact,
it can be called an American Exergy Unit (AEU); the
ambient air one is cold dead, absolutely moribund,
with respect to its environment. It can do no work and
has an AEU content equivalent to zero. The AEU can do
everything a BTU can, but not vice versa. The Second
Law denies the interconvertibility that the First Law
admits of.

Belief that energy is indeed exergy is perhaps the most
widespread, popular delusion of our times. Consumers
really purchase energy for its exergy content.

Consider an electric arc furnace melting steel. A
classical energy efficiency of 75% might be found on
a BTU basis. An exergy efficiency of 52.4% would be
found on an AEU basis. The moot financial question is:
What portion of electricity costs should be attributed
to the steel melting process, 75% or 52.4%? Should 25%
or 47.6% be assigned to losses? The proper thermo-
economics reside with exergy. Energy management would
have its cost accounting treat dollars as indiscrimi-
nately as it treats BTU's.

In the midst of the American productivity crisis, the
luxury of utilizing incomplete thermodynamics, and
"generally accepted accounting principles" on the
wrong energy commodity, the BTU, is untenable. The
penalties for competitive failure in the marketplace
are, to say the least, severe. Energumens of the First
Law will surely perish there.

Appreciative of needs of the marketplace, exergy
management has emerged. It does not treat energy in
sterile isolation from the other elements of productivity, but rather seeks eclectic synergism with them
in the interest of productivity and enhanced competitiveness.

Exergy management views energy conservation without
prejudice and dispassionately, certainly not as an end
in itself. Instead, it proceeds by assessing all the
degrees of freedom open to a business organization and
seeks an optimal solution through judicious synergism
within existing boundary conditions.

Exergy management considers enhanced productivity as
the universal prize. Mundane conservation savings will
openly be sacrificed to gain greater savings or profits
among the elements of productivity. This energy leveraging for the common good never candidly appeared in
energy conservation circles, even though much energy
savings was achieved by production and quality circles'

leverage into the energy area.

Beyond D.O.E.'s dismantlement, there are recurring
tales of energy management functions being dissolved,
disbanded or dismissed because energy savings couldn't
be documented. If rightly so, so be it. However,
consider the possibility that genuine energy leveraging
occurred. There would be no savings in energy dollars
or BTU's, but the effect would be reflected in the
elements of productivity. Many traditional accounting
systems lack the sophistication to document an energy
induced savings which is distributed in other areas.
An exergy accounting system would be equal to the task.

Exergy efficiencies are equal to or less than energy
efficiencies. The energy viewpoint makes things look
much more efficient than they really are. A domestic
gas or oil fired water heater may have an energy
efficiency of 75% while operating. The exergy efficiency is 4.5%. By taking an optimistic view of things,
many investment opportunities are bound to be overlooked and some of the investments indicated may be
directed at symptoms rather than root causes.

By operating in accordance with at least two Laws,
exergy management avoids many of the myopic or over-
generalized visions energy management can easily fall
prey to. It provides a clear map of the elevations
and depressions on the true surface of things. With
its superior insight, it is less error prone and
more able to proceed along the path of least resistance
towards enhanced productivity.

Exergy management will be the manager of necessity in
tomorrow's productive ventures. Being soundly backed
by the Laws of Thermodynamics, it can supply a
lucidity previously unseen in the marketplace. The
productive managers on manager's row will see the light
and proceed accordingly.

A stir, perhaps, a present shock may accompany the
acceptance of exergy management. History has a
propensity to repeat itself, historians say. Perpetual
motion devices of the second kind were not denied by
the First Law, which also admitted of chemical
processes, though inventive, were impossible. Energy
management is the latest repetition, and, as the First
Law, is a necessary, but insufficient mechanism in the
marketplace.

The wastage of labor, capital and manpower in trying
to invent perpetual motion devices and commercializing
impossible processes must have been exceedingly great.
An insufficient mechanism operating in today's marketplace can be outright ruinous.

Exergy management is the complete energy mechanism
needed. To determine current postures, exergy assays
will be undertaken, not at the level of naivete of
energy audits, but with all the intensity of an in-
depth quality audit planned against definite thermodynamic criteria and with correct cost analysis. The
findings will not be treated in isolation.

Identified problem areas will be prioritized and
solutions sought with complete thermodynamic analysis,
correct thermoeconomic costing and due consideration
for the other elements of productivity. The chosen
solutions will be optimized with respect to productivity within an entity's existing degrees of freedom.

In the coming productivity competitions in the marketplace, the fate of industrial empires will be decided.
The persevering captains of industry will need the best
stereographic charts that exergy management can provide
to navigate the productivity terrains and seas of the
future.

Chapter 41

INDUSTRIAL ENERGY MANAGEMENT . . . LOOKING BEYOND THE PLANT GATE

E. M. Loyless

I would guess that almost everyone in this room is attending this conference and this particular session because you are responsible for energy management at a large industrial facility. As the emphasis on controlling the energy portion of manufacturing costs has grown, your area of concern has probably grown to include anything and everything within the plant gates which might affect energy consumption. Today, I hope to persuade you that there are more factors outside your plant gates than within them which affect your energy budget, that chief among these factors is government and government's driving force - public opinion, and that as Energy Managers you must deal with these "outside factors" in order to do your job properly. I'll explain what I mean by that shortly, but first let's examine where we are today regarding energy supplies.

AN ENERGY CRISIS?

Beginning in the early seventies, we mounted an energy conservation effort in this country which has been very successful. It probably helped us avoid a short-term disaster. In the long run, however, it would be dangerously naive to believe that energy use in this country can decline while we maintain a healthy, growing economy. We must redouble our conservation efforts so that we can produce the maximum amount of goods and services for each unit of valuable energy we consume, but we must also recognize that if we are to produce at a rate which will provide our children with an expanding rather than a shrinking future, our energy requirements are going to increase. The Electric Power Research Institute (EPRI) has pointed out that in order to avoid a net decline in the real life-time earnings of U. S. workers over the next two decades, the U. S. economy must grow at an annual rate of 2.5 percent. According to EPRI, this rather modest growth rate would require a 50 percent increase in energy use by the end of the century. This increase includes a doubling in the use of that most highly refined, most expensive form of energy - electricity. Can we meet these future energy needs or are we faced with an energy "crisis"? I am among those who believe that we do have enough raw energy resources and that we do have the potential technology and the national will to develop those resources. But raw energy resources do not run our factories. Energy must be recovered from increasingly difficult natural states. It must be refined and converted into forms that the end users, such as yourselves, can use. It must be distributed to the places where it is to be used. The cost of this process will reach staggering dimensions. The capital cost of the facilities to gather, convert, and distribute the energy we need in the next two decades will dwarf the total value of all the energy facilities currently in place. The electricity supplier to our plant in Florida, for example, recently announced that its capital spending budget for the next five years will be more than its total capital spending in its entire 82 year history to date.

We will have usable energy available only to the extent that the energy suppliers can acquire the very expensive capital necessary to gather, refine, and deliver the energy to us. When they are able to do this, the huge capital cost will be reflected in the price of that energy. The situation is this: Although we have plenty of energy resources in this country, the energy you need to run your plants - in the form you can use it effi-ciently, in the place you want to use it, and at a price you can afford to use it - is going to be very scrace indeed. You may want to call that an energy crisis. I prefer to call it an energy cost crisis.

CONSERVING ENERGY DOLLARS

Of course, rapidly increasing energy costs are not a new experience to industry in the United States. Beginning in the early seventies, we began to reduce our unit energy consumption almost as a matter of survial. Industrial energy managers can be proud of a job well done. Energy savings in the industrial sector have far exceeded those in the residential, commercial, or governmental sectors. U. S. industry has reduced its energy consumption per unit of product by 18 percent since 1972. I am proud to say that the Chemical Industry has reduced its unit energy consumption by 23 percent since 1972, and is well on its way toward its 1985 goal of a 30 percent reduction in energy use per unit of product. We are saving huge amounts of energy. I cal-culate that our Florida plant saved over 200 billion BTU of energy last year. My salary, however, is not paid in British Thermal Units. I have insisted on being paid in dollars, and conversely my employer expects to see results in dollars saved in energy costs. I expect you have a similar arrangement with your employer. The job, then, is to reduce energy cost in dollars. Energy cost is the product of two components - energy consumption and energy price. We have done a good job so far on containing energy consumption and we must continue to do so. Unfortunately, we have already done the easy, inexpensive things to reduce consumption. Further reductions will be at an ever diminishing rate of return on our energy conservation investments. Furthermore, rising energy prices have offset the dollar savings from those energy conservation measures we have already taken. In short, we have already "taken up the slack" associ-ated with inefficient industrial processes, and we are still faced with energy costs which are projected to escalate more rapidly than any of our other cost factors. While we continue our efforts to reduce energy consump-tion, we must attack the other half of the energy cost formula - energy price. Up until ten or fifteen years ago, the objective of industrial energy management was to assure an adequate supply of energy to operate our various processes. The nature of the process determined how much energy, and little effort went into reducing the amount. Then economic survival dictated that we reduce the energy consumption of each process, so the role of energy management expanded to include energy conservation. Some outside factor determined the rapidly escalating energy price and little effort went into containing it. Now economic survival dictates that the role of energy management expand to include some effort to slow the rapid increase in energy prices to industry.

THE ENERGY PRICE MECHANISM

Is it possible for us as buyers to have any effect on energy prices? If we were setting out to get a firmer grip on the price of some other feedstock material for our industrial process, we could institute aggressive purchasing policies and use the mechanism of a free and competitive market to assure that we secured the best possible price. Energy prices, on the other hand, are largely determined by governmental regulation - an artificial pricing mechanism designed to simulate a free market. Recent history suggests that this regulated pricing mechanism is far less than perfect. The artificial pricing of petroleum products below their real cost led precisely where any freshman economics student could have predicted - a petroleum supply crisis that would have to occur eventually with or without the OPEC cartel. More recently, regulatory bodies such as the Federal Energy Regulatory Commission and the various state public service commissions have recognized that electricity and natural gas suppliers must receive substantially higher prices for their products if they are to be able to make the necessary investments for future growth. These are the huge investments we discussed earlier, and I believe almost everyone accepts the painful fact that energy prices must go up if we are to have adequate supplies to support a healthy economy. It is not, however, either desirable or necessary for energy prices to industry to escalate as rapidly as they have in the past.

THE REGULATOR'S DELIMMA

The various commissions charged with regulating energy prices are characteristically political bodies. The members are either elected directly or appointed by other elected officials. When faced with the unpleasant necessity of raising energy prices, there is a significant tendency to avoid large increases in energy bills to homeowners by placing a larger than justified portion of the increased price burden on industrial energy users. As engineers, we like to quantify things, so I would suggest a handy index which helps explain the dynamics of regulated energy pricing. Calculate the energy used in your industrial operation and count the number of people directly concerned with energy costs. Calculate the energy use in your home and neighborhood and count the number of people directly concerned with that energy cost. A comparison of this "votes per BTU index" explains the tremendous pressure on regulatory commissions to depart from a proper allocation of energy costs among all energy users.

There are many apparent and some not so apparent reasons that the unit cost of energy delivered to a large, continuous industrial operation is less than that delivered to a smaller, more sporadic user. That analysis is the subject of another paper, but suffice it to say here that this difference in cost is not fully recognized in the energy price regulatory process. In electricity pricing, for example, the Electricity Consumers Resource Council (ELCON) has analyzed the cost of service studies of 700 regulated electric utilities and found that in 80 percent of the cases, the utility's industrial customers were paying substantially more than their fair share of the costs. In other words, the utility was earning a larger than average rate of return on its cost to serve industry

in order to support a smaller than average rate of return on its cost to serve residential customers. It is estimated that industry in the United States pays $1-1/2 billion each year over and above its fair cost of electricity in order to subsidize residential electric rates. The names and techniques for effecting this subsidy are many and, on the surface, complex. Such terms as social ratemaking, capital substitution, marginal cost pricing, etc. simply obfuscate a conscious decision to protect homeowners from the real cost of energy at the expense of U. S. industry.

THE CONSEQUENCES

Is protecting homeowners such a bad thing? I'm a homeowner. I don't want to pay any more for the energy I use than is absolutely necessary. Let's look at some of the consequences of continuing to subsidize residential energy costs at the expense of industry:

First and most importantly, there simply is no free lunch. Additional costs borne by industry will eventually have to be paid by the consumer. Those businesses which are competitively able will pass the increased cost along in the price of their products. Those which cannot will eventually fail - leaving fewer suppliers, less competition, and the same result - higher consumer prices.

A second consequence of allowing industrial energy prices to subsidize other energy users is that we risk killing the goose that lays the golden eggs. Industry provides the large dependable base load that allows regulated utilities to recover the cost of such capital intensive facilities as pipelines and power plants. If increasingly high prices drive the industrial users to non-regulated energy sources, such as oil to replace natural gas or self-generated electricity, those users remaining on the system will be left to support the cost of the pipelines or power plants and energy prices would have to increase dramatically.

The third consequence of inequitable pricing of energy is that it causes an improper allocation of scarce resources. At a time when we should be conserving energy we are pricing the energy to heat swimming pools below its actual cost and pricing the energy to build American automobiles above its actual cost.

A fourth consequence of pricing energy to industry on political rather than cost-to-serve considerations is that it creates uncertainty as to future energy supplies and prices. Such uncertainty limits expansion of industrial facilities and costs our economy sorely needed new jobs and new markets.

WHAT WE CAN DO

It's one thing to point out the problem, but what about solutions? Is there anything we, as energy managers, can do to affect the regulated price of the energy we purchase? There are four things that I believe you not only can but must do if you are to properly do your job in managing energy costs:

The first thing is to educate yourself on the issues involved. I would recommend the Electricity Consumers Resource Council (ELCON) as an excellent source of facts in this area.

Second, educate the public. Public opinion is what finally determines what direction the various regulatory commissions take and this is as it should be. We should never blame the politicians for doing what the voters want. That is what democracy is all about. Miss no opportunity to speak on energy matters. The public should be made aware of industry's contribution to energy conservation and of the grave economic dangers of continuing to allow industry to subsidize residential energy costs. Most people, you will find, are more economically aware and wiser than for which some give them credit. With the facts properly placed before them, people will let the regulators know that they prefer the long range benefits such as jobs and upward mobility that a growing economy can provide more than the short range benefit of artifically low energy bills.

Third, involve yourself in the regulatory process. This is where your energy costs are being set, and you should be there to protect your interests. The law provides that any action by the various regulatory commissions must be preceeded by a public hearing and that any party whose interests are substantially affected by the outcome may intervene in the process. It is often advantageous for several industrial energy users with similar interests to intervene in regulatory matters as a group so that they can present a united view and share the costs of retaining attorneys and expert witnesses. For example, the Florida Industrial Power Users Group routinely intervenes in electric ratemaking procedures on behalf of its members. Every dollar my company has spent on these efforts has returned at least ten dollars in annual energy cost savings. This is a considerably higher return than we can earn through investing in energy conservation measures although we will, of course, continue to do both.

The fourth thing which you can do about regulated energy prices is to be prepared to adopt alternatives if the regulated price becomes unreasonable. Non-regulated fuels may be substituted for natural gas in many applications. I am sure you are aware that federal regulations arising from the National Energy Act of 1978 make it much easier now to generate your own electricity if you can do so at a cost less than the regulated utility price.

SUMMARY

To summarize, I have tried to make three points:

 I. If we are to effectively manage energy costs, we must be concerned with energy prices as well as energy consumption.

 II. The current system of regulated energy prices causes industry to pay more than its fair share of the cost of providing energy.

 III. Industrial Energy Managers can work to combat the rapid escalation of energy prices by:
   1. educating yourself
   2. educating the public
   3. intervening in the regulatory process
   4. being prepared with alternative energy sources

I'll close by telling you that at our Florida plant we store a large amount of #5 fuel oil. If one of our storage tanks began to leak at the rate of ten barrels per hour, my employer would not call that a ten barrel per hour leak or even a six million BTU per hour leak. He would call that a $300 per hour leak, and if that leak were ignored, he would quickly find someone else to look after his energy concerns. A proposed electricity rate increase currently being considered by the Florida Public Service Commission threatens to cost my

company about $300 per hour in increased energy costs. To ignore that threat and make no effort to avoid the increased cost would be just as much dereliction of duty as to continue to allow that ten barrels of oil per hour to be wasted.

Chapter 42

CORPORATE ENERGY MANAGEMENT AT WESTERN ELECTRIC

T. A. Mulhern, L. A. Elder

INTRODUCTION

This paper presents the Energy Management Program developed by the Western Electric Company. The program includes managerial accountabilities; administrative and technical activities; and a recently implemented comprehensive five-year energy planning process. The planning process is designed to (1) provide increased Corporate energy direction; (2) assure that all Company locations have comprehensive energy management programs; (3) improve communications to reduce duplication of effort; (4) provide engineering effort and capital required for energy related projects, and (5) provide increased energy conservation input to the Corporate Business Plan.

THE COMPANY

Western Electric is the manufacturing and supply unit of the American Telephone and Telegraph Company (AT&T). Western Electric owns or leases over 65 million square feet of space in 450 locations throughout the United States. The approximately 150,000 employees are involved in the manufacture, installation, service and repair of a wide variety of telecommunications products. These products include telephones, electronic components, cable and wire, switching equipment and transmission systems.

MANAGEMENT COMMITMENT TO ENERGY CONSERVATION

The implementation of an efficient Energy Management Program requires top management commitment to establish an environment in which energy is a prime consideration in the planning, evaluation and administration of all Company operations. Mr. Donald E. Procknow, President of Western Electric, was quoted recently in Electronic Engineering Times: "Energy conservation must be integrated into a company's normal business operations. It's just good business for a company to encourage its employees to be energy savers. Because there is a direct contribution to the bottom line, it's to our advantage in the long run to make every effort to conserve energy."

Western Electric has emphasized an energy consciousness among all levels of management in all segments of its operations. Most important, is the leadership provided at the Corporate level to guide locations in establishing their own objectives and programs that support the Company's goals.

ENERGY MANAGEMENT STRUCTURE

In 1973, the Corporate Energy Management Organization was established within the Corporate Engineering Division. This organization is responsible for developing the Corporate Energy Management Program and to coordinate its implementation.

Initially, to develop a communications network, an energy coordinator was appointed at each of our 75 major facilities. Activities were focused on monitoring immediate energy conservation efforts such as reduced space temperature, lighting restrictions and air flow improvements. With its network of local coordinators, the Corporate organization endeavored to develop an energy ethic among all employees.

Today, each of our major facilities has a middle level manager designated as the Energy Manager for that location. He is authorized by the location's top executive to review and recommend action in all areas of energy usage. This includes product design and production as well as building operations and motor pools.

Local energy committees are a very important part of the energy management structure. The purpose of the energy committee is to assist the Energy Manager in the development and implementation of the energy programs. The committee's objectives include the promotion of energy awareness among employees, the monitoring of areas for compliance with energy conservation practices and the identification of new conservation opportunities. Each energy committee is comprised of representatives from organizations such as plant and factory engineering, production, maintenance and public relations with the Energy Manager serving as Chairman. By having representatives from such diverse organizations, a broad based knowledge is provided of all major building systems and manufacturing and repair operations. Through scheduled monthly meetings, formal agendas and minutes of meetings, energy programs can be effectively developed, communicated and implemented.

CORPORATE ENERGY MANAGEMENT PROGRAM

Working through each location's Energy Manager, the Corporate Energy Management Organization provides both administrative and technical support to all locations. The activities such as those outlined below are ongoing:

Energy Awareness Programs

Our Awareness Programs are designed to keep energy conservation in the forefront of day-to-day activities and are vital to the success of conservation efforts. The concept that "no saving is so small it should not be considered" is emphasized. For example, the simple action of turning off lights when work areas are unoccupied for short periods of time contributes to the conservation effort.

Effective use is made of energy related posters, pamphlets and technical magazines. Articles are developed for use in Company publications and trade journals. Employees are also provided home energy-saving tips and are encouraged to participate in voluntary car and van pools. At the end of 1981, approximately 40 percent of Company employees were either participating in car and van pools or making use of public transportation to and from work.

Poster Contest

The Corporate Energy Management Organization recently sponsored a company-wide, energy poster contest open to all employees and their children. The contest was designed to increase interest in energy conservation and to generate energy-saving ideas for use in conservation awareness programs. Poster ideas stressed energy savings on the job, in the home, in the community and while driving.

Each location judged the entries submitted by adults and by children and selected the three top entries from each age group. These selections were sent to the Corporate Energy Management Organization for consideration in the company-wide contest. A total of 140 local winning posters were submitted from 28 locations of which three adult entires and three children's entries were chosen as overall Company winners.

Composite ideas of the winning posters were incorporated into four posters which were distributed to all locations. In addition, reduced copies of the winning posters were published in an issue of a Company magazine that is distributed to all employees.

Video Tapes

Heating, ventilating and air-conditioning (HVAC) videotapes pertaining to the maintenance of boilers, refrigeration systems and dampers have been produced by our Corporate Plant Service Organization. The tapes are for use at Company locations and emphasize that proper HVAC equipment maintenance will achieve higher operating efficiency and lower energy costs.

Technology Transfer Conferences

Within the past four years two Technology Transfer Conferences were conducted to provide a forum for the exchange of technical ideas. These conferences were attended by 170 engineers and other energy management personnel from throughout the Company and featured energy conservation techniques and methods of obtaining optimum equipment performance.

Energy Consumption Forecasting and Reporting

Energy consumption data is maintained for all major locations from 1972 onward. Since 1974 local energy consumption data has been reported to Corporate on a monthly basis using forms. In 1975, we began inputting this data to an outside computer time-sharing system to aid in the data analysis and the preparation of reports.

We have recently transferred our energy data to AT&T's Energy II System specifically designed for energy information management. Rather than sending data forms to Corporate as in the past, personnel at each location are responsible for inputting their monthly consumption data to the Energy II System. They also have the capability of immediately running reports on their data. Higher level reports, i.e., division and company totals and location comparisons can be run after all location data has been entered.

The primary advantage of the Energy II System is the direct accessibility by each location reducing by at least one month the turnaround time necessary for a location to obtain a report of its results. Each location has the capability to tailor reports to their own requirements and to use the system to perform special types of analyses on its energy data.

Once a year each location inputs its forecasted energy usage in the same manner monthly data is entered. Monthly reports utilize this information by comparing actual usage with that forecasted.

Energy Surveys

An Energy Survey Program was recently introduced to assist the locations in the development of energy programs and to assess the adequacy of energy conservation efforts. In order to make the most efficient use of manpower within Corporate Engineering, two types of surveys are being performed - an Energy Management Survey and an Energy Conservation Survey.

The Energy Management Survey is basically a walk through type survey. The survey provides a broad examination of major energy consumption areas, a review of the existing energy management program and identifies potential energy conservation measures which merit consideration. Generally, it takes two engineers approximately three days to perform this type of survey at a typical one million square foot manufacturing plant.

Either on request of local management or as a result of the Energy Management Survey findings, it may be decided to perform a more in-depth survey at a location. The Energy Conservation Survey encompasses the Energy Management Survey and also provides a detailed evaluation of the building's mechanical and electrical equipment. Specific recommendations are made and a cost analysis is performed to show the cost and savings to be expected from implementing each recommendation. Recommendations to improve energy efficiency may include such actions as improved equipment maintenance, equipment modifications and the replacement of equipment.

Members of our Corporate Plant Service Organization perform the mechanical and electrical equipment evaluations and make the financial analyses. Generally, four engineers are used to perform the mechanical and electrical equipment evaluations and the amount of time required depends on the complexity of the building systems. On site review normally takes one week.

Formal reports are prepared for both surveys and copies are provided to local management and engineering personnel.

Eleven Energy Management Surveys and two Energy Conservation Surveys were performed in 1981. Thirty Energy Management Surveys and two Energy Conservation Surveys are scheduled for 1982.

Contingency Plans

During 1980, each of our major facilities developed an energy contingency plan for use in the event of energy shortages or forced curtailments. For planning purposes, energy reduction conditions of 5-10 percent, 10-20 percent and greater than 20 percent were used.

For each energy reduction condition, specific actions were identified to reduce the location's energy consumption and to minimize the detrimental effects of energy shortages. Each action required, and the impact of the action on the location, the customers, and the employees was described in the plan.

The development of the plans revealed shortcomings in energy supply situations at certain locations that required improvement. Plans are reviewed and updated at least annually or at more frequent intervals if warranted by changing operating conditions or energy supply.

CORPORATE RESULTS

Western Electric's energy requirements in 1981 were 20 trillion BTU's (equivalent to 3.5 million barrels of oil) with resultant costs of $135 million.

Since the inception of the Energy Management Program in 1973, total energy consumption as of the end of 1981 had been reduced by over 11 percent. During this period, production levels increased by 26 percent, the amount of space occupied increased by 11 percent and a highly energy intensive resource recovery facility came on line.

As shown in Figure 1, the Energy Management Program has resulted in $250 million in cost savings between 1973 and 1981. Energy conservation measures requiring minimal engineering effort and low capital expenditures such as reducing space temperatures, lowering lighting levels and improving air flow efficiencies have been implemented at most locations. The savings in energy consumption in 1981 was equal to approximately 10 percent of the Company's net income.

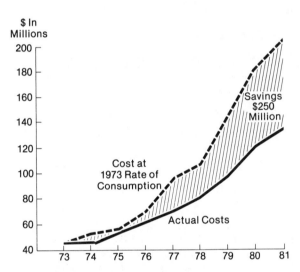

FIGURE 1. ENERGY COST AVOIDANCE

The area of greatest energy savings has been with general plant systems, i.e., HVAC equipment, lighting and reductions in building losses. On a BTU per square foot basis, energy usage since 1973 has been reduced by approximately 48 percent.

Energy usage by our manufacturing facilities since 1973 has been reduced by 16 percent. On a BTU per dollar of production (adjusted for inflation) the reduction amounts to 35 percent.

Although our energy conservation results to date are substantial, opportunities exist for additional energy savings through a formalized energy planning process that will ensure that energy programs are implemented on a timely and cost-effective basis.

ENERGY PLANNING

The energy situation in our Company changed considerably between 1973 and 1981. Although substantial progress was made in reducing energy consumption, many of the easy to perform and inexpensive energy conservation measures had been implemented. Additional energy savings were requiring increased capital investment.

To obtain capital funding during this period of tight money required that the economic rate of return of proposed energy related projects be evaluated against that of production projects. In addition, local energy conservation measures were being applied on an inconsistent manner between locations. A means was needed to ensure from a Corporate view that the most potentially cost effective energy conservation measures were receiving priority consideration at the local, divisional and corporate levels.

In 1981, Western Electric instituted the energy planning process as part of its five-year Business Plan. Energy planning is a formalized process whereby you analyze your existing energy situation, decide what type of energy situation you desire and select strategies to be implemented to achieve the desired situation. Energy planning is performed on the location, division and corporate levels and results in the development of plans to be implemented over a six year period - the present year and the succeeding five years.

Energy planning is designed to fulfill two needs. First it assists in the establishment of an ongoing Energy Management Program. Secondly, for those locations with a program already in existence, it provides a means of assessing and measuring its performance. Locations have invested considerable effort over the past eight years to improve the energy efficiency of their operations and facilities. Planning provides an improved means of formalizing this task and streamlines communications.

Planning is not new to our Company. A formalized five-year Business Plan has been in effect since 1977. However, only the most prominent energy related considerations were being included in the development of business plans. A separate energy planning process was needed to accomplish our energy goals at the local level and to provide increased energy input to the business planning process.

Energy Planning Guideline

Our energy planning process was initiated by the development of a 150 page, Energy Planning Guideline which provides information to assist in the preparation of plans and programs.

The guideline also serves to consolidate much of the Company energy related programs and procedures such as instructions for energy consumption reporting and forecasting, instructions for preparing contingency plans and forecasts of energy inflation rates for use in economic evaluations of energy related projects.

The guideline includes the following major sections:

    1. Corporate Energy Goals: As described below, general energy goals have been established which are to be achieved through implementation of the location plans:

        Comply with all laws and regulations governing

the consumption and conservation of all forms of energy.

· Continuously strive to improve energy conservation by limiting our energy use to the minimum practical level that our operations require.

· Employ available and new energy-efficient technologies and equipment, as the needs of the business permit, continuously improving the Company's effective use of energy.

· Reduce our indirect energy consumption by recycling and reusing the by-products from our operations and maximizing our use of recycled materials.

· Include energy conservation considerations among the criteria by which all phases of our operations are planned, administered and measured.

· Develop an awareness in our employees, suppliers, customers and community of the need for adherence to sound energy conservation practices.

Specific energy consumption goals have also been established for both manufacturing and non-manufacturing locations. The manufacturing goal is to reduce energy consumption per dollar of merchandise produced by 15 percent during the planning period (1982-1987). The non-manufacturing goal is to reduce energy consumption per square foot of floor area by 15 percent during the planning period.

2. Division Level Input: Space is provided for staff personnel in each division to advise their locations of energy related information. This could include such things as production forecasts, new planned operations and division energy goals.

3. Energy Price and Availability: Tables and charts reflect how energy prices of coal, electric power, natural gas and oil are expected to change over the planning period. The predicted availability of each fuel type is also analyzed. This information enables the locations to estimate the cost benefits of energy projects and to plan for the most cost effective fuel supplies.

4. Energy Management Program: The minimum elements and procedures which should be included in a location's Energy Management Program are outlined:

· Management responsibilities.

· The identification and quantification of present energy usage.

· The identification of potential energy conservation measures.

· The priority assessment of energy conservation measures.

· The performance of energy surveys.

· The preparation of energy contingency plans.

· The maintenance of appropriate record and reference files.

5. Energy Consumption Reporting and Forecasting: Instructions are provided for use in the annual

forecasting and monthly reporting of energy consumption.

6. Sample Energy Plans: Sample energy plans assist the locations in the preparation of plans and ensure that a common plan format is used.

7. Situation Analysis Checklist: A checklist was developed to assist each location in assessing their energy situation. The checklist is highly beneficial in providing corporate direction to the locations as to the type of energy conservation measures that should be considered. A portion of the 17 page checklist is shown in Figure 2.

Is Heating and Cooling Equipment Properly Operated and Maintained?	OK	Action Needed
A. Room Temperatures Within Limits of Guidelines. No Heating Above 65°F and No Cooling Below 78°F.	_____	_____
B. Boilers:		
– Tubes Clean and Unobstructed	_____	_____
– Firing Rate Proper	_____	_____
– Preheated Fuel Oil at Proper Temperature	_____	_____
– Blowdown Frequency Not Excessive	_____	_____
– Piping Insulated for Low Heat Loss	_____	_____
– Steam Traps Properly Maintained	_____	_____

FIGURE 2. SITUATION ANALYSIS CHECKLIST

Energy Planning Process

Figure 3 depicts the energy planning process. Central to this process is location and division planning guided by Corporate goals and functional guidelines. Each division develops an energy summary from its location plans which is reviewed by the Corporate Energy Management Organization for incorporation into a Corporate Energy Summary. This summary is then used as input to the Corporate Business Plan.

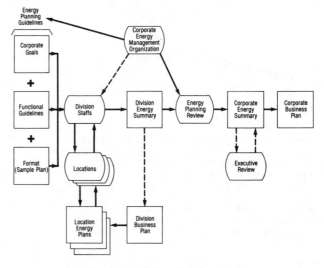

FIGURE 3. ENERGY PLANNING PROCESS

The first step in the planning process is the development of preliminary location plans. Normally, the sequence described below is used in the plan development.

1. Analyze the Existing Energy Situation: The strengths and weaknesses of existing energy programs are determined.

2. Establish Objectives: Broad statements are prepared that support the Corporate goals and that guide and direct the plans.

3. Choose Strategies: One or more strategies are chosen to achieve each objective.

4. Prepare Major Action Programs: A Major Action Program is prepared for each identified strategy. The expected benefits of the strategy are described and each step on the way to implementation is stated. The following information is also provided for each strategy:

- The organization responsible for implementation.

- The approximate implementation schedule.

- Expenditure and cost saving data.

- Incremental engineering manpower requirements.

Review of Plans

The Division and Corporate level reviews include a determination of the following:

- Does the plan support the Division and Corporate Goals?

- If a Corporate energy survey has been made at the location, does the plan include actions recommended in the survey report?

- Can sufficient funding be obtained to implement the strategies?

- Will the implementation of the strategies provide a sufficient rate of return on investment?

- Are there duplications of effort? For example, two or more locations may be planning to investigate or test the same new energy saving technology when action by a single location or a joint effort would be more practical.

The development of the plans is by no means the end of the planning process. Periodic reviews are made to determine the status of plan implementation and the plans are updated at least annually.

The review of location plans for the 1982-1987 planning period was recently completed. We are very pleased with the quality of the plans and the obvious seriousness that the locations have given to the energy planning effort.

CORPORATE ENERGY SUMMARY

Division Summaries of location plans have been incorporated into a Corporate Energy Summary. Many energy saving opportunities have been identified. From a Corporate view, the strategies identified below offer the potential for the most cost effective energy savings and in general will be given priority attention.

The institution of an annual energy survey by local personnel.

- The implementation of programs to increase employee energy awareness and encourage their participation in energy conservation measures.

- Increased use of permanent and portable instrumentation in order to quantify the energy usage.

- The installation of new energy management systems and the expansion of the capabilities of existing systems.

- Improved maintenance of heating, ventilating, and air-conditioning systems.

- The installation of new controls on heating, ventilating and air-conditioning systems.

- Increased use of energy efficient motors.

- The purchase and installation of incinerators for waste disposal and heat recovery.

- Waste heat recovery from boiler flues and manufacturing processes.

- Increased use of high pressure sodium lamps.

CONCLUSION

Our Energy Management Program has been successful in reducing energy consumption and holding down energy costs. Many of the programs that have been successful in the past will continue to yield savings in the future.

From a business standpoint, energy costs remain a significant part of our operating expenses and it is important that reductions in energy usage continue.

The Energy Planning Progress has been successful in identifying many energy conservation opportunities. The implementation of new energy conservation measures will require increased capital investment. We feel that the planning process best ensures that new energy programs are implemented on a timely and cost effective basis.

The energy problems of the past, whether they are shortages or ever increasing prices, will be with us for many years. Therefore, we are making energy conservation an integral part of our planned business activity. Based on our own experiences, we feel that energy conservation and energy management are just plain good business.

Chapter 43

CONSERVATION OF ENERGY
AT WESTINGHOUSE R & D

P. Norelli, V. Roy

Presented herein is the history of "Conservation of Energy at the Westinghouse R&D Center" from the time of the OPEC crisis to the present and forecasts of our intended programs for the next few years. The energy and fuel saved from 1973 to 1977 are credited to the Westinghouse Corporate Task Force recommendations and to our own informal conservation programs. The natural gas shortages of early 1977 and the national coal strike of early 1978 resulted in extreme gas and electric energy savings but not without some loss of comfort and research. More modest savings are attributed to the computerized Energy Management System consisting of a central Westinghouse 2515 Minicomputer Control Center and five (5) remote NumaLogic Microprocessors -- but without loss to comfort or to research. This report also presents an economic analysis of the investment in the current R&D installation of a waste heat recovery system with TEMPLIFIER heat pumps. Some current projects such as a new boiler-burner, dual fueled, with oxygen sensor controller added -- and a new insulated glass system "piggy-backed" on existing window panes -- are also presented.

INTRODUCTION

The Westinghouse R&D Center is located on a 142 acre site about 12 miles east of the Pittsburgh "Golden Triangle" business district. This research center

has slightly over one million square feet of gross area contained in five main buildings plus several smaller ones. These five main buildings are shown in Figure 1; like your buildings, they are identified by numbers. Buildings 401 and 801 are served by common boiler and mechanical rooms - as are 501 and 601. Building 701 has its own heating and cooling equipment room. Employees total 2400 on site. Our annual visitors number almost ten times as many. The research programs are coupled to the World of Science for its professional people and to the World of Business and Government for money and facilities as shown by the following diagram:

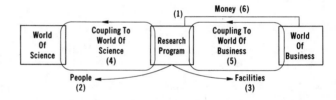

It is the management and conservation of fuel and energy to operate these facilities which we will now describe.

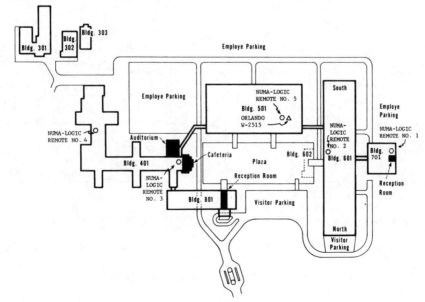

Fig. 1

EARLY HISTORY OF ENERGY CONSERVATION

This is a review of "where" we have been, "where" we are now, and "where" we intend to be - as far as energy consumption at Westinghouse R&D is concerned. Figure 2 is a history of electrical usage from 1964 to the present time. Note the first peak demand (9580 kVA) was reached in July 1970 and then again (9600 kVA) in July 1973. The nameplate capacity of the substation was and still is 10,000 kVA. Our supplier, Duquesne Light Company, recommended that we should plan a new 15,000 kVA substation at a cost of $300,000 (1971 dollars). While contemplating their advice we drew the graph of Figure 2 and we initiated a conservation program. This conservation program - turning off lights and turning off idling equipment - enabled us to occupy two new buildings (701 and 801) without increasing demand or consumption of electricity (see Figure 2).

The first OPEC oil embargo in 1973 caused Westinghouse to take a strong position to emphasize conservation of fossil fuels in all of its physical plant facilities including R&D. Figure 3 is a history of our natural gas usage and clearly indicates

wasteful consumption during the winters of 1969 through 1973. The conservation philosophy is evident from 1973 to the present time - until 8/1/77 - its program simply involved thermostatic and ventilation cut-backs as well as fume hood exhaust reduction. This same conservation program was available for immediate use during the natural gas shortage of 1977 and the electrical power curtailment of 1978.

ADVENT OF IDEA -- "HEAT PUMPS FOR R&D"

Exactly one year later, on 1/17/78, after eleven weeks of nation-wide coal strike, our customer's representative at Duquesne Light Company called by phone to ask for a voluntary cut-back of electrical usage - a voluntary cut-back of 25 percent - meaning that we should reduce our daily electrical usage from 106,000 kW hours to 79,500 kWh -- as can be seen from Figure 4 which also shows that we achieved the necessary reduction even before it was made mandatory on 2/15/78. Again it was simply a matter of reducing lighting levels, searching out idle equipment to turn off - but this time, we also had the help of our computer controller energy management system consisting

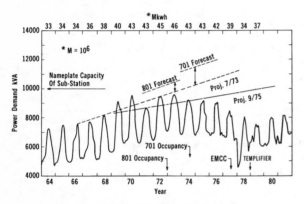

Fig. 2 - Electrical Power Demand Chart
Westinghouse R&D

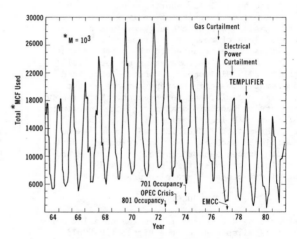

Fig. 3 - Gas Usage Chart At Westinghouse R&D

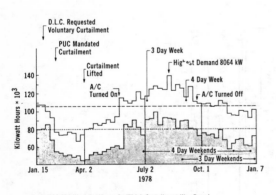

---- 106,000 Weekday Usage On Which Curtailment Was Based
········· 79,500 Weekday &75%) Limit On Usage

Shading = Avg. Daily Usage On Weekend
Shaded + Plain = Avg. Daily Usage During Week

Fig. 4 - Average Daily Kilowatt Hours Used

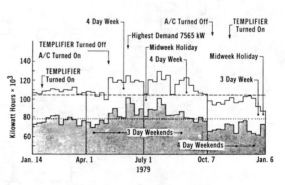

---- 106,000 Weekday Usage On Which Curtailment Was Based
········· 79,500 Weekday &75%) Limit On Usage

Shading = Avg. Daily Usage On Weekend
Shading + Plain = Avg. Daily Usage During Week

Fig. 4A - Average Daily Kilowatt Hours Used

of a Westinghouse 2515 Central Unit plus five NumaLogic remote stations. With this system, we were able to increase our schedule of load shedding.

Of course another reason for our curtailment success was the cooperation received from R&D personnel; special R&D Newsletters kept all employees informed and involved in the common effort.

What else can we learn from Figures 4 and 4A? We can determine that the a/c load appears to be 30,000 kW hours/day, 20,000 kW hours/day for lighting, 30,000 kW hours/day for ventilation, 40,000 kW hours/day for laboratory process power (based on the difference between daytime and night-time operation).

MORE RECENT CONSERVATION HISTORY

Figure 5 indicates the history of "Energy Conservation" on site from the time of the OPEC "crisis" to the present time and forecasts some of our intended programs for the next couple years. Recommendations emanating from corporate seminars plus our own conservation program produced the significant saving from 1973 to 1977. The extreme natural gas savings

for 1977 was obviously the result of the 1977 natural gas crisis which began on the −17°F day of January 17, 1977, and which continued into March of that year.

Figure 5 also shows that we invested $173 K for our computerized energy management system, $250 K for a closed-to-the atmosphere recirculating process cooling water system and $600 K for a waste heat recovery system utilizing a Westinghouse heat pump known as a TEMPLIFIER, $100 K for a new dual-fueled burner and combustion control system with a Westinghouse oxygen sensor controller, and $210 K for a 28,300 sq. ft. of thermal add-a-pane insulating glass system for buildings 501, 601, 602 and pedestrian walkways.

This chart also indicates our intention to use a TEMPLIFIER heat pump for waste heat recovery in other buildings on site, providing the installation can be justified. One investment not indicated in Figure 5 is the installation of 1267 storm windows in Building 401 to reduce heat loss and air infiltration. This investment, like the other committed ones, passed the 15 percent ROR after taxes test in order to qualify for the energy conservation special program.

| Year | Compared to Previous Year | | Description of Investment For Energy and Fuel Savings | Dollar Value of Investment |
	Natural Gas Savings (MCF)	Electric Power Savings (kW hours)		
1973	7,799	-76,000*		$ 1,500
1974	24,500	3,520,000	Revisions of Temperature and Damper Control Systems.	$ 1,500
1975	3,638	48,000	New Time Clocks	$ 1,500
1976	-691*	1,016,000		$ 1,500
1977	42,589	2,366,000	Computerized Energy Management System Installed	$173,000
1978	- 5,718*	5,490,000	Recirculating Process Water System Installed	$250,000
1979	12,653	-2,816,000*	TEMPLIFIER Waste Heat Recovery System Installed	$600,000
1980	- 1,554	-1,712,000*	New Dual Fired Burner and Combustion Control System With Microprocessor Based Oxygen Trim Installed	$100,000
1981	6,300 Estimated	35,000 Estimated	Thermal Addapane System Of Insulating Glass For Buildings 501, 601, 602 And Pedestrian Walkways	$210,000
1982	5,000 Estimated	500,000 Estimated	Waste Heat Recovery Systems for Buildings 501, 601, 701	$750,000

* The Increase in Electric Power in 1973 is Attributed to the Occupancy in January 1973 of a New 100,000 Square Foot Administration Building; While the Increase in Natural Gas in 1976 is Due to the Weather Which Was 9 Percent Colder Than the Previous Ten-Year Average; and the Increase in Natural Gas in 1978 is Due to the Extreme Savings in 1977 as a Result of the Natural Gas Crisis Which Began in January 1977 and Ended in March 1977; and the Increase in Electric Power in 1979 is Due to the Extreme Savings in 1978 as the Result of the Nation-Wide Coal Strike Which Caused the Electrical Power Curtailment that Began in January 1978 and Ended April 1, 1978; and the Increase in Natural Gas and Electric Power in 1980 is Due to the Weather which was 9 Percent Colder in the Winter and 30% Warmer in the Summer than the Previous Ten Year Average.

Fig. 5 - Energy Conservation History At Westinghouse R&D

THE TEMPLIFIER HEAT PROGRAM

Figure 6 is a schematic diagram of our TEMPLIFIER Waste Heat Recovery System including the 401 chiller bank which serves our winter cooling load while providing source heat from its condenser.

Another element of the source loop is the heat recovered from the new process cooling water system serving the 401 and 403 buildings. The dotted lines in Figure 6 indicate possible extensions of source and sink loops

Calculations show that it is difficult to justify the cost necessary to recover the 401 lab exhaust air heat. However, it was relatively easy to justify extending the utilization loop to pick up more (all) of our Building 401 lab air supply.

The service hot water load, shown here in common with the other heat loads, is really on a separate utilization loop as required by the potable water standards.

Figure 7 compares the operating costs of the new TEMPLIFIER heating system with the previous gas fired steam heating situation.

For the twelve-year economic life of the new system - Plan B - the total savings escalates to more than 1-1/2 million dollars.

The twelve-year annual average of $143,000 simulates a uniform annual series which can be handled mathematically as an annuity. Figure 8 indicates corporate fuel cost escalation factors issued in March of 1977 for use in evaluating ROR.

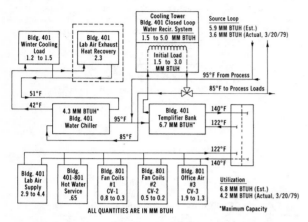

Fig. 6 - Templifier Heat Recovery Concept - Bldgs. 401-801

Energy	Percent Increased	
	Yrs. 1 to 5	Yrs. 6 to 10
Electrical	7.7	6.3
Natural Gas	12.5	9.2
Fuel Oil	6.6	5.5
Propane Gas	10.8	8.8
Coal	6.6	4.9

By plotting our water and sewage costs back to 1969, we were able to ascertain an average escalation of 7 percent per year. This is also reflected in Fig. 7

Fig. 8 - Fuel Cost Escalation Factors

EVALUATION OF RESULTS

How can we evaluate our efforts regarding waste heat recovery? Figure 9A summarizes the analytical data recorded since the start-up in January 1980 thru December 1980 of the Building 401/801 Heat Recovery System. Some interesting points to review are:

The Domestic Hot Water System

o The ratio of heat pump to water pump operating hours.
o The cost to generate a gallon of domestic hot water.
o The equivalent natural gas savings.
o The dollar savings.
o The system efficiency.

Figure 9 is formed to account for depreciation, investment tax credit, and income tax.

Basically, we are investing P now at interest rate i, getting back n equal amounts R at equal intervals. This is called an annuity.

Interpolating indicates a rate of return on investment equal to 15.7 percent. The latest Westinghouse study forecasts escalation rates of 7,10 and 20 percent for general costs, electricity and natural gas costs respectively. These new escalation rates would estimate to a new ROR of approximately 18.5 percent after payment of income taxes on the savings.

	Plan A (Costs)		Plan B (Costs)	Savings
	Prior System		New System	
Year	Water, Sewage and Electrical*	Gas	Water, Chemicals, Electrical Power	(A - B)
0 - 1978 -	Capital Investment Expenditure Equals $600,000			
1	58,700	74,100	57,600	75,200
4	71,900	105,500	70,200	107,200
8	94,300	154,600	90,800	158,000
12	123,600	219,800	116,000	227,400
Total	1,050,000	1,675,300	1,010,000	1,715,400

Average savings for 12 yr. economic life = $143,000
* Electrical and maintenance for two series connected 30 h.p. pumps equals $10,000 for year 1

Fig. 7

Years	Cash Flow Before Taxes	Write-off or Depreciation	Anual Taxable Income	Annual Cash Flow for Income Taxes	Cash Flow After Taxes
0	- 88,000 (expense)		-88,000	+ 40,480	- 47,520
0	-600,000 (capital)			+120,000 (tax credit)	-480,000
1 to 12	+143,000	-50,000	+93,000	- 42,780	+100,220

P = Present Worth = -47,520 -480,000 = -527,520
R = Income after Taxes = 100,220

$$P = R \left[\frac{(1+i)^n - 1}{i(1+i)^n} \right] \qquad \left[\quad \right] = \text{Present Worth Factor/PWF}$$

PWF = P ÷ R = 527,520 ÷ 100,220

= 5.264

i = 15.7% = ROR

Fig. 9 - Estimation Of Cash Flow

The Space Heating System

o The ratio of heat pump to water pump operating hours.
o The cost per million Btu generated.
o The equivalent natural gas savings.
o The dollar savings.

The above characteristics for both systems are also shown in graph form in Figure 9B and Figure 9C.

At this time, we believe our Building 401/801 Heat Recovery System with TEMPLIFIERS is operating as intended; however, we plan to continue monitoring it to insure maximum system operating efficiencies.

DOMESTIC HOT WATER SYSTEM - 1980

Month	Operating Hours Heat Pumps	Operating Hours Water Pumps	City Water Used (10³ Gallon)	Btu (10⁶) Required to Heat City Water	Btu (10⁶) Input to Domestic Hot Water Syst.	Electrical Power Consumed kWh (10³)	Electrical Power Cost ($)	Electrical Power Cost/Gallon City Water ($)	Equiv. Natural Gas Saved (MCF)	Equiv. Natural Gas Cost ($)	Savings ($)	System Efficiency (%)
Jan	(1) 676	318	142.2	87.1	121.9	11.7	434	0.0031	174	612	178	71
Feb	724	319	190.8	105.4	130.3	12.4	460	0.0024	186	655	195	81
Mar	729	318	190.4	97.4	131.0	12.5	462	0.0024	187	658	196	74
Apr	636	319	181.3	92.6	102.2	10.1	403	0.0022	146	514	111	91
May	(2) 484	330	150.3	64.4	79.2	8.2	328	0.0022	113	398	70	81
June	(3) 376	269	114.5	46.2	61.7	6.4	251	0.0022	88	310	59	75
July	412	320	90.9	37.3	68.1	7.2	260	0.0029	97	343	83	55
Aug	421	329	112.2	38.8	69.6	7.4	295	0.0026	99	350	55	56
Sept	270	317	103.9	29.2	46.4	5.4	215	0.0021	66	247	32	63
Oct	370	330	124.8	42.2	61.9	6.7	276	0.0022	88	333	57	68
Nov	478	294	142.4	56.4	77.8	7.9	324	0.0023	111	418	94	72
Dec	567	342	149.5	61.7	92.0	9.3	392	0.0026	131	494	102	67
Totals	6143	3805	1693.2	758.2	1042.1	105.2	4100	0.0024	1486	5332	1232	73

(1) Lowered Set Point From 130 to 105°F on 1/14/80.
(2) Cleaned Heat Pump Condensers on 5/13/80.
(3) Heat Pumps Down From 6/2/80 Thru 6/5/80 For Annual Storage Tank Cleaning.

SPACE HEATING SYSTEM - 1980

Month	System Operating Cond. Heat Pump Hours	System Operating Cond. Heat Pump On Time (%)	System Operating Cond. Water Pumps (Hours)	Btu Generated (10⁶)	Electrical Power Consumed kWh (10³)	Electrical Power Cost Per kWh ($)	Electrical Power Consumed ($)	Cost Per 10⁶ Btu ($)	Equivalent Natural Gas Saved (MCF)	Equivalent Natural Gas Cost/MCF ($)	Equivalent Natural Gas Saved ($)	Savings ($)
Jan	698	97	720	1584.2	133.8	0.037	4954	3.13	2263	3.52	7966	3012
Feb	733	99	743	1659.1	139.3	0.037	5154	3.11	2370	3.52	8342	3188
Mar	524	73	720	1353.8	118.0	0.037	4366	3.22	1934	3.52	6808	2442
Apr	385	53	721	989.0	93.7	0.040	3747	3.79	1413	3.52	4974	1227
May	(1) 303	55	552	758.3	71.8	0.040	2872	3.79	1083	3.52	3812	940
June	—	—	—	—	—	0.039	—	—	—	3.52	—	—
July	—	—	—	—	—	(2) 0.036	—	—	—	3.52	—	—
Aug	—	—	—	—	—	0.040	—	—	—	3.52	—	—
Sept	—	—	—	—	—	0.040	—	—	—	3.72	—	—
Oct	(3) 115	97	118	275.2	22.9	0.041	937	3.40	393	3.76	1478	541
Nov	513	76	673	1444.2	122.3	0.041	5013	3.47	2063	3.76	7758	2745
Dec	703	91	769	1746.3	146.1	0.042	6137	3.51	2495	3.76	9380	3243
Totals	3974	79	5016	9810.1	847.9	0.039	33180	3.38	14014	3.60	50518	17338

(1) Partial Month. Stopped Heat Pump 5/19/80. Total Elapsed Time for May Was 552 Hours.
(2) Reflects $11.565 Credit for Over Charge in April and May.
(3) Partial Month. Started Heat Pump 10/23/80.

Fig. 9A - INITIAL OPERATIONAL RESULTS: TEMPLIFIER SYSTEMS

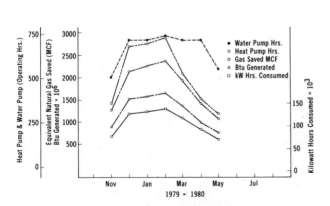

Fig. 9B - Operating Results Of Templifier Space Heating System

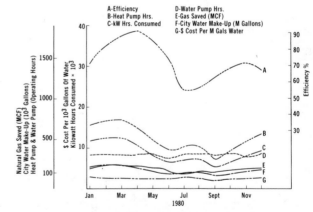

Fig. 9C - Operating Results Of Templifier Domestic Hot Water Heating System

MEASURING CONSERVATION - ENERGY AUDITS

What criteria can we develop to measure achievements and establish goals? Figure 10 shows a range of energy requirements for schools in Northern United States of America in terms of M Btu/sq.ft./year.

Figure 11 compares the energy requirements of our major R&D buildings with the new ASHRAE 90-75 standards and with Dr. Rudoy's record of energy needs of school buildings.

The R&D buildings show a significantly large number in the space heating column and for ventilation and make-up air. Figure 11 reveals that the R&D Center required less energy for space heating than for heating make-up air needed for fume hoods.

* Energy Conservation in School Buildings — by Dr. W. Rudoy,
May 1974 — Univ. of Pittsburgh

Space Heating	Btu/ft²/°Day	M Btu/ft²
"Typical Construction"	12 - 16	70 - 110
"Tight" Thermal Building	6 - 7 1/2	35 - 45
Electrical Lights and Power	3 - 6 kWh/ft²	10 - 20
Hot Water (gas input energy)		5 - 9
Incineration and Miscellaneous (gas input)		5 - 7

"Typical" Construction - Uninsulated Building with Single Glass Windows

"Tight" Construction - Building with Well Insulated Walls and Roof and Insulated Glass

Fig. 10 - Energy Requirements By Schools In Northern USA

(M BTU/FT²/YEAR)

	Space Heating Typical	Tight	Electricity Lights and Office Power	Ventilation and Make-up Air Heating	Cooling[x]	Service* Hot Water	Totals[φ]
Dr. W. Rudoy	70/110	35/45	10/20			(5-9)	
Pre-ASHRAE 90-75	82.5		53.5	22.2	16.0		174.2
ASHRAE 90-75		20.5	28.3	17.3	6.4		73.0
(W) R&D Bldgs. 401-801	50.0		28.0	57.5	17.0	(6.0)	152.5
(W) R&D Bldgs. 501-601		31.0	28.0	65.0	17.0	(6.0)	141.0

x Electricity = 3413 Btu/kWh
* Omitted from totals

φ Does not include process power[Δ]

Δ Process power ≈ 35 M Btu/ft²/year for R&D Labs.

Fig. 11 - Energy Requirements Of Certain Buildings In Northern USA

TO SUMMARIZE

These conservation moves reduced our annual electrical consumption from 46 million kW hours to 39 million kW hours per year as seen in Figure 12 which plots kilowatt hours against time (in calendar years). This graph also indicates the occurrence of various significant events such as occupancy of Building 801 and Building 701 as well as the first OPEC crisis and the two major curtailments, natural gas in early 1977 and electricity in early 1978. Note also the advent of EMCC - electrical management computer controller and the TEMPLIFIER heat pump.

During the same period, the annual gas usage was reduced by 89 million cubic feet per year, as can be seen in Figure 13 which plots million cubic feet of gas against calendar years. Again note the occurrence of the same significant events. It appears that the projection is for an early plateau.

Figure 14 combines gas and electrical consumption in terms of Btu/sq.ft./year and indicates its conservation from 1972. The five-year projection in Figure 14 represents a practical goal. While generating Figure 14, the thought occurred to express in terms of persons served as well as area served. The result is Figure 15 showing energy usage from 1972 in terms of Btu per employee per square foot per year. Given the fact that existing laboratory facilities are being used about one-third of the time -- 90 Btu per employee per square foot per year appears to be a reasonable energy usage rate plateau.

Figure 16 is intended to provide a reasonable separation of building energy needs and laboratory process needs. Some elementary calculations would show that 40 percent of the R&D energy usage is for

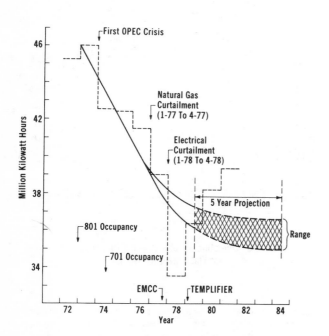

Fig. 12 - History Of Electrical Usage At Westinghouse R&D

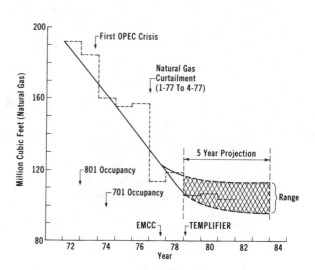

Fig. 13 - History Of Natural Gas Usage At Westinghouse R&D

building needs such as space heating, lighting and office power, comfort cooling and service hot water. The other 60 percent is split 33 percent for make-up air heating and 27 percent for laboratory process energy. Obviously the big target for energy conservation continues to be make-up air heating and we do plan to continue to tackle that problem, especially the off-hours idle operation of our exhaust air systems.

It appears that 66 M Btu/sq.ft./year of lab process requires a physical plant which utilizes 165 M Btu/sq.ft./year including make-up and exhaust air requirements. How do we improve the ratio of lab to plant utilization of energy? One apparent answer is multiple shift operation of the laboratory process functions. Perhaps the research function within the World of Science should adopt the efficient World of Industry practice of multiple shift operations.

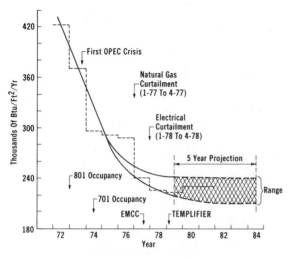

Fig. 14 - Historical Audit Of Energy At Westinghouse R&D

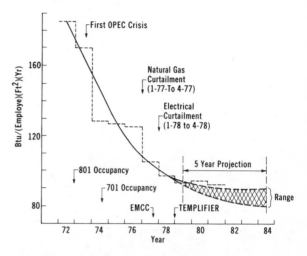

Fig. 15 - Historical Audit Of Energy Usage At Westinghouse R&D

Year	Total Energy M Btu/Ft^2/Yr	Building Energy Needs M Btu/Ft^2/Yr				Lab. Process Needs M Btu/Ft^2/Yr		
		Space Heating	Lighting and Office Power	Comfort Cooling	Service Hot Water	Total	Make-Up Air Heating & Fan Energy	Lab. Process Energy
1970	400	100	40	20	10	170	120	110
1971	397	98	40	19	10	167	120	110
1972	422	112	44	20	14	190	122	110
1973	371	90	40	20	12	162	110	99
1974	296	57	36	20	12	125	100	71
1975	292	55	34	21	12	122	100	70
1976	289	55	34	20	10	119	100	70
1977	240	44	30	18	8	100	75	65
1978	225	40	28	17	6	91	74	60
1979	223	38	28	17	4	87	74	62
1980	231	39	29	18	5	91	74	66
1981	230	37	30	19	5	91	73	66

Fig. 16 - Audit Of Energy Usage At Westinghouse R&D Building Energy vs. Lab. Energy

SECTION 14
HEAT RECOVERY

Chapter 44

DEVELOPMENT OF GENERIC DESIGN GUIDELINES FOR CONDENSING HEAT EXCHANGERS FOR RESIDENTIAL FURNACES AND COMMERCIAL BOILERS

J. J. Lux, Jr., R. D. Fischer

SUMMARY

This paper presents both analytical and experimental results of an initial investigation into the development of generic design procedures that may be used in designing condensing heat exchangers. During this study, research was conducted to determine the current state of knowledge of heat-transfer with condensing vapors in the presence of a noncondensible gas, typical of conditions found in a furnace flue. Based on this research, an analytical design procedure was developed and adapted to selected heat exchanger geometries typical of those that would be considered for use in a residential condensing furnace. Experimental heat exchangers were then designed, constructed, and tested in the laboratory using flue gas from a commercially available oil-fired furnace. The laboratory experiments were designed to provide information that could be used to refine and validate the analytical design procedure.

INTRODUCTION

Sensible and latent heat are extracted from flue gases in processes involving independent and/or simultaneous heat and mass transfer. Flue gas is a mixture of noncondensing gases including nitrogen, carbon dioxide, oxygen, and carbon monoxide, and condensing vapors of water and sulfur dioxide. Within the heat exchanger, flue gas is initially cooled through a sensible heat exchange. The gas will continue to lose heat in this manner until the cooling surface temperature drops below the dewpoint temperature of the mixture. At this point a fraction of the water vapor will condense on the cool surface through a latent heat exchange while the remaining gas mixture continues to lose heat through sensible heat transfer. This process of simultaneous latent and sensible heat exchange for the gas to the cooling surface will continue until the bulk mixture temperature approaches the dewpoint temperature. From this point to the outlet of the heat exchanger the mixture remains saturated.

Figure 1 shows the variation of available heat as a function of temperature for a typical flue gas. The portion of the curve from points A-B characterizes the heat transfer by sensible heat exchange for the noncondensible gases and a desuperheating of the water vapor. Once the gas contacts a wall temperature at or below the dewpoint (Point B), part of the water vapor condenses while the remaining fraction continues to desuper-

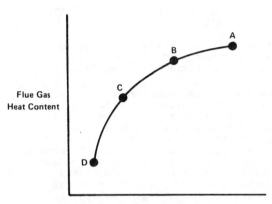

FIGURE 1. VARIATION OF AVAILABLE HEAT FOR A TYPICAL FLUE GAS

heat and the noncondensible gases continue to lose sensible heat. At point C the bulk gas temperature is at the dewpoint and the water vapor loses heat only through condensation. Again the noncondensible gases continue to lose heat by a sensible heat exchange. It should be noted that the dewpoint of the gas mixture at point C is lower than the dewpoint between points A and B because the quantity of water vapor condensed between points B and C depresses the dewpoint by decreasing the water content of the gas mixture.

ANALYTICAL APPROACH

Energy balances may be used to determine the flow of heat from the flue gas mixture to the cooling surface for each of the three heat transfer areas discussed earlier. The sensible heat exchange that occurs in the first region (points A-B) may be written as follows:

$$q_S = \varepsilon \, (\dot{m}C)_{min}[T_{FGI} - T_{AI}] \qquad (1)$$

The heat exchanger effectiveness ε is a function of the overall heat transfer coefficient U between the flue gas and the cooling fluid, the heat transfer area A, and $(\dot{m}C)_{min}$. Values of ε for numerous generic heat exchanger geometries may be found in Reference 9. The temperature profile for the gas mixture in this region is shown in Figure 2. There is a temperature drop in the flue gas near the heat exchanger wall. This temperature drop exists because heat is transferred from the flue gas to the cooler wall. Be-

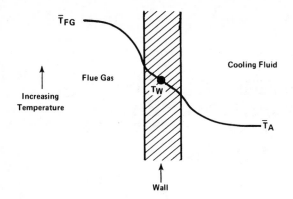

FIGURE 2.   DRY HEAT TRANSFER TEMPERATURE
PROFILE

cause flue gases are cooler near the wall,
condensation occurs before the overall flue
gas temperature reaches the dewpoint.

The process of condensation and the
presence of condensate on the cooling surface
complicates the heat transfer analysis for
the process from points B-D in Figure 1. A
typical cooling wall section in the condensing region is shown in Figure 3. The condensate is shown as a thin film on the surface
of the wall. The noncondensible portion of
the flue gas exchanges heat based on the difference between the bulk gas temperature
$\overline{T}_{FG}$ and the condensate-gas interface temperature T_I and not the wall temperature as
in a noncondensing mode. The equation for
this sensible heat exchange may be written as
follows:

$$q_S = hA[\overline{T}_{FG} - T_I] \tag{2}$$

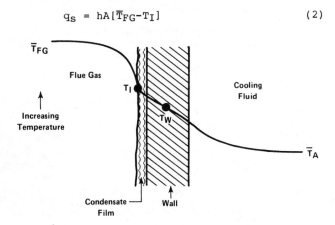

FIGURE 3.   CONDENSATION HEAT TRANSFER
TEMPERATURE PROFILE

The heat transfer process for the water vapor
condensing on the cool wall surface is more
complex. The molecules must move through the
noncondensible gas to reach the cooling surface. For the vapor molecules next to the
cooling wall, little or no movement is necessary. However, as the water molecules condense, a partial pressure gradient is established between the liquid at the cooling surface and the vapor molecules in the bulk gas
stream. This pressure gradient provides the
driving force for moving the water molecules
to the cooling surface. The latent heat re-

moved from the water molecules may be expressed as:

$$q_L = K_g M_V L(\overline{P}_V - P_I) \tag{3}$$

The latent and sensible heat exchange could
be analyzed separately if there was no interaction between the two processes. In reality
this is not the case. As water molecules mi-
grate toward the cooling surface, molecules
of noncondensible gas are displaced from near
the surface. The movement of noncondensible
gases near the cooling wall and the presence
of a convective boundary layer at the condensate interface establishes a region of high
resistance to the latent heat transfer and
complicates the determination of gas interface temperature. In effect, the presence of
noncondensibles reduces the rate of condensation heat transfer.

In the past, numerous studies have evaluated methods that may be used for determining this gas/condensate interface temperature
and the coupled sensible and latent heat exchange of a condensing vapor in the presence
of noncondensible gases (1-6). One of the
earliest studies was performed by Colburn and
Hougen in 1934 (7). This method involves an
iterative approach for the calculation of the
gas/condensate interface temperature and the
heat transfer area required to achieve an assumed overall drop in mixture temperature due
to cooling. Later, a number of investigators
modified the method. These modifications include defining an enthalpy potential as the
driving force for the heat transfer and various graphical techniques for the determination of the condensate interface temperature.
Some simplify the calculations of condensing
heat transfer with a penalty of reduced accuracy. The iterative Colburn-Hougen method is
more complicated, but is easily adaptable for
use with computer techniques. During this
project, a computer program based on the
Colburn-Hougen method was developed.

The Colburn-Hougen method involves
establishing energy balances for discrete
segments of a condensing heat exchanger. A
unique feature of this method is that the
heat exchanger is divided into segments based
on flue gas temperature drop increments as
opposed to increments of area. For example,
to design a condensing heat exchanger section
having an inlet flue gas temperature of 130 F
and a desired final flue temperature of 90 F,
the heat exchanger could be divided into segments having equal flue temperature drops of
5 F for a total of 8 segments. The area of
each segment is unknown, and will likely be
different because the rate of heat transfer
changes with flue gas temperature, moisture
content and cooling fluid temperature. The
purpose of applying this method to condensing
heat exchanger design is to determine the
heat exchanger area required to achieve the
desired overall flue gas temperature drop.

The method applies equally well to cases
involving condensation with saturated or unsaturated gas/vapor mixtures. The simplest
case involves the condensing of a saturated
mixture (Process C-D, Figure 1). Figure 4
shows a segment of a condensing heat exchanger. Saturated flue gases enter with a mass

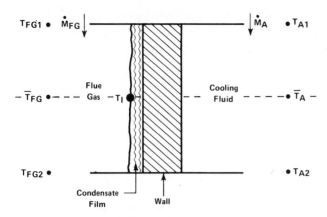

FIGURE 4. HEAT EXCHANGER SEGMENT USED TO
EXECUTE THE COLBURN-HOUGEN METHOD

flow rate of $\dot{m}_{FG}$ and at temperature T_{FG1}.
The flue gases are assumed to leave the segment (and enter the succeeding segment) at
T_{FG2}. Within the segment, the average flue
gas temperature, $\overline{T}_{FG}$, is the algebraic
average of T_{FG1} and T_{FG2}. The cooling
fluid enters the heat exchanger segment at a
known temperature T_{A1}, and leaves at an unknown temperature T_{A2}. The mass flow rate
for the cooling fluid, $\dot{m}_A$, is known. Within the segment, the average fluid temperature, $\overline{T}_A$, is also unknown. The condensate/
gas interface temperature shown in Figure 4
reaches a value at T_I which is somewhat
below $\overline{T}_{FG}$. For this reason, the wall
temperature, T_W, is at or below the flue
gas dewpoint.

Using the Colburn-Hougen method, an energy balance relating heat transfer through
the wall can be expressed in terms of flue
gas properties, gas/condensate interface temperature and average cooling fluid temperature. Heat transferred from the flue gas to
the gas/condensate interface is a combination
of latent and sensible heat.

A second energy balance relating the net
loss of energy by the flue gas to the energy
increase of the cooling fluid completes the
necessary energy balances. The energy balance
equations are expressed on a unit heat exchanger area basis because the actual segment
area is unknown.

The energy balance equations are as
follows:

Interface Energy Balance

$$h(\overline{T}_{FG}-T_I) + K_g M_V L(\overline{P}_V-P_I) = U(T_I-\overline{T}_A) \quad (4)$$

where:

$h(\overline{T}_{FG}-T_I)$ = sensible heat loss from flue
gases (Eq 2)

$K_g M_V L(\overline{P}_V-P_I)$ = latent heat loss from the
flue gas (Eq 3)

$U(T_I-\overline{T}_A)$ = heat transferred through the
condensate film and heat
exchanger wall

U = overall heat transfer coefficient
due to conduction through the condensate, conduction through the
wall, and convection in the cooling
fluid stream.

Gas Stream Energy Balance

$$\dot{m}_{FG}(H_1-H_2) = \dot{m}_A C_A(T_{A2}-T_{A1}) = q_T \quad (5)$$

where:

$\dot{m}_{FG}(H_1-H_2)$ = heat lost by flue gases in
the segment

$\dot{m}_A C_A(T_{A2}-T_{A1})$ = heat gained by the cooling fluid in the segment

Flue gas enthalpy is determined by temperature and moisture content. Because the
flue gas conditions in terms of temperature
and moisture content are known, the enthalpies are also known. Equation (5) can be
solved to determine T_{A2} using the known
value of T_{A1}.

Equation (4) can now be solved because
the only unknowns are T_I and P_I. The
partial pressure of the vapor at the gas/
condensate interface, P_I, is a function of
the temperature T_I because of saturated
conditions. In effect, there is only one unknown in Equation (4). To solve this equation, a value of T_I is assumed. The value
of T_I is iteratively adjusted until Equation (4) is satisfied and convergence is
achieved.

Because the total heat transferred
between the flue gas and cooling fluid is
known, and the interface temperature, T_I,
for the segment is known, the segment area
can be calculated from the relation

$$A = \frac{q_T}{U(T_I-\overline{T}_A)} \quad (6)$$

The exit conditions for this segment are
used as inlet conditions for the next segment
and the solution proceeds until the final
segment is completed resulting in the desired
overall flue gas temperature drop. Segment
areas and heat transfer rates are summed
resulting in overall heat exchanger size.

For a process involving unsaturated
cooling (Process B-C) another equation must
be introduced in the calculation routine
because the segment flue gas outlet conditions may not be saturated. The method for
this analysis is the same as discussed previously with the addition of the following
equation for vapor pressure at the segment
exit flue gas temperature conditions (8):

$$\ln\left[\frac{P_T-P_I}{P_T-\overline{P}_V}\right]\frac{(P_{V2}-P_{V1})}{(P_T-\overline{P}_V)} = \left[\frac{T_{V2}-T_{V1}}{\overline{T}_V-T_I}\right]\left[\frac{N_{PR}}{N_{SC}}\right]^{2/3} \quad (7)$$

Equation (7) involves three unknowns (P_{V2},
P_I, and T_I). The heat transferred within
the heat exchanger segment and the value of
the interface temperature T_I must be determined simultaneously for each step of the

iteration. Equation (7) also shows that the flue gas vapor pressure at the exit of the heat exchanger segment (and hence the total heat transfer) is a function of the interface temperature.

LABORATORY HEAT-EXCHANGER EVALUATION

In order to validate the analytical design procedure, laboratory tests were performed on various condensing heat exchanger configurations. One of the heat exchangers used vertical tubes in crossflow with circulating air on the inside of the tubes. This heat exchanger shown in Figures 5 and 6 has a total of 225-1/2 in. O.D., 18 gauge, 304 stainless steel tubes in a 4.51 in. x 4.75 in. duct. One side of the heat exchanger was replaced with a glass panel to allow visual determination of the condensation pattern.

This configuration may not be optimum from a manufacturing standpoint, but is useful for determining intermediate temperatures within the heat exchanger for comparison with the calculated results, and to visualize the point at which condensate first begins to form. Three tests at various levels of excess air and inlet temperature were run for this configuration. An additional test was performed with the downstream half of the heat exchanger fitted with PTFE Teflon coated tubes. For each of these tests, data recorded included the inlet and outlet flue gas and

circulating air temperatures, and intermediate flue gas temperatures at various locations within the heat exchanger. Also, excess air level and the flow rates of flue gas and circulating air were observed. Steady state condensation rates for the same flue gas conditions (inlet temperature and excess air) were recorded for the cases of coated and uncoated tubes. A comparison of these data is presented in Figures 7, 8, 9 and 10.

Figure 7 shows the effect of heat transfer area and excess air on flue gas temperature. At high excess air levels, temperatures within the heat exchanger were higher. Inlet flue gas temperature was also higher for cases involving increasing levels of excess air. These higher inlet temperatures cause the heat exchanger outlet temperature to rise because the heat transfer rate (and hence, condensation rate) is highly dependent on inlet temperature. For the range of excess air levels shown, the general shape of the temperature profiles in Figure 7 is the same, indicating that the heat transfer rate is not strongly dependent on excess air level. Viscosity and thermal conductivity of the gas is also not appreciably influenced by excess air levels.

When condensation first occurs within the heat exchanger along the flue gas flow path, only a portion of each tube is wetted. Because the cooling air entered the top of the tube bank in crossflow relative to the

FIGURE 5. LABORATORY HEAT EXCHANGER USED FOR MODEL VERIFICATION

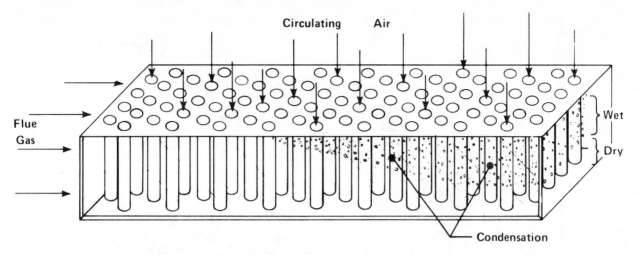

FIGURE 6. CONDENSING HEAT EXCHANGER TEST CONFIGURATION SHOWING AIR AND FLUE GAS FLOW PATHS
AND TYPICAL OBSERVED CONDENSATION PATTERN

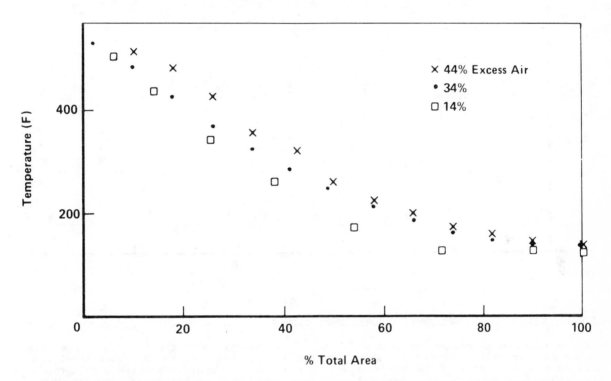

FIGURE 7. TEMPERATURE VS. % TOTAL HEAT EXCHANGER AREA AT VARIOUS EXCESS AIR LEVELS

flue gases, the upper regions of each tube
are cooler than the rest of the tube. A
typical condensation pattern is shown in
Figure 6. As the flue gases become progressively cooler after crossing each successive
tube bank, the length of the condensing region on each tube increases. Figure 8 shows
the length of the wetted portion of the heat
exchanger tubes throughout the heat exchanger
for various excess air levels. The data show
that higher inlet gas temperatures reduce the
condensation regions of each tube. For each
case shown in Figure 8 the condensation profiles indicate that the heat exchanger area
experiencing condensation decreases with increasing excess air levels due to changes in
flue gas dry bulb and dewpoint temperature.
Figures 7 and 8 show that beyond the point of

initial condensation, the change in flue gas
temperature through the heat exchanger is
reduced. This characteristic is due to the
average flue gas temperature approaching the
dewpoint. In this region, latent heat transfer is a large percentage of the total, and
significant heat is removed with a small drop
in overall flue gas temperature. In the dry
region, the heat transfer is sensible and
results in a large change in flue gas
temperature.

Figure 9 shows the effect of a Teflon®
coating on the performance at a condensing
heat exchanger. Materials similar to Teflon®
are hydrophobic or nonwetting (10). This
characteristic promotes the formation of
water droplets instead of a condensate film.

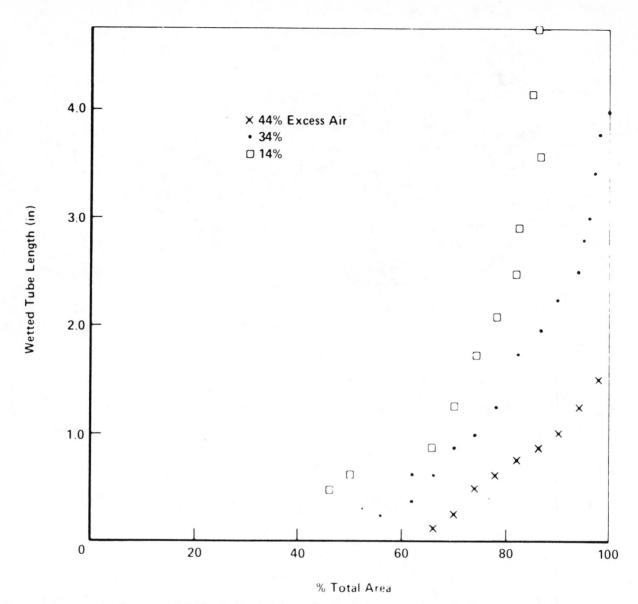

FIGURE 8.   WETTED TUBE LENGTH VS. % TOTAL HEAT EXCHANGER AREA AT VARIOUS EXCESS AIR LEVELS

Dropwise condensation increases the heat transfer by exposing more tube surface area and limiting the resistance of the condensate film layer that is present in film condensation. The general shape of the two curves in Figure 9 is almost identical down to the point where condensation occurs. Beyond this point, the superior heat transfer of dropwise condensation results in a greater reduction of flue gas temperature. The greater condensation rate from the coated tubes results in a net increase in heat recovery and potential furnace efficiency. Further evidence of this fact is provided by the larger quantity of condensate collected from the coated heat exchanger. The condensation profile for coated tubes is shown in Figure 10. These data were estimated because the formation of water droplets is more difficult to observe than a condensate film.

A comparison of the flue gas temperature profile generated by the computer model and the actual temperature profile within the laboratory heat exchanger is shown in Figure 11. The experimental data are identical to the data presented in Figure 7 for 34% excess air. The computer model used an assumed value for the dewpoint of the flue gas because little data were available in the literature relating flue gas dewpoint to excess air levels and sulfur content of distillate fuels.

In the future it is planned to extend the design procedure to additional geometries and to package the procedure in a form that can be readily used by manufacturers.

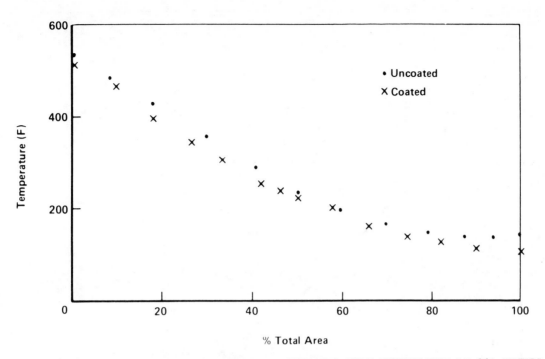

FIGURE 9. TEFLON COATED TUBE TEMPERATURE VS. UNCOATED TUBE TEMPERATURE AT 34% EXCESS AIR

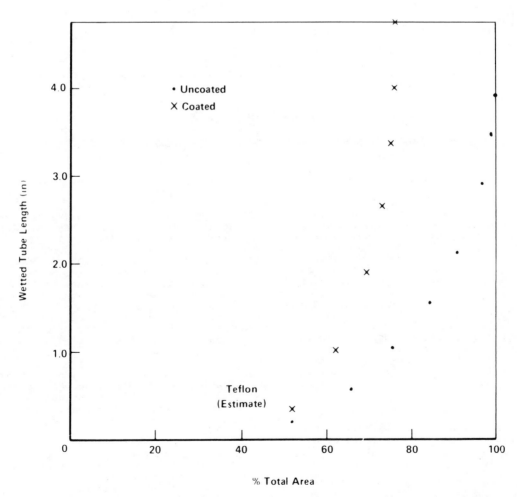

FIGURE 10. UNCOATED WETTED TUBE LENGTH VS. TEFLON COATED TUBE LENGTH AT 34% EXCESS AIR

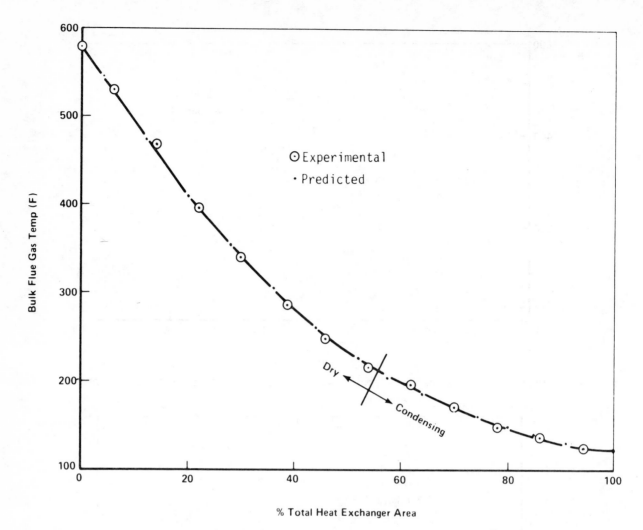

FIGURE 11. CONDENSING HEAT EXCHANGER TEMPERATURE PROFILE: EXPERIMENTAL VS. PREDICTED DATA
AT 34% EXCESS AIR

CONCLUSIONS

The cooling of a flue gas in a condensing heat exchanger may be predicted by use of a dry heat transfer model for regions of no condensation and by the Colburn-Hougen method in regions of saturated condensation. In conjunction with Equation (7), the Colburn-Hougen method may be used in regions of unsaturated condensation. The performance of a condensing heat exchanger is strongly influenced by the inlet gas temperature and only slightly dependent on the excess-air level. Materials that exhibit hydrophobic (nonwetting) properties provide significant enhanced heat transfer rates due to the promotion of dropwise condensation and the minimization of condensate film thickness.

REFERENCES CITED

1. F. Votta, Jr. and C. A. Walker, "Condensation of Vapor in the Presence of Noncondensing Gas", AIChE, Vol. 4, No. 4, p. 413, 1958.

2. G.H.P. Bras, "A Graphical Method for the Calculation of Cooler-Condensers", Chemical Engineering Science, Vol. 6, p. 277, 1957.

3. F. Votta., Jr., "Condensing from Vapor-Gas Mixtures", Chemical Engineering, Vol. 71, p. 223, 1964.

4. E. M. Sparrow and S. H. Lin, "Condensation in the Presence of a Noncondensible Gas", Journal of Heat and Mass Transfer, ASME Transactions, Vol. C86, p. 430, 1963.

5. A. P. Colburn and T. B. Drew, "The Condensation of Mixed Vapors", AIChE Transactions, Vol. 33, p. 197, 1937.

6. D. R. Webb and R. G. Sardesai, "Verification of Multicomponent Mass Transfer Models for Condensation Inside a Vertical Tube", International Journal of Multiphase Flow, Vol. 7, No. 5, p. 507, 1981.

7. A. P. Colburn and O. A. Hougen, "Design of Cooler Condensers for Mixtures of Vapors with Noncondensing Gases", Industrial and Engineering Chemistry, Vol. 26, p. 1178, 1934.

8. A. P. Colburn, "Problems in Design and Research on Condensers of Vapors and Vapor Mixtures", Proceedings of the Institute of Mechanical Engineers, Vol. 164, p. 448, London, England, 1951.

9. U. M. Kays and A. L. London, "Compact Heat Exchangers", McGraw Hill Book Co., New York, New York, 1964.

10. J. G. Collier, "Convective Boiling and Condensation", McGraw Hill Book Company, New York, New York, 1981.

NOMENCLATURE

A   Area (ft^2)

C   Specific heat $\left(\dfrac{BTU}{lbm \cdot F}\right)$

D   Diffusivity $\left(\dfrac{ft^2}{hr}\right)$

h   Convective heat transfer coefficient $\left(\dfrac{BTU}{hr \cdot ft^2 \cdot F}\right)$

ħ   Enthalpy $\left(\dfrac{BTU}{lbm}\right)$ (from psychrometric relations for air)

k   Thermal conductivity $\left(\dfrac{BTU}{hr \cdot ft \cdot F}\right)$

K_g Mass transfer coefficient $\left(\dfrac{mol}{hr \cdot ft^2 \cdot atm}\right)$

L   Latent heat of vaporization $\left(\dfrac{BTU}{lbm}\right)$

m   Mass flow rate $\left(\dfrac{lbm}{hr}\right)$

M   Molecular weight

P   Pressure (atm)

P_{LM} Flue gas mixture log mean pressure

q   Heat flux $\left(\dfrac{BTU}{hr}\right)$

T   Temperature (F)

U   Overall heat transfer coefficient $\left(\dfrac{BTU}{hr \cdot ft^2 \cdot F}\right)$

Subscripts

ABS   Absolute

A   Cooling fluid

AI   Cooling fluid inlet

FG   Flue gas

FGI Flue gas inlet

I   Flue gas-condensate interface

L   Latent

M   Flue gas-water vapor mixture

min Minimum

S   Sensible

V   Vapor

W   Wall

T   Total (sensible + latent)

1   Heat exchanger segment inlet

2   Heat exchanger segment outlet

Greek Letters

ε   Heat exchanger effectiveness (from Ref. 9)

ρ   Density

μ   Dynamic viscosity

Superscripts

$\bar{x}$ Arithmetic average or bulk

Equations

N_{PR} Prandtl No. $= \dfrac{C\mu}{k}$

N_{SC} Schmidt No. $= \dfrac{\mu}{\rho D}$

$$D = \frac{(6.234 \times 10^{-5}) T_{ABS}^{3/2}}{P_{ABS}}$$

$$K_g = \frac{h}{M_M C_M P_{LM}} \left[\frac{N_{PR}}{N_{SC}}\right]^{2/3}$$

$$P_{LM} = \frac{(P_T - P_I) - (P_T - \bar{P}_V)}{\ln[(P_T - P_I)/(P_T - \bar{P}_V)]}$$

Chapter 45

HEAT RECOVERY BOILERS IN INDUSTRY

R. R. Cutler

ABSTRACT

This paper outlines the implementation of Heat Recovery Projects in the glass and plastic industry and the results therefrom.

Described herein are the preliminary investigations, application studies, project economics with justification through fuel savings, system description, boiler design, installation and operational experience.

INTRODUCTION

Our firm is multi-purpose, providing services through boiler sales, rentals, repairs and engineered turn-key plant installations.

We are employed by the entire industrial spectrum to assist with their utility requirements. Our experience extends from small package boilers through large industrial units be it oil-gas or coal fired. Today I wish to describe our involvement in the design of Waste Heat Boilers and their installation for generation of steam only - or steam for co-generation.

It is not necessary for me to reidiorate the impact of fuel costs for our group, only to say that energy intensive industries are finding a return on an investment in Heat Recovery Projects between one (1) and two (2) years. We are confident that companies will strive harder to effect energy conservation measures to reduce or minimize inflationary pressures on the cost of their products.

PART I - EXPERIENCES IN THE GLASS INDUSTRY

The first area I will discuss is our experience with glass plants in which heat is recovered from high temperature flue gases generated in the glass tank or furnace.

In the glass forming process sand (Silica), sodium carbonate, lime and some small amounts of other constituents are thoroughly mixed (Batch) and fed into the melting area of the glass tank. The melting of glass creates a homogeneous composition with various gases being given off, depending on the particular "fining section" in which gas bubbles and final dissolving occurs. The glass then flows to the forming machinery, be it molding machines, flat glass or other.

On Figure 1 - REGENERATIVE FURNACE we have shown a typical furnace in the silicate industry. Since 1920 the industry has utilized regenerators, constructed of massive brick settings, to capture a large portion of the high temperature flue gases. The regenerators cool the exit flue gases down to 900 - 1200°F, in the process the regenerator is heated to about 1600°F, a reversing valve then rotates, directing the cold combustion air through the heated regenerator for preheating combustion air. The flow of flue gases is reversed about every 20 minutes. Our project utilized the flue gases after the reversing valve, before the stack, with gas temperatures of 900 to 1200°F being available for heat recovery.

SELECTION OF THE HEAT RECOVERY SYSTEM

Certain constraints were agreed upon as necessary for a successful Heat Recovery Operation, including the following:

1. Relatively large gas passages to minimize fouling of heat transfer surfaces.

2. Design and construction to prevent cold end corrosion.

3. Any device installed in the flue system shall not upset furnace pressure control.

4. Minimize horse-power for the handling of the flue gases.

Consideration was given to air-preheating, economizers and the Heat Recovery Boiler. Based on the constraints mentioned above and on some previous experiences the Heat Recovery Boiler for steam generation was selected for economic reasons and also fulfilling the constraints discussed earlier.

PRELIMINARY REPORT

For each project it was required to make preliminary designs of the boiler installations with the associated items of equipment, piping, foundations-structural and electrical work to enable reasonable cost estimates to be obtained. Specifications and drawings were issued to manufacturers and contractors to obtain project estimates.

Capital recovery was based on 15% interest, 20 year life. The savings from the Waste Heat Boiler was established as the cost of equivalent process steam from a fuel fired boiler.

Project cost estimates varied from $500,000 to $1,000,000. The results of our studies were presented in a report which showed a payback of one (1) to two (2) years. Accordingly, the different projects were approved for final design-procurement and construction.

To date the projects with which we have been involved have utilized the steam generated for process requirements. We are now looking at a number of potential possibilities where the generated steam would be utilized first for electrical generation and then for process.

WASTE HEAT BOILER ARRANGEMENT

With variations to accommodate plant arrangements the boilers are tied into the plant steam distribution utilizing the existing auxiliaries of fuel fired boilers such as softeners, deaerators and boiler feed pumps.

While the boiler is a highly reliable piece of equipment it was felt justified to provide for the by-pass of the boiler if desired. The by-pass arrangements were made by use of dampers, either manual or motor operated in flues and duct work. We found the by-pass operation helpful in furnace start-up or shut-downs, which we will comment on later.

Figure 2 - Schematically outlines the Flue Gas and Steam System Arrangement.

In light of previous difficulties certain features were incorporated in the design:

a. Tube Cleaning Facilities. It is inherent that particulates and condensing vapors deposit out on the inlet tube sheet and tubes depending on glass composition. Previous practice had been to hand lance the tubes with air or steam through the front end. A vertical soot blower using steam as a blowing media is provided in the new arrangement. The blower upon operator initiation will traverse the tube sheet during the blowing sequence. The lance cross arm has a series of nozzles which coincide with boiler tube rows which effectively blow the tubes.

b. Tube Sheet Protection. Periodic cleaning of tubes of long term build-up by caustic washing has been required in the past. The washing process had resulted in corrosion of rear tube sheets as the solution ran over the sheet during cleaning. We have provided an acid resistant cement over the rear sheet surface to prevent such corrosion.

c. Controls. The items requiring automatic control are feedwater and draft control on furnace.

1. Feedwater - due to sharp demands in plant steam usage the use of a two-element feedwater control was utilized.

2. Draft Control - the I.D. fan is equipped with an outlet damper that is automatically operated to maintain required pressure in the glass furnace.

OPERATIONAL RESULTS

The boiler operations have all been successful. In fact the steam output from the recovery boilers is greater than that predicted in most cases, as high inlet temperatures are occuring.

The problem of tube fouling has been controlled by the regular blowing of the tubes.

As important to us, is that we have found a good correlation of heat transfer necessary for the selection of heating surface for future projects.

PART II - PLASTIC INDUSTRY

The other application which I wish to describe is the recovery of heat from a thermal fume incinerator.

In the manufacture of plastic gears our customer first impregnates a cotton canvas with plastic binders. To cure the binder the coated fabric is passed through a curing oven. The solvents which are driven off during curing were identified by E.P.A. to be hazardous, and accordingly a direct fired incinerator was installed.

A few years after the incinerator installation we made a "Steam Utilization Study" of the plant operation and identified the potential for heat recovery, again by the installation of a boiler on the outlet of the incinerator. It was estimated that the generation of steam from a Waste Heat Boiler would effect a fuel savings of $45,000/year.

WASTE HEAT BOILER ARRANGEMENT

On Figure 3 - INCINERATOR WAST HEAT BOILER we have shown the schematic arrangement of the system.

The installation of the boiler, ductwork, fan with piping and electrical work was about $65,000. Accordingly, the simple pay back was identified as approximately 18 months.

OPERATIONAL RESULTS

This system was installed the middle of 1981 and has been on line since then. The original data showed a 900°F exhaust from the incinerato and a gas flow capable of producing up to 2500 lbs./hr. of 150 psig steam.

The results have been more than satisfactory, with up to 3000 lbs/hr. of steam being produced at various times.

The operation has not created any production problems with only a minimum attention from the boiler operators for periodic blowdown.

SUMMARY

The results of these and other projects have assured us of our design approach, as the realized performances have met-or-exceeded the anticipated results.

In some applications the Heat Recovery Boiler has been arranged as a gas cooler for a tail end pollution control system.

Energy recovery projects require an evaluation of not only the recoverable energy but also the utilization of recovered BTU's and how to integrate a new system into an existing situation. Each project requires a thorough evaluation and study to effect a successful Heat recovery installation.

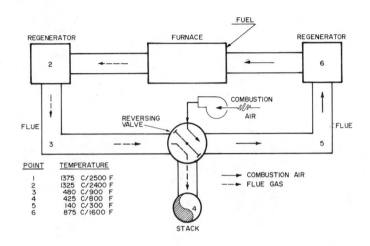

POINT	TEMPERATURE
1	1375 C/2500 F
2	1325 C/2400 F
3	480 C/900 F
4	425 C/800 F
5	140 C/300 F
6	875 C/1600 F

FIGURE I - REGENERATIVE FURNACE

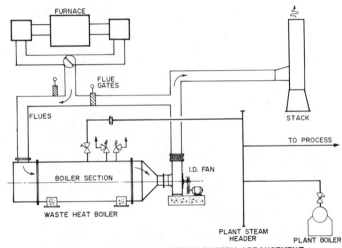

FIGURE 2 - FLUE GAS AND STEAM SYSTEM ARRANGEMENT

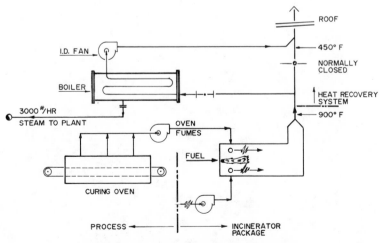

FIGURE 3 - INCINERATOR WASTE HEAT BOILER

Chapter 46

ENERGY CONSERVATION WITH RADIANT TUBE SYSTEMS

R. B. White

The Radiant Tube System is an Industrial and Commercial Heating System, which was developed in the United Kingdom seventeen years ago, is operating successfully throughout many countries of the world and has now been introduced into the United States. The arrival has been well timed because of the national interest in energy savings, and the system can be described simply as expanding on the well-known qualities of infra-red heating.

It is a system with greater capacity than conventional systems and is most suitable for buildings with large air volumes and multiple air changes. Typical applications would be high bay industrial buildings, aircraft hangars, and loading bays or dispatch docks.

FIGURE 1. A TYPICAL HANGAR INSTALLATION USING ONE HEATER ONLY.

The System consists of a direct or indirect fired heater connected to a galvanized spiral duct layout, which consists of parallel banks of tubes up to three feet in diameter mounted at high level. Heated air is circulated around this closed loop, so continuously supplying heat to the tubes and creating a powerful source of radiation. Heater sizes vary from 150,000 Btu's/hr. to 6,000,000 Btu's/hr. and, tube lengths can be installed 600 feet from the heaters, which minimizes the number of units required.

Natural gas is the energy source with the direct fired system, but there is a choice of fuels with the indirect fired, which utilizes a heat exchanger to raise the tube air temperature. Dual burners can be employed with the indirect system.

Tube surface temperatures are much lower than other radiant systems and, depending on building design, the flow air temperature is approximately 450° F. dropping to 250° F. on the return, providing an average tube surface temperature of 300° F. Again, depending on the tube size design, air speed averages 2,000 ft./minute, the volume of air passing along the tubes varies from 1177 to 12,563 SCFM, and the heat emission varies from 1360 to 14,310 Btu's/ft./hr. Tube diameters are from 9" to 36", the lightweight components provide minimal loading to the roof structure, with the weight of the total tube installation varying from 10 to 53 lbs./ft. in total.

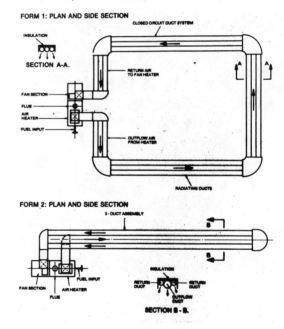

FIGURE 2. DIAGRAM SHOWING TWO ALTERNATIVE LAYOUTS.

Combustion efficiency of the heater is 95% for direct fired, radiant efficiency is 85%, and there is total fail-safe protection with a sophisticated control package. The heater can be located on the floor or elevated in the roof trusses, and the heater can also be placed outside if this is desirable.

However, the most impressive feature of the Radiant Tube System, lies in it's energy saving potential.

Special foil-backed insulation is used on top of the tubes to restrict heat loss rising, and suppressor side plates avoid tube surface temperature reduction, due to air currents. Insulation is used instead of reflector plates, because it is believed these become less effective through dirt attraction after a short period and they act as excellent conductors of heat. This method minimizes heat stratification within the building, there is no overheating of the roof or troublesome condensation and 85% of the radiant heat is directed to floor level.

A significant advantage of a radiant system lies in the fact that comfortable working conditions can be obtained at a lower air temperature than with a convected warm air system. It is possible to feel warm out of doors on a cold day, when standing in the clear sun, because the sun's radiant temperature compensates for the low air temperature. The sun's heat may warm by radiation being absorbed either directly or indirectly on adjacent surfaces re-radiating their own heat. Radiant tube heating provides a comfortable environment within buildings in the same way.

Comfort depends as much on the radiant temperature of the surroundings as on the air temperature. There are, therefore, two measures of warmth: the air temperature measured by an ordinary mercury thermometer, and the radiant temperature, measured by a globe thermometer. A increase in the temperature of one kind means that to keep the same feeling or warmth, the other can be decreased by the same amount. The comfort level is, therefore, the average of the air temperature and mean radiant temperature.

FIGURE 3.   THIS LARGE BUILDING IS 660 FT. LONG, 96 FT. WIDE AND 72 FT. HIGH, WITH TUBES MOUNTED AT MAXIMUM HEIGHT SO THAT NATURAL LIGHTING IS NOT OBSCURED.

Most surfaces of and within the building absorb heat, and in turn re-radiate it. In a building with radiant heating, warmth is absorbed by these surfaces directly and a high mean temperature is therefore radiated. Typically, the mean radiant temperature rises by 11° F. above the air temperature and can sometimes be as much as 20° F.

Heat is lost from a solid floor only through a narrow strip around the perimeter. 95% of the heat returns to the space by re-radiation from the floor surface, which can rise as much as 12° F. above air temperature. This heated floor effect contributes further to the feeling of comfort and there are no drafts or cold spots as with blown air heaters.

As the system does not rely on raising air temperatures, there are substantial energy cost savings when installed in buildings with multiple air extractions

or frequent large door openings, such as hangars or loading bays.

Heating costs are reduced considerably due to the rapid heatup from a cold start and it is not necessary to heat the premises at night or weekends when the work-force is not present and there are no freezing risks with the equipment.

Probably the most effective aspect of the system, is the fact that the heated air is re-circulated within the enclosed loop. The air heating process utilizes the return air at a temperature of 250° F. instead of ambient or external air, and only the combustion air is taken from outside and exhausted at the end of the loop.

There are no physical height limitations to the system. There is presently a system operating at over 90 ft. above floor level and, in fact, the higher the installation, the greater the energy savings and benefits over other heating methods.

Among the other advantages of Radiant Tube Systems is the fact that it is not necessary to heat the whole building: for instance, if one area only requires additional heating, such as a dispatch dock, this can be covered with the tubes in that zone only.

Dew point is encouraged by convector heaters but prevented by radiant heat which warms everything in it's path. This has reduced considerably the rejection rate in plants where materials such as rolled sheet metal or glass are stored. As there is no air movement within the building, installations have proved most effective in manufacturing processes such as foundry's where dust is a major problem.

FIGURE 4.   IN THIS MODERN COLD ROLLING WORKS RADIANT TUBE KEEPS THE STOCK FREE FROM CONDENSATION AND PREVENTS CHILLING FROM THE FLOOR SO THAT REJECTS ARE REDUCED AND PLANT EFFICIENCY IS INCREASED.

In large industrial buildings and aircraft hangars it is not unusual to find savings of up to 50% of fuel bills when the Radiant Tube System is installed. The

initial installation costs is sometimes more than
other systems, but the payback is substantial, plant
room and attending labor are unecessary, the plant
life is considerably longer than most competitive
systems and maintenance costs are negligible.

Heat Recovery

It must be realized that the Radiant Tube System
requires only an adequate heat source. Therefore,
waste heat of suitable temperature and capacity from
furnaces, ovens or boiler house flue stacks, can be
utilized, provided there are no corrosive constituents. This application of the system does not re-
circulate the captured heat, which is exhausted to
atmosphere at the completion of the tube run.

By using insulated duct work, it is possible for
the tube runs to be some distance from the heat source
and automatic dampers are used to mix in ambient air
to reduce the higher temperatures to those required
for the system. The advantages mentioned beforehand
are identical for this heat recovery system, plus the
additional benefits that the energy source comes from
previously waste heat and therefore provides cost-
free operation.

FIGURE 5. A TYPICAL WAREHOUSE INSTALLATION.

Radiant Tube Systems provide a custom made
design/build service and all materials are locally
manufactured.

Our policy is for regional Representatives to
provide contact with customers, we design a system
for each individual plant and arrange for installation
by local contractors. We are responsible for the correct operation of the design temperature from the
specifications provided by the client. Retrofit installations are just as suitable as new buildings and
installation is simple and quick.

Chapter 47

PYROLYTIC INCINERATION OF SOLID WASTE FOR ENERGY RECOVERY

P. J. Ollman

The East Moline Plant of International Harvester is primarily an assembly plant for Axial Flow Combines, Cyclo Planters, Tractor Control Centers, and attachments for the combines, consisting of grain platforms and corn heads.

The facility utilizes a total area of 156 acres with approximately 52 acres, or 2.2 million square feet, under roof. In addition to the assembly function, the plant houses machining, welding, heat treating, stamping, receiving and shipping departments. Normal employment at the facility is around 4,000, however, streamlining of operations will probably reduce that number to around 2,200 in the future.

One of the problems facing our operation was the ever-increasing volumes of solid waste to be removed from the site. As production increased, and we relied more on outside suppliers, solid waste volumes increased as well.

For several years we knew that something had to be done about this problem, but nothing was accomplished until 1975. In 1975 we began seriously studying the problem and found that there was justification for incineration equipment to reduce disposal costs and provide for energy offsets as well.

An appropriation request was prepared and submitted to our Corporate Headquarters detailing the project we wanted to fund. The project would cost $320,000 with a discounted payback of two years-11 months, and provided an internal rate of return of 43.48 percent, based on 1976 economics. The ROI was based on the need to dispose of more than seven million pounds of combustible waste annually.

After consideration of each of the options available for solid waste disposal, it was concluded that a system using a pyrolysis approach would best serve the needs of our plant. A pyrolytic incineration system offered pollution-free operation, a relatively simple design, heat recovery options, and little modification to adapt to the East Moline Plant waste stream.

Pyrolytic reduction has advantages over direct combustion methods in that it can operate pollution-free without scrubbers and collectors. It also requires less input fuel because the process becomes self-sustaining after reaching operating temperatures. The pyrolytic process is the termal degradation of solid material in an oxygen-lean atmosphere to yield gaseous products and a carbon-rich char. This is a two-step process which, when properly operated, results in pollution-free incineration. (See Figure #1 on page 2.)

In the thermal reactor which is the second step of the process, the gaseous products are reduced to a clean, effluent stream of carbon dioxide, water vapor, and recoverable heat energy. Left behind is an ash residue which is roughly five percent of the original volume and about ten percent of the original weight.

THE SYSTEM

In June, 1976, East Moline Plant placed its order for two Kelley Model 1280 Pyrolytic Incinerators equipped with 2.6-yard hydraulic ram feeders and a cart-tipping mechanism adapted to our plant trash cart which was already in service. Complimenting this equipment are two York-Shipley Waste Heat Boilers (3-pass fire tube) which generate up to 5000 pph saturated steam at 125 psig each from the hot flue gases from the incinerators.

This equipment is housed in a 66-foot square building which provides adequate space for all the above equipment as well as the water treatment equipment for the boiler feed water.

The project was to be a turnkey project for the two Model 1280 incinerators, feeders, and York-Shipley waste heat boilers. Once installed, we would have the capacity to consume over 10 million pounds of solid waste annually and produce more than 36 million pounds of steam for production use.

The Kelley system was to be compatible with our present trash hauling system; two-yard carts pulled in trains by a lawn-garden tractor and emptied by a tipping mecnanism. By integrating the new incinerators into the existing trash system, we felt we could operate the incinerators with the personnel assigned to haul the solid waste. We have since learned that to realize maximum benefit from the incinerators a full-time operator must be assigned to each operating shift. All too often, the trash haulers would overload the incinerators, reducing their efficiency and causing incomplete burning.

In addition to attending to and feeding the incinerators, the operator is able to perform routine maintenance, sort and examine loads for improper materials, and keep an almost-constant eye on boiler output.

The incinerator-boiler equipment is operated a minimum of 16 hours per day, consuming approximately 1,000 pounds of solid waste per hour per unit. The incineration of an estimated 3,500 tons of combustible material annually produces the equivalent of approximately 22.9×10^9 BTU annually and a savings of more than $71,000 in natural gas costs at 1981 prices. Total savings from reduced hauling costs and fuel savings are more than $350,000 annually.

While incineration equipment is inherently self-destructive, project life is anticipated to be eleven years on the incinerators proper. Other attendant equipment, with reasonable maintenance, will last somewhat longer.

Loads of waste brought to the incinerator site are checked for non-combustible materials, as these items tend to cause deterioration of the refractory lining of the incinerators. Should a load contain more non-combustible than combustible materials, the contents are emptied into a 44-yard closed compactor container and compacted to help keep disposal costs in line for the non-combustibles. By doing so, incinerator life is protected and the maximum amount of non-combustible material is removed in each trip to the landfill. We are attempting to eliminate virtually all open-top hauling from our plant site. We plan to use only two such containers, where we had up to ten before.

ENERGY RECOVERY

Available energy in the form of hot flue gases from the combustion process aids in the conservation of natural gas. The natural gas conservation is accomplished through the use of heat recovered from the pyrolysis process to heat two paint drying ovens in our Control Center manufacturing area. By synchronizing the burn time on the pyrolyzers with the operation times of the ovens, we are able to utilize the recovered heat energy to fully operate the ovens. Existing burners in the ovens remain to serve as stand-by units. We feel that this combination of pyrolytic incineration with production equipment like paint ovens represents the best approach to maximum return on investment and maximum recovery of the energy content of the combustible solid waste stream.

The paint drying ovens normally operate on a two-shift basis, for the full production year, thereby providing constant demand for the heat recovered from pyrolysis of the waste generated during the same time period. Generally, most heat-using production operations provide satisfactory demand for recovered heat energy.

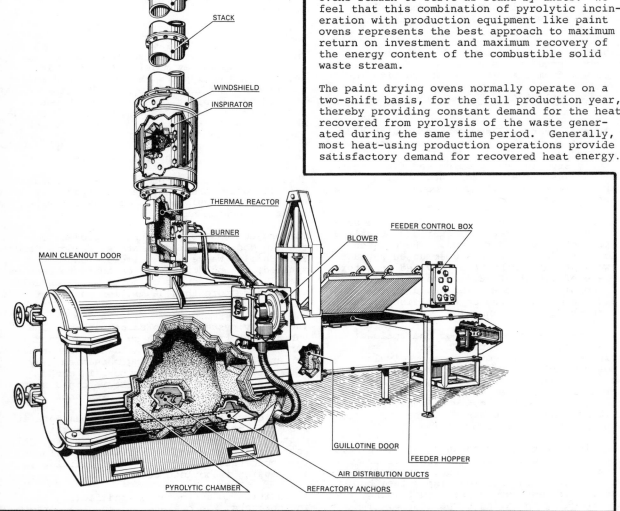

FIGURE #1 - PYROLYTIC INCINERATOR

Seasonal applications (i.e., space heating) should be considered only if no process operation is available or adaptable. Seasonal applications do not normally recover as much of the available heat energy as would production applications which use recovered heat energy over a longer time span.

We are currently investigating the expansion of the steam system to include a seven-stage parts washer and a tube washer, both located in the Control Center manufacturing area as well. If capacity allows, we will also provide some limited space heating in this area at doors to the outside storage areas.

We originally considered air-to-air heat exchange to heat the paint ovens, but changed when we realized the lack of flexibility presented by an air-to-air system. Steam, on the other hand, is extremely flexible and adaptable to a much larger variety of end-use applications than hot air. As we view the overall system today, we are pleased with the decision to use steam.

The surplus steam available today is a result of a major change in the Control Center paint line. We installed in 1980 an "E-Coat" Paint System which requires oven temperatures greater than can be achieved by our steam system. Consequently, we temper the make-up air to the oven, and the gas burner heats the air up to the required temperature of 390°F. The other oven is fully supported by the steam system for heating requirements, with the existing gas burner standing by in case of steam failure. By retaining the burners, the ovens can be easily operated at times when the incinerators may not be operating; and, of course, the incinerators can be operated without having to raise steam should the need arise.

PROBLEMS/SOLUTIONS

Now we must consider other important factors associated with the thermal reduction of combustible solid waste. This is the experience we have gained through the start-up and actual operation of this equipment.

At the very outset of a project like this, you must plan for a combustible waste stream. Foreign material (non-combustibles) will destroy the thermal reduction process and create untold handling and feeding problems. You must be able to insure a combustible waste stream, reasonably free of material detrimental to the incineration process. A problem that became very apparent early in our experience with the equipment was that hardware in the waste stream caused untold problems with the feeding mechanism. The hardware would get caught under the feeder ram and cause it to bind and jam.

An effective solution to the hardware jamming problem has been installed and we have no more difficulties in that area, although hardware items are still a part of our waste stream to the incinerators.

A second problem encountered within the waste stream is that of metal banding used to secure a variety of items for transport from place to place. This material balls up in the incinerator main chamber, making clean-out a major

problem. We are working with plastic strapping material which is combustible to effect a solution to this problem. Both in-house and vendor use is being considered so the proposed solution is a total rather than a partial one. In addition, the operators sort any major amounts of metal banding from loads. The banding is then chopped and sold as part of our miscellaneous scrap metal.

Also of benefit in a problem such as this one is to invite all shop management to tour the incinerator facility and explain to them how it works and what causes problems in the system. With their support and cooperation, many potential waste stream problems can be minimized and/or eliminated. Here is where a well-trained operator becomes a great asset in the overall operation of the equipment by explaining to the shop people what they can do and how it helps the incinerators operate efficiently. While planning your incineration project and investigating suppliers, you should keep in mind that no matter how nice the advertising photos may show incineration installations to be, dealing with the disposal of solid waste is a dirty job, so prepare for the worst in this regard. People are required to make this system function properly. Be sure you have them "on board" and let them know the contribution they are making is helping us all. If these people are not for you, your project will be in trouble from the start.

Be sure to plan equipment to handle needs as far in the future as you dare project. Solid waste streams can vary greatly, so don't get caught short. Consideration of several smaller units may be advantageous over one or two large units to allow for variations and all-important maintenance. When shopping for this kind of equipment, everyone puts on the best show possible, and as a result, all looks simple and straightforward. However, when the equipment is installed and operating in your plant, all is subjected to the hard scrutiny of reality and you begin to realize that dealing with a problem as complex as solid waste disposal is not a simple, straightforward situation solved with quick, easy decisions. In fact, many agonizing hours will need to be spent in fully adapting any equipment to your facility's needs.

Another item to keep in mind is that incineration equipment is inherently self-destructive. Temperature cycles, abrasive materials, etc., all contribute to the degradation of the feeding units and the combustion areas.

A good maintenance program will pay big dividends. The program should heavily involve the operator as the first-line preventative maintenance individual. Items beyond his ability or classification should be handled by the maintenance department. Daily checklist inspections will prevent minor problems from becoming major ones.

As one might expect, maintenance costs will generally increase as the equipment becomes older. Our experience indicates that maintenance has cost about $75,000 for five years of operation. Included in this cost are extensive modifications and repairs being done this year. This program will cost an estimated $35,000 and will bring our units back to first-class operating condition.

CONCLUSION

In spite of all this, we feel that the real long-range solution to our ever-growing solid waste disposal problem lies in thermal reduction of this waste material.

All one needs to do is consider the ever-increasing solid waste disposal cost, continually raised by the complex myriad of federal, state, and local regulations, together with the spiraling cost of energy coupled with potential supply interruption, and one can see the significance of such a program.

European countries have long disposed of domestic and hazardous combustible wastes by incineration techniques. They were forced to this technology years ago because they simply did not have land to waste for landfill operation. Consequently, they have little worry about ground water contamination by materials leached from inadequate landfills which is becoming a major concern in our country.

Incineration breaks the "cradle to grave" hold of the RCRA regulations because the remaining ash is, for the most part, not hazardous. Freeing your company from this kind of liability for the future certainly merits strong consideration in the overall evaluation of thermal reduction of solid waste.

I would encourage you to consider this method of dealing with solid waste problems. If you approach it with as much knowledge as is possible of all aspects and the potential ramifications, your project will be highly successful. Aspects of a pyrolytic incineration tion which provide attractive "fringe" benefits are:

1. Decreasing dependence on landfill disposal and reduction of land pollution.

2. Reduction of trash back-ups around in-plant contrainers which create potential health and pollution problems.

3. Elimination of scrap paints and oils requiring special handling and permits; and

4. The creation of a positive environmental action program for industry that may be used to improve our image in the community; a worthwhile public relations endeavor.

In summary, the struggle and frustration involved in bringing a project like this on-line is worthwhile when all the ramifications for positive environmental and economic impact are considered.

Pyrolytic incineration is a "solution to pollution."

Chapter 48

KILN WASTE HEAT RECOVERY

M. A. Mozzo, Jr.

INTRODUCTION

Sanitaryware (ceramic) tunnel kilns have for many years been outstanding candidates for heat recovery technology. Traditional firing of such product was done primarily in muffle-fired (indirect radiation) tunnel kilns. Such kilns were terribly inefficient with at most 5% of the input heat, and more likely only 1% of the input heat actually used to vitrify the product. The remainder 95 to 99% of the heat input is lost forever through exhaust stacks, furnace wall and ceiling radiation and transmission, and kiln cars and furniture. Some attempts were made at heat recovery, primarily in the form of steam generation, but the low price of energy prior to the mid-70's insured that waste heat recovery was a secondary aspect in the manufacture of sanitaryware.

Since the mid-70's three things have affected waste heat recovery technology for ceramic sanitaryware kilns. First, the price of energy has risen sharply affecting production costs. The price of natural gas delivered to our U.S. plants has risen from $1.65/MCF to $3.73/MCF during the period of 1976 to 1981, a rise of 126%. During the same time frame, fuel oils delivered to our U.S. plants have risen from 31.3¢/gallon to 80.8¢/gallon, a rise of 158%. Such increases have made heat recovery projects much more economically feasible.

The second major event which has impacted heat recovery technology is closely related to energy prices. With the huge increases in energy costs, we have seen more types of heat recovery equipment being marketed. Now open to the engineer is a choice of steam, hot water, hot air, and even cogeneration heat recovery equipment. The engineer now has a wider choice of what type of heat recovery equipment he desires to use.

And lastly, the kiln manufacturers now have a market for newer, more efficient (in terms of energy) kilns. With rising energy costs contributing more to cost-of-goods sold, sanitaryware manufacturers are paying more attention to kiln design, discarding old tried-and-true methods of firing, for newer, more efficient methods. These new kilns still exhaust hot gases from which waste heat can be recovered, but because of increased energy efficiencies, the potential savings are not as great as previously and the heat recovery engineer must work harder to recover this energy.

This paper discussed the steps used in evaluating and engineering potential heat recovery projects for sanitaryware tunnel kilns. Also reviewed are case studies of heat recovery projects which have been installed in our plants or were seriously considered by American-Standard for installation.

PHASES OF A WASTE HEAT RECOVERY PROJECT

Typical waste heat recovery projects can be divided into three phases: (1) Preliminary or Concept Engineering, (2) Design Engineering, and (3) Construction and Startup. While all phases are important, it is my opinion that the first phase is the most important. Proper background work must be accomplished in this phase to determine the technical and economic feasibility of the project.

PRELIMINARY ENGINEERING OF A WASTE HEAT RECOVERY PROJECT

The primary objective of this phase is to evaluate the potential of a waste heat recovery project. Proper attention must be made to all major aspects of the project before the decision to commit resources for design engineering is made. Some of the steps which should be considered in this phase are:

1. Determine the actual amount of the waste heat available. This involves a physical measurement of the volume and temperature of the exhaust stream.

2. Determine the dollar value of the waste heat being rejected. This will allow the engineer an initial opinion as to the economic feasibility of the project.

3. Examine the various means available to recover the waste heat. Table One shows the method of heat recovery and their criteria, which we have used.

4. Determine uses for the waste heat in whatever form it is recovered. Distances from source to use is obviously a factor as well as hours of operation of the source and the use. Cascading use of the waste heat may be applicable.

5. Select a waste heat recovery system that is technically feasible for your operation. Determine the preliminary costs and savings of the system to resolve the economic feasibility.

6. Insure that the system will meet local codes, especially codes requiring attendant operations.

TABLE ONE

HEAT RECOVERY TECHNIQUES

Techniques	Criteria
Steam Generation	Requires a temperature generally greater than 600°F.
	Temperature leaving the waste heat boiler will be 50-100°F higher than the steam temperature.
Hot Water Generation	Good technique if the waste heat temperature is less than 600°F.
Air-to-Air	Primarily used at low temperatures (below 400°F), but there are recuperators for temperatures above this limit. Extensive ductwork could be a cost factor if distance between source and use is great.
Cogeneration	High temperatures (800°F+) required for steam turbines.
	Can use lower temperatures (600°F+) if an Organic Rankine Cycle is used.

DESIGN ENGINEERING AND CONSTRUCTION PHASES

Once a concept for the waste heat recovery project has been selected and determined to be both technically and economically feasible, funds must be approved for design engineering and construction phases. These two phases are important in their proper execution. Primarily, good engineering practices must be followed in the design and construction of a waste heat recovery system.

CASE STUDIES

The following are examples of waste heat recovery which we have installed or are considering installing in our plants worldwide.

Steam Generation

A facility located in a northern climate with five (5) kilns and available waste heat as follows:

        Kiln #1 - 19,625 lb/hr @ 950°F
        Kiln #2 - 22,774 lb/hr @ 920°F
        Kiln #3 - 15,736 lb/hr @ 970°F
        Kiln #4 - 10,701 lb/hr @ 655°F
        Kiln #5 - 9,010 lb/hr @ 546°F

        Total = 77,846 lb/hr @ 858°F

Steam is used by the plant for both heating and process loads with demand varying from a minimum of 9,000-13,000 pounds per hour (summer) to a maximum of 21,000-26,000 pounds per hour (winter). Steam generation was selected because of favorable exhaust temperatures and the need for year round steam.

All kilns were ducted to one waste heat boiler sized to generate 125 PSIG steam at a rate of 9,724 lb/hr. Annual cost savings will be $567,300 at a cost of $926,000 or a simple payback of 1.7 years.

Hot Water Generation

Two (2) low temperature kilns in a midwest environment have the following qualities:

        Kiln #1 - 11,688 lb/hr @ 550°F
        Kiln #2 - 5,122 lb/hr @ 685°F

Steam is used in the plant for heating and process. A hot water system for process requirement was installed about five years ago. The hot water system is used year round. Because of the low exhaust temperatures, the proximity of the source to the hot water system, and the year round use of the hot water system, hot water was selected as the best heat recovery technique. Because of legal codes, separate hot water coils were placed into each stack, and the coils were piped into the existing hot water system. Savings and costs for each kiln are:

        Kiln #1: Savings = $ 18,759
                  Cost   = $143,149
                  Payback = 7.6 years

        Kiln #2: Savings = $ 14,564
                  Cost   = $103,651
                  Payback = 7.1 years

Air-to-Air Heat Recovery

This plant has three (3) direct fired tunnel kilns with exhaust temperatures from several gaseous streams in the range of 165°F to 350°F. Existing space heating and process heating is accomplished by gas-fired heaters and a ductwork system. By recovering the waste heat through air to air recuperators and ducting to the existing system, a savings of $216,000 per year can be achieved. At an investment of $250,000, a simple payback of 1.2 years is realized.

Cogeneration

This plant is located in the Los Angeles area. The facility had four (4) kilns firing 24 hours/day generating waste heat as follows:

        Kiln #1 - 12,540 lb/hr @ 760°F
        Kiln #2 - 19,440 lb/hr @ 900°F
        Kiln #3 - 18,900 lb/hr @ 900°F
        Kiln #4 - 19,320 lb/hr @ 700°F

        Total = 70,200 lb/hr @ 820°F

Steam generation appears to be the best waste heat recovery technique; however, steam requirements at this facility are low year round. A waste heat boiler would generate more than enough steam, with still more waste heat nonrecoverable.

This led to the conclusion that cogeneration using an Organic Rankine Cycle was the best technique. With over 70,000 lb/hr of 820°F exhaust gases, we could generate 515 KW (net) of electrical power. The best arrangement with the electrical utility was a simultaneous sell-buy-back arrangement. By selling our generated power at the utility's avoided cost, we calculated savings of $324,825 per year. At a project cost estimate of $1,176,000, we had a simple payback of 3.6 years.

CONCLUSIONS

Waste heat recovery projects in the sanitaryware industry are feasible, even with modern direct fired kilns. A variety of techniques is available to the engineer. Careful evaluation in the preliminary engineering or concept development phase is required to insure project success.

Chapter 49

BOROSILICATE GLASS -
ITS USE IN WASTE HEAT RECOVERY

M. J. Ruston

INTRODUCTION

The addition of borax to silica glass formulations, to produce a new family of glasses, namely the borosilicates, was probably one of the most significant advances in glass technology since antiquarian times.

Work was begun by Otto Schott of Jena, in Germany, in the early 1890's. He found that adding borax to a silica glass melt produced a material with significant improvement in thermal shock resistance and a low coefficient of expansion. Based upon this pioneer work and spurred on by serious failure problems with the glass then used for lantern globes, chimneys and signal lamp lenses, Arthur Day and Charles Houghton began work in the United States in 1908. Helped later by William Taylor and Eugene Sullivan, their efforts bore fruit in 1912 with the introduction of Nonex, the firest mass produced borosilicate glass formulation.

Although Nonex solved the problems of shock and expansion found in silica glasses, it was extremely susceptible to moisture attack.

Research continued until 1914, when a formulation which also had excellent chemical resistance was found. This material was called Pyrex[R], a name which is now familiar in almost every household. Since that time, low expansion borosilicate glass has found many uses in the home with oven and stovetop ware, in science for laboratory and optical equipment, in industry for large scale chemical plant, gauge glasses and incandescent lamps.

All of these end uses are based upon the material's basic properties; low coefficient of expansion, high thermal shock resistance, excellent chemical resistance and transparency.

These features also make low expansion borosilicate glass an ideal material to consider for the heat transfer surface in heat recovery applications from exhaust gas streams. Particularly where the gas stream is corrosive and contains particulate matter which could foul the heat exchange surface.

PROPERTIES OF LOW EXPANSION BOROSILICATE GLASS

Approximate Composition (wt. %)

SiO_2	81
Al_2O_3	2
B_2O_3	13
Na_2O	4

Physical Properties

Density (lb./ft.^{3})	139.3
Young's Modulus (x10^6 psi)	9.1
Thermal Expansion (x10^{-7} in./in./^{o}F)	18.1
Upper Working Temperature (^{o}F)	
- Normal Service	446
- Extreme Service	914
Lower Working Temperature (^{o}F)	-300
Thermal Shock (^{o}F)	320
Specific Heat (BTU/lb. ^{o}F)	0.233
Thermal Conductivity (BTU/ft.^{2} hr.^{o}F/in.)	8.1

Chemical Resistance

Borosilicate glass is inert to almost all materials with the exception of hydrofluoric acid, phosphoric acid and hot concentrated caustic solutions. Of these, hydrofluoric acid has the most serious effect and, even when present in a few parts per million, attack will occur.

Phosphoric acid and caustic solutions cause no problems when cold, but at elevated temperatures corrosion will occur.

Borosilicate glass has been the chemical industries standard for many years for use in highly corrosive applications, both for laboratory ware and large scale chemical plant operations.

The graph shown in Figure 1 shows the general effect of pH on corrosive attack.

Although borosilicate glass is attacked by the listed chemicals, they are unlikely to be found in the exhaust gas streams being expelled from the majority of plant equipment involved in combustion and drying operations. Fluorides can be present in calcining operations.

The overall excellent corrosion resistance of borosilicate glass, therefore, makes it ideal for situations where acidic condensation can occur. Exhaust gases can be cooled below their dew points with no concern for corrosive attack. Cooling below the dew point increases savings due to sensible heat recovery and also adds latent heat.

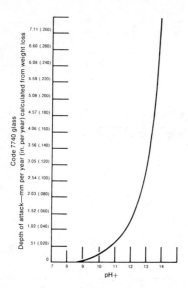

FIGURE 1: Effect of pH on corrosive attack.

Surface Finish

A common definition of "glass", a general name for a whole family of materials is, "Glass is an inorganic substance in a condition which is continuous with, and analogous to, the liquid state of that substance, but which, as a result of a reversible change in viscosity during cooling, has attained so high a degree of viscosity as to be, for all practical purposes, rigid".

In simple terms, glass is not a crystalline material and therefore has no discrete grain boundaries. It is a supercooled fluid which has an ultra-smooth surface. The quality of the surface finish is much higher than that which can be achieved in the manufacture of metal tubing.

This ultra smoothness, therefore, does not promote fouling or contamination of the heat transfer surface. If contamination does occur, it can be removed by a simple washing operation. Efficiency of operation is usually maintained for long periods before routine maintenance is required.

This makes a borosilicate heat transfer surface ideal for "dirty" applications where particulates are contained in exhaust gas streams.

An additional advantage is, borosilicate tubing is a "plain" tube and has no fins or extended surfaces which can act as deposition sites for fouling materials. This also means there is no possibility of heat exchanger blockage in heavy fouling situations.

HEAT TRANSFER COEFFICIENT

Probably one of the greatest concerns of any potential user of a heat recovery system, which utilizes a transfer surface made up from borosilicate glass tubing, is the thermal efficiency of that system.

One generally considers glass a poor conductor of heat, it should follow, therefore, that any such system would have a lower thermal efficiency than a metal tubed system. This, however, is not the case.

In a gas/gas heat exchange situation, and to a slightly lesser degree in a gas/liquid system, the major resistance factor which controls the rate of heat transfer is the gas film which exists at the tube wall. The thermal conductivity of the tube material, therefore, has very little effect on the overall rate of heat transfer and thus the thermal efficiency of the system.

The table below (Table 1) demonstrates this phenomenon for a range of tube materials. The system described is for sensible heat exchange only, i.e., dry coefficients only, for two gas streams. Clean surfaces, with no fouling are also assumed.

The system is described by Figure 2, which shows a section through a typical heat exchanger tube.

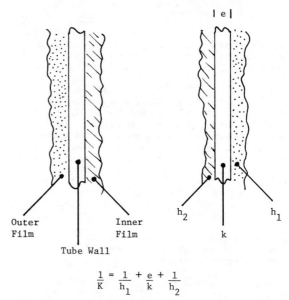

$$\frac{1}{K} = \frac{1}{h_1} + \frac{e}{k} + \frac{1}{h_2}$$

FIGURE 2: Heat Transfer System

TABLE 1: Overall Heat Transfer Coefficient

Variable	Tube Material			
	Boro-silicate	SS 316	Aluminum	Copper
h_1 Outer Coefficient	11.4	11.4	11.4	11.4
h_2 Inner Coefficient	9.4	9.4	9.4	9.4
k Thermal Cond.	0.66	9.4	128	225
e Material Thickness	0.0078	0.0016	0.0016	0.0016
K Overal Coeff.	4.88	5.15	5.15	5.16

h_1, h_2: Film coefficients, BTU/hr. ft.^{2} ^oF
k : Thermal conductivity, BTU/hr. ft.^{2} ^oF/ft.
e : Material thickness, ft.
K : Overall heat transfer coeff., BTU/hr. ft.^{2} ^oF

As the same system is considered for all tube materials, and the tubes are assumed to be clean, the inner and outer gas film coefficients will be the same.

The significant difference in thermal conductivity between tube materials is lost when the overall heat transfer coefficient is calculated, even when material thickness is taken into account.

The difference between borosilicate glass and the other tubing materials range between 5.2% and 5.4%. This initial difference between overall heat transfer coefficients would soon be compensated for by the continuing pristine condition of the glass surface. Whilst the metal tubing will tend to foul and corrode.

Practical experience has shown that where glass has been used side by side with metal tubed systems, there is no perceptible difference in performance during the initial running period. As running time continues, the glass tubes began to show a definitely superior performance as that of metal tubes drop due to corrosion and fouling.

EARLY USES IN HEAT RECOVERY

The first uses of borosilicate glass tubing as a heat transfer surface in heat recovery systems occurred in the late 1950's. These early applications were in air preheater systems recovering heat from stack gases by preheating combustion air for oil and coal fired boilers.

Generally, with the high sulfur content of the fuels used and the dirty exhaust gases, severe corrosion and fouling problems existed with the carbon steel and cast iron systems then operated, particularly in areas where acidic condensation occurred.

These original systems were generally installed in power stations and oil refineries all examples of large boiler operations. Usage soon afterwards expanded into marine applications on ships boilers.

Tubing used for these early systems was typically 2" in diameter with a wall thickness of 1/8". Lengths of up to 40' were common.

These systems were usually so massive that they were built on site. Tubes were mounted vertically being suspended from an engineered glass bead (Figure 3). When mounted horizontally, the tubes were held in position by a clamping plate (Figure 4).

Tube to tube plate sealing was affected by using a specially woven asbestos or fiberglass doughnut type bushing or seal. This allowed tube movement for thermal expansion and contraction, and gave a reasonably good gas tight seal.

These early installations were extremely successful giving longer operating times between scheduled maintenance, significantly reducing down time, and solving the corrosion and fouling problem.

However, problems did occur with these early systems after long service periods. Many of these problems were associated solely with material perception and operator education. During maintenance and cleaning operations, operators tended to use rods, shot blasting and brushing techniques, rather than the recommended water washing, which resulted in tube breakage. In some cases, where water washing was used, thermal shock failure occurred as the maximum recommended limits of temperature difference was exceeded.

It is believed the major problem with these early systems existed in the area of the tube to tube plate seal. Although the tubes and sealing gaskets had excellent corrosion resistance, the carbon steel tube sheet was unprotected.

When contact occurred over long periods between acidic condensate and the seal area, age hardening of the seal occurred, plus corrosion of the tube sheet. This resulted in the impregnation of the woven seal with metallic salts.

These two phenomena caused a reduction in the cross-sectional area of the tube plate hole, and effectively bonded or locked the tubes in position. Tube movement was prevented, resulting in abnormal stresses and eventually tube breakage.

FIGURE 3: Vertical Mounting

FIGURE 4: Horizontal Mounting

307

All of these early problems were associated with, either material perception, or the mechanical sealing techniques then used. Since that time the wider range of seal materials available, improvements in sealing methods, and the use of PTFE type bushings and elastomeric materials have cured the mechanical problems of the early days.

Borosilicate glass tubed heat recovery systems, therefore, can now be installed and operated with the degree of confidence associated with any metal tubed system.

SYSTEM FLEXIBILITY

The final design and choice of any heat recovery system, particularly when retrofitting to an existing process or operation, is a trade-off between a variety of factors.

Of all the considerations for any heat recovery project, those shown below are possibly the most important.

The Project Must Be Economically Viable

The return on capital investment must fall within an acceptable period. This is dependent upon:

- Capital Cost

- Thermal efficiency of the system

- Energy Savings

Increased Pressure Resistance

The additional pressure drop imposed by the heat exchanger must not disturb the operating balance of the process. It should, if possible, fall within the capacity of existing fans, avoiding the need for replacement or uprating.

Space Limitations

The unit should be compact, avoiding the need for large structural alterations which can often out weigh the cost of the heat exchange surface.

Borosilicate glass, in common with other tubular type recouperation systems have a relatively simple form of construction. Essentially they are banks of tubes which are located in supporting tube plates, by a seal system.

This approach gives this type of unit a large degree of flexibility in meeting design restraints, matching process and operator requirements to give an optimized system, at an economic price.

For each application an almost "customized" unit can be provided, without incuring unfavorable variances in manufacturing costs, and difficulties in manufacturing techniques.

A wide range of tubing diameters can be obtained from 0.08 inches O.D. up to 7 inches O.D., with wall thicknesses ranging from 0.006 inches to 0.13 inches. Tubing lengths of up to 12 feet can be obtained as a standard, longer lengths can be specially fabricated.

With the development of numerically controlled manufacturing equipment, punching or drilling highly accurate tube plates is also achieved easily and at low cost. This means tube pitch can, therefore be varied.

This flexibility can be further extended if a building block approach is adopted. Each heat exchanger assembly can be built from a multiple of standard modular units. This reduces mechanical handling problems during manufacture and shipping, and also gives greater flexibility in the siting and layout of the heat exchanger.

FIGURE 5: Modular Construction

SYSTEM DESIGN

Definitions

Thermal Efficiency: The generally accepted definition of thermal efficiency, often called the temperature exchange efficiency, is:

$$\eta = \frac{t_2 - t_1}{T_1 - t_1}$$

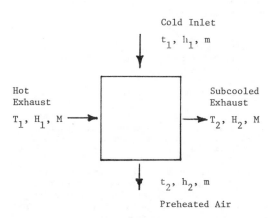

Where:

T_1 = Hot Exhaust $\qquad$ t_2 = Preheated Air

T_2 = Subcooled Exhaust $\qquad$ M, m = Mass Flow Rates

t_1 = Inlet Air $\qquad$ H, h = Enthalpies

The mass flow rates are not incorporated, so the expression does not give a precise value of the heat transfer efficiency. But as a comparative value the thermal efficiency is acceptable and is easily calculated.

Heat Transfer Efficiency: A more precise expression for the heat exchanger efficiency is given by:

$$\eta = \frac{Q\ actual}{Q\ theoretical}$$

Where:

Q actual = $M(h_2 - h_1) = M(H_1 - H_2)$

Q theoretical = the greater of $M(H_1 - H'_2)$

and $m(h'_2 - h_1)$

h'_2 is the enthalpy of m at temperature T_1

H'_2 is the enthalpy of M at t_1

Sensible Heat Recovery: Heat which is transferred due to change in temperature only.

For the cold stream q = $mCp(t_2 - t_1)$

Which is equivalent to Q = $MCp(T_1 - T_2)$ for the hot stream.

The above expressions can also be written as:

$$q = m(h_2 - h_1)$$

$$Q = M(H_1 - H_2)$$

Latent Heat Recovery: Heat which is transferred due to phase change, e.g., condensation of a vapor to give a liquid. When this occurs, a quantity of heat equivalent to the heat required to initially vaporize that condensed mass of fluid is given up. For an exhaust gas stream which is not at its dew point as it enters the exchanger, sensible heat must first be removed, cooling the gas to its dew point. On further cooling of the gas, condensation will occur thus making latent heat recovery possible. If the exhaust gas is at its dew point on entering the exchanger condensation will occur immediately, latent heat will be recovered throughout the heat exchanger.

The equation for heat added to the inlet air for the inlet gas remains as:

$$q = mCp(t_2 - t_1) = m(h_2 - h_1)$$

The equation for heat removed from the exhaust stream becomes:

$$Q = MCp(T_1 - T_2) + \lambda w = M(H_1 - H_2) + \lambda w$$

Where:

λ = latent heat of vaporization (condensation)

w = mass of liquid condensed

SYSTEM PERFORMANCE

Tubular recuperation systems by virtue of their construction are restricted to simple cross flow arrangements. One gas stream flowing through the tubes, the other on the shell side of the unit, outside the tubes.

In its basic form the tubular system is designed as a single pass for each stream. As short path lengths are involved this approach will give a lower pressure drop (Figure 6 and Figure 8).

However, it can restrict efficiency through temperature crossover considerations between the preheated air temperature, t_2, and the temperature of the subcooled exhaust, T_2.

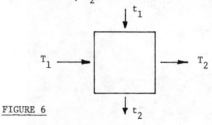

FIGURE 6

As a result, this type of flow pattern will generally limit thermal efficiencies to between 50% and 60% for sensible heat recovery. It can achieve efficiency levels of up to 80% where the hot exhaust gas is near its dew point on entering the heating exchanger, making latent heat recovery possible, from condensing vapor.

An extention of the single pass unit is to adopt multipass configurations for the exhaust or inlet gas streams. Pressure drop penalties must be accepted, as pressure drop will be higher with the increase in path length for the gas stream. However, higher efficiencies can be attained by the preheated inlet air stream, t_2, achieving a closer approach temperature to that of the hot exhaust, T_1.

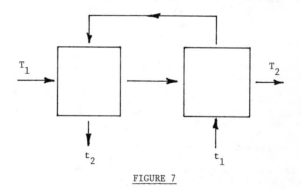

FIGURE 7

In a sensible heat recovery situation efficiencies between 60% and 75% can be achieved. For sensible plus latent heat recovery an 80% efficiency still remains the upper limit.

A multiple pass configuration for the exhaust stream can be a distinct advantage where particulate matter is contained in the exhaust, and cooling the exhaust gas below the dew point is feasible. The first section or pass of the heat exchanger can be used to remove the heat required to cool the gas down close to its dew point. The second pass of the heat exchanger can then be used to cool to the dew point and beyond. This approach will contain particulate matter and condensate in a specific area of the heat exchanger where it can easily be handled.

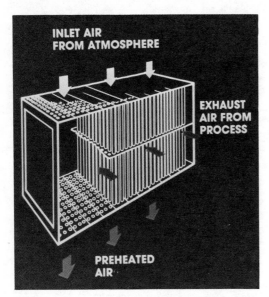

FIGURE 8: Simple Crossflow System

PRACTICAL APPLICATIONS - OPERATION EXPERIENCE

One of the leading sectors of industry in the field of energy conservation in the United States, Canada and Great Britain is that of the malting companies.

Malt production is a complex process. Starches contained in barley grains are converted to fermentable sugars for use in brewing and distilling. The process begins with the saturation of barley grains with water to stimulate germination. After saturation, the grains are transferred to germination boxes where this process continues until germination is complete. The final processing step in malt production is to dry the malted barley.

This drying phase serves several functions - to complete and then terminate biochemical conversion, to add color to the product and to reduce moisture content in the grain to a level acceptable for storage purposes. This phase of production is carried out in a fixed bed dryer and is essentially a batch process. It is at this phase of malt production that there are enormous opportunities for heat recovery.

A malt drying kiln is essentially a large rectangular box with a slotted wire mesh floor, onto which the saturated barley is loaded. The flooring mesh retains the barley while allowing the hot, drying air to pass through.

Batch sizes for typical malt drying operations can range between 40,000-90,000 lb. of "green" malt, with drying air flow rates between 49,000 and 1,000,000 scfm.

Exhaust air cannot be recirculated because it contains particulate material and corrosives, and in the early phases high moisture levels. To recover heat some form of heat exchanger is required which prevents cross contamination between exhaust and inlet air streams.

The drying process is split into usually four distinct phases, lasting a total of approximately 20-22 hours.

Phase 1

The free drying stage where moisture is removed from the surface of the grain, duration about 12 hours. Exhaust air temperature remains steady at 82-90° F, relative humidity, 90-100%.

Phase 2

Stages 2 and 3, where the drying air temperature is progressively increased to remove moisture from within the grain itself, duration about 7-8 hours. Exhaust air temperature averaging about 140° F, relative humidity, 36%.

Phase 3

The curing period, where color is put into the dried malt, duration about 2 hours. Exhaust air temperature approximately 185° F, relative humidity, 5%.

The greatest potential for heat recovery exists during the first stage of the malt kiln cycle. Although the exhaust temperature is low, the high humidity level enables high heat recovery to be achieved. This is possible because the exhaust gases are near their saturation points. Condensation will occur immediately the exhaust gas enters the heat exchanger.

The heat recovery systems described in the following text were built and installed for Bass Maltings, Ltd., a member of the Bass-Charrington organization, one of the largest brewing companies in Great Britain. The first system was installed at their Alloa Maltings, where a heat exchanger system was designed as an integral part of a modernized malting facility.

The second systems were retrofitted on two existing kilns at the Mirfield Malting facility. Space was not available within either of the existing buildings; the heat recovery systems were, therefore, sited on external support structures.

FIGURE 9: Mirfield Drum Plant Recovery System

The prime considerations for all three systems were: absolute corrosion resistance, a non-fouling surface, and the pressure resistance imposed by the heat exchangers must be kept to a minimum without affecting heat exchanger efficiency.

Heat recovery systems using a borosilicate glass heat transfer surface were chosen.

The operational and design parameters for the Mirfield Drum Plant recovery system are detailed, e.g., flowrates, temperature, humidity, to give an indication of process conditions.

For the other installations, as for any malting plant process, conditions other than air flow rates are similar, only the measured fuel savings are given.

BASS MIRFIELD - DRUM PLANT

Operational Conditions

Exhaust Conditions:

Flow Rate:	71,108 acfm	
	5,097 lb./min.	
Temperature:	86° F	
Humidity:	95% R.H.	

Inlet Conditions:

Flow Rate:	64,389 acfm
	4,999 lb./min.
Temperature:	50° F
Humidity:	80% R.H.

Drying air temperature: 151° F

Heating System: Heat Load 7.2MM BTU/hr.

Direct fired process air heating.

3% Sulfur Fuel Oil.

Sulfur burnt during process.

Measured Performance

Preheated Inlet Air:	79° F
Subcooled Exhaust:	78° F
Water Condensed:	1458 lb./hr.
Heat Recovered:	2.07MM BTU/hr.
Thermal Efficiency:	80%
Heat Saving:	28.6%
Total Pressure Drop:	0.7 inches W.G.
Heat Exchanger Surface Area:	24,640 ft.^{2}

Fuel Savings (Therms)

	1980	1981	SAVINGS
August	34,720	24,304	10,416
September	30,700	23,100	7,600
October	34,247	24,722	9,525
November	33,395	23,870	9,525
December	34,400	24,930	9,470
January	34,800	25,200	9,600

1980 Estimated Annual Usage:	404,524 Therms
1981 Estimated Annual Usage:	292,252 Therms
Estimated Annual Saving:	112,272 Therms
% Fuel Saving:	28%

BASS MIRFIELD - SALADIN PLANT

Operational Conditions

Exhaust Air Flow Rate: 35,000 acfm

Direct fired process air heating

3% Sulfur Fuel Oil

Sulfur burnt during process

Measured Performance

Thermal Efficiency:	79%
Heat Saving:	28.4%
Total Pressure Drop:	0.68 inches W.G.
Heat Exchanger Surface Area:	14,080 ft.^{2}

Fuel Savings (Therms)

	1980	1981	SAVINGS
August	21,615	15,190	6,425
September	20,950	14,700	6,250
October	21,600	15,190	6,410
November	22,000	16,275	5,725
December	22,700	15,723	7,427
January	22,157	15,190	6,967

1980 Estimated Annual Usage:	262,044 Therms
1981 Estimated Annual Usage:	184,536 Therms
Estimated Annual Saving:	78,412 Therms
% Fuel Saving:	29.9%

Total savings for Drum and Saladin units:

190,684 Therms

Capital Cost of Systems

Approximate cost of heat exchanger surface and immediate ducting:

£ 70,000.00

Total capital cost of installed systems, inclusive of
support structure, ducting, insulation, etc.:

£ 175,000.00

Annual value of fuel saved (1980 prices):

£ 59,684.00

Simple Payback: 2.9 Years

At current U.S. dollar values, and U.S. fuel prices:

     Capital Cost: $332,500

     Fuel Savings: $ 92,653

     Simple Payback: 3.6 Years

BASS ALLOA

Operational Conditions

Exhaust Air Flow Rate: 220,000 acfm

Direct fired process air heating

Natural Gas, with 3% Sulfur Fuel Oil as standby.

Sulfur burnt during process.

Measured Performance

Thermal Efficiency: 82%

Heat Saving: 27.5%

Total Pressure Drop: 0.60 inches W.G.

Heat Exchanger Surface Area: 69,725 ft.^{2}

Fuel Savings (Therms)

	1978/79	1981/82	SAVINGS
August	94,860	61,659	33,201
September	87,210	59,670	27,540
October	85,374	56,916	28,458
November	89,505	61,965	27,540
December	99,231	66,402	32,829
January	97,232	66,402	30,830

1978/79 Estimated Annual Usage: 1,106,824 Therms

1981/82 Estimated Annual Usage: 746,028 Therms

Estimated Annual Savings: 360,796 Therms

Capital Cost of System

Approximate cost of heat exchanger surface and immediate ducting:

£ 115,000

Capital cost of installed unit:

£ 180,000

Annual value of fuel saved:

£ 108,239

Simple Payback: 1.67 Years

At current U.S. dollar values, and U.S. fuel prices:

     Capital Cost: $342,000

     Fuel Savings: $175,311

     Simple Payback: 1.95 Years

Direct comparisons between "before and after" fuel
savings for this particular unit are difficult to
make, as the whole malt production facility under-
went extensive modification. Some portion of the
savings could, therefore, be due to higher burner
efficiencies, improved air distribution, etc.

However, comparison between capital cost and payback
for the Mirfield and Alloa units does indicate the
vast differences in expenditure between retrofitting
a system and including a heat exchanger as an
integral part of a new project.

FIGURE 10: Mirfield Drum Plant
Recovery System During Operation

SUMMARY

Borosilicate glass has been a proven engineering
and construction material for more than sixty five
years. Applications for its use have ranged from
oven ware to telescope mirrors, laboratory apparatus
and chemical plant.

As a heat transfer surface in exhaust gas energy
recovery applications, it will operate as efficiently
as any metal surface.

As a material of construction in fouling and corrosive environments it has a distinct advantage over other materials.

Using modern thermoplastic and elastomeric materials, the problems of the late 1950's and early 1960's associated with sealing and breakage no longer exist.

A borosilicate glass system can, therefore, be installed with confidence in the knowledge that it will give many years of efficient trouble free service.

The installation of what is believed to be the largest borosilicate glass heat exchanger in the world, has recently been completed for a mid western brewery. This unit contains over 550 miles of tubing, has a heat transfer surface of 518,000 ft.^{2} and is designed to handle an exhaust flow rate of 1,000,000 scfm.

Acknowledgements:

Mr. R. H. Smeaton, Managing Director, Bass Maltings Ltd., Burton-on-Trent, England, for his kind permission to use the Alloa and Mirfield Heat Recovery Systems as examples of operational installations.

Corning Process Systems, Corning, New York.

Mrs. N. Robinson, for her patience.

313

Chapter 50

SUCCESSFUL INSTALLATION OF A TRASH INCINERATOR/WASTE HEAT BOILER - A CASE STUDY

A. J. Gallo

INTRODUCTION

With energy costs continually rising above the general inflation rate, incineration of industrial trash (wood, paper and cardboard) becomes more attractive as a means of energy cost reduction.

This paper discusses a case history of the successful installation of an incinerator/waste heat boiler in the power house of our Administrative complex in Corning, New York.

Various aspects of the project will be discussed from the early decision factors through a short discussion of potential design/operating problems.

KEY DECISION FACTORS

Before you can actively consider the installation of an incinerator and waste heat boiler - you must have a sufficient quantity of combustible material and an ultimate use for the resulting steam. Ideally, steam use should be a continuous - year-round requirement. Rising boiler fuel costs are obviously a major factor in the decision to pursue an incinerator boiler project. However, disposal costs and landfill availability can also enter the picture as decision factors.

In Corning, we had enough combustible trash and the use for the steam. The key factor that triggered our decision to pursue this project in 1979 was rapidly escalating boiler fuel costs. Between 1972 and 1980 our boiler fuel costs increased by a factor of 5 from $.60/MCF to $3.00/MCF. This does not include the impact of Phase I incremental pricing which approximated $220M additional cost for our Corning boiler complex in 1981.

Hauling and landfill costs added to our justification since they were approaching $120M/year.

Supplementing these obvious financial justifications was a strategic concern - where we foresaw our local industrial landfills filling up in the next five years forcing us to look elsewhere for suitable disposal facilities. All of these factors were instrumental in the decision to pursue incineration of combustible materials.

SITE LAYOUT

Our main Corning complex spans approximately two miles and is bisected by the Chemung River. The core of that complex is comprised of 700,000 Ft² of administrative area, two

manufacturing facilities and a large museum that are supplied with steam by our Main Plant Power House. Steam is supplied by two water tube boilers at 150 p.s.i. and 50° superheat. Boiler No. 1, sized at 80,000 lbs/hour capacity operates predominantly in the winter season and is augmented by a No. 2, 60,000 lbs/hour, boiler when necessary. During the summer season, our smaller process loads were supplied by the No. 2 boiler only.

Figure 1 illustrates the seasonal steam demand satisfied by the power house boiler in a typical year. Typically, our year-round process load averages 15,000 lbs/hour while our winter heating demand averages 60,000 lbs/hour and can go as high as 120,000 lbs/ hour on a bitterly cold day. The vast majority of our steam is utilized for heating in the winter season. This demand curve plus an accurate inventory of our available trash formed the basis for our decision on the size of the incinerator/waste heat boiler and the design of the complex.

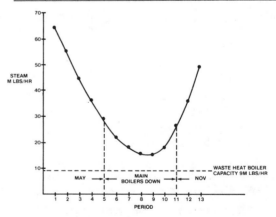

Figure 1
Seasonal Steam Demand - Corning, New York

The major factor that determined the size of the unit was the amount of available combustible trash. You must have enough data to ensure that you do not oversize the capacity of the incinerator and the waste heat boiler. Our studies showed we had 24 tons/day of trash, enough to justify six days/week - 24 hours/day operation at 9,000 lbs/hour steam maximum. Although this value was less than our typical year-round process load, we designed around the 9,000 lbs/hour maximum since it was compatible with our available trash level and since we had visions of lowering our year-round process load with other energy conservation projects. The horizontal dotted line on the chart illustrates the year-round anticipated steam supply. This base load for year-round operation is important - both when sizing and justifying the unit.

With the short summers in the northeast, the spring and fall steam requirements are very large. We actually only have two or three months of flat process load with no heating demand. This led to inefficient operation of our 60,000 lbs/hour boiler when it had to operate at very low steam production levels in the spring and in the fall. In fact, when summer demand dipped below 15,000 lbs/hour we often had to manually tend the boiler. We investigated retrofitting different burners on the boiler and found that the costs would exceed the cost of a new smaller boiler. Taking all of these factors into consideration we decided to add a 20,000 lbs/hour auxiliary gas-fired boiler to the incinerator/waste heat boiler system. By installing this auxiliary boiler in the system, we were able to:

1. Augment the incinerator/waste heat boiler output of 9,000 lbs/hour to meet the summer load when necessary.

2. Satisfy the heating requirements between May and November, thereby allowing our large boilers to shut down for this period of time.

3. Obtain higher efficiency of steam generation during this time period due to higher loading of the auxiliary boiler vs. low percent loading of the No. 2 boiler.

INCINERATOR OPERATIONS

Figure 2 depicts a schematic layout of the complex. The main steam producing components of the installation are:

1. The incinerator which will burn up to 2,000 lbs/hour of trash.

    Sunbeam Corp/Comtro Div. A-50 Combustion Unit

    2,000 lb/hr capacity - 8,500 BTU/lb material

2. The heat recovery boiler (fire tube) rated at 9,000 lbs/hour of saturated steam at 200 p.s.i.g.

    Eclipse model 5HR waste heat boiler single pass

3. The auxiliary water tube boiler rated at 20,000 lbs/hour at 160 p.s.i.g. and 100° superheat.

    Cleaver Brooks Model D-42-S-260

4. The two main water tube boilers (80,000 lbs/hour and 60,000 lbs/hour capacity) operating at 150 PSIG and 50° superheat.

The superheat is required for a steam turbine driven air compressor and pumps located in the power house.

Since we did not have space in our existing power house to house the incinerator/waste heat boiler and auxiliary boiler, we had to construct a separate building. We wanted it near the existing boilers to maximize utilization of existing operators and equipment. Construction was industrial type poured concrete foundation and floor, structural steel frame with standard corrugated metal siding on the building.

The combined steam output from the waste heat boiler and the auxiliary boiler are fed through an overhead insulated line to the power plant's existing steam manifold.

Three new stacks were constructed for the complex. The large exhaust stack from the incinerator has a diameter of 52", is 50' high and is refractory lined. This stack is equipped with a cap that closes under normal operations to divert the gases into the waste heat boiler. In the event that the steam demand falls below the capacity of the heat recovery boiler, or it is necessary to shut the boiler down, the hot gases can be bypassed by opening the cap. The other two stacks are standard metal stacks that provide the exhaust from the waste heat boiler and the auxiliary boiler.

Since a high percentage of our trash was scheduled to originate from our glass manufacturing plants, we realized that it would be very difficult to keep all glass out of the incinerator by segregation alone. To minimize chances for slagging on the hearth by entrained glass, we included a design feature whereby we circulated the incoming boiler feed water through piping enclosed in the refractory lining of the incinerator hearth. With this feature, we felt the hearth would be kept cool enough so that slag material would not adhere and would be subsequently pushed out by the plow with the other non-combustible materials. Another positive feature was additional energy anticipated to be obtained by pre-heating the incoming boiler feed water from 240°F to 300°F prior to entering the boilers.

Figure 2
Incinerator - Heat Recovery Boiler Complex

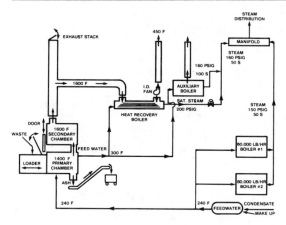

The incinerator is a dual chamber type with ignition burners in both chambers and excess air capability in the secondary chamber. Thermocouple sensors monitor the temperatures in both the ignition and combustion chambers with operating temperatures adjustable through use of automatic temperature controllers. Combustible material is placed in the loading chamber of the ram feeder by a front end loader. The operator fills the loader hopper with solid waste and presses a button to close the hopper cover. A second pushbutton activates the automatic loading sequence. The guillotine door opens and the ram moves forward charging the waste into the primary combustion chamber. The ram then moves to its full open position, the guillotine door closes, and the operator pushes a button to open the hopper lid.

On initial start-up, primary burners ignite the waste. After the primary chamber temperature reaches 1400°F, the primary burner shuts off. The burners will reignite if the primary chamber temperature falls below the preset level. In the primary chamber, the waste is oxidized with less than stochiometric air. This pyrolysis action produces a clean gaseous hydrocarbon fuel. This gaseous fuel enters the secondary chamber where it is mixed with additional air. The secondary burner ignites this mixture completing the combustion process. Temperature controls regulate the amount of combustion air introduced and modulates the secondary burner firing rate to maintain a chamber temperature of approximately 1600°F. On leaving the secondary chamber, the clean products of combustion enter a fire tube boiler where the temperature drops from 1600°F to approximately 450°F producing an average of 9,000 lbs/hour of steam. An induced draft fan exhausts the flue gas into the atmosphere. If the boiler is not in use, the incinerator can still be fired with the flue gases directly entering the atmosphere through a bypass stack. When the combustion air is properly adjusted, the discharge plume meets New York State opacity standards, thereby negating need of a pollution control device.

Instruments and safety circuits monitor every aspect of the system. Temperature controls regulate the burning rate and prevent overheating by adjusting the overfire, underfire and secondary air automatically. Pressure sensing instruments control the draft within the system. The safety circuits initiate alarms when any equipment fails or reaches critical status.

PROJECT JUSTIFICATION

Figure 3 elaborates on the financial factors mentioned earlier that influenced our decision to proceed on this project.

Figure 3
Project Justification Factors

Fuel Savings

- Incinerator
- Efficiency Improvement

Figure 3 (Con't)

Disposal Costs

- Hauling
- Landfill

IRB Financing

Tax Credits

- Investment
- Energy

Obviously the major savings area is from the fuel saved by the operation of the incinerator. However, another factor that contributed to the justification was the equivalent fuel savings due to efficiency gains of operating the auxiliary boiler at higher efficiencies for six months/year.

Hauling and landfill costs were quantified. Financing and tax credit impacts certainly contributed to our justification. We chose to finance our investment with industrial revenue bonds. Once we did that, we were only allowed to claim one half of the applicable 10% energy tax credits. I should also mention that by qualifying this investment under the recycling portion of the Federal energy tax law, we found we were able to qualify more equipment for the tax credit than if we had qualified it under the waste heat boiler section of the tax law.

Figure 4 summarizes a full year savings and costs for all of the categories mentioned. The assumption behind the fuel savings value is 7500 hours/year operation of the system at 9,000 lbs/hour production rate and $4.40/MCF boiler fuel costs.

Figure 4
Project Savings & Costs

Project Savings		1982 $M
· Fuel		$363M
· Efficiency		25
· Disposal		120
		$508M/Yr
	IRB - $169M	
	ITC - $137M	
	ETC - $ 50M	

Operating Costs	1982 $M
· Labor	$111M
· Fuel & Power	10
· Maintenance	15
	$136M/Yr

Capital = $1,472M

In Corning, we use a discounted cash flow analysis of after tax savings to justify our investments. It is a fairly standard calculation that we have modified to take into consideration the above normal inflationary rates for energy. Figure 5 shows our expected rate of returns for the project. For a capital investment of $1,472M, the savings outlined in Figure 4, and with our assumptions for the inflationary rate for natural gas, our most likely IRR was 30% with a payback of four years three months.

Figure 5
Discounted Cash Flow Results

Savings Level	20% Less	Most Likely	20% More
15 Yr IRR	24%	30%	36%
Payback	4.8 Yrs.	4.3 Yrs.	3.9 Yrs.

As anyone knows who tries to predict energy prices, it's a risky business. I have therefore shown the results of a sensitivity analysis of this investment where we calculated the impact with 20% less energy savings and 20% more energy savings. This actually covers two situation, if gas prices don't rise at the forecasted level or if you don't generate steam at the assumed rate.

POTENTIAL PROBLEM AREAS

In the design of a system of this magnitude, we found certain design areas to be very important.

Overall site layout and trash storage considerations are extremely important. Thought should be given to how the trash will be delivered, unloaded and subsequently processed. Will the trash be loose or compacted? Will any final segregation be needed? If delivery is not continuous, is there enough storage area to support continuous - 24 hr/day - incineration? Your design goal should be to provide the material handling capability to address these questions with minimal labor costs.

The relationship between the size of the loading chamber, the size of the primary combustion zone and the average BTU content of the trash is critical. The loading chamber size must be matched to the other two parameters such that a reasonable time period exists between charges. For example, if the loading chamber is too small - charges could be so frequent that additional shift labor could be required.

Additionally, frequent loading cycles could introduce excessive quantities of air into the pyrolytic zone - upsetting the pyrolytic process.

If you have trash that has greater than 5% by weight of glass or other slagging materials, emphasis on segregation is required. Excessive slagging can hinder operations in two ways - spalling of castable refractories or potential jamming of the ram feed mechanism. In our case, we conducted extensive training sessions of plant personnel and provided a system of specially marked trash containers to minimize this problem.

We are pleased with our system. In recent months we have made rapid progress on the "learning curve" and are sure that our forecasted return will be realized.

I would recommend that any facility having sufficient volume of combustible waste products and a use for the resultant energy seriously consider this opportunity for energy cost reduction.

Chapter 51

WASTE TO ENERGY - STRATEGIES AND PAYOFFS

J. S. Gilbert

ABSTRACT

Many industrial firms are taking a hard look at converting waste to energy. The opportunities for positive significant operational impact are not without large capital outlays. Past experiences indicate that an understanding of the basic alternatives, strategies, and typical economic performance can go a long way in directing corporate efforts, and in engineering an economically viable project.

This paper addresses boiler and engine-based systems, their performance, operating advantages and disadvantages, and the economic performance of each of the major hardware alternatives. This formulation and decision process for actual waste-to-energy projects is examined with sample energy and economic examples. These guidelines should assist the energy manager in deciding between waste-fired cogeneration or conversion of the powerhouse to alternative fuels.

INTRODUCTION

Why mix together the concepts of cogeneration and waste to energy? Because they represent what we think are going to be the "dynamic duo of the 80's." As a matter of definition, waste is normally construed to mean something thrown away, rejected, or declared to have little or no value. Ironically, most wastes now are valuable in a reverse way since disposal costs are at unprecedent levels. Historically, natural gas was considered a waste product of oil well and refinery operations. As a matter of fact most combustible waste products are generally quite valuable when you consider the combination of their Btu content and the cost of disposal.

But why cogenerate? Isn't it good enough to merely burn the wastes to make steam? The answer is that while it is true that under some circumstances the cost of the electric generation equipment would not be marginally attractive, the majority of American industry should include power generation in the design. This is very easy to see when you consider that every million Btu's converted to elecricity can usually produce about 265 kWH(assuming a 95% efficient generator and 95% efficient mechanical systems). Given that 265 kWH at 4c kWH amounts to $10.58, where the fuel burned is waste, this electricity is a more valuable energy form.

This is not meant to imply that conventional fuels are less amenable to cogeneration. We have just seen that the economics for waste fired cogeneration are often even better. Conventional fuels are defined here to include natural gas, carbon monoxide, LPG, fuel oils, coal, and wood. The candidate wastes that we will highlight are municipal waste, refuse derived fuel(RDF), industrial waste, wood residues, and bio-gas. The common denominators are that they are at least somewhat non-homogeneous, usually have relatively low Btu/lb heat content, and represent a disposal problem(opportunity).

Actual Btu contents are quite variable even within these candidate fuel types. Normally the variability is strongly dependant upon moisture content. Typical values range from 3-8,000 Btu/lb. Industrial wastes can actually have Btu values higher than primary fuels. Values of 16-20,000 Btu/lb are not that unusual, especially among the petroleum and chemical industrial wastes.

There are basically two types of cogeneration systems; boiler and engine based systems. Each of the common engine types (gas turbine and diesel)will be presented. It may be obvious at the start that a certain system is not feasibly fired on a given fuel. For instance, current technology does not permit a gas turbine to be fired with RDF. It may be possible in future years. The emphasis then will be to describe a conceptual framework that logically investigates waste to energy and cogeneration potentials for your facilities.

Putting First Things First

Always start out by carefully defining your current energy requirements. For instance, what is the use of electricity, gas, oil, and other fuels on an absolute and relative basis. This should be developed for an individual plant broken down by time of day, season, product mix, and any other index that can be directly tied to energy consumption. Many building loads will correlate well with heating and cooling degree days. Firgure 1 would represent the minimum amount of data to define use. It could be very misleading without a 24 hour breakout and a daily variation.

Example Energy Requirements

Month	Heating Degree Days	Cooling Degree Days	Days of Production	Electric kWh	Steam Mlb	Gas Mcf	Oil Gal	Coal Tons
Jan.	1,320	0	20	1,231,000	7927	2,444	43,130	0
Feb.	865	0	19	1,203,000	5,989	6,350	1,950	0
Mar.	831	0	22	1,164,000	6,075	1,607	44,820	0
Apr.	465	0	21	1,142,000	3,786	2,610	9,960	0
May	214	24	20	1,069,000	3,621	0	26,380	0
June	18	116	22	1,227,000	3,520	0	24,800	0
July	0	286	12	871,000	2,024	0	17,910	0
Aug.	0	346	20	1,511,000	2,908	1,224	17,720	0
Sept.	41	148	21	1,404,000	3,835	1,778	13,490	0
Oct.	297	6	22	1,231,000	4,226	1,608	27,470	0
Nov.	646	0	18	1,284,000	6,451	1,770	30,510	0
Dec.	1,038	0	20	1,262,000	7,336	1,796	42,350	0
Total	5,735	926	237	14,599,000	57,698	21,187	300,490	0

Fig.1 Example Energy Requirements 813395

Without detailed energy consumption data your early estimate of cogeneration economics could be seriously in error. We have found that these variations typically make or break the cogeneration project, and that the hourly and daily variations are often more exaggerated than anticipated.

The most common situation is that both heat and kWH consumption at the industrial complex in question can be reduced significantly. It would be poor management to design and build a cogeneration system to provide energy that is known to be wasted. Even though the energy produced in the cogenerator may be relatively inexpensive, remember that the capital required to build the cogeneration system is almost directly proportional to this waste.

One other frequent situation which would invite agressive conservation is when the amount of waste material is low relative to the energy requirements of the plant. Now reducing the energy requirements permits a relatively larger impact to be made on the bottom line of plant operations.

Energy requirements can be reduced through heat loss reduction, recovery, and conversion. Most plant energy managers are working on the heat loss recovery concepts of heat exchange, waste heat boilers and heat pumps, and reducing heat loss through insulation upgrade or minor process revamp.

Heat pumping is especially attractive where electricity is relatively inexpensive and waste heat is attractively close in quality and quantity to heating requirements. It can also represent an excellent way to reduce the heat to power balance in the plant by cutting thermal energy requirements and raising the plant electrical load.

It is also possible to convert waste heat to electricity through the use of organic and conventional Rankine systems. Electricity, compressed air or other mechanical work can be generated.

These conservation and conversion strategies can have a significant impact on plant energy consumption, improve cogeneration feasibility, and also provide acceptable returns on investment. Where corporations have payback horizons that tolerate 3 and 4 year returns, it is often possible to cut energy use by 20-40%, through a combination of the heat loss reduction, recovery and conversion technologies.

Some Waste-Fired Cogeneration Basics

Most of the wastes considered here would either be used directly in a waste-fired boiler to generate steam or would be gasified to fuel an existing boiler, process heaters (not a cogeneration situation), gas turbines, or diesel engines. As mentioned earlier, three systems are generally considered: boiler, gas turbine, and diesel based systems. Each of these will be described briefly.

Steam boiler based cogeneration systems have been around a long time. Paper makers have used the wastes of wood residue, bark, and black liquor to generate steam at higher than required pressure, expanding the steam through back pressure turbines to generate electricity. Some plants generated so much power through cogeneration and hydro that they become power companies also. The hardware involved in this type of cogeneration system has been around for a long time.

These boiler based systems have only one "undesirable" feature. They generate the lowest amount of electricity relative to thermal output. Generally less than 75 kW can be generated from every million Btu's of fuel burned using back pressure turbines. Even less power than this is usually generated in small plants because of the historical low efficiency of these back pressure devices (35-40% in many instances). While not a cogeneration system, it is possible to generate more power through condensing turbines.

In any event, these boiler based systems are very flexible in the fuel they can utilize. Boilers can be designed to simultaneously burn solid, liquid, and gaseous fuels. Plant chilling requirements in these types of systems are often met using absorption (Lithim-Bromide) machines. Plant space heating and process heating requirements are normally met by condensing the steam in the appropriate heaters.

Figure 2 is a schematic of a hypothetical boiler based system indicating typical strategies. In industrial practices, boiler pressures over 600 psi can be achieved and may be appropriate to improve the relative electrical to thermal split for these systems. The most common situation is that the plant will have to purchase electricity from the utility to meet plant electrical requirements.

Example of Boiler-Based System

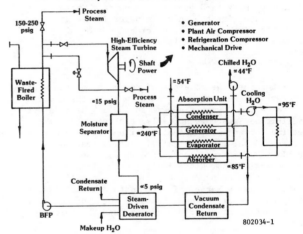

Figure 2

There has been a tremendous interest over the past few years in gas turbine based systems. Aircraft derivative engines are now available that can rival utility electric generation heat rates. We have even heard of hardware designed to run on gasified wood waste. A typical system schematic is shown in Figure 3. While the advantages to a gas turbine are significant, pay special attention to fuel quality. Where the gas turbine manufacturer will warrantee the engine in gasified fuel service, you can rest somewhat easy. The gas turbine driven generator can also be designed to achieve a heat rate better than a diesel through the use of a supplemental steam cycle fired by the exhaust of the gas turbine. The common name for this type of system is a combined cycle power generation system. An example is shown in Figure 4.

Gas-Turbine-Based System Schematic

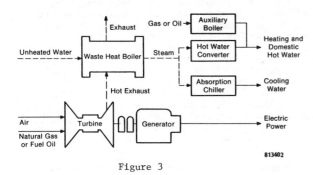

Figure 3

Diesel cogeneration has received a great deal of press in recent years. This is partly due to the fact that they do represent a simple cogeneration system(when the jacket water can be used for heating). Many firms, especially hospitals and universities, already have them for back up power. Where the local utility company charges a major part of the electric bill as a "demand charge," these diesels may also be profitably operated to lower these charges through automatic controls. Figure 5 is a diesel-based system diagram.

Diesel-Based-System Schematic

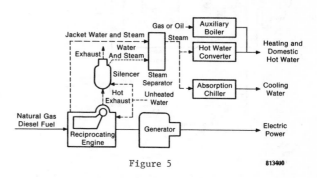

Figure 5

Schematic of Combined-Cycle System with Two-Pressure Steam Cycle

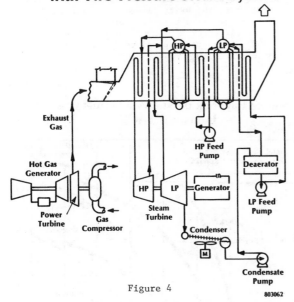

Figure 4

Another advantage of diesels is that they can generate electricity at an efficiency of 30-35% even in the 300-1000 kW range. The basic heat recovery characteristics is that the jacket water and exhaust are each approximately equivalent to the electrical generation level (on a Btu equivalent basis). The exhaust of the diesel can be recovered to produce steam and even supplemental electrical power as shown in Figures 6 and 7.

Diesels have the greatest fuel flexibility of the internal combustion engines. They can be designed to run on almost any liquid fuel, gas, and the products of gasifying coal, wood, or other candidate fuels.They have even been built to run on coal dust. This is not meant to imply that diesels should be considered best engine for utilizing waste fuels. In general, boilers can be designed to burn waste fuels with much higher reliability and fuel flexibility.

General Arrangement of MTI Binary Rankine Cycle System for Waste Heat Recovery/Electric Power Generation

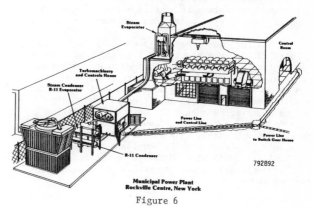

Figure 6

Cycle Schematic for the MTI Binary Demonstration System

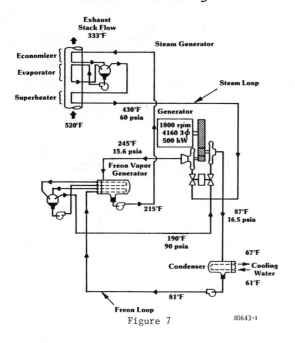

Figure 7

80643-1

The Economics of Waste Fired Cogeneration

MTI recently generated a bill of materials and costing for a 500-shaft HP waste-fired industrial system. A steam turbine condenser was included in the analysis to consider the possible redundancy that may be required to allow the absorption unit to be serviced while the turbine is operating. The data are based on steam turbine throttle conditions of 150 psig and 450°F and an exhaust of 30 psig. The chiller would be capable of delivering approximately 850 tons of refrigeration at these conditons.

Steam Turbine(500 HP)	$100,000
Absorption Unit(850 tons)	212,500
Waste-Fired Boiler(16,000 lb/h)	360,000
Cooling Tower (25 x 10^6 Btu/h)	33,750
Deaerator/Pumps (16,000 lb/h)	16,000
Steam Turbine Condenser(15 x 10^6 Btu/h)	18,750
TOTAL CAPITAL COST	$741,000

Simple paybacks were calculated (See Figure 8) using a conservative assumption that the installed cost for the complete waste-fuel-fired TAC^{TM} system would equal twice the above total capital cost(calculations were based on varying waste fuel and purchased electricity costs). The data reflect the savings using waste fuel at a rate of 3 ton/hour versus nearly 1 MW of purchased electricity for shaft power used for chiller drives and other plant equipment. Assuming installation at twice the equipment cost, the waste-fired system can show attractive returns. If the system were credited with displaced electric motor equipment costs(1000 kW costs about $50,000) and waste disposal costs, the payback would be accelerated further.

How To Change The Steam To Electric Balance

The most common problem with potential cogeneration at a site is that the steam or heat requirements are too low in relation to electrical demands. Part of the cogeneration implementation then is to bring these utilities into a stable and attractive relationship. This does not normally mean that you should use steam turbines instead of electric drives, since the only steam turbine that would qualify economically is a condensing steam turbine where the exhaust energy is wasted. This is not cogeneration. However, it may be economically attractive to run a condensing steam turbine as part of a large cogeneration project, especially as an alternative to venting steam, if that is a frequently encountered operational characteristic.

Process cooling can be accomplished using LiBr steam absorbers rather than electric driven chillers. Sometimes cooling can be supplied through heat pumping and thereby also providing heat elsewhere in the process. As an aside, heat pumping is more frequently attractive where you want to move away from an overly high steam to electric balance. The best way to improve the ratios is always through conservation and heat recovery where possible. In situations where electricity is valuable (over 6¢/kWH) and large quantities of waste heat are available (30 million Btu/hr and above at 250°F or higher), power generation using Rankine cycle machinery may prove economic.

Simple Payback *versus* Waste Fuel Cost, Industrial TAC™

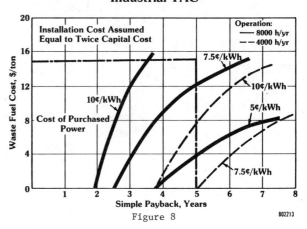

Figure 8

802213

CONCLUSIONS

The opportunities for converstion of combustible waste to valuable energy forms have never been greater. The hardware is available now to convert these wastes to steam, electricity, chilled water, and other plant utility requirements while meeting economic constraints.

Chapter 52

LOW ORDER HEAT RECOVERY WITH WATER SOURCE HEAT PUMP

R. D. Allen

Those of us who have been working in energy conservation are constantly on the lookout for opportunities to use low order heat to save energy. There are many situations where the heat source is available but there is just no economical way to recover and use it. Consequently, when we find a source and a place to use it, we like to share the find. Hence this paper.

This application is a solution to an energy conservation project at the Dworshak Fish Hatchery at Ahsahka, Idaho. The Corps of Engineers and the U.S. Fish and Wild Life Service who maintain and operate the facility became concerned with the rising costs to operate due to excalating cost of fuel oil. There is no natural gas now available there and electrical energy, though plentiful, is becoming more expensive as the grid system mix changes. The hatchery has a need to maintain a constant temperature in the fish rearing ponds which translates into a need for heating during the fall, winter and spring. The facility now uses hot water boilers, both electric and oil-fired, and wastewater heat exchangers as heat sources and as pointed out

earlier, costs are rising.

DESCRIPTION OF THE HATCHERY

Dworshak National Fish Hatchery is located below the Dworshak Dam between the North Fork of the Clearwater River and the Clearwater River in Idaho. (Figure 1) The mechanical features, so far as this paper is concerned, include a main pumping station which pumps water from the North Fork into the hatchery, a water treatment facility for aeration, rearing ponds where the fish are raised, filtration and water treatment areas for the recirculated water from the ponds, and mechanical buildings for the treatment and conditioning of make-up water to the ponds. Figure 2 is a schematic flow diagram of the water used in System 1.

PROCESS OPERATION

1. Existing System Operation

   A. System 1

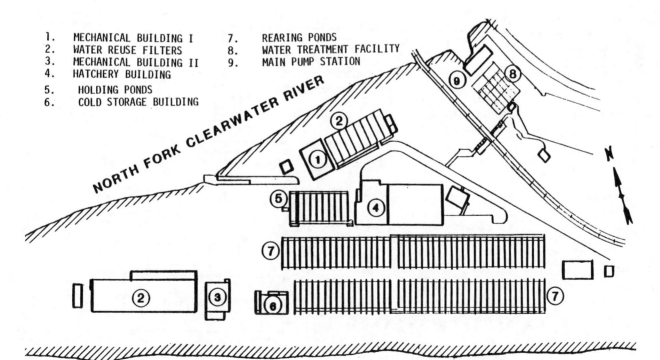

1.	MECHANICAL BUILDING I	7.	REARING PONDS
2.	WATER REUSE FILTERS	8.	WATER TREATMENT FACILITY
3.	MECHANICAL BUILDING II	9.	MAIN PUMP STATION
4.	HATCHERY BUILDING		
5.	HOLDING PONDS		
6.	COLD STORAGE BUILDING		

NORTH FORK CLEARWATER RIVER

CLEARWATER RIVER

FIGURE 1 FACILTY LOCATION PLAN

(1) General - System 1 is the original rearing pond installation at the hatchery. The main components are a 25 pond reuse area; a water treating area for System 1 reuse water; a mechanical equipment building housing filters, sterilizers, boilers, chillers and heat exchangers for treating and conditioning make-up water for System 1, incubator water, and heated water for general plant use; and associated water distribution piping. Each of the 25 rearing ponds is approximately six feet wide by 32 feet long with three feet depth of water. Initial fill is from the North Fork of the Clearwater River. The water is pumped through an aeration facility to the ponds and is recirculated from the ponds to System 1 water treating area. There it is clarified and treated before being returned to the rearing ponds. During normal operation, 15,000 gallons per minute of water is circulated in this system. Ten percent or 1500 gpm is blown down or wasted from the system. Make-up water is provided from the river. All make-up water can be processed in System 1 mechanical equipment building, designated Mechanical 1, where the make-up water can be filtered, sterilized, heated or cooled as required before mixing with the bulk of the reuse water being circulated. Figure 2, Flow Diagram System 1, is a schematic diagram of this system.

(2) Temperature Control - Temperature of the reuse water in the ponds is controlled by the temperature of the reuse make-up water. Reuse make-up water is pumped from the main pumping station to Mechanical 1 where the water is filtered, sterilized and either heated, cooled or left unconditioned for mixing with the reuse water. The original conditioning system provided heat exchangers where the make-up water could be cooled using chilled water from any of four large (nominal capacity - 390 tons each) chillers or heated with hot water from either of two oil-fired boilers

(nominal capacity - 13,400 Btu/hr output). Modification to this system, now being installed, provides two waste heat exchangers whereby the wastewater from System 1 is used to preheat the make-up water. Wastewater is pumped through the exchangers before being discharged to the river. Figure 3, Flow Sheet Mechanical 1, is a schematic flow diagram of this system.

PROCESS WATER HEATING REQUIREMENTS

The desired rearing pond water temperatures determined by the U.S. Fish and Wildlife Service are:

Period	Pond Water Temperature
September through November	54°
December through March	56°
April through June	Ponds Inactive
June through August	52°

During the summer period, June through August, the pond water temperature is approximately the same as the river water temperature, so that little or no conditioning is required. River water temperature can be controlled through selective take off from the dam reservoir.

The temperature of the North Fork of the Clearwater River varies through the fall, winter and spring as shown in the tabulation of average monthly river water temperatures:

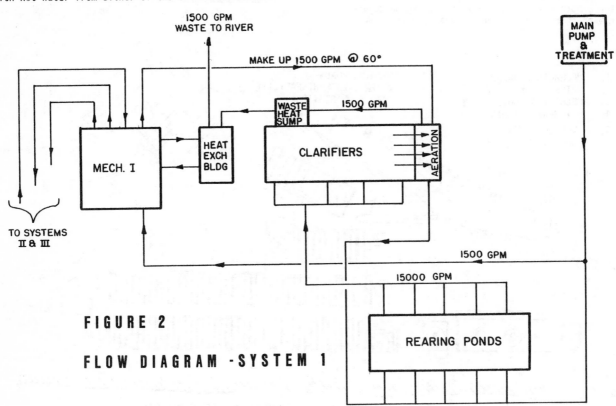

FIGURE 2

FLOW DIAGRAM - SYSTEM 1

Month	Temperature
November	46°
December	41°
January	38°
February	38°
March	39°
April	41°
May	45°

For purposes of this study, the river water temperature for the period September through November was assumed to be 43°F and for the period December through March, to be 38°F.

Since the bulk of the water in the rearing ponds is recirculated, it has been assumed that the temperature of the recirculated water after leaving the water treatment area and before being mixed with the reuse make-up water is 2°F colder than the desire pond water temperature. This assumption is based on a loss of about 1°F during the time the water is in the rearing ponds and about 1°F during processing through the clarifiers or filters and other treatment procedures in the Water Treatment areas.

According to information received from the U.S. Fish and Wildlife Service, operators of the hatchery, 10 percent of the circulating water through the rearing ponds is blown down or wasted to the river and 90 percent of the water is reused after being treated. Consequently, 10 percent of the flow is made up from the North Ford of the Clearwater River. In order for the temperature of the recirculated water to be at the desired pond water level, the reuse make-up water is heated. Accordingly, the temperature of the heated make-up water has been estimated to be:

Period	Temperature
September through November	72°F
December through March	73.8°F
April through June	Ponds Inactive
July through August	52°F

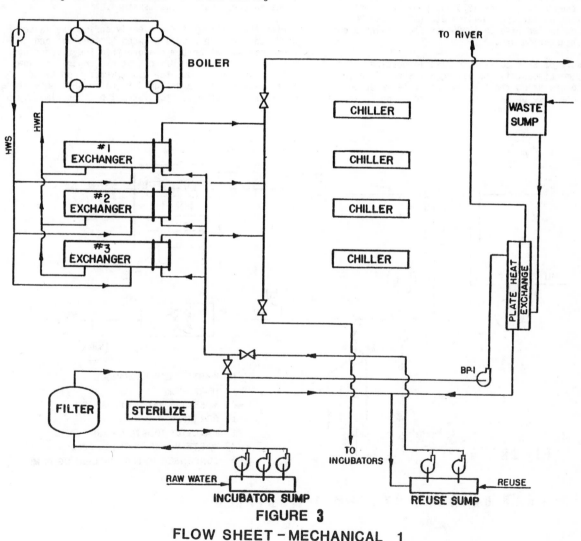

FIGURE 3

FLOW SHEET – MECHANICAL 1

325

The temperature of the pond make-up water may vary due to ambient weather conditions, degree of clarification and type of water treatment in use. Therefore, the means available to control the heat in the make-up must have some flexibility to allow for anticipated temperature of the make-up water ranging from 75°F to 70°F. This flexibility is required principally during the coldest periods in December through March.

PRESENT SYSTEM PROBLEMS

In the original concept of System I, temperature control in the reuse circulating water system was maintained by heating or cooling the make-up water to the system by means of heat exchange equipment located in Mechanical 1. According to information provided by the U. S. Fish and Wildlife Service, there has been no need for chilled water to cool the reuse water for the past several years. Heating, when required, is obtained through the use of hot water, heated in two oil fired hot water boilers which is circulated through heat exchangers raising the temperature of the make-up water as required.

The cost of operating the oil fired boilers on #2 fuel oil has been rising since 1973 and presently costs about eight dollars per million Btu ($8.00 per MMBtu) with an estimated rise to $9.00/per MMBtu being predicted in the near future.

As a result of this continuing rise in heating costs, hatchery management is installing a system of waste-heat recovery using the heat in the waste water to preheat the make-up water before introducing it to the reuse circulating water system. However, the amount of heat to be recovered is far short of that desired, so additional means of providing the supplementary heat without using the oil fired boilers is required or desired.

There is a secondary problem in the use of the waste water as a heat source. This water retains some of the residue from the rearing tanks which creates a potential fouling problem in any heat exchanger used to extract the heat, consequently rate of flow must be reasonably high and flow surfaces smooth to reduce the possibility of fouling the heat exchange surfaces. There is also the probability of bacterial growth in the waste water piping and on heat exchanger surfaces which requires chlorine or other anti-bacteria agents to be used for cleansing the exposed surfaces. Materials of construction of exchangers and other associated equipment such as pumps should be resistant to chlorine and organic attack.

ADDITIONAL WASTE HEAT RECOVERY

So, we decided to investigate the possibility of using So, we decided to investigate the possibility of using existing chillers installed in Mechanical 1 as water source heat pumps. The water source considered is the waste water being discharged from the rearing ponds in these water reuse systems.

During the fall, winter and early spring the waste water temperature ranges between 52° and 56° and the reuse make-up water temperature from the river ranges between 38° and 46°. Waste heat exchangers are now being installed to use the waste water from System I. These exchangers are designed to raise the temperature of reuse make-up water to System I by approximately 8°F.

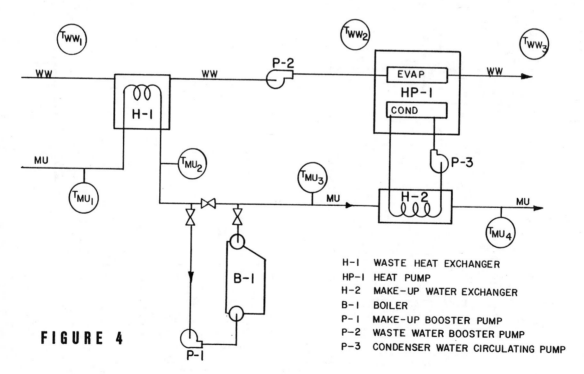

FIGURE 4

H-1   WASTE HEAT EXCHANGER
HP-1 HEAT PUMP
H-2   MAKE-UP WATER EXCHANGER
B-1   BOILER
P-1   MAKE-UP BOOSTER PUMP
P-2   WASTE WATER BOOSTER PUMP
P-3   CONDENSER WATER CIRCULATING PUMP

HEAT BALANCE SYSTEM 1 MODIFIED

The existing chillers in Mechanical 1 have the capacity to extract an additional 8°F of temperature from waste water and transfer that heat to the make-up water, thereby raising the make-up water temperature approximately 8°F. The performance of the chillers as heat pumps is based on the total heat (enthalpy) available in the waste water streams.

The efficiency of the chillers in Mechanical 1 is estimated to be .80 kW/ton which means an expenditure of 2730 Btu of electrical power for every 12,000 BTU of heat extracted or added in the chiller. The efficiency of an electric boiler is approximately 97 percent which means that 3518 Btu must be expended for 3413 Btu added to the system. If oil fired boilers are used, their efficiency is estimated to be 85 percent which means 4015 Btu must be expended to add 3413 Btu to the system.

It is logical to assume, based on the figures shown above, that any heat that can be added to the reuse make-up water by use of heat pumps is more economical than that added through use of an electric boiler. It is also more economical to operate the electric motor driven heat pump than to use the oil fired boiler at the fuel cost rates prevalent at the hutchery.

There are two ways where heat pumps can be used in this system. The first, and most logical one is to heat the reuse make-up water before mixing with the re-circulated reuse water to the ponds. The second way would be to preheat boiler feed make-up water which will reduce the amount of heat required in the hot water boilers. The first option is most attractive because of the volume of water being heated. The volume of boiler feed make-up is quite low and would not utilize the potential heat pump capacity to best advantage. Consequently, we concentrated on the concept of heating the reuse make-up.

The four existing chillers are identical Chrysler Air Temp. centrifugal water chillers, model C2HK668 with a nominal capacity of 390 tons each. They use R-11 (Trichloro monofluoro methane) as a refrigerant. Limiting temperature in the condensers is about 130°F. Limiting temperature in the evaporator/cooler is about 40°F. Flow in the condensers and evaporator tubes is 1440 gpm and 1425 gpm respectively. Design flow rate is 10 feet per second. At these flows the design range of the condenser is 10°F and the evaporator is 9 - 12°F. Tube sizes within the heat exchangers are large enough to accommodate anticipated flow of the waste water and the reuse make-up water at 1500 gpm. There is a manufacturers restraint which requires that the leaving condenser water be at least 25°F above the leaving chilled water temperature.

The existing chillers can be modified for use as water source heat pumps by providing piping changes which permit the flow of waste water through the evaporator or cooler where heat is extracted from the waste water. The reuse make-up water is circulated through the condenser bundle where the heat is transferred to the make-up water. Because of the temperature level of the waste water and make-up water, and the temperature restriction regarding the leaving condenser water, it is necessary to provide a secondary heat transfer loop on the condenser side of the chiller.

Figure 4 is a flow diagram showing the proposed change.

MAKE-UP FLOW #/HR. W_{mu}	RAW WATER TEMP. T_{mu1} °F.	MAKE-UP WATER TEMP. T_{mu2} °F.	DISCH. FROM BOILER T_{mu3} °F.	COLD REUSE WATER TEMP. T_{cr} °F.	COLD REUSE FLOW W_{cr}	REQUIRED POND WATER TEMP. T_p	POND WATER FLOW #/HR. W_p	
750,000	43°	48°	67°	52°	6.75x10⁶	54°	7.5x10⁶	

WITH HEAT PUMP			WITHOUT HEAT PUMP					
ENTHALPY OF HEATED M.U. h_{mu4} BTU/#	TEMP. OF HEATED M.U. T_{mu4} °F.	HEAT REQ. FOR HEATED M.U. Q_{mu3} MMBTU/HR.	ENTHALPY OF HEATED M.U. h_{mu4} BTU/#	TEMP. OF HEATED M.U. T_{mu4} °F.	HEAT REQ. FOR HEATED M.U. Q_{mu4} MMBTU/HR.	WASTE WATER FROM PONDS T_{ww1}	WASTE WATER FROM H-1 T_{ww2}	WASTE WATER FROM HEAT PUMP T_{ww3}
40.08	72.04°	14.23	40.08	72.04°	15.75	54°F.	46°F.	41°F.

SYSTEM 1 HEAT BALANCE VALUES

TABLE 1

The most advantageous use of heat pumps in System I is to provide supplemental heat to the reuse make-up water using residual heat in the System I waste water as a heat source. During normal operation of the reuse water circulating system to the System I rearing ponds, 1500 gallons per minute (gpm) of water is replaced with fresh water taken from the North Fork of the Clearwater River. The waste water is released from the system at temperatures which range from 52°F in the fall months, to 54°F in the winter months. The make-up water ranges in temperature from an average of 43°F in the fall to 38°F in the winter. The waste water passes through two (2) waste water heat exchangers which lower its temperature approximately 8°F, releasing its heat to the reuse make-up water. The waste water is then released to the river and the heated reuse make-up water is routed to the reuse water circulating system. If additional or supplemental heat is required to provide sufficient heat to the reuse make-up water to maintain pond temperature, then the oil fired hot water boilers in Mechanical 1 can be used in conjunction with existing heat exchanger.

We propose to use the capacity of the four chillers now installed in Mechanical 1, as heat pumps, extracting additional heat from the waste water to further heat the reuse make-up water. Figure 4 is a schematic heat balance of the method proposed and Table 1 is a computation of temperatures and other pertinent data which describe conditions affecting the heat balance during average conditions which are estimated to exist during the fall period of the year and during the winter period. Under average temperature conditions, it is estimated that the reuse make-up water temperature should be 72°F at point of injection into the reuse water being recirculated to the ponds in order to produce a pond temperature of 54°F. The heat content of the reuse make-up water at that temperature is 40.08 British thermal units per pound (Btu/lb) or 30.06 MMBtu/hr. Heat available in the reuse make-up water after preheating in the waste heat exchangers and in the heat pumps is 15.83 MMBtu/hr. This condition means that 14.23 MMBtu/hr. must be provided from another source to satisfy the need. During the fall period, there is excess capacity, estimated to be approximately 10.2 MMBtu/hr. available from the electric boiler in Mechanical 2. If this capacity is used then the 4.03 MMBtu/hr. still required could be supplied by the oil fired boilers in Mechanical 1. If the oil fired boilers were not used, then reuse make-up water temperature would have estimated value of 68.2°F and pond water temperature would not be maintained at the desired level of 54°F.

During the winter months, December through March, under average temperature conditions existing during that period, reuse make-up water temperature is estimated to be 73.8°F in order to maintain pond temperature of 56°F. Heat content required in the reuse make-up water 31.5 MMBtu/hr. Heat available in the make-up water after processing through the waste heat exchangers and the heat pumps is 15.72 MMBtu/hr. During the winter months, 15.78 MMBtu/hr. must be supplied from a supplemental source to satisfy the heat requirement. During this period of the year, there is no excess capacity available from the existing electric boilers in Mechanical 2, therefore, all of the supplemental heat would have to be supplied by the oil fired boilers in Mechanical 1.

Use of the chillers at heat pumps, will reduce the supplemental heat requirement provided by the boilers from 18.01 MMBtu/hr. to 14.23 MMBtu/hr. in the fall months and from 19.52 MMBtu/hr. to 15.78 MMBtu/hr. in the winter months. This is an average reduction of approximately 3.75 MMBtu/hr. during the heating seasons.

If the oil fired boilers are used in conjunction with the heat pumps, there is an estimated savings of $28.00/hr. in fuel costs during the fall period compared with the cost of providing all supplemental heat with oil fired boilers alone. If a combination of electric boilers, oil fired boilers and Mechanical 1 heat pumps is used, there is a rise in operating cost of $4.50/hr. over the cost of operating oil fired boilers alone.

Operating cost savings based on fuel or energy cost only, during the winter period, if heat pumps and oil fired boilers are used is $39.80/hr. compared with use of oil fired boilers alone.

Estimated annual savings based on 2184 hours operation during the fall period and 2904 hours during the winter period is $176,876. Estimated implementation costs for change if only oil fired boilers are used is $144,750. Using uniform capital recovery over a period of 20 years with a 7.625 percent interest rate, the annual cost is $14,502 and the benefit/cost ratio is 12.20.

CONCLUSIONS AND RECOMMENDATIONS

The existing Chrysler Air Temp. chillers, installed in Mechanical 1, can be adapted for use as heat pumps. The heat source will be waste water from System I. Heat will be transferred to the reuse make-up water before mixing with reuse water recirculating back to rearing ponds.

There is excess heat pump capacity in Mechanical 1. System I requirements can be met with two of the four existing chillers. Under present operating conditions, use of the chillers as heat pumps is limited by the heat available in the waste water stream from System I.

During the fall heating period, September through November, with the heat pumps operating, approximately 14.25 MMBtu/hr. of boiler capacity is required to heat the reuse make-up water. There is excess capacity in the two electric boilers installed in Mechanical 2 sufficient to provide about 10.0 MMBtu of this requirement. However, based on energy cost criteria established for this report, it not economical to use the electric boilers for this purpose. During the winter heating period, there is no excess boiler capacity available in Mechanical 2 boilers.

There is sufficient oil fired boiler capacity in Mechanical 1 to provide supplementary heat to reuse make-up water in System I, II and III during winter heat period, assuming that all waste heat exchangers in Mechanical 1 and 2 are operational and the heat pump options are implemented.

If heat pump conservation is implemented, it will be necessary to treat the waste water flowing through the evaporator section of the chiller/heat pump with chlorine or other anti-bacterial substance to reduce probability of organic residue build-up in the evaporator bundle.

By using the existing chillers as heat pumps to heat reuse make-up water to System I, energy savings of an estimated 18,500 MMBtu/year can be realized. In System I, the primary energy sources are electrical energy and fuel oil. Estimated annual savings due to reduction in boiler heating demand is $177,000 using the oil fired boilers in Mechanical 1 to provide supplemental heat when required. Implementation costs, using Mechanical 1 oil fired boilers is estimated to be $144,750. Annual cost based on 20 year investment period and a capital cost of 7.625 is $14,502. Benefit/cost ratio is 12.2.

The savings described here are very favorable due in part to the presence of the large chillers. If new heat pumps had been required, the added cost of purchase would have to be included in the implementation costs, reducing the cost benefit ratio. Even so, this situation is made to order for their use as heat pumps as we have the heat source, the heat sinks and a means to transfer the heat economically.

SECTION 15
LIGHTING UTILIZATION

Chapter 53

IES ILLUMINANCE RECOMMENDATIONS AND U.P.D. LIGHTING POWER BUDGET PROCEDURES

W. S. Fisher

PART I.   ILLUMINANCE RECOMMENDATIONS

The Illuminating Engineering Society of North America has been the authority for recommendations on quality and quantity of illumination for many decades. It has sponsored research programs for much of its existence and has tried to place its recommendations on a scientific basis, whenever possible.

The Society has also provided forums for discussion among its broadly-based membership (and outside its membership, as well) so that any recommendations not based in science might be based on consensus agreements.

In 1975, the IES Board of Directors passed a resolution which sought a more flexible method of illuminance specification than the single number target values which had been used for sometime. These numbers· had been based on an averaging of assumptions regarding worker age, eyesight, task difficulty, level of performance, etc. The principal responsibility for proposing a new illuminance system fell on the IES Technical Committee for the Recommendations on Quality and Quantity of Illumination (RQQ).

This proved to be an exceptionally difficult task and it was June of 1979 when the RQQ Committee's work resulted in a proposal for a new system that the IES Board of Directors approved unanimously. RQQ Report #6[1] is now the basic policy of the Society regarding task illuminance recommendations.

Following the adoption of this Report by the IES Board, the technical committees of the Society began the formidable process of changing all the task/area illuminance recommendations over to the new system. After this, the new illuminance values were published in the Sixth Edition of the IES LIGHTING HANDBOOK, available in March 1981.

The new system has ranges of illuminance rather than single number target values as before. The ranges correspond to those which appear in the International Commission on Illumination's (CIE's) Report #29[2]. These values and the type of activity for each illuminance range are summarized in Table I.

In the IES LIGHTING HANDBOOK, Sixth Edition[3], several hundred tasks/areas are classified into one of nine letter categories, A through I, each letter representing a specific range of illuminances as seen in Table I.

It is important to read all the copy of Table I for a full understanding of its implications and how the illuminances should be applied.

The illuminance ranges for each letter category are very broad, two to one, or greater. What illuminance value should one choose within each range? The answer involves raising two or three questions about the people and tasks involved in· each application, and obtaining the correct answer to each question. This will generate two (or three) weighting factors which will determine a specific illuminance value according to Table II. Again, it is vital to read all the descriptive matter surrounding this table so as to apply it correctly.

It is believed by many that this procedure is a better method for keying the illuminance of a job to the people and tasks than any available previously. It puts more responsibility on those who design and those who buy lighting systems than before. This is where the responsibility should lie.

TABLE I — ILLUMINANCE CATEGORIES AND ILLUMINANCE VALUES
GENERIC TYPES OF ACTIVITIES IN INTERIORS

| | Illuminance Category | Ranges of Illuminances | | Reference Work-Plane |
		Lux	Footcandles	
Public spaces with dark surroundings	A	20-30-50	2-3-5	
Simple orientation for short temporary visits	B	50-75-100	5-7.5-10	General lighting throughout spaces
Working spaces where visual tasks are only occasionally performed	C	100-150-200	10-15-20	
Performance of visual tasks of high contrast or large size	D	200-300-500	20-30-50	
Performance of visual tasks of medium contrast or small size	E	500-750-1000	50-75-100	Illuminance on task
Performance of visual tasks of low contrast or very small size	F	1000-1500-2000	100-150-200	
Performance of visual tasks of low contrast and very small size over a prolonged period	G	2000-3000-5000	200-300-500	
Performance of very prolonged and exacting visual tasks	H	5000-7500-10000	500-750-1000	Illuminance on task, obtained by a combination of general and local (supplementary lighting)
Performance of very special visual tasks of extremely low contrast and small size	I	10000-15000-20000	1000-1500-2000	

These target values represent maintained values.

TABLE II   WEIGHTING FACTORS TO BE CONSIDERED IN SELECTING SPECIFIC ILLUMINANCE
WITHIN RANGES OF VALUES FOR EACH CATEGORY

a. For Illuminance Categories A through C

| Room and Occupant Characteristics | Weighting Factor | | |
	-1	0	+1
Occupants ages	Under 40	40-55	Over 55
Room surface reflectances*	Greater than 70 per cent	30 to 70 per cent	Less than 30 per cent

b. For Illuminance Categories D through I

| Task and Worker Characteristics | Weighting Factor | | |
	-1	0	+1
Workers ages	Under 40	40-55	Over 55
Speed and/or accuracy**	Not important	Important	Critical
Reflectance of task background***	Greater than 70 per cent	30 to 70	Less than 30 per cent

*Average weighted surface reflectances, including wall, floor and ceiling reflectances, if they encompass a large portion of the task area or visual surround. For instance, in an elevator lobby, where the ceiling height is 7.6 meters (25 feet), neither the task nor the visual surround encompass the ceiling, so only the floor and wall reflectances would be considered.

**In determining whether speed and/or accuracy is not important, important or critical, the following questions need to be answered: What are the time limitations? How important is it to perform the task rapidly: Will errors produce an unsafe condition or product? Will errors reduce productivity and be costly? For example, in reading for leisure there are no time limitations and it is not important to read rapidly. Errors will not be costly and will not be related to safety. Thus, speed and/or accuracy is not important. If however, prescription notes are to be read by pharmacist, accuracy is critical because errors could produce an unsafe condition and time is important for customer relations.

***The task background is that portion of the task upon which the meaningful visual display is exhibited. For example, on this page the meaningful visual display includes each letter which combines with other letters to form words and phrases. The display medium, or task background, is the paper, which has a reflectance of approximately 85 per cent.

PART II UPD LIGHTING POWER BUDGET PROCEDURE

For the past decade the IES has been in the vanguard of the program developers who seek to save energy in all possible applications:

1. The well-known "12 points....."[4] appearing in 1972 pointed out how to obtain needed illumination with less energy use.

2. The lighting section of ASHRAE 90A-80 (IES/LEM-1P)[5] and its revisions, developed by the IES, has become a model for the energy codes of many states in the construction of new buildings.

3. The lighting sections of the ASHRAE 100 Series documents also point the way to energy conservation by holding up the lighting power limits of ASHRAE 90A-80 (new buildings) as the model for existing buildings.

ASHRAE 90A-80 (IES/LEM-1P) outlined a procedure for computing a lighting power budget for a building which assumed that efficient light sources, luminaires and application techniques would be employed for the illumination. The procedure is simple for anyone familiar with the basic lighting design routine. It provides for the consideration of many special and important aspects of lighting design and application such as veiling reflections, glare, color rendering, the effect of room size and surface reflectance. Such considerations are vitally important to the maintenance of quality in illumination, a factor that is lacking in all-too-many of today's lighting systems -- and in many of the proposals being put forward for limiting lighting power or energy use.

The procedure can also be set up for use with a programmable calculator and magnetic card. In the opinion of many consultants who have used it, the procedure requires only a few additional man hours beyond that required to do the actual lighting design for a building. And its comprehensive basis helps to insure that a building's lighting power budget is neither too generous nor too skimpy.

It must be emphasized that the power budget procedure is not a design procedure. It merely provides an upper limit for a building's lighting power that the actual design may not exceed.

Despite the many good features of the IES power budget procedure (and its simplicity for anyone who knows lighting) there are some who want an even simpler method. For this audience, the IES Energy Management Committee has now provided a Unit Power Density (UPD) Procedure which greatly reduces the computations required, yet maintains in its system as much as possible of the technology of lighting that will allow for good design and maintain quality.

The procedure supplies tables of pre-calculated watts/sq.ft. for various tasks performed in a building or for areas of buildings without specifically defined tasks such as corridors, elevators and toilets. The procedure appears in the LEM-1P publication of the IES and also in the IES LIGHTING HANDBOOK, Sixth Edition, Applications Volume, previously referenced.

The UPD procedure provides a work sheet, Form 1 (Fig. 1) and two tables, A and B, (Figs. 2 and 3). The power budget for each space is determined by multiplying four factors together as follows:

$$P_r = P_b \times A_r \times RF \times TAM$$

Where: P_r = Room Power Budget in watts

P_b = Base unit power density (UPD) from Table A

A_r = Area of room from building plans

RF = Room Factor, from Table B

TAM = Task Area Multiplier, from Form 1

All identical rooms (same tasks and areas) can be totalled on the same line of Form 1. The procedure continues for each individual room type until all those included in the building have been done.

While the rote described above can be followed by anyone capable of doing arithmetic, a brief discussion of what is involved in each step of the process is desirable for a better understanding of the procedure.

The base UPD is a watts per square foot value appearing in Table A for a specific task or area. It is the unit power necessary to satisfy the lighting requirements for a particular task, assuming that the power is utilized effectively in a large, unobstructed room with recommended reflectances on room surfaces and furnishings.

Task areas are determined by identifying the number of work stations at which workers perform tasks. At each work station, for lighting power determination purposes, a task area of 4.6 m^2 (50 square feet) is assumed. If the task area at a work station is greater, the actual task area should be used.

In rooms where there is more than one type of task or area, the base UPD is determined by taking the weighted average of the individual task/area UPD's.

The Room Factor (RF) is a multiplier varying between 1.00 and 2.00 which adjusts the base UPD for rooms of various dimensions. Small rooms utilize light much less efficiently than large ones and the 2:1 range for the Room Factor is considerably less than that experienced in practice.

The Task Area Multiplier (TAM) is a factor ranging between 0.4 and 1.0 which adjusts the room power downward when the area assigned to visual tasks (A_t) is less than 50% of the total room area. In the majority of well-utilized space, half, or more, of the total area is task area (4.6 m^2 or 50 sq.ft. assigned to each work station) and the TAM is 1.0. Multipliers of less than one may be encountered in large executive offices having a generous allotment of space, in lobbies where there may be an isolated work station, or in spaces that are underutilized.

The following problem will illustrate some of the principles involved:

Space: 40' x 60' with 9' ceiling height

Tasks: 12 people doing conventional drafting

         4 people doing computer-aided drafting with CRT's

From Table A (Fig. 2)

Base UPD for conventional drafting: 4.7 w/sq.ft.

Base UPD for computer drafting: 1.7 w/sq.ft.

Weighted average UPD:

$$4.7 \times \frac{12}{16} + 1.7 \times \frac{4}{16} = 3.95 \text{ w/sq.ft.}$$

$$P_r = P_b \times A_r \times RF \times TAM$$

$$P_r = 3.95 \times 2400 \times 1.05 \times 0.75$$

$$P_r = 7466 \text{ watts}$$

The Room Factor (RF) above came from Table B (Fig. 3) and the Task Area Multiplier (TAM) came from Form 1 (Fig. 1). Figure 1 shows how Form 1 should be filled out for this problem.

Note on line "f" of Form 1 that, for an existing building, the allowable power budget is 1.2 times that for an existing building. Owners and operators of existing buildings are encouraged to have their lighting power conform to that of new buildings. However, it is recognized that existing buildings may have older lighting equipment that is less efficient than currently available and is impossible to retrofit. While it is recognized that capital investment in new lighting technology can have attractive paybacks, industrial and commercial business may have alternative investments for their capital with even greater return on investment; or, in some years, there may be no funds available for capital improvements or investments.

If a particular task/area is not listed in Table A, then it will be necessary to use the full procedure.

Values of lighting power obtained with the UPD procedure compare very well with those obtained by using the original lighting power budget procedure. So, the UPD procedure is now available as an alternate to the originals for guidance in design energy-conserving lighting systems.

REFERENCES

1. "Selection of Illuminance Values for Interior Lighting Design (RQQ Report #6)", A report of the Committee on Recommendations for Quality and Quantity of Illumination of the IES (RQQ), Journal of the IES, April 1980, pp. 188-190.

2. "Guide on Interior Lighting", published by The International Commission on Illumination (CIE), Report #29 (TC-4.1), Bureau Central De La CIE, 4, Av. DuRecteur-Poincare, 75782 Paris CEDEX 16, France.

3. IES Lighting Handbook, Sixth Edition. Published by the Illuminating Engineering Society of North America, 1981. 345 East 47th Street, New York, N. Y. 10017.

4. Kaufman, John E., "Optimizing the Uses of Energy for Lighting", Lighting Design and Application, October 1973, pp. 8-11.

5. ASHRAE 90A-80 (IES /LEM-1P), "Energy Conservation in New Building Design". Cosponsored by the American Society of Heating, Refrigerating and Air Conditioning Engineers, Inc. and the Illuminating Engineering Society of North America.

LIGHTING POWER BUDGET

PART 1A – BUILDING INTERIORS/EXTERIORS

PROJECT __AAE DEMONSTRATION PROBLEM__

UPD

PAGE ____

BY ____

1

A	B	C	D	E	F	G	H·G/E	I	J= C·E·F·I	K	L= J·K	M
		P_b	ROOM DIMEN. (FT.)	A_r	RF	THESE STEPS SHALL BE USED ONLY WHEN $A_t < 50\% A_r$			P_r	NO. OF IDEN. RMS.	POWER BUDGET FOR ALL IDENTICAL ROOMS (W)	N O T E
RM. NO.	ROOM IDENTITY AND/OR TASK DESCRIPTION	BASE UPD (W/SF)	L x W LUM. H.	ROOM AREA (S/F)	ROOM FACT-OR	NO. OF WORK STA. X 50 SF	% OF TASK AREA	% / TAM: 40/.85, 30/.70, 20/.55, 10/.40	ROOM POWER BUDGET (W)			
1	OFFICE 2 TYPES DRAFTING	WT. AV. 3.95	40x60 9'	2400	1.05	800	33%	0.75	7466	1	7466	
—	DRAFTING –CONVENTIONAL	4.7	40x60 9'		—		→	—	— — —	→	— — —	
—	DRAFTING –COMPUTER	1.7	40x60 9'		—		—	—	— — —		— — —	
	REMAINING SPACES WHICH DO NOT HAVE SPECIFIC VISUAL TASKS											
a	SUB-TOTAL (THIS PAGE)											
b	SUB-TOTAL (PREVIOUS PAGES)											
N O T E S					c	SUPPLEMENTARY LIGHTING: 3% OF (a+b)						
					d	FACADE LIGHTING: 2% OF (a+b)						
					e	EXTERIOR OR SPECIAL: (SUBMIT SUPPLEMENTARY)						
					f	TOTAL	NEW BLDG: (a+b+c+d+e)					
					g	PROJECT	EXIST. BLDG: 1.2 x f					
					h	BUILDING UPD, W/SF						
j	GROSS BLDG. AREA ($\sum$ E·K)				i	PROJECT UPD, W/SF						

UPD-4

FIGURE 1. This is Form 1 of the UPD on which the basic data for the project is entered and computations made, resulting in a lighting power budget for each room. Results for individual rooms are totalled along with figures for supplementary, facade and exterior lighting to obtain the building total.

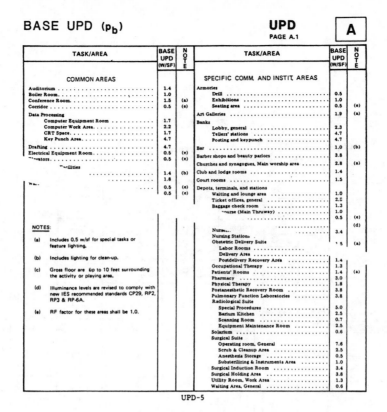

UPD-5

FIGURE 2. Table A of the UPD provides the base power densities for specific tasks and areas that will, when efficiently applied, result in illumination that is appropriate in both quality and quantity for the application.

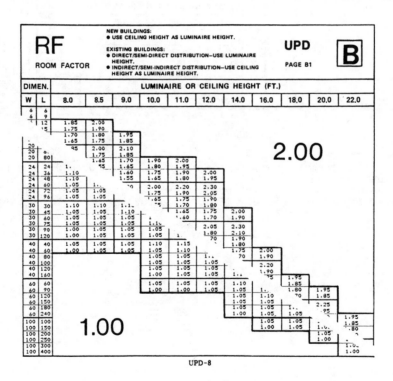

UPD-8

FIGURE 3. Table B of the UPD provides a Room Factor (RF) for modifying the base UPD obtained from Table A. This takes account of the effect of room size on the utilization of light.

338

Chapter 54

LIGHTING CONTROL: APPLICATIONS AND DIRECTIONS

C. P. Farnham

OVERVIEW

With the advent of a new lighting ethic - utilizing light where and when it is required, the proper application of lighting controls becomes of paramount importance in the energy management program. Recent studies have shown that upwards of 30% of the total lighting kilowatt hours can be saved through the proper application of controls.

Control techniques have been with us since the first lamp - in the form of on-off switching. Stage lighting requirements began the utilization of dimmers, primarily of the resistance type, in the 1920's. Fluorescent dimming systems came into some use during the 1950's with the development of the rapid start circuitry. In the 1960's, the solid state incandescent "wave chopper" began to be marketed. This device, which can be easily installed in a wall switch box, popularized low cost dimming in many residences.

The 1970's brought a tremendous explosion in solid state technology, and with it wide and varied innovations in lighting control.

The very recent emphasis in energy control and management has flooded the market with lighting control products. Today, users are demanding an effective systemized approach to controlling lighting levels, coordination with natural daylight and burning hours, along with other features.

TYPES OF CONTROLS

Generally, control systems available today can be broken into two catagories: centralized controls and localized controls. Centralized control systems can include a centralized computer through a simple time clock, with many variations. The only requirement is that it control a large area, from a single floor to an entire building. The localized system can control a single small office up to a part of a floor, i.e. the perimeter. To further cloud the issue, a centralized system can be utilized to activate local controls. The matrix (1) below outlines the functions required, with the type of control designed to perform that function or functions. For obvious reasons, no attempt is made to differentiate the actual technology utilized by manufacturers. See Figure 1.

PLANNING THE CONTROL SYSTEM

Application

The first step in planning an effective control system is to zone each area, and develop control options. Some general possibilities are listed below.

1. Industrial

   a. Photo electric sensors

   b. Time of day; clocks, micro-processors

   c. Ambient light control

   d. Selective on/off

   e. Remote control

   f. Individualized on/off

2. Offices/Schools/Institutions

   a. Photo electric sensors

   b. Time of day; see above

   c. Selective on/off

   d. People sensors

   e. Remote control

   f. Individual on/off

3. Commercial (Retail)

   a. Photo electric sensors

   b. Time of day; see above

   c. Selective on/off

   d. People sensors (warehouse)

Typical Plan

The plan view of a typical "open plan" office, with varied tasks, is shown below with control recommendations (2). See Figure 2, 3.

Zone 1 - Perimeter (15'): This area can utilize natural daylighting to augment the existing lighting. Based on nominal requirements (70 FTC), daylight can provide from 35% to 50% of the total lighting requirements.

LIGHTING CONTROLS

Type of Control	Total Building Control	Localized Control	Programmable	On/Off	Manual Override	Telephone Control	Time of Day	Holiday/Weekend Schedule	Dimming	Co-ordination with Daylight	Remote Control	Co-ordination with HVAC	Over-the-wire Control	Low Voltage
Centralized Controls:														
Computer Operated	X	X	X	X	X	X	X	X	X	X	X	X	X	X
Over-the-wire Controller Clock	X	X	X	X	X		X	X			X		X	X
Programmable Control	X	X	X	X	X		X	X			X	X		X
Localized Controls:														
Daylight Control (dimmable)[1]		X		X					X	X				X
Daylight Control (non-dimmable)		X		X						X	X			X
In Fixture Daylighting Control		X		X					X	X	X			
Solid State Dimming (no dimming ballast)		X	X	X					X	X			X	
People Sensors		X		X	X									X
Ambient Light Co-ordinators[2]	X	X	X						X	X			X	
Timer Control	X	X		X	X		X						X	

[1] Requires dimming ballast in each luminaire
[2] Compensates for light loss factors
* Many functions can be added, check with manufacturer

Figure 1 - Lighting Controls Matrix

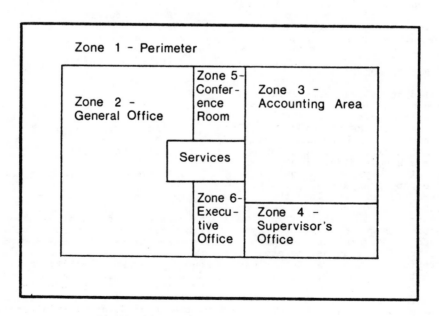

Figure 2 - Typical "Open Plan" Office

Zone 2 - General Office (typing pool, etc.):
Open office space. Task lighting to 75 FTC.
Time of day control, since working hours are
fairly well set. Recommend on/off time control, with additional feature of adjusting
the ambient lighting level to the 50% to 33.3%
level for off line periods and janitorial services.

Zone 3 - Accounting Area: Made up of small
cubicles - 6' high partitions. Task lighting
level - 100 to 125 FTC. Recommend time of
day control, with option for localized lighting for off hours work. This would require
sub-zones within the zone. Options could be:
localized switching, telephone control, or
timer switches. The off hour cleaning, maintenance feature should be retained here.

Zone 4 - Supervisor's Offices: Small offices,
general lighting level - 75 FTC. Recommend
individual controls - people sensors, automated on/off devices. Lights on only when
space is being utilized.

Zone 5 - Conference Room: Varying lighting
level, to serve many varied tasks. Recommend
both on and off capability, i.e. automated
switches with dimming controls to vary lighting levels.

Zone 6 - Executive Office: Working lighting
level - 75 FTC, with emphasis on aesthetics,
subzoned for both working and aesthetic system. Recommend people sensors, for working
area, wall switching for aesthetic zone.

Service Area: Minimal lighting. Recommend
timer switching, to insure no lighting is left
on. Not cost effective for people sensors.

Summary: By analyzing and zoning your building, adequate illumination levels can be maintained, while energy consumption is reduced.

Based on average usage as spelled out in our
example, total energy savings should be within the following range. See Figure 3.

Notes: Estimated reductions are based on a
nominal annual use of 3,200 hours plus 600
hours of off time use - cleaning, maintenance
and after hours work.

Adequate lighting for security and safety purposes should be maintained at all times.

EFFECTIVENESS ANALYSIS

An analysis of various control functions
should include some measure of their effectiveness, as it applies to a particular operation. A graphical analysis of the three control functions utilized in the above example
points out the time of day savings.

Time of Day Controls

Time of day savings utilizes either a time
clock or micro-processor to turn the lighting
on at 7:30 a.m., off at 5:00 p.m. and on for
one hour of cleaning. Additional lights may
be turned off during the noon hours. See
Figure 4.

Daylighting

The coordination of natural daylight and artificial (man-made) lighting can result in savings from 35% to 50% on the perimeter 15'.
Several types of low cost Photo Electric cell
controlled systems are available. They are
listed below.

1. Dimmable controllers - controlling a maximum of 8 lamps.

2. Dimmable controllers - controlling an entire circuit - wiring must be designed so
   that the perimeter lighting is on one circuit.

3. Non-dimmable controller - utilizes relays
   within individual luminaires. Does not require any wiring changes.

The chart below shows that the maximum effect
from daylighting occurs from 10:00 a.m. to
6:00 p.m. (3). However, this will vary by
latitude and building orientation. Other factors which will affect daylighting economics
are listed below. See Figure 5.

1. Room dimensions

2. Wall and ceiling reflectivity

3. Glass area - % of room wall

4. Transmittance - % of light transmission

5. Window light control - blinds, louvers,
   drapes, etc.

6. Ground reflectance - material, grass, concrete, etc.

7. Overhang - length of overhang of window

Motion Detectors

Smaller office areas, such as Zone 4 and Zone
6, should utilize individualized controls.
Motion detectors have proven to be most effective in this situation. Manufacturers tests
show that 40% savings are not exceptional (4).
Note that the effective area coverage is fairly small, 300 to 500 square feet, thus a large
area should utilize several such detectors or
other means.

Timer Switches

Service areas are generally little used, and
exotic sensors cannot be justified economically. A simple inexpensive timer switch will
suffice in these areas.

Total Control Effect and Analysis

The graph below depicts the effect of the entire control network. In large buildings,
utilizing H.V.A.C. controls, a centralized
system may be justified, whereas in smaller
areas individual systems are more cost effective. See Figure 6.

Zone	Type of Control Action	Estimated Reduction (%)	% of Total Lighting	Estimated % of Energy Savings
1	Perimeter - Co-ordina-tion with natural daylight	35%	41.7%	14.6%
2	Timer control, central-ized, to co-ordinate with working hours	25%	22.1%	5.5%
3	Timer control, central-ized, to co-ordinate with working hours, localized control	25%	22.9%	5.7%
4	Localized control, dimming	15%	1.5%	₂2%
5	People sensors	45%	6.3%	2.8%
6	People sensors	45%	4.7%	2.1%
	Service area	50%	.6%	.3%
	Total Estimated Percent (%) of Savings			31.2%

Figure 3 - Composite Control Effect

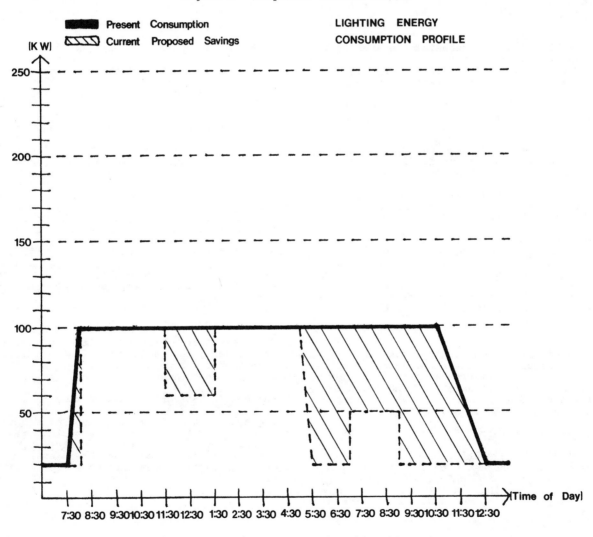

Figure 4 - Time of Day Switching Effect

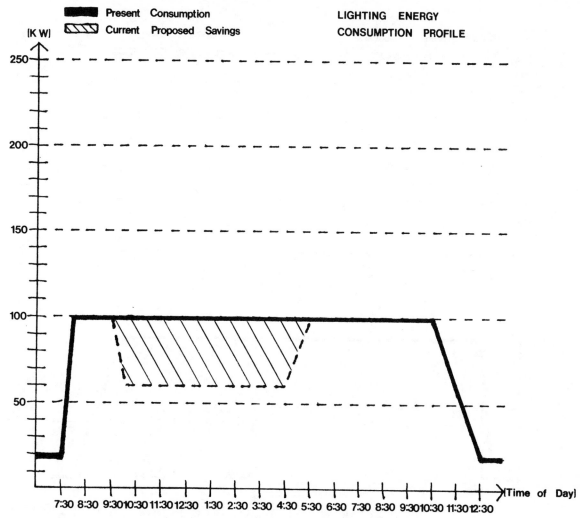

Figure 5 - Daylighting Control Effect

SUMMARY

The planned application of lighting controls offers the user the most effective low cost method of tailoring their lighting system to individual employee requirements. The control system can be as complex or as simple as desired, and can possibly be installed utilizing existing wiring. A few advantages of a planned lighting control system are shown below.

1. Maximum energy savings while maintaining adequate task illumination.

2. Flexibility to meet variances in working hours and locations.

3. Coordination with other building systems.

4. Maximum use of natural daylight.

5. Minimum lighting for cleaning and other maintenance functions.

6. Manual control of little used areas, janitor closets, mechanical rooms, and maintenance areas, with provisions for turning off after use.

7. Coordination with the H.V.A.C. system, thereby augmenting the heating system, and not overloading the air conditioning.

8. Maintains existing lighting geometry and aesthetics.

REFERENCES

(1) "Lighting Energy Management", 1981, Lighting Technology, Inc., Bellevue, WA.

(2) IBID

(3) "Automatic Lighting Control For Energy Conservation", #EC-803, B.C. Hydro and Power Authority, Vancouver, B.C.

(4) "Light-o-Matic Sales and Installtion Manual", Novitas, Inc., 1982.

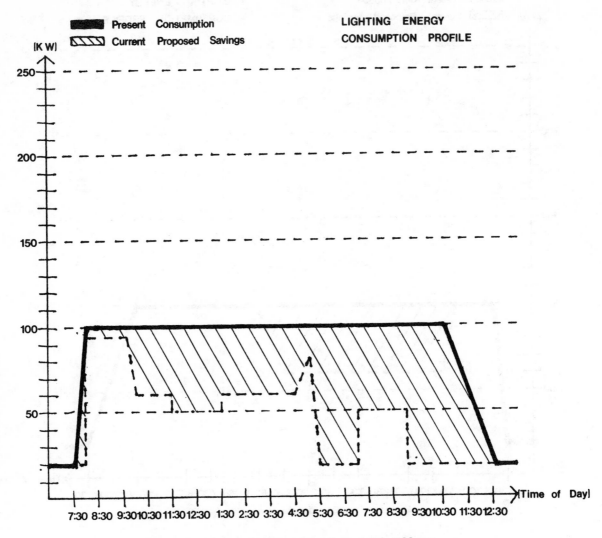

Figure 6 - Composite Result of Control Effect
Including Daylighting, Time of Day and People Sensors

SECTION 16
RESIDENTIAL ENERGY AUDITS

Chapter 55

A COST EFFECTIVE ENERGY STRATEGY
FOR NEW HOUSING

M. LaRue

INTRODUCTION

The optimum thermal design of a structure minimizes the sum cost of the structure and the energy required to heat and cool it. This best design is achieved by individually optimizing each of the thermal elements of the structure. These thermal elements include the components of the envelope (such as windows, walls and ceilings), air infiltration, and the efficiencies of the heating and cooling equipment.

The optimization procedure compares the value of the savings resulting from a contemplated improvement with the cost of that improvement. It then accepts those improvements whose value is greater than cost. Thus, for example, a thicker wall would be selected if its incremental cost were $.25 per square foot and the present value of its future energy savings were $.30 per square foot.

With few exceptions it is possible to optimize the elements of a residence on a unit area basis. Since the dimensions of the residence need not be known, the thermal design can be defined while the structure is still in the concept stage.

The value of a contemplated improvement depends on a number of factors, among which are fuel and electricity prices, the cost of money, the thermal characteristics of the element, heating degree-days, cooling hours, and the future owner's tax bracket. When passive solar inputs are possible, there are the additional factors of orientation, latitude and climate.

In the case of custom housing, where the residence is being designed and built to order, the techniques presented here are of even greater value. The thermal elements can be selected with the future owner in attendance, with the thermal design being tailored on the spot to his estimates of future energy prices, his tax bracket, and his mortgage terms.

THE PRESENT VALUE OF FUTURE SAVINGS

The value of the savings from a thermal improvement is found from Equation 1:

1) $$\text{Value} = \frac{S(1 - (1 + i)^{-n})}{i} \text{ , where}$$

S = savings realized each time period,
n = number of time periods,
and i = interest rate per period, expressed as a decimal.

The Present Value After Taxes

It is often convenient to compare the value of a thermal improvement with the after-tax value of a tax-able alternative investment. Since the savings from a thermal improvement are not taxable, their value for comparison purposes is:

2) Value (compared to a taxable investment) $= \dfrac{100(\text{value})}{100 - \text{tax bracket \%}}$

THE DECISION METHOD

"S" in Equation 1 is also the periodic payment which amortizes an equal-payment principal-plus-interest loan. A thermal improvement will be profitable when its periodic savings are greater than the amortization payments, and this desirable situation occurs whenever value is greater than cost.

Cost effectiveness decisions are made as follows:

a. Determine the periodic net savings which will be realized from a contemplated improvement.

b. Note the length of time over which the savings will be realized.

c. Use Equation 1, and Equation 2 if applicable, to find the present value of the periodic savings.

d. Note the net cost of the improvement (after subtracting tax credits or other incentives).

e. Include the improved element in the design whenever value is greater than cost.

f. Seek a further improvement, and repeat the procedure until value no longer exceeds cost.

THE THERMAL DESIGN PROCEDURE

The thermal design of a new dwelling in climates where heating is the greater expense begins with optimizing its orientation, glazing and solar shading for winter solar gains while minimizing summer gains. Then each of the conductive elements is examined in turn, using the above decision method to maximize cost effectiveness. Infiltration is examined next, then the efficiencies of the heating and cooling equipment. Obvious losses are corrected along the way, such as those from ducts and pipes carrying conditioned air or fluids through unconditioned space.

ANNUAL WINTER SOLAR GAINS

Solar Considerations

Direct gain passive solar techniques which involve only structure orientation, glazing and shading are relatively inexpensive and thus highly cost effective.

In cooler climates, where the predominant use of energy is for space heating, the long axis of the dwelling should run east and west. One of the long sides of the structure then faces southerly, and it is on this face, of course, that the majority of the glazing is placed. Properly sized and placed overhead projections reduce or eliminate solar entry in the summer while allowing maximum entry in winter. Proper shading is aided by avoiding tall windows. The projections should shadow the upper half of the windows at spring and fall equinoxes. Additional window area may be left unshaded in cooler climates, since there is less danger of appreciable overheating in early autumn.

The resulting reduction in annual heat loss from just these few things can be considerable, and can easily reduce the heating bill of a midwest home by 20%.

Estimating the Annual Solar Gain Through Glazing

The net annual direct solar gain through glazing depends on the glazing direction, number of glazing layers, latitude, and the modifying effects of the local climate. Basic solar data is available from the U.S. Department of Commerce. Better data for the purpose may be available from a neighboring university. An example of such local data is that contained in "Technical Note 14" published by the Small Homes Council of the University of Illinois. Such data can be used to derive "Solar R Values" which simplify the estimation of solar gains.

Basic heat loss equations can be manipulated to obtain the following relationship:

$$3) \quad R = \frac{(Degree\text{-}days)(24)}{\substack{Annual\ heat\ gain\ (or\ loss) \\ per\ hour\ per\ square\ foot\ of\ glazing}}$$

Equation 3 allows "Solar R Values" to be derived when the net annual BTU gain (or loss) through glazing is known. These R values are then used in place of conventional R values to estimate the annual gain or loss through southerly windows. In Equation 3, a net gain is represented as a negative quantity.

The values in Table I were derived from University of Illinois data.

TABLE I

SOLAR R VALUES FOR SOUTHERLY GLAZING IN THE CHICAGO AREA
(Appropriate shading above the glazing)

Glazing layers	Southeast	South	Southwest
1	1.7	3.8	1.7
2	5.7	-8.0	5.7
3	-7.0	-3.0	-7.0

Negative solar R values indicate a net annual solar gain. The values in Table I are not for use in sizing heating or air conditioning equipment.

OPTIMIZING THE STRUCTURE SHELL CONDUCTIVE ELEMENTS

The conductive elements of the shell include

    Exterior walls
    Windows
    Exterior doors
    Exterior ceilings
    Exterior floors
    Basement walls
    On-grade slabs
    Chimney masonry

The average monthly cost of the heat loss through one square foot of a conductive shell element is

$$4) \quad \$ = \frac{(Annual\ degree\text{-}days)(\$\ per\ therm)}{(50,000)(e)(R)}, \quad \text{where}$$

R = "R" value of the building section, and
e = heating plant efficiency, as a decimal.

From the above it may be shown that the monthly savings per square foot from an increase in the section R value is

$$5) \quad \$ = \frac{(Annual\ degree\text{-}days)(\$/therm)(1/R_1 - 1/R_2)}{(50,000)(e)}$$

where $1/R_1$ and $1/R_2$ are the present and contemplated R values, respectively.

In Equations 4 and 5, it is suggested that the efficiency of a normal furnace be assumed, say in the range of .55 to .65. The cost effectiveness of higher efficiencies is explored after the shell has been optimized.

Optimizing a Conductive Element

The method of optimizing a conductive element is as follows:

a. Consider an alternate construction of the element which has a higher R value.

b. Determine the monthly savings per square foot of the element if the alternate construction were used, from Equation 5.

c. Calculate the value of the savings from Equation 1, and Equation 2 if applicable.

d. Compare value with the incremental cost. If value is greater than cost repeat steps a through d.

e. Incorporate the construction with the highest "R" whose value exceeds cost.

This sequence continues to add improvements whenever their return on investment exceeds that selected in Equation 1.

Improvements are usually possible only in fairly large discrete jumps, such as with alternate wall sections or the number of glazing layers. An exception is with blown ceiling insulation where any desired depth is possible within the limitations of the roof design. In this case, the depth may be explored inch by inch until the value of the last inch of thickness just equals its cost, thereby squeezing out the last possible penny of profit. Such a computation is best made with a programmable calculator.

Basement Walls

Studies made by the Dow Chemical Company indicate that 20% of the heat loss in a conventional two story home is through the walls of a full basement. A full depth concrete wall typically averages R = 5, with the upper four feet averaging R = 2.5. These values include the adjacent earth.

Foundation insulation will generally be found to be cost effective in the central midwest and northerly. Placing the insulation on the outside of the wall (provided it is suitable for underground use) is preferable, since this adds the foundation wall to the thermal mass of the structure. Cost effectiveness is explored as for other conductive elements, using the initial R values given above.

On Grade Floor Slabs

The reduction of losses from floor slab edges is explored after the initial R value is found, then proceeding as with other conductive elements.

Chimney Masonry Losses

Conduction losses through chimney masonry are usually overlooked, perhaps because of the difficulty of determining them. A typical interior masonry fireplace structure, with a five by seven feet cross-section, will conduct about two million BTU annually from a Chicago area residence. An exterior chimney will easily lose twice this much. This is in addition to losses from loosely fitting dampers and from building a fire in the fireplace. While the dollar value of these losses is not large at today's fuel prices, they do cause discomfort and complaints of "cold masonry".

Some inventiveness is needed in the construction of fireplaces and chimneys. One considerable improvement for interior structures is to create a thermal stop in the portion of the masonry which passes through the upper ceiling and its insulation. The thermal stop can be several courses of a low-conduction ceramic material such as sintered glass. Cost effectiveness of alternate constructions is explored for the masonry as a unit. Calculate the total heat loss before and after the contemplated change, find the fuel savings, then apply Equation 1 and Equation 2 if applicable.

INFILTRATION

Infiltration is usually the single greatest loss from a dwelling, and particularly so from a single family home. One air change per hour in a typical two story home causes about 38% of the total loss, and most homes have about this amount of infiltration. It can be reduced considerably, and is as low as .05 air changes per hour in some recently built Canadian homes. However, this can result in high interior humidities and danger of air contaminants build up, with air-to-air heat exchangers then being necessary to supply fresh air without a commensurate heat loss. The objective is to raise the ventilation rate to about .3 air changes per hour while using fuel to reheat perhaps .1 air changes per hour.

Infiltration can be measured (though at considerable cost) in an existing home, and can only be estimated in the planning stages. Our experience has shown that incorporating items (a) through (f) from the list below reduces air infiltration to about .5 air changes per hour, and that these additions are cost effective in climates as warm as 5000 degree-days.

The following infiltration improvements are given in the approximate order of their cost effectiveness;

a. Adequate vapor barrier
b. Vapor transmitting infiltration barrier
c. Outside combustion air intakes
d. Low infiltration windows
e. Sill sealer
f. Well installed weatherstripping
g. Caulking
h. Vestibule entrances
i. Weatherstripped and insulated attic access
j. Gasketted electrical boxes

The savings resulting from infiltration improvements can be found from Equation 6.

6) $$\text{Monthly savings} = \frac{(V)(.018)(ac1 - ac2)(\text{degree-days})}{50,000\,(e)},$$

where V = volume of the structure in cubic feet
ac1 = existing air changes per hour
ac2 = contemplated air changes per hour
e = furnace seasonal efficiency

AIR CONDITIONING

Most thermal improvements cause additional savings when the structure is air conditioned, and may make further improvements cost effective. The use of NAHB data can simplify estimates of air conditioning savings. The NAHB, in their publications "Thermal Performance Guidelines", uses annual cooling hours instead of annual cooling degree-days. Since cooling hours represent the annual full load operating hours of the cooling equipment, it is not necessary to make separate computations of sensible and latent heat.

Table II shows the annual cooling hours for certain cities, abstracted from the NAHB "Thermal Performance Guidelines", 1978.

TABLE II

ANNUAL COOLING HOURS FOR CERTAIN CITIES

CITY	COOLING HOURS
Anchorage, Alaska	40
Phoenix, Arizona	2010
Los Angeles, California	530
Denver, Colorado	750
Miami, Florida	3250
Atlanta, Georgia	1320
Honolulu, Hawaii	3950
Chicago, Illinois (S)	790
Des Moines, Iowa	810
Topeka, Kansas	1030
New Orleans, Louisiana	2090
Boston, Massachusetts	660
Detroit, Michigan	710
Minneapolis, Minnesota	640
Kansas City, Missouri	1180
Omaha, Nebraska	860
Las Vegas, Nevada	1760
Albuquerque, New Mexico	1120
New York, New York	850
Charlotte, North Carolina	1250
Portland, Oregon	340
Philadelphia, Pennsylvania	920
Nashville, Tennessee	1210
Houston, Texas	2060
Richmond, Virginia	1090

Equation 7 may be used to find the monthly savings in air conditioning (per square foot of a conducting element) from a thermal improvement.

7) $$\$ = \frac{(\text{Cooling hours})(\$/\text{KWHr})(1/R_1 - 1/R_2)}{900\,(\text{COP})},$$

where

COP = coefficient of performance of the air conditioning equipment.

These are the monthly savings averaged over the entire year, as are those in previous equations.

Equation 7 applies to heat loss elements whose thermal characteristics are described by R values. Appreciable air conditioning savings, however, do not result from on-grade or below grade thermal improvements such as foundation wall insulation.

Air conditioning savings from reductions in infiltration can be most conveniently estimated by a proportion method. Select some conductive element (such as walls) for which both heating and air conditioning savings have been calculated, then apply Equation 8.

8)
$$\text{Air conditioning infiltration savings} = \text{Heating infiltration savings} \times \frac{\text{Conductive AC savings}}{\text{Conductive heat savings}}$$

EFFICIENCY OF HEATING AND COOLING EQUIPMENT

After the total annual fuel and electricity usage has been found, as will be discussed in the following, the value of increases in heating and cooling efficiencies can be explored. The monthly average savings from an improved furnace are,

9)
$$\text{Monthly savings, improved furnace} \quad \$ = \frac{\text{Annual heating cost}}{12}(1 - e1/e2),$$

where e1 = present furnace seasonal efficiency and e2 = contemplated efficiency.

Similarly, the monthly savings from improved air conditioning equipment are,

10)
$$\$ = \frac{\text{Annual cooling cost}}{12}(1 - COP1/COP2)$$

Heat Pump

A heat pump may be explored by computing its annual cost of electricity with Equation 11, and subtracting this from the present heating cost to find the annual savings.

11)
$$\text{Annual heat pump electrical costs} = \frac{\text{Annual therms (e)(\$/KWHr)}}{.03415\ (COP)}$$

As before, find the value of the savings by applying Equation 1, and Equation 2 if applicable.

If a suitable water supply should be available, a water-to-air heat pump might be found to be desirable because of its appreciably higher coefficient of performance. While the system cost may be higher, particularly if a second well is needed, tax credits may reduce the net cost of the system. Additional savings will result if the system is also used for cooling or domestic hot water production.

Caution is indicated with any heat pump to assure that it has adequate capacity. Undersized systems will require more costly supplementary resistance heating during severe weather.

REMARKS ABOUT THE BOTTOM LINE

Figure 1 illustates the overall value versus cost relationship as more and more thermal improvements are added to a structure. The more promising improvements are added first, in the neighborhood of point A. These improvements add much more to value than they do to total cost, and the curve rises more rapidly than the "value equals cost" line.

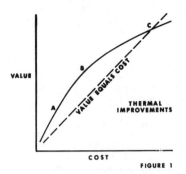

FIGURE 1

As good, but less valuable, improvements are added, the slope of the curve decreases until at point B the added cost equals the added value. Past point B, successive improvements cost more than they are worth. The curve becomes less steep, and finally meets the "value equals cost" line at point C. Further improvement causes the total value to become less than the total cost.

The greatest net return on investment is obtained at point B. It is at this point where our thermal design strategy refuses further investments and thus maximizes return on investment.

THE PERFORMANCE OF THE STRUCTURE

The annual cost of heating and cooling can be found when the final form of the structure is known. Figure 2 is a convenient calculation sheet for the purpose. Figure 2 also shows the characteristics of a large two story residence as it was originally intended to be built.

RENCON • 22 INDIAN TRAIL ROAD BARRINGTON, ILLINOIS 60010

BUILDING ELEMENT	NOTES	"R"	A/P/V	UNIT LOSS Winter	Summer
Wall type A		19.0	2903	2074	2074
Wall type B					
Wall type G	Adj. to garage	17.1	140	8.2	8.2
Ceiling type A		19.0	2226	117.2	117.2
Ceiling type B		11.0	40	3.6	3.6
Ceiling type C		19.0	633	33.3	33.3
Doors type A		3.0	57	19.0	19.0
Doors type B		1.8	27	15.0	15.0
Glazing (1) N E W (2)		1.0 / 1.8	32 / 272	32.0 / 151.1	32.0 / 151.1
Glazing (1) S ( ) (2)		3.8 (1) / -2.0 (1.8)	11 / 11 / 85 / 85	2.9 / -10.6	11.0 / 47.2
Glazing ( ) S (2) (shaded) ( )		1.8	40	22.2	22.2
Basement wall		5.0	1880	376.0	—
Slab			P		
Floor	Over passage	19.0	A 275	14.5	14.5
Masonry				50.0	50.0
Infiltration	(1.0 ac/hr)(.018)		V 48.3K	869.4	869.4
Total Unit Loss. Winter (WTUL), summer (STUL)				1911.2	1601.1

Heating: $\frac{(1911.2\ WTUL)(24)(6530\ DD)}{(100{,}000)(.55e)}$ = 5446 therms at ($.40/therm) = $ 2178.35

Cooling: $\frac{(1601.1\ STUL)(790\ CH)}{(75)(2.1\ COP)}$ = 8030 KWHr at ($.076/KWHr) = $ 610.31

TOTAL ANNUAL HEATING AND COOLING = $ 2788.66

L40-44 1281

ESTIMATED ANNUAL COST OF HEATING AND COOLING

Client: Noyer

Structure: Lot 16 Farm Trails

Location: Lake Barrington

Notes: Per original design

Estimated by: MAL

Date: 5/22/81

RENCON Residential Energy Conservation

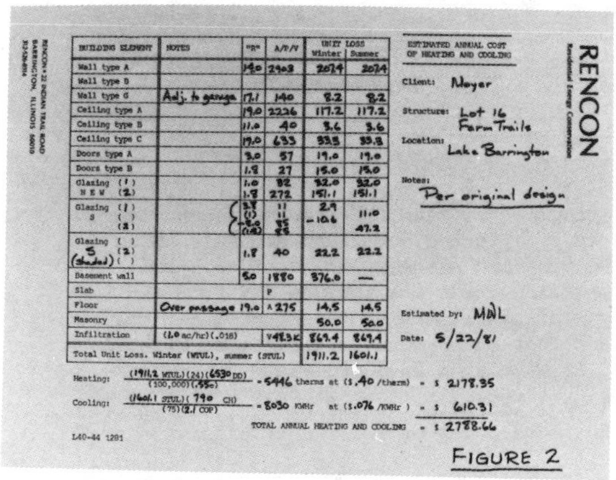

FIGURE 2

The described thermal strategy was then applied, using the following parameters specified by the client:

Natural gas at $.75 per therm
Electricity at $.10 per kilowatt-hour
16% interest rate
25 year mortgage
45% owner tax bracket

Changes in orientation and glazing were specifically not permitted.

Figure 3 shows the characteristics of the same residence after optimization. The additional construction cost was approximately $4500. A comparison of Figures 2 and 3 shows annual savings of $1547 at the then existing energy prices, a tax free return on investment of 34%. The return under design conditions would be 61%!

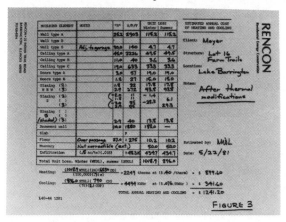

The actual fuel consumption of this home has been measured over one heating season, and was within 5% of the estimate in Figure 3 despite initial difficulties with defective windows. The home, shown in Figure 4, has 4600 square feet of living area plus 1900 square feet of heated basement. Its natural gas consumption is 2290 therms annually (6500 degree-days), or .35 therms annually per square foot of heated floor area.

FIGURE 4

It is interesting to compare this with a somewhat smaller home to be constructed. This home will take full advantage of direct gain solar inputs. A model of it is shown in Figures 5 and 6. It has 2900 square feet of

FIGURE 5

FIGURE 6

living area plus 1600 square feet of heated basement. Natural gas consumption will be 756 therms annually, or .17 therms per square foot of heated floor area. This prediction does not include the beneficial effects of internal gains or window coverings. Occupancy will materially increase its performance. Without solar inputs, its natural gas consumption would be 1116 therms annually, even with triple glazing.

CONCLUSION

It is evident that there is always some least cost for a given structure and the energy required to heat and cool it. The energy strategy which has been presented here simplifies the process of closely approaching that least cost. The principles used are those of life cycle costing, but simplified for easier application to the specific problem of minimizing the total cost of home ownership.

In causing the structure to dissipate less energy, it should be noted that it is also made more comfortable by reducing internal temperature differences.

Solar gains are usually ignored in the design of new housing. Perhaps the availability of solar assisted designs has been delayed by the difficulty of accurately predicting solar performance. If so, the "solar R values" method which was described may help to hasten the development of workable solar-assisted housing. Similarly, the estimating method which was shown simplifies the prediction of heating and cooling costs while in the design stage, and thus enables rapid comparisons between alternate designs.

While it was not emphasized, it should be noted that the elements of the structure were purposely optimized in a specific sequence. In particular, the last elements to be considered were the heating and cooling equipment. This is of importance. If the equipment **is** considered earlier in the sequence, much of the financial incentive to reduce the thermal losses of the structure can be lost, and the lowest minimum cost is not then achieved. It is much more effective to reduce the need for fuel first, and then to look for greater efficiency in burning what fuel must still be used.

Chapter 56

MASS-SAVE: AN AGGRESSIVE APPROACH TO REGULATED RCS PROGRAMS

D. Parisi

Introduction

Each electric and gas utility company that is covered by
Title II of the National Energy Conservation Policy Act
(NECPA) of 1978 has had to face the issue of how to meet
NECPA's mandates. With regulatory signals from DOE
being unclear and signals from the Reagan administration
being clearly non-supportive, many utilities have taken
a wait-and-see attitude toward implementation of a
program.

Mass-Save, Inc. was formed as an aggressive response to
both NECPA and a state law, Massachusetts C.465, by the
state's covered utilities. The aggressive approach has
resulted in a highly successful visible program which
can perhaps provide substantial benefits to the member
utilities in addition to the conservation actions effected by the program.

This paper is a description of Mass-Save as it functions
in a complex organizational and regulatory climate, as
well as a description of some of its special programs
that are manifestations of its aggressive approach to
performing a residential conservation service (RCS)
program.

Genesis of Mass-Save

Mass-Save, Inc. was chartered in the Commonwealth of
Massachusetts on September 4, 1980, for the purpose of
delivering the Residential Conservation Services man-
dated by the National Energy Conservation Policy Act of
November 1978 and by Chapter 465 of the Massachusetts
Acts of 1980.

Mass-Save, Inc. came to evolve into its present form as
a result of a combination of legislation, regulation,
innovation, and good management. Mass-Save is a private, non-profit corporation with 48 of the state's 59
electric and gas utilities and municipal light departments as participants. It is the largest utility
consortium formed to date to deliver the RCS program.

As an initial response to NECPA, 11 of the state-
regulated utilities formed a pilot program, known as
SAVE* (Save All Valuable Energy), that ran from September 1979 to June 1980. Based on the SAVE* experience,
but incorporating the regulations from Chapter 465,
Mass-Save, Inc. was established with the state's
investor-owned utilities as members with municipal
light departments participating as contract utilities
(Figure 1).

Structure of Mass-Save

The concept of a utility consortium to deliver the mandated RCS program began to evolve in Massachusetts in
1978. The primary motivating factors were cost economies that were anticipated to accrue from a consolidated
management effort, and the ability to deliver a uniform
product in all service territories of the Massachusetts
utilities.

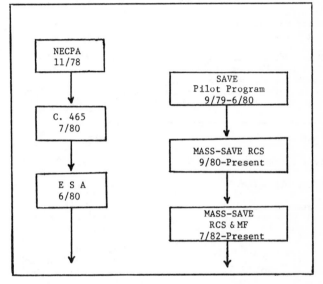

FIGURE 1. GENESIS OF MASS-SAVE

Another motivating factor was the advent of Chapter
465. The Massachusetts state law both mirrors and goes
beyond NECPA. It requires that all Massachusetts
utilities, including municipal light departments, provide RCS services to their customers. That provision
brought many of these small municipal light departments
into the Mass-Save fold because of the magnitude of the
task of providing a full RCS program and the threat of
the imposition of economic penalties by the state.
Many municipal light departments remain upset that the
state brought them under its regulatory arm. Nevertheless, they have chosen Mass-Save as a cost-effective
way of providing the mandated service.

Throughout the formative period, however, an unusual
degree of cooperation was engendered. Given the dis-
parity of the various entities involved--large vs.
small utilities, gas vs. electric utilities, regulated
vs. unregulated untilities, regulatory agencies vs.
regulated entities--the extent of progress and the
degree of commitment to getting Mass-Save started
within a few months of the promulgation of the enabling
Massachusetts legislation is an outstanding achievement
by the Massachusetts utilities.

Developing an implementation plan that was responsive
to NECPA and Chapter 465 and yet satisfactory from a
policy perspective to 48 unique utility companies was
no mean feat. The continued successful operation of
Mass-Save as the largest utility consortium nearly two
years and 100,000 audits later is a tribute to the
cooperative yet aggressive posture of the members of
Mass-Save toward delivering the RCS program.

A Board of Directors represents the utility members. The 15-member Board is comprised of five members of the gas utilities, 5 members of the electric utilities, and five representatives of the municipal light departments.

Reporting to the Board of Directors is the Executive Management Committee, an 11-member utility body including five Board members and six non-Board members, whose function is to interact directly with the management team of Mass-Save and make recommendations to the Board on policy and budgetary issues.

An early decision of the Board was that there would be no employees of Mass-Save. That is, all functions would be contracted, including overall managemeent of the program. Through a competitive bid process, management of the Mass-Save program was awarded to Peter Merrill Associates, a small Boston-based consulting firm.

Peter Merrill Associates, under the guidance of the Executive Management Committee and Board of Directors, formulated the delivery structure of Mass-Save and proceeded with a complex bidding process to secure vendors for each of the services listed in Figure 2.

Contractor	Function
Volt Energy Systems, Inc.	Audit Delivery
DMC Energy, Inc.	Audit Delivery
Strategic Information and Energyworks, Inc.	Single Family Audit Algorithm and Computer Processing
Strategic Information and Xenergy, Inc.	Multi-Family Audit Algorithm and Computer Processing
Harold Cabot & Company, Inc.	Advertising
Costello, Sullivan & Hammer	Legal
Arthur Andersen & Company	Accounting
Peter Merrill Associates, Inc.	Program Management

FIGURE 2. MAJOR CONTRACTORS FY 82

Between September 1980 and March 1981, all vendors were chosen, six offices were established, 200 auditors were trained, and audits were being delivered to an eligible customer base that represents approximately 95% of the entire state geography. As of September 1982, over 100,000 residential energy audits (buildings with four units and under) have been delivered.

Regulation of Mass-Save

The relationships diagrammed in Figure 3 indicate the complexity of the regulatory process to which Mass-Save must respond. Mass-Save's position is such that if any of the relationships become adversarial it can have its functional life squeezed out of it. That same position, however, allows it to be a buffer and a facilitator between the utilities and the various regulatory bodies around them. To date, the regulatory climate has been primarily cooperative. Mass-Save and its member utilities have tended to stay abreast of the regulatory process in its activities and therefore are able to contribute actual experience to the formula-

tion of regulations. As a result, the regulations have tended to be functional and thus encourage compliance on the part of Mass-Save.

The key features of this relationship are:

- The provision in Chapter 465 allowing utilities forward cost recovery based on a budget approved in advance by the DPU. This feature is vital to the conduct of an aggressive yet cost-effective program, and eliminates apprehension and uncertainty on the part of member utilities.

- A state law which goes beyond NECPA in committing all utilities to the same quality of service. Chapter 465 is indicative of a strong state commitment to an aggressive program. It also required a multifamily audit service to be provided by January 1982.

- Goals specified in the state regulations required that utilities provide seven percent of their eligible customer base annually with some sort of RCS service. Four percent must be class A audits and three percent may be other services such as arranging for contracting or financing.

- A cooperative attitude between both the regulatory bodies and the utilities. While the issues to be resolved have been complex, the process of resolving them has been consistent, fair, and cooperative.

Audit Delivery

A graph of audit uptake vs. delivery (Figure 4) shows that Mass-Save experienced an overwhelming initial response to its program announcement, outstripping its ability to train, certify, and field adequate numbers of auditors to meet the demand. At its peak, requests of over 2,000 audits per week were being received and waits of three to four months were required for some customers to be served. Over 200 auditors, performing nearly 2,000 audits per week, remained active through the summer and fall, achieving a production of 60,000 audits in our first calendar year, including a three-month start-up phase. Mass-Save should perform its 100,000th audit by September 1982, representing about seven percent of the total eligible customer base, making it one of the largest single RCS programs in the country.

Audit delivery costs for a large scale program fluctuate significantly with audit delivery rates, as Figure 5 demonstrates. In the months of June and July 1981, when audit delivery was at its peak rate, per audit costs dipped to about $90. In contrast, when backlog declined and audit delivery rates slowed in November and December 1981, per audit costs climbed to nearly $130 per audit (Figure 5).

There are several contributing factors to the increased costs, but the major one is that the marketing and administration expenses are fairly fixed and sizeable in a program operating out of six offices. Hence, the effect of audit delivery rates on unit costs is significant. The net result is that "more is cheaper," which argues for increasing marketing efforts. The difficulty is that there is precious little marketing history for the product and a multiplicity of influences that makes the point of diminishing returns on audit uptake for dollars invested in marketing efforts impossible to determine. So we are continuing to "feel our way" trying to keep per audit marketing costs low, and the rate of uptake stable.

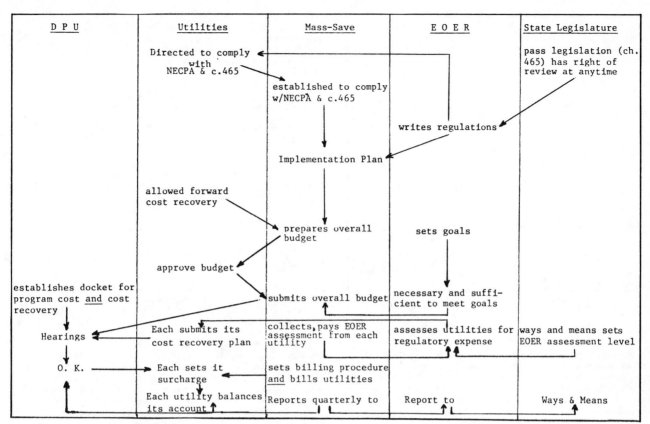

FIGURE 3. REGULATION OF MASS-SAVE

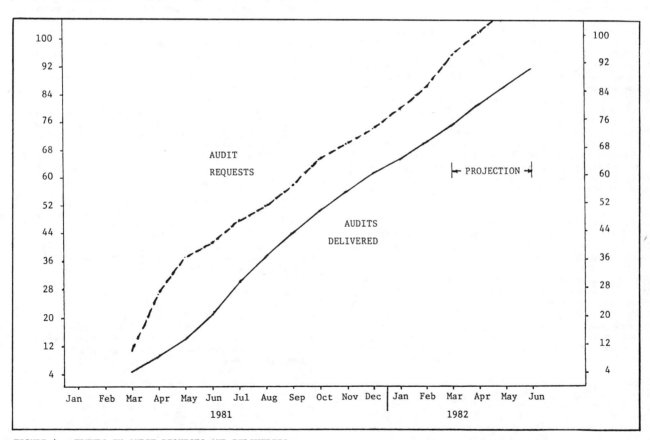

FIGURE 4. TRENDS IN AUDIT REQUESTS AND DELIVERIES

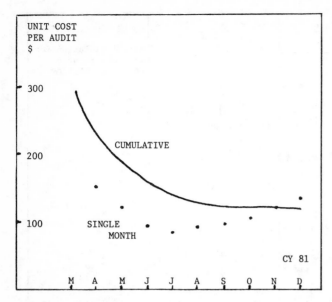

UNIT COST
PER AUDIT
$

300

CUMULATIVE

200

100   SINGLE
       MONTH

                                    CY 81

M   A   M   J   J   A   S   O   N   D

FIGURE 5. TRENDS IN UNIT AUDIT COSTS

Foreign Language Program

In the first several months of Mass-Save's operation,
it became apparent that low income and ethnic minority
populations were not adequately represented in the
population of audited customers. Several ethnic minority population loci were easily identified. They
included substantive numbers of Portuguese-speaking
residents of Fall River and New Bedford, Massachusetts,
and concentrations of Spanish-speaking residents in
Boston and Springfield, as well as others.

Fall River was selected as the area for a pilot program
because over two-thirds of the population of Fall River
is of Portuguese descent (approximately 60,000) and
many of those are non-English-speaking (approximately
35,000). Further, most of the non-English-speaking
community are lower income apartment dwellers, to whom
Mass-Save felt it could provide some special education
concerning energy conservation.

The problems to be overcome by a utility mounting an
effort to reach a low income, ethnic minority community
are manifold, especially if the customers are non-
English-speaking. Mass-Save chose to work through a
lead agency to reach its target audience. An agency
with established credibility, a proven track record,
and a widely read publication was approached, and it
agreed to work with Mass-Save. We had learned that our
name, Mass-Save, and our product, an energy audit, were
not recognized or understood in the Portuguese community. The agency became the key through which to reach
the Portuguese community.

The Portuguese Program objectives were threefold: to
offer the Mass-Save Energy Audit to Portuguese-speaking
people in their native tongue, to conduct an aggressive
effort to reach a low income population which would
normally be only marginally receptive to a utility-
sponsored energy audit program, and to take advantage
of the publicity attending the program to boost Mass-
Save visibility and overall audit requests.

We modified the product by translating all audit materials and promotional materials into Portuguese, paying
attention even to the correct dialect of the local
population. Since we already had Portuguese-speaking
auditors and schedulers staffing the Regional Office we
did not need to do any special training of staff.

From the Mass-Save and utility perspective, the program has been a success. Over 1,000 Portuguese program audits have been done and more are scheduled.
The program has been expanded to reach the Portuguese
populations in New Bedford and, to a lesser extent,
in Cambridge.

Publicity surrounding the program was good and media
coverage consistently laudatory. Response from consumer groups and state legislators was supportive,
and it appeared that general audit uptake may have
also improved.

Attempts to initiate a Spanish program have been less
successful. Although materials have been translated
and Spanish-speaking staff are available, an agency
comparable to the one described in the Portuguese
program has not yet been found. Since Mass-Save as
an entity has little or no proven credibility in the
Spanish-speaking community in Boston or Springfield,
it may remain hamstrung until such an agency is found.

Community-Save Programs

Mass-Save has several efforts under way designed to
provide a direct linkage between audits and weather-
ization work done. One such effort is the Community-
Save program.

The process is based on a Mass-Save audit in which the
auditor fills out a detailed work order for attic
insulation and ventilation. The work orders are
batched by the community and contractors are allowed
to bid on groups of 10 houses at a time. The program
takes advantage of the contractors' normally slow
season by stimulating demand when the contractor most
needs work. Contractors bid in competition with each
other but invest nothing in marketing other than the
bid preparation itself. The work is localized, which
saves the contractor time and travel.

The participating community awards the package of
houses to the lowest qualified bidder and inspects the
work to insure compliance with DOE standards when the
job is done and before the homeowner makes payment.

The program is successful and growing. Three such
communities participated in Community-Save programs in
1981, and ten or more communities will participate in
1982. As of May 1982, about 500 homes have had attic
insulation installed under the program and another
500 to 1,000 are anticipated for the remainder of 1982.

Elementary School Program

As part of its ongoing general energy awareness program, Mass-Save conducted an energy education program
for fourth grade students in 15 elementary schools in
Fall River. Taking advantage of an already established and politically neutral character, the DOE energy
ant, Mass-Save conducted its version, "Annie the
Energy Ant" with DOE's blessing.

From the perspective of generation of audit uptake,
the program was somewhat of a disappointment. It was
difficult to document audits generated as a result of
Annie's efforts. From the perspective of the fourth
grade students of Fall River, as well as from Mass-
Save's public relations perspective, however, the
program was an enormous success.

Annie, a Mass-Save promotional staff member, along
with a local state representative, visited each of
the 15 schools to give presentations on energy conservation, hand out Energy Ant membership cards,
award Energy Ant tee shirts to Energy Essay Contest
winners, and tape student interviews on energy

356

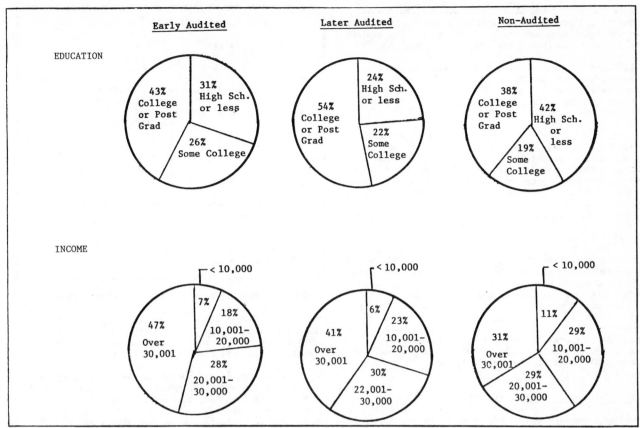

FIGURE 6. DEMOGRAPHIC CHARACTERISTICS OF CUSTOMERS: EDUCATION AND INCOME

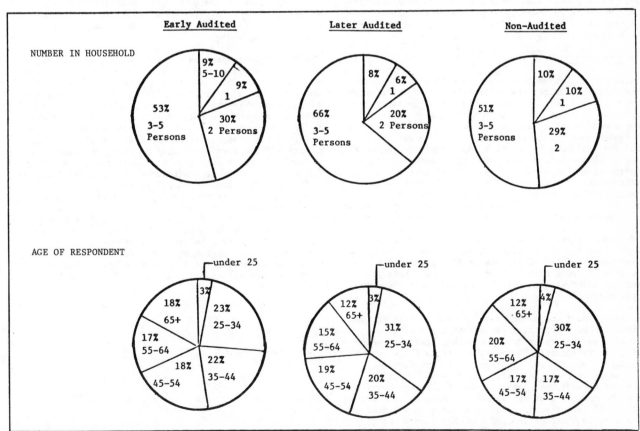

FIGURE 7. DEMOGRAPHIC CHARACTERISTICS OF CUSTOMERS; NUMBER IN HOUSEHOLD, AGE OF RESPONDENT

conservation for the local radio station that helped
promote the program. The station sponsored free rides
at the local amusement park for all Energy Ant membership card holders when the program was completed.

The interest, publicity, and good will generated by
"Annie," as well as the conservation message brought
home by the fourth graders, made the substantial
effort to produce the Energy Ant program well worth
the investment.

In addition to the three programs highlighted here,
Mass-Save is conducting literally dozens of special
efforts with communities across Massachusetts and
coordinating our audit program with community development block grant programs, housing authorities, home
care corporations, community action agencies, and many
other programs. Approximately 10,000 special program
audits have been performed to date.

Program Impact

In order to assess the impact of the Mass-Save program,
Mass-Save conducted two customer surveys, one in June
1981 and one in December 1981. A total of 500 residents was surveyed in June and 600 in December. The
second survey was designed in part to follow up on the
earlier one by including 200 residents chosen at random from the earlier group. A research firm, Decision
Research Corporation, was retained to conduct the
second survey, which included 200 earlier survey residents, 200 later audited residents, and 200 who had not
yet had an audit.

Profile of Audited Customers

Audited Mass-Save customers tended to have a higher
proportion of highly educated individuals and of persons earning more than $30,000 annually. While eligi-
ble customers included apartment dwellers in buildings
of four units and under, 99% of audited customers were
homeowners (Figures 6 & 7).

Conservation Actions Taken

There is a relationship between the measures taken and
length of time since the audit. Early audited customers spent 17% more on conservation than the later
audited group and 40% more than the unaudited group.
The amount spent by all groups for the time period represented seems quite significant, the average for all
audited customers being about $650 spent over an aver-
age of 6½ months and about $500 for non-audited cus-
tomers spent "over the last year." (Interestingly,
information from other sources in our program seems to
indicate that contractors are not experiencing an
increase in conservation business. The do-it-yourself
suppliers, however, appear to be benefiting somewhat.)

Homeowners were very positive about their Mass-Save
audit. Nearly all customers found it "very useful" or
"somewhat useful." This data is also borne out by the
results of audit satisfaction cards mailed back from
audited customers, the results of which are shown in
Figure 8.

An anomalous issue arose from the data that is of some
significance. Approximately 85% of the audited customers credited the audit with either "greatly influencing" or influencing "somewhat" their decision to take
conservation actions. Customers also indicated a
strong perception that energy conservation measures
pay for themselves. Yet in spite of a strong commitment to action, a strong belief in the economic advisa-
bility of taking action, and an apparent commitment of
funds on the part of audited customers, only one in

Were you satisfied with your audit?	
● yes	97%
● no	3%

How would you rate your auditor?	
● excellent)	
● good)	97%
● fair)	
● poor)	3%

Do you plan to implement the recommended measures?	
● yes	94%
● no	6%

FIGURE 8. RESULTS FROM AUDIT SATISFACTION CARDS

ten expected to take out a loan to complete weatheriza-
tion activity. Our experience with several banks who
have attempted to market energy conservation loan
packages bears out the veracity of the data.

While we do not fully understand what is causing the
barrier to financing, we think it is a combination of
the existing high home improvement rates and an inherent resistance to, or fear of, the process of
financing (or of obtaining a contractor, for that
matter). To that end, Mass-Save is developing a
financing booklet and worksheets designed to be used
by the auditor with the homeowner, as an enticement to
additional action on major investment items through
the financing route.

The Mass-Save program as a regulated RCS program represents a marvelous opportunity for the utility companies of Massachusetts. The issues of cost economies
and consistency of program delivery have been substan-
tiated by nearly two years of successful cost-effective operation. The relationship to regulatory bodies
is well orchestrated and uniform in spite of its complexity, resulting in a cooperative regulatory climate
toward RCS programs. Our audited customers felt very
strongly in our survey that the service was a valuable
one and appropriate to be offered by the utilities,
indicating the significant public relations value to
Massachusetts utility companies. There is a message
here that grows louder as time passes and more audits
are done: in spite of the past history of development
of the RCS program, there are some substantial benefits that can accrue to utilities who conduct an
aggressive conservation program. Improved public
image, more cooperative regulatory climate, and improved community relations may even be the least of
these.

Chapter 57

AN RCS AUDIT FOR A HAND-HELD PROGRAMMABLE CALCULATOR

M. D. Beanland

Abstract

An RCS Class A residential energy audit methodology has been designed, developed, and implemented by the City of Palo Alto Utilities
Department and Northern California Power Agency
member utilities. The methodology is designed
for use on hand-held programmable calculators.
The Palo Alto version has been accepted by the
Department of Energy (DOE) and the California
Energy Commission (CEC). The method's low cost
of implementation and its flexibility are ideal
for a small municipal utility or community
based organization. The procedure comprises a
repertoire of programs that can be used in
almost any sequence or by-passed if not applicable. The procedure provides the homeowner
with accurate information about energy and cost
savings. The repertoire structure allows for
easy updating of program as parameters change
or analysis techniques improve.

Background

The City of Palo Alto, when faced with RCS,
chose to develop its own plan rather than fall
under the State of California RCS plan. This
commitment to the development of a plan included the need to create a field audit methodology. The City had exposure to central-compu-
terized energy audits but found the lack of documentation, the difficulty of field use, and
the high costs of use were not to their liking.

At this time, energy surveys were being done
using the HP-67 programmble calculator. There
were problems extant, keeping track of where
you were in the program or correcting entry
mistakes for example, but these did not seem to
be insurmountable. The City decided to develop
a Class A RCS audit for the HP-67. The calculator's low cost and multiple uses more than
offset the lack of hard copy output and limited
memory.

The City of Palo Alto developed the necessary
software and firmware to implement a Class A
RCS audit. Partial funding of the development
was supplied by the California Municipal
Utilities Association (CMUA) under a contract
with the California Energy Commission (CEC)
using federal funds. Subsequently, the
Northern California Power Agency (NCPA) contracted with the author to expand the audit
software to allow its use by the member utilities. The expansion refined many calculations,
added a few, and made the audit more generic.
NCPA standard ized on the more powerful HP-41CV
calculator. The resultant audit is being used
by six NCPA cities in northern and central
California. The users feel very comfortable

with the audit and many have developed utility
and climate-specific improvements. This flexibility and ease of modification are major benefits of an audit based on a programmable calculator. The remainder of this paper gives an
overview of the audit software developed.

Input/Output Format

The two major drawbacks of a programmable calculator over a computer are limited memory and
lack of hard copy. The limited memory requires
the development of modular software. The lack
of hard copy requires the development of a
foolproof input/output process.

The I/O drawback was overcome by the use of
numerical prompting. In this technique a
numerical prompt (1,2,3...) appears in the calculator display. This prompt is keyed to a
data sheet which the auditor has used to collect relevant data about fuel use or the house
being audited. The calculator prompts the auditor for the next sequential piece of data.
The data sheets were developed to act as worksheet, permanent record, calculator input form,
and results presentation form. Palo Alto uses
a two-part form when analyzing an energy saving
measure: the top sheet is designed for the
auditor's ease in executing the audit; the
second part acts as a presentation sheet to
give results of the audit to the homeowner.
Only the relevant information is transferred
from the first sheet to the second by a carbon
paper strip. Standard outputs were selected
as being important to give the homeowner.
These typically are: energy savings in the
first year; the dollar value of this first-year
savings; the net present worth of the savings
over the expected life of the measure; an estimate of the do-it-yourself cost to implement
the measure; an estimate of the contractor-in-
stalled cost of the measure; and the contractor-
installed cost after available tax credits are
taken. If, as a result of an RCS audit calculation, a measure pays for itself over its
lifetime, it is eligible for a state tax credit
in California. This had a great deal to do
with the choice of outputs generated. For
better focus on the salient facts, we generally
try to minimize the information given the homeowner.

Organization

The overall organization of the calculation
methodology is broken into three main sections.
The first is an analysis of the customer's
utility bill to determine the heating and/or
cooling load, base load, and savings rate. The
second part analyzes the building itself. This

359

is a preparation for part three, which is a repertoire of calculation procedures used to analyze the audit measures. The repertoire system allows an almost random choice of any of thirty or more audit measure calculations. It also allows unnecessary calculations to be by-passed. Analyses can also be conducted several times using different inputs.

Billing Analysis

In order to provide the customer with realistic estimates of energy savings, it is necessary to start from actual heating and/or cooling fuel use. When monthly (or bi-monthly) utility bills are available, they are the best source of this information.

We have found that the easiest way to estimate heating/cooling loads is to graph a year's data.

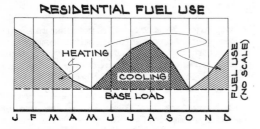

RESIDENTIAL FUEL USE

From the graph one can see the fuel used for heating or cooling. The minimum use is the base load. This is deducted from the total load to give the heating/cooling fuel use. These data are used in the Building Analysis to provide realistic, rather than theoretical, estimates of energy savings.

Savings Rate

To estimate the value of any energy savings accurately, it is important to know accurate fuel costs. When fuel rates are flat (no change on increased consumption) it is obvious that any fuel savings will occur at that flat rate.

When rates are tiered and monthly use varies from tier to tier, the average rate for saved fuel is not so obvious; this is why the concept of Savings Rate was developed. The Savings Rate is the average value of fuel saved per unit of fuel ($/KWH, etc.).

It is computed by comparing a customer's fuel cost before conservation with his expected cost after a reduction due to conservation. The amount of reduction is chosen based on experience. Palo Alto assumes that a 25% reduction in space heating fuel use can be obtained from reasonable conservation.

A calculator program is used to estimate a homeowner's fuel bill over a heating or cooling season. The load is based on the previously estimated heating or cooling load and the base load. Monthly fuel bills are based on a degree-day proration of the heating or cooling load. A season bill at no conservation is computed; then a bill at 25% load reduction is computed; and the Savings Rate is given by:

$$\text{S Rate} = (Bb - Ba)/25\% \times \text{Load}$$

where:

    S Rate = Savings rate
        Bb = Bill before conservation
        Ba = Bill after conservation
      Load = Heating or cooling load

What this means is that with Palo Alto's heavily tiered rate ($.40, $.65, $.80/therm), a high use customer will have a high savings rate and higher incentive to conserve. The Savings Rate is used to give value to energy savings estimates generated in the repertoire of measure analysis programs.

Basic Analysis

The Building Analysis program is in many ways the heart of the energy audit. In this routine, the details of the house structure are input and analyzed to generate a series of factors. These five factors relate actual heating and cooling energy use, determined in the billing analysis, to the theoretical heat gains and losses of the structure. Using this approach avoids the problem of the "saving-more-energy-than-last-year's-bill" syndrome.

The five factors are: (f) heating conductive loss; (m) heating infiltration loss; (r) cooling radiation gain; (s) cooling infiltration gain; and (c) cooling conduction gain. (When originally trying to settle on letter symbols, there was much discussion about what letters to use. This was done very scientifically: r obviously stands for radiation, c for cooling, s just sounded nice, f stands for "fudge factor", and m was a totally random choice.)
These factors rely on an accurate estimate of the theoretical total heat loss and heat gain. The total heat loss is the sum of the loss coefficients (U) multiplied by their applicable area (A) for ceiling, walls, floor, slab perimeter, windows, doors, duct work (adjusted for increased losses when operating), and infiltration. This total heat loss UA is generated in units of (Btu/hr-°F).

The total house heat gain UA includes the above UA's for the structural elements, adjusted for expected operating conditions during the cooling season.

The adjustments for duct heat loss when operating add a loss element due to the increased differential between duct and ambient temperature and, for cooling, also considers whether the ducts are located in an attic or a basement/crawl space.

The heating conductive loss factor (f) is given by:

$$(f) = \frac{\text{Avg. HDD} \times \text{AHFU}}{\text{Act. HDD} \times \text{Total HUA}}$$

where:

    Avg. HDD = Average annual heating
               degree-days
    AHFU   = Actual heating fuel use, last 12
               months
    Act. HDD = Actual heating degree-days last
               12 months
    Total HUA= Total heat loss factor for the
               house

This is a weather-normalized ratio of the actual fuel use to the heat loss coefficient. The units are therms or KWH per (Btu/hr-°F). This factor is useful in that a reduction in the total house heat loss (UA) reduces fuel use by (f) times the change in UA. For example: if insulating an attic reduces the theoretical house heat loss by 100 Btu/hr-°F, then [100 x (f)] is the expected fuel savings.

The factor (f) is also a behavioral indicator. Two identical houses will have different f-values depending on the way each house is used. The warmer a house is kept, the higher the f-value.

The f-value is used when analyzing for energy savings from any measure which alters the conduction heat losses of the house (e.g., attic, floor, wall, duct insulation).

The heating infiltration loss factor (m) is given by:

$$(m) = [GFA \times (f) \times (ACH-1) \times 8 \text{ feet} \times .03 \text{ Btu/Cu ft-°F}] / \text{Total Crack}$$

where:

GFA = gross floor area of the house
(f) = heating conductive loss factor
ACH = existing air change rate of the house
8 feet = average ceiling height
$\frac{.03 \text{ Btu}}{\text{Cu ft-°F}}$ = an average enthalpy change infiltration must undergo from ambient to indoor conditions
Total Crack = sum of all cracks producing infiltration

This term assumes that the infiltration of a house can be brought down from its current rate (ACH) to a final value of 1 air change per hour if all sealable cracks are sealed. The units of (m) are therms or KWH per linear foot.

The Total Crack includes window and door perimeters, sills (measured as house perimeter) and other cracks (e.g., masonry/wood interfaces). An estimate of existing ACH must be made depending on house construction and location. Typical values range between 2 and 3 ACH.

Factor (m) is used to estimate the energy savings for door and window weatherstripping and caulking.

The cooling radiation gain factor (r) is a ratio of the actual fuel used for cooling (KWH) and a theoretical measure of seasonal cooling load. It is like the f-value but must consider radiation heat gain as well as gain by infiltration and conduction. It is computed by:

$$(r) = \frac{ACFU}{\text{Total CUA} \times 24 \text{ hr/day} \times \text{Act.CDD} + Rad}$$

where:

ACFU = annual cooling fuel use (KWH)
Total CUA = total cooling heat gain UA
Act. CDD = the acutal cooling degree-days accumulated over the last cooling season
Rad = radiation heat gain attributable to windows.

The radiation heat gain is estimated by summing the incident radiation over the cooling season on the house's windows. The size, orientation, and shading coefficient of the windows are used. The units of (r) are KWH per Btu. The (r) factor is used to estimate expected energy savings due to improved shading of glazing. A reduction in cooling season radiation heat gain produces (r) times that in fuel savings. Note that the EER of the air-conditioner is implicitly included in the (r) factor.

The cooling conduction gain factor (c), like the (f) factor, provides a measure of expected energy savings due to added insulation. The (c) factor is generated by using the (r) value times an estimate of the conduction induced cooling load.

$$(c) = (r) \times \frac{24 \text{ hr}}{\text{day}} \times \text{Avg. CDD}$$

where:
Avg. CDD = cooling season average degree-days

Expected energy savings during the cooling season would be estimated as a change in UA times (c).

The final factor used is (s), the cooling infiltration heat gain factor, which is identical to (m) except that the average enthalpy change is based on the cooling season conditions.

The factor is given by:

$$(s) = [(c) \times GFA \times 8 \text{ feet} \times (ACH-1) \times .018 \text{ Btu/Cu ft-°F}] / \text{Total Crack}$$

where:

$\frac{.018 \text{ Btu}}{\text{Cu ft-°F}}$ = the average enthalpy change outdoor air would undergo during cooling to room conditions

The enthalpy factor is applicable for a dry air cooling climate as in California. Higher humidity areas would use an appropriately higher factor.

The calculator program which generates these factors functions in two parts. The first part is the input routine. Here the calculator prompts for the data about the various structural elements and stores the data in registers. No analysis is done until all data are stored.

If a mistake is made during input, the auditor can correct it by manually storing the correct value in the proper register. This "register map" is part of the data collection and input worksheet.

Once all necessary data have been input, the auditor directs the program to run the analysis by pressing a single key. The analysis runs, storing all of the intermediate results and the factors for later use.

Measure Analysis

Once the utility bill analysis and the building analysis have been performed, the analysis for the expected energy savings of the conservation and renewable resources begins. Measure analysis is done on a repertoire basis. Each measure to be analyzed has its own independent analysis routine. In this way an audit can be built from any sequence of measures. New measures can be added simply by creating software for the new item; none of the existing software would need to be modified. As improved methods of analysis are developed, they can be integrated easily without requiring a major comprehensive revision.

Methods of analysis calculation and calculator software were developed for the following items:

    Insulation
        Attic/Roof Insulation
        Wall Insulation
        Floor Insulation
        Slab Edge Insulation

    Infiltration Reduction
        Weatherstripping
        Caulking

    Heat Distribution
        Duct Insulation
        Hydronic Pipe Insulation

    Windows
        Storm/Thermal Windows
        Window Thermal Controls
        Window Sunlight Controls

    Heating Plant
        Clock Thermostat (Heating & Cooling)
        Replacement Furnace or Boiler
        Replacement Oil Burner
        Flue Opening Modifications
        Replacement Heat Pump

    Air Conditioning
        Replacement Central Air Conditioner
        Whole House Fan
        Evaporative Cooler

    Water Heating
        Water Heater Insulation
        Heat Pump Water Heater

    Swimming Pools
        Swimming Pool Cover
        Active Solar Pool Heater

    Active Solar
        Solar Domestic Water Heater
        Combined Space and Water Heater
        Space Heater

    Passive Solar
        Direct Gain
        Trombe Wall
        Water Wall
        Thermosyphon Air Panel
        Greenhouse/Sunspace

Each analysis in the repertoire is based on standard and readily available resources. Many measures are based on the DOE Model Audit and the ASHRAE Handbook of Fundamentals. The solar measures are based on the F-Chart method and LASL's work on passive analysis. The bibliography at the end of this paper gives a detailed list.

The single biggest challenge in applying these resources to the audit methodology was developing simple algorithms which could be easily executed in a programmable calculator. Much of the DOE Model Audit and LASL's work rely on graphs or tables to supply results. Having the auditor carry a collection of graphs into the field or putting tables into calculator memory are not practical. To make the analysis compact, tables and graphs were analyzed using curve-fitting techniques to generate algorithms.

For any graph or table, several curve fits were tried until an accurate but simple algorithm was found. Some of the final algorithms were logarithmic or quadratic in form.

It is not practical in this paper to discuss the analysis of all of the measures listed above or how their algorithms were derived. (The NCPA audit documentation is more thn eighty pages long without calculator program listings!) Reviewing a few of the measure analysis techiques will give a feel for the overall methodology. The most interesting and useful ones for presentation are the basic insulation analysis, the clock thermostat analysis, and the passive solar analyses.

Insulation

The insulation (attic, floor, wall, etc.) analysis makes use of the previously calculated (f) and (c) factors. The factors give energy saved for a change in UA.

Given that the area and existing U-value are already in calculator memory, the auditor simply enters the added insulation R-value. The sequence of analysis runs as follows:

(1) Compute final U-value:

$$Uf = 1/(1/Ue + R)$$

where:

> Ue = existing U-value
> R = added R-value
> Uf = final U-value

(2) Compute change in UA:

$$UA = Area \times (Ue - Uf)$$
where:
> Area = area of surface treated

(3) Compute heating and cooling energy savings:

$$Qh = (f) \times UA$$
$$Qc = (c) \times UA$$

(4) Compute the dollar value of the energy savings using appropriate savings rates:

$$\$ = Qh \times SRate\ h + Qc \times SRate\ c$$

(The savings rates may be different depending on the fuel used for each.)

(5) Estimate the present value of dollar savings over the life of the measure:

Life $ = PWF x $

where the Present Worth Factor (PWF) is given by

$$PWF = \frac{Rate}{Rate - 1} \quad (Rate^{LT} - 1)$$

where:

    Rate = real fuel cost
         escalation factor
         (i.e., 5% real = 1.05)

(6) Compute an estimated cost of installation:

Cost $ = Area x (A + B x R-value)

where:

    A = base insulation cost
        in $/sq ft

    B = variable insulation cost
        in $/sq ft -R

The values for A and B are computed by doing a linear curve fit of expected price per square foot versus R-value. For example, in Palo Alto, A is $.05/sq ft and B is $.015/sq ft -R.

(7) Compute the cost after tax credits are applied assuming a 40% net tax credit is available. (The State of California offers a 40% conservation tax credit. The credit is so structured that the combined state and federal credits total no more than 40%.)

    ATC$ = Cost $ x .6

where:
    ATC$ = after tax credit
           cost

The various results generated above are manually recorded on a record sheet and given to the customer. This record gives the customer the information necessary to accurately assess the personal cost-effectiveness of this measure.

Clock Thermostat

Heating Setback: A clock thermostat (or manual thermostat setback) can save large amounts of energy in California's mild climate. Because of this, however, a simple rule-of-thumb of "percent per degree" will not provide accurate results.

To develop a better estimating technique, a thermostat simulation program was developed and executed using an Apple II desk top computer. The program used local climate data (monthly average maximum and minimum temperatures and degree-days) to simulate an hour-by-hour outdoor temperature profile over the heating season. Then different thermostat setting profiles (daytime setting, nighttime setting; 6:00 a.m. set up and 10:00 p.m. set back times were held constant) were compared to the outdoor profile to give an hour-by-hour temperature difference. This degree-hour temperature difference was accumulated over the heating season. All such degree-hour totals were normalized against actual degree-days using a degree-hour total based on a constant 65°F indoor temperature.

Simulations were run with various daytime and nighttime thermostat settings. These runs generated a table of effective seasonal degree-days for different thermostat settings.

To convert this table (some 400 values) into something usable in a calculator took two steps. First, the data were plotted by the computer for visual interpretation. Plots of degree-days versus daytime setting and versus setback (daytime minus nighttime setting) were made on the computer video screen. An inspection of this data showed that for the most part, over the thermostat settings of interest, the effective degree-days could be represented by an equation of the form:

HDD = A + B x TS + C x SB

where: HDD = effective heating
                degree-days
           TS = Daytime thermostat setting
           SB = thermostat setback
        A,B,C = constants of propor-
                tionality

In the second step, the constants A, B, and C were generated using a multivariate regression routine. This routine generates values of A, B, and C which model HDD to the highest degree of accuracy. This equation can then be implemented in the programmable calculator.

In Palo Alto, the values for A, B, and C are:

A = -8310, B = 167.8, C = -67.2

The B value says that an increase of one degree thermostat daytime setting adds 168 degree-days; C means that one degree increase in nighttime setback substracts 67 degree-days.

A savings in heating energy can be estimated by comparing the effective heating degree-days at the homeowner's current thermostat settings to the effective heating degree-days at alternate settings. The savings are given by:

$$Qf = (1 - HDDn/HDDc) \times (f) \times THUA$$
where:
           HDDn = effective heating
                  degree days at new
                  thermostat settings
           HDDc = effective heating
                  degree-days at current
                  settings
            (f) = heating conduction
                  loss factor
           THUA = total heating UA

(Note: (f) x THUA is a weather normalized seasonal heating load)

From this the cost savings can be estimated using the Savings Rate.

Cooling Setup: The same kind of analysis can be run for thermostat changes during the cooling season. An increase in indoor temperature, however, will reduce only that part of the cooling load induced by conduction and infiltration heat gains. Radiation heat gain is not affected.

As in the heating setback, a computer analysis was run comparing indoor temperature profiles with outdoor profiles and accumulating cooling degree-days. On visual inspection of the data, a non-linear relationship between degree-days and indoor temperatures was found. The algorithm that gave the best curve fit was found to be:

$$CDD = EXP (D + E \times TS + F \times SU)$$

where:

CDD = effective cooling degree days

EXP = exponentiation to base "e"

TS = thermostat nighttime setting

SU = daytime thermostat set up (noon to 9 p.m.)

D, E,& F = constants of proportionality

A multivariate regression was applied to generate the best values for D, E, and F.

Because of the non-linear nature of the algorithm, the values for D, E, and F do not have an intuitive meaning. For example, Sacramento, California has:

$$D = 14.49, E = .1143, F = .0593$$

This algorithm is used as in the heating setback case (now using (c) and total cooling UA) to derive cooling fuel savings and cost savings.

Passive Solar

The analysis technique developed for passive solar retrofits (direct gain, water wall, Trombe wall, greenhouse, and thermosyphon air panel) is derived from the excellent work done by Los Alamos Scientific Laboratories (LASL).

The DOE Model Audit uses a technique of Collector Load Ratio (CLR) where LASL uses Load Collector Ratio (LCR). The techniques are identical. LASL's LCR was chosen because the available documentation is much more comprehensive.

In LASL's work, the LCR is defined as the ratio of load induced by all but the south wall to the collector aperture. Dealing as we are with existing houses, trying to remove the south wall load is impractical, especially since only a small portion of the wall is usually subject to passive retrofit.

For the purposes of the audit, LCR was defined as:

$$LCR = \frac{\text{Heating Load (Btu)}}{\text{Degree Days} \times \text{Collection Area (sq ft)}}$$

To implement this in the methodology requires an expanded form. Here heating load is given by:

$$HL = (f) \times THUA \times FEF \times SE$$

where:

HL = heating load (Btu)

(f) = heating conduction loss factor (units/UA)

THUA = total heating UA

FEF = fuel energy factor (Btu/unit)

SE = system conversion efficiency--fuel to heat

Since solar access can be a serious obstacle to retrofit in an established area where trees have grown, the Prime Solar Fraction (PSF) as defined in the DOE Model Audit is also included

$$LCR = \frac{HL (Btu)}{\text{Degree-Days} \times \text{Collection Area} \times PSF}$$

Tables of Solar Savings Fraction (SSF) were developed by LASL for various locations, types of passive systems, and values of LCR. It can be found by curve fitting these tables of SSF vs LCR, that curves of the following form occur:

$$SSF = A - B \times Ln (LCR)$$

where:

Ln = natural logarithm

A, B = constants of proportionality

LASL provides tables for night-insulation and no night-insulation cases. For the audit algorithms, an average of the two was used. This was a tradeoff between detailed accuracy and simplicity; simplicity won! The above equation can be easily implemented in a calculator program.

For any city A and B values for each passive measure can be developed based on LASL's work. The field audit is executed by having the auditor select from the repertoire which passive retrofit is to be analyzed. This selection puts into memory the appropriate values of A and B, as well as values for do-it-yourself and contractor costs and estimated measure life. The software then prompts the auditor for the prime solar fraction and area (aperture) of retrofit, and then proceeds with the analysis.

Experience

Palo Alto has been using this audit methodology in the field for almost two years. The expanded methodology has been used by the NCPA cities for several months. After the first initial debugging phase, everyone has found this methodology, based on a programmable calculator, to be very workable.

In the typical audit, the auditor first collects all of the necessary information about the house on the standard data sheets. From the house measurements, total window areas by orientation; gross and net wall areas; and ceiling and floor areas are computed. Furnaces and water heaters are inspected, usually with the homeowner present to discuss possible improvements. The auditor then discusses with the homeowner any applicable low- or no-cost practices or standard maintenance measures recommended. Normally by this time an hour to an hour-and-a-half has elapsed in the two-hour audit. The auditor then begins the analysis using the calculator. For a typical house where ten measures from the repertoire are analyzed, the analysis takes only thirty minutes. Homeowners are often very interested in this process and request extra analyses. For example: the homeowner might want an analysis for R-22 and R-30 attic insulation. The repertoire structure makes running the extra insulation analysis quite simple.

The auditors have found that if they make an error of data entry (or want to double-check the results), rerunning a measure from a repertoire is very easy and consumes little time. The auditors have indicated that this gives them confidence that the results presented are correct.

The results of the audit are presented to the customer on the presentation form which discusses the "what and why" of each measure.

Updating of software as errors are found, outputs added (the audit had been in use one year before tax credits were added), or price changes have proven relatively simple. Again the repertoire organization makes this a simple action. All that need be done is: the individual program is called up, changed, and rememorized on its magnetic card. A typical price change takes five minutes.

Future Plans

New rules coming out of DOE are allowing audits much reduced in scope to be performed. Rather than having to consider the applicability of thirty measures on every audit, only ten or fifteen might be needed. Palo Alto, working on this basis, has segregated its audit into a minimum audit and optional packages (e.g., air-conditioning, passive retrofits, swimming pools). No change in the methodology or software was required; only a change in the data analysis and presentation forms was made.
A version of the NCPA expanded audit is being developed for the Apple Computer to help with the Class B audit program. In this audit, information collected by the homeowner will be analyzed in the utility office rather than in the field.

Summary

The choice of using a hand-held programmable. calculator to conduct Class A RCS audits was an excellent one for Palo Alto and the NCPA cities The low cost of these calculators and the flexibility of the methodology have proven a boon to the smaller municipal utility that cannot justify the expense of central computer interties. As the need for energy audit services expands in the multi-family and small commercial areas, this type of audit will continue to serve a useful purpose.

Bibliography

. DOE Model Audit, Oak Ridge National Laboratories

. Passive Solar Building Analysis, by Dong Balcomb, LASL, January 1980, DOE/CS-0127/2

. Performance Estimates for Attached-Sunspace Passive Solar Heated Buildings, R.D. McFarland and R.W. James, LASL, LA-UR-80-1482

. HUD Intermediate Minimum Property Standards, Solar Heating and Domestic Hot Water Systems, Appendix A, 1977 Edition

. Residential Conservation Service Hand-held Programmable Calculator Methodology Documentation, Michael D. Beanland, Northern California Power Agency, 8421 Auburn Blvd., Suite 160, Citrus Heights, CA 95610

SECTION 17
INNOVATIVE FINANCING
OF ENERGY
CONSERVATION PROJECTS

Chapter 58

THE APPROACH TO FINANCING ENERGY CONSERVATION PROJECTS AT JOHNSON & JOHNSON PRODUCTS, INC.

H. Pearson

A. Commitment To Energy Conservation

At Johnson & Johnson, we are committed to reducing energy consumption and we have strong management backing to carry out energy programs. With energy cost escalation considered, our management will support projects with an 8 year payback. Our goal is to reduce energy 20% or more in our world wide operations and some of our plants have achieved that goal.

B. Return On Investment

Both the Standard and Energy model return on investment are available to determine the benefits of implementing a capital project with energy reduction. At Johnson & Johnson, each plant is operated independently and therefore, can choose the ways and means to implement their own energy savings programs. The assumptions used for the energy saving model are as follows:

1) Equipment has 8 year life and uses sum of years digits - method of depreciation.

2) Energy UNIT PRICE ESCALATION

   10% per year for electric power
   15% per year for fuel oil
   20% per year for natural gas

3) Tax Rate = 46%.

4) Investment Tax Credit & Energy Credit = 20% - Additional 15% credit for solar and wind power projects.

   (Amortized over life of asset for P & L effect fully utilized in first year for cash flow)

Obviously, the Energy saving model will increase the return on investment and make the energy project more attractive. In actuality, the energy saving return or investment is often not used because the return on investment stands on it's own using the standard return or investment analysis. As an example, an economizer installed on our boiler at our Chicago plant in 1979 produced an estimated return on average investment of 61% and a payback of 3.5 years. The appropriation was written for $85,500, $70,000 capital and $15,500 expense with an annualized savings of $44,652. This savings was based upon 22¢ per therm for natural gas, at our 1982 gas rate of 40¢ per therm; the savings is $81,185 annually.

C. Highlights Of Boiler Room Improvements

The boiler room is not normally the highlight of the plant and therefore, can be overlooked for energy conservation. At our Chicago plant, the boiler room funnels through it over 40% of the energy used at this facility, therefore it deserves most close attention. In 1977, our Chicago boiler room had 3 boilers fired up, 2 on standby ready to take on the steaming load in case the first went down. This energy wasting procedure was stopped and more attention given to insuring that the one boiler carrying the load did not falter and cause process shutdowns. In 1978, our one main boiler was operating at 71% estimated efficiency. Presently, that same boiler is operating at 71% estimated efficiency. Presently, that same boiler is operating at 83 to 85% efficiency. This efficiency improvement is a result of:

1) Improve burner tuning   (2%)

2) Boiler economizer on exhaust stack (6%)

3) Blowdown heat recovery unit   (3%)

4) Excess oxygen automatic controls (2%)

5) Automatic blowdown control   (1%)

D. Capital And Maintenance Type Programs

Many energy savings programs have been successfully carried out at our Chicago manufacturing facility and our Lemont, Illinois warehouse operation. At our warehouse, electricity and natural gas use have each been reduced 57% since 1977. Mercury type lights were used at this facility and the first step was to turn half of them off reducing the lighting costs from $84,000 annually to $42,000. The next step was to replace the mercury lights with high pressure sodium lights cutting the lighting cost again in half to $21,000 annually. Air conditioning at this warehouse was originally installed in 1968, as a means to control humidity and keep our 30 foot high stacking height from toppling over. With the low cost of energy the initial reasoning was soon forgotten and temperatures were controlled at 70°F in this 400,000 square foot facility. In 1978, humidistats were installed and tied into our Honeywell Energy Management System to maintain 58% relative humidity and thus keep the 30 foot of stacking height from toppling. As a result, the air conditioning systems rarely are needed.

With these energy projects and others, our electrical and gas use dropped dramatically at the Lemont warehouse. Our electricity supplier sensing something was wrong and therefore, checked the electric meter 3 times and finally replaced the meter. This was of no consequence because this did not change the fact that the electricity use had been greatly reduced and electricity charge was indeed correct.

E. Selling Energy Projects

Numerous energy projects and programs have been carried out at both our Lemont warehouse and our Chicago manufacturing facility. They range from the complex Honeywell Delta 1000 energy management control with over 100 air conditioning control points to the simple removal or disconnect of 2 or 3 lights in an office. Successfully implementing an energy program requires top management support with the dollars backing up both maintenance and capital programs. I believe the most important ingredient is to have the right person or persons implementing energy reduction programs. The right ingredience includes:

1. Enthusiasm
2. Perseverance
3. Dynamic individuals
4. Good selling ability
5. Merchandising
6. Good understanding by plant management

Often as not, reducing energy is not popular with plant operating people because it can cause discomfort or more careful operating to reduce waste. Therefore, the top level support and understanding is a requirement in selling Energy Conservation.

Chapter 59

DETERMINATION OF AN ENERGY BASELINE AND ADJUSTMENTS FOR A SHARED SAVINGS FINANCING PROGRAM

M. R. Castonguay

ABSTRACT

Many energy financing projects utilize an "energy baseline" approach for allocating future shared savings. Establishment of a reasonably accurate baseline value, as well as various other adjustment factors, is crucial for both parties in the financing program. This paper details the process utilized in baselining heating energy requirements for a 300,000 sq. ft. plastics manufacturer. The information presented is instructive for any general energy baselining approach, but is particularly useful for analysis of buildings with substantial internal heat gains.

The paper discusses these key factors:

o   Analyzing historical energy data
o   Calculating normalized energy usage
o   Establishing the energy baseline
o   Predicting adjustment factors for variations in weather

INTRODUCTION

With high interest rates and a shortage of capital, many firms are seeking alternative methods of raising the necessary funds for energy savings projects. Given the new tax laws and energy credits, many investors see financing of energy reduction projects as profitable. As a result, the two groups have come together to produce "shared-savings" programs.

Under a typical agreement, investors provide the working capital for energy-related improvements in return for a share of future energy savings. The building owner or corporation, on the other hand, realizes a reduction in operating costs, physical improvements in the facility, and often increased productivity without the need for raising capital.

A central feature of these arrangements is how future savings are measured or determined. Many projects utilize an "energy-baseline" approach whereby future energy consumption is to be compared against an established "normal" energy consumption. Establishment of a reasonably accurate baseline value, as well as various adjustment factors, is crucial for both parties in the energy financing program.

This paper details the process utilized in baselining heating energy requirements and future weather-related adjustments for a 300,000 square foot plastics product manufacturer.

ESTABLISHING THE ENERGY BASELINE

Analyzing Historical Energy Data

The first steps are to determine which energy types will be significantly affected by the proposed conservation measures, and to collect historical usage on these fuels. Subsequently, usage must be distributed among all of the end uses for a particular fuel.

For this plastics manufacturer, both No. 6 oil and natural gas are used to produce steam for space heating equipment. Natural gas is also utilized for domestic hot water production and a cup sealing process; the No. 6 oil has no other end use. Because of its minor impact, changes in electrical consumption of boiler-related equipment are not included.

Process gas consumption per case of product is available from measurements, and domestic hot water requirements are estimated based on summer natural gas consumption patterns. Net natural gas heating consumption is then obtained after subtracting out these values.

Table 1 presents the resulting net heating energy use at the facility for the last four heating seasons.

TABLE 1. BASELINE DATA - HEATING ENERGY USAGE

	1977-1978		1978-1979		1979-1980		1980-1981	
	Oil (gals)	Gas (ccf)	Oil (gals)	Gas (ccf)	Oil (gals)	Gas (ccf)	Oil (gals)	Gas (ccf)
Sept.	--	--	--	--	--	--	--	966
Oct.	--	7,000	6,905	--	--	8,800	--	7,006
Nov.	6,809	300	6,819	--	--	16,700	--	23,482
Dec.	20,432	--	27,282	--	6,925	24,400	--	38,752
Jan.	34,113	--	20,526	00	6,826	20,100	--	34,118
Feb.	28,261	5,400	34,322	2,200	5,851	34,400	--	27,491
Mar.	20,498	5,300	6,920	20,000	--	21,900	--	28,111
Apr.	13,616	--	11,695	8,000	--	14,300	--	8,019
May	6,801	--	--	700	--	1,800	--	1,766
TOTALS:	130,531	18,000	114,469	30,900	19,602	142,400	--	169,712

Normalizing Energy Consumption

Next in the baselining process is the normalization of historical energy consumption.

For most purposes, a degree-day base is sufficiently accurate. Each year's heating BTU consumption is divided by the corresponding heating degree-days. The result then provides an historical indication of space heating efficiency. Any anomalies observed should be investigated and a satisfactory explanation obtained.

In Table 2, normalized heating energy consumption is listed for the manufacturing facility. As can be seen, energy consumption decreased steadily from 1977 to 1980, with a more significant drop in 1981. The slight decrease in the first three years correlates strongly with roof insulation measures undertaken at that time. The almost exclusive use of a single gas-fired boiler, rather than use of the two oil-fired boilers and the natural gas boiler as in previous years, accounts for the 1980-81 drop in consumption.

TABLE 2. NORMALIZED YEARLY HEATING ENERGY USAGE

	1977-1978	1978-1979	1979-1980	1980-1981
Total Million BTU	20,662	19,631	17,072	16,971
Heating Degree Days Sept-May	6,236	6,028	5,485	6,014
Million BTU per Degree Day	3.31	3.26	3.11	2.82

Establishing the Energy Baseline

Since the data available presents no substantial inconsistencies in consumption patterns, it is reasonable to apply the most current usage figures as the energy baseline. In this way, current heating system operation and performance form the basis for future energy reductions and savings calculations. If anomalies had been present, it would have been necessary to devise appropriate adjustments or perhaps to utilize values averaged over several years. In some cases, an energy baseline might not even be possible, and other methods of measuring savings would be required.

ADJUSTMENTS TO BASELINE VALUES

Determining the Appropriate Degree-Day Base

Having established the 1980-81 heating energy consumption as the baseline for calculating future savings, it now becomes necessary to assess the impact of potential differences in heating-related factors - most importantly, weather.

Although the 65°F base degree-day is the most readily available heating-related weather data, other factors can be important. In some cases, wind or hours of sunshine may have a substantial effect on heating requirements. More importantly, for the 65°F base degree-day, little or no internal heat gains are assumed. Clearly, such is not the case for a plastics manufacturing process.

As internal heat gains become important, a larger percentage of the heating space requirements shift to the colder months, and less heating is required during the spring and fall. When a base temperature lower than 65°F is used, the same shift in relative monthly degree days as a fraction of the yearly total occurs. By choosing the appropriate degree-day base, the shape of actual fuel consumption patterns can be matched more closely to the degree-day profile.

The National Climatic Center has compiled 30-year normal degree-days to selected bases, ranging from 25°F to 70°F. Monthly degree days as a percent of yearly total are plotted in Figure 1 for various bases. Also shown in the same figure is the relative heating fuel consumption for each month during the 1980-81 season. In the case of the plastics facility, the 57.5°F base (an average of 60°F and 55°F values) provides reasonable agreement with fuel consumption, as Figure 1 illustrates.

FIGURE 1.

MONTHLY DEGREE DAYS AND FUEL USE

(PERCENT OF YEARLY TOTAL)

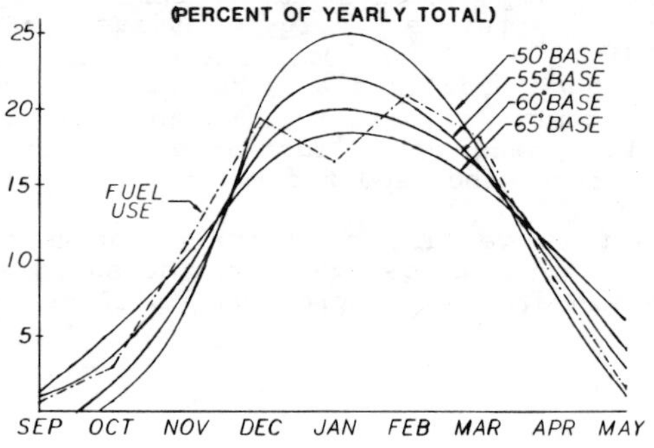

Adjusting Available Weather Data

If actual degree-days to selected bases were readily available on a month-to-month basis, the normalized fuel consumption could be calculated for the appropriate degree-day base, and a degree-day adjustment factor obtained (i.e., total annual fuel consumption divided by yearly degree-days at 57.5°F base). Unfortunately, such monthly degree-day information is not easily obtained.

Adjustments to the heating energy baseline would be greatly simplified by use of a 65°F base degree-day data. This approach requires that the annual adjustment factor (65°F base) be weighted by the relative percentage of degree-days occurring in a given month, using the 57.5°F thirty-year normal data. Table 3 lists the adjustment factors calculated for the plastics plant.

TABLE 3. CALCULATION OF WEATHER-RELATED ADJUSTMENT FACTOR

$$\frac{\text{Annual Consumption}}{\text{Annual 65°F Base Degree-Days}} \times \frac{\text{Monthly Degree-Days (57.5°F Base)}}{\text{Annual Degree-Days (57.5°F Base)}} = \text{Adjustment Factor}$$

Sept. 1980	28.22 CCF/°Day	x	.71	=	20.04 CCF/°Day
Oct. 1980	28.22 CCF/°Day	x	.74	=	20.88 CCF/°Day
Nov. 1980	28.22 CCF/°Day	x	.88	=	24.83 CCF/°Day
Dec. 1980	28.22 CCF/°Day	x	1.07	=	30.20 CCF/°Day
Jan. 1981	28.22 CCF/°Day	x	1.10	=	31.04 CCF/°Day
Feb. 1981	28.22 CCF/°Day	x	1.11	=	31.32 CCF/°Day
Mar. 1981	28.22 CCF/°Day	x	1.04	=	29.35 CCF/°Day
April 1981	28.22 CCF/°Day	x	.90	=	25.40 CCF/°Day
May 1981	28.22 CCF/°Day	x	.76	=	21.45 CCF/°Day

ALLOCATION OF ENERGY SAVINGS

Thus far we have examined the 1980-81 heating energy usage and adopted it as the energy baseline; adjustment factors for variations in weather have also been derived. This information, along with the 1980-81 heating degree-days, provides a method for allocating future energy savings between the building owner and financing company.

Once energy conservation measures have been implemented, gross energy consumption for each month can be adjusted as required for production usage, non-heating related usage, and weather variations, to yield net heating energy usage. Comparison with the baseline value will quantify the savings achieved.

SECTION 18
STEAM SYSTEM UTILIZATION

Chapter 60

INDUSTRIAL STEAM SYSTEM OPTIMIZATION

W. L. Viar

ABSTRACT

Industrial steam systems provide unique opportunities for the development of low cost mechanical shaft power. A properly applied back-pressure turbine can deliver power to a machine at 40 percent of the cost of an equivalent motor operating on purchased electricity. Conversely, the cost of power from a turbine exhausting to the atmosphere will be in the order of six times greater than that provided by an efficient motor. Therefore, the relative costs of fuels and purchased power in your area, coupled with designs of merit and proper operating practice, will determine the success you experience when you "power up your steam system!"

STEAM SYSTEM ANALYSIS

Purpose

The orderly study and evaluation of an industrial steam system should be the result of a planned program with well defined objectives. The ultimate goal must be to save fuel and purchased power. The first step of such an analysis should be to determine the current status of the system. There will be a need to know where you are now, how you stand, and what are the prospects of significant savings? Objectives will include the improvement of performance of existing equipment to reduce losses; a search should also be made for favorable design changes, possibly leading to retrofits. One of the most fruitful objectives will be to maximize the cogeneration potential of the plant steam system.

Learn The Steam System

Once the objectives of a steam system analysis are established, the natural question will be, "who will conduct this study?" The answer will vary from company to company, from division to division, and from plant to plant. Most often the selection of technically competent people to conduct such an elaborate study will be made from (available) in-plant supervisory or engineering staff members, corporate central engineering staff, or qualified outside consultants. The individual, or team members, must be well trained and skilled in the analytical techniques required to organize a thorough study, to know where and how to search for significant savings opportunities, to be aware of the economic feasibility of

apparently desirable changes, and to be able to locate and assess system limits. In short, he must have the necessary complement of powers of observation gained from long experience in the field.

To initiate the analysis, the system must be defined, unit by unit, component by component, element by element. All steam sources must be identified as to types, capacities, normal loads, applied fuels, steam conditions, and relative efficiencies. Power units must be located and listed as to output ratings, throttle conditions and flows, steam rates or efficiencies, exhaust conditions, mechanical states, and various machine characteristics.

The distribution of steam must be well understood. Line sizes, mechanical conditions, pressure-temperature levels, meter locations, trap types, and insulation adequacy should be noted. All connected steam loads and the characteristics of these loads should be established and recorded. For example, it must be learned whether the steam mixes with the process fluid or is kept separate by closed heat exchanger surface; what methods are used to control steam and condensate flows? Are the steam requirements well defined, measurable, and can they be determined readily?

The condensate return system must be examined to determine its extent, adequacy, state of repair, and prospects of stream contamination. Cold water make-up requirements, its source, pumping, and treatment for scale and corrosion control must be studied.

Learning the steam system requires a thorough search of available power system and process area flow diagrams and schematics. A first hand examination of system components and interconnecting lines (some line tracing); and extensive discussions with unit managers, engineers, and operators must be conducted. Pertinent historical operating records and logs, along with manufacturers' specifications and data sheets must be gathered for subsequent evaluation.

Preliminary Mass Balance

As progress is made in the intensive steam system learning process (and emerging evaluations), data must be gathered to determine flow quantities used throughout the plant. The application of necessary steam properties at points of use must be verified. That is, if actual throttle steam conditions vary from those established by design and set by turbine manufacturers, the variances should be noted and taken into account. Further, if a closed steam heat exchanger was designed to operate on dry saturated steam at 100 psig, but is, in fact, operating on wet steam at 80 psig, the prospects of poor heat exchanger performance must be examined. Conversely, if such a heat exchanger is connected to a 300 psig steam header having 100°F of superheat, this misapplication must also be assessed.

The preferred way of establishing the true distribution of steam mass throughout a system is by the extensive use of fluid flow meters. Although strategically located and properly calibrated metering will lead to the best possible steam mass balance, closure with zero error is not a realistic goal. All too often, meter installations are too few in number and proper maintenance has been deferred.

Therefore, other methods have to be employed to determine flows to, or through, individual components. For example, reasonable condensate flows can be collected and weighed, or collected in vessels with known volumes while being timed. The liquid temperature should be noted and flash vapor should be minimal. Process steam flows can be determined indirectly by sufficient knowledge of heat exchanger characteristics and process energy demands.

Unmeasured steam flows to a turbine can be estimated reasonably if the driven machine load can be characterized. Alternatively, determine throttle and exhaust steam conditions and assume a reasonable steam rate or expansion efficiency based on observed turbine size, speed, valving, and apparent maintenance and operating practices. On some machines, measured first-stage or ring pressure can be used in conjunction with manufacturers' performance curves to approximate steam flows and corresponding shaft power output.

By the judicious use of these techniques, a steam mass balance diagram must be developed that describes the current system status. Steam headers should be shown schematically with straight lines to the extent that they pass through the plant. For example, if boiler steam is distributed throughout the plant, this header would be represented by a full length line near the top of the diagram. The next lower pressure header can be laid off two or three inches below the top line. Lower level lines of this type should be used to represent the remaining steam lines extending through all, or part, of the plant.

All steam generators should be shown connected to their proper headers as system inputs. Turbines, pressure reducing valves, and steam loads that are connected to the several headers should be drawn in as loads or outputs from the headers. Exhaust steam or drained condensate should be connected to the appropriate lower pressure header.

After all components have been laid out schematically in the diagram, the mass flows and approximate steam properties should be labelled at the proper places on the diagram. Each component should be identified by name and/or equipment piece number, in keeping with established plant identification nomenclature. To complete the mass balance throughout the system, every flow entering the header must be summed up and compared with the total of all flows leaving that header. The absolute values of individual components will invariably differ from those that are metered, otherwise measured, or estimated. However, with the very careful and deliberate gathering of information and data that has gone on before, the steam balance so established should be reasonably accurate and adequate for most engineering evaluations utilizing the diagram. Pertinent pressures and temperatures should be noted on the schematic.

When the preliminary balance is prepared, it should be reviewed with plant personnel who have specific knowledge about their equipment and who have contributed to the development of the steam balance, so that they may assess the results and make necessary revisions. With these procedures completed, the steam mass balance will represent normal plant operation, and will be extremely useful in planning that leads to steam system optimization. Keep in mind that as the study progressed, opportunities for improved performance will have been noted, acted upon, or established as management objectives.

STEAM SYSTEM EXAMPLE (Figure 1)

Steam Generation

A preliminary mass balance has been established for the hypothetical industrial steam system illustrated in Figure 1. The highest pressure steam is generated in the combination of multifuel-fired boilers. The steam conditions are 850 psig and 750°F. Additional steam is generated at the 400 psig pressure level, with a substantial fraction of this steam being supplied by heat recovery boilers. These HRB's utilize high temperature energy from process exothermic reactions and fired process furnaces, that would otherwise be rejected to the atmosphere. The remaining steam generated at 400 psig is from fuel-fired boilers which establish the incremental cost of steam at that pressure. Clearly, the HRB's are extremely important since they offset the firing of fuels in the system.

Power Turbines

There are substantial electrical power and mechanical shaft loads on this plant that are driven by a mix of power turbines and purchased electricity. Some of these units drive generators, others are directly coupled to the driven machines.

From original plant designs and subsequent facility additions, turbines have been selected and placed between pressure levels that contributed to overall steam system balance at the times of installation. There are straight back-pressure machines, one extraction/back-pressure unit, and one turbine operating condensing. The throttles are at high pressure, intermediate, and low pressure.

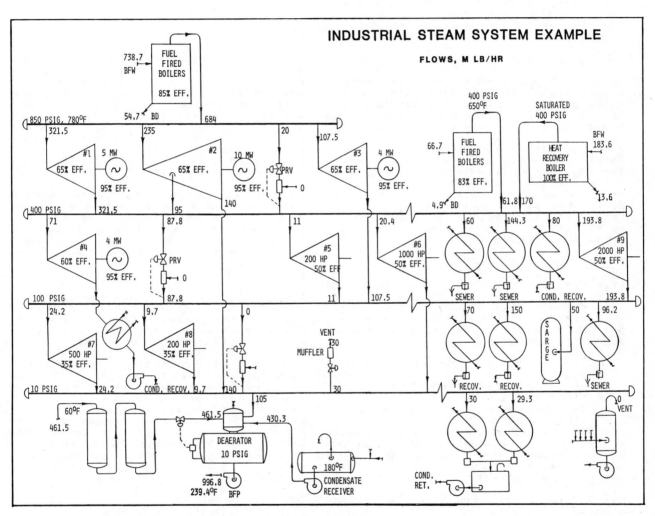

Figure 1

Pressure Reducing Stations

Operating in parallel with the turbines between each header set, pressure reducing valves (PRV's) regulate pressure in the downstream headers. The PRV's throttle the high pressure steam in response to the lower pressure sensors. Desuperheater stations are shown in place but inactive.

Boiler Feedwater System

To the extent that it is recovered, condensate is pumped to the deaerating heater, along with cold treated water makeup that is regulated by level in the deaerator. Steam at 10 psig is condensed in this open heater to heat and degas the incoming water. Deaerated water is stored and then pumped on demand to the several boilers.

Steam Distribution

Steam is exported from the power house to widely distributed process loads on this multi-product plant at the levels of 400, 100, and 10 psig. As shown, the 400 psig boilers are located in production areas.

Although not shown in the schematic, a few meters provide useful operating information for the plant staff. The recorded flow quantities are used for accounting and accountability purposes. Line sizes, pressure drops, and heat losses are not shown in the schematic, although the latter two items can be significant. The large number of distribution steam traps are not shown.

Process Steam loads

At times, the types of process steam loads appear to be limitless. There are an enormous number of ways used in industry to apply this very common heat transfer and working medium. Generally, steam loads can be classified in several ways:

- By pressure and temperature requirements.
- As motive steam or heating steam.
- Magnitudes, large or small.
- Kept separate or mixed with process fluids.

If steam loads are satisfying thermal requirements, they can be determined by process heat duties, temperature differentials, fluid properties, heat exchanger surface areas, space heating needs, etc.

Where steam is mixed with the process streams, it may be for heating or cooling (temperature control), sparges, agitation, dilution, or partial pressure control. Mass flow rates are often constant and manually controlled.

Fundamental cogeneration principles dictate that process loads be valid and useful applications of steam in systems such as this. The electrical power that is generated, as well as the shaft mechanical power, must be functions of these process heat loads, and not be determined solely by power requirements. It simply is not economic to generate power while exhausting steam to the atmosphere.

Condensate Recovery

Steam or condensate that is mixed with process streams will be unavailable for recovery and return to the boiler feedwater system. All other condensate is subject to recovery and reuse. The heat content represents the value of the condensate. High purity is good, but it is secondary. Water scarcity can, of course, change the relative importance of these two characteristics. In this sample example 40 to 45 percent of the boiler feedwater requirements are satisfied by returned condensate.

COMPUTER ASSISTED ANALYSIS (Figure 2)

Why?

With a steam system mass balance in hand similar to the one described above, some technique must be used to evaluate effects of the changes that appear to be needed to improve system performance. It is well known that an alteration in one part of a steam system affects virtually the entire system. Because of this fact and the many iterations required to evaluate desired changes, longhand calculation methods make it impractical to identify all the mass and property variations.

It has become essential in the study of complex steam systems to utilize computer routines developed for these purposes. Without the use of an accurate computerized simulation technique, it is nearly impossible to reliably analyze any but the most simple industrial systems. One such computer program is known as SYNTHA III (1).

Properly used, the prepared model faithfully reproduces the actual steam system in its entirety and computes the mass flow, pressure, temperature, enthalpy, quality, and entropy of each flow stream. It also computes and summarizes boiler fuel consumption, horsepower developed by direct-drive turbines, and electricity produced by turbogenerators. Having established a set of baseline conditions, a change (or a set of changes) can easily be imposed on the model and it will calculate and tabulate all of the effects of the changes. The use of such a tool in the analysis of steam systems not only improves the detail, the accuracy and reliability of the calculations, but will also speed up the computations immeasurably.

Computer Simulation
Of The Preliminary Mass Balance

The example steam system has been modified schematically into a model or numerical adaptation which complies with the computer program. See Figure 2. The components which represent boilers, turbines, generators, PRV's, process heat loads, flash tanks, and feedwater treatment are recognizable since they are quite similar to the preliminary schematic diagram counterparts. In addition to these familiar items, the steam headers (in particular) are fitted with a number of splitters and mixers that separate or join two streams. Despite their extreme importance in this modeling technique, they appear to be no more than pipe tees in the

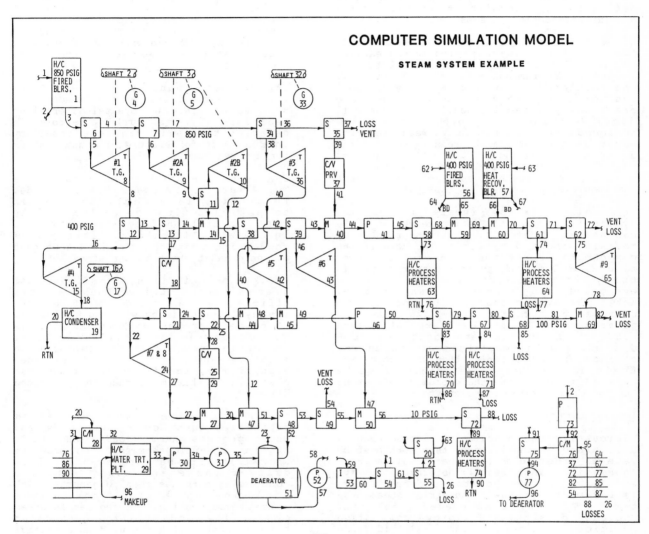

Figure 2

system. A number of priorities can be established with the mixers and splitters, and in the program run, mass and energy balances are made around each of them repeatedly as the properties, or flow rates, of the streams being joined or split vary.

As can be seen in the diagram, every component has been assigned a number, which need not be in order, but none of which may be repeated. Similarly, every stream has been identified by number, and none of these may be repeated. This is the method used to define the system with its many components and interconnecting streams.

After the above format has been established, the operating data gathered in the earlier stages of the study can be fed into the computer program and a run made. Typically, the tabulated results are carefully reviewed and compared with values established in the preliminary steam balance. Variations will occur because of improved accuracy now achieved, but discrepancies resulting from input data errors can be readily changed and a corrected run made. This will produce the "Base Case" for the plant steam system. It will represent the best available information and knowledge pertaining to the system.

The engineer is now prepared to test the system for apparently desirable changes. He can answer "what if" questions promptly by revising input data for one or more of the components or streams as needed to reconfigure, and making another computer run. In turn, the results of each test case are evaluated, and successive runs are made, until the system analysis is complete. The changes that occur in fuel-firing rates and power generation are evaluated after each run. If the changes are positive as predicted, and the magnitude of the energy savings are acceptable, then unit costs for fuels and purchased electricity must be applied to determine the economics of making the indicated system improvements.

Implementation

Ultimately, the practical and sound changes determined in analyses of this sort **must be made** in order to achieve the actual savings. Aggressive managers will determine the ways and means to make positive, constructive improvements in their industrial steam systems. They will save energy dollars, increase profits, and improve competitive position. The managers who persist in finding reasons not to do things will fail to take advantage of powerful analytical techniques such as this, and their positions will not improve.

SAVINGS OPPORTUNITIES

Example System

In plants where steam is the dominant medium, it is not uncommon to find that savings in fuels and avoided power purchases can range from 10 to 40 percent of annual costs. The example system has significant potential for savings, some of which are listed. Refer again to Figure 1.

1. Improve boiler efficiencies. This includes complete combustion with low excess air, and additional heat recovery.

2. Maximize generation from HRB's. This will reduce load on the 400 psig fuel-fired boilers.

3. Recover heat from the boiler blowdowns:
   - Flash and recover steam at the 400, 100, and 10 psig levels -- cascade system. Most of this flashed steam will then be expanded usefully through turbines instead of being throttled wastefully to the atmosphere.
   - Recover heat from the 10 psig saturated liquid by exchange with some sink, such as treated cold water enroute to the deaerator.

4. Improve performance of turbines. Check for leaks, throttling at part-loads, open hand valves, internal bypassing, erosion, seal losses, etc. Turbine performance deterioration will manifest itself by increases in temperature of exhaust steam for a given load.

5. Challenge the economics of operating the No. 4 turbine in the condensing mode. This unit, in effect, operates in competition with the public utility central station from which power is purchased. Like the utilities, this condensing unit rejects the latent heat of the exhaust steam to the atmosphere.

   Unlike the utilities, this industrial system cannot achieve Rankine cycle efficiencies in the order of 35 percent. Its cycle efficiency will more likely be 20 percent or less. Economies of scale do not favor the small industrial condensing power turbine. Depending on relative costs of fuels and purchased electricity, the in-plant kilowatt can cost 1.2 to 1.6 more than the purchased kilowatt.

   In plants where power generation capacities equal or exceed loads, and where process thermal loads cannot fully utilize exhaust steam from back-pressure units, it will be more economical to deliver the shaft power with the condensing turbines than to generate with back-pressure units that cause atmospheric venting. In the sample steam system the modest sized condensing turbogenerator might be used to advantage in avoiding high demand of purchased power, particularly where demand charges are high and ratcheted. At all other times, steam flow through the condensing unit should be kept to minimum (tail-end cooling) flow rates.

6. In plants where condensing turbines are operated for whatever reason -- unusual economics, reliability, or habit -- they should be made to operate with highest possible efficiency. In the example, turbogenerator No. 4 is shown operating to an exhaust pressure of 17 in. Hg A. This is an unusually high condenser pressure and will exaggerate most actual cases. However, there are units that operate this poorly, and the problem is

caused by inadequate cooling of the condenser or by an excessive infiltration of air into the system. Regardless of cause, either of which should be corrected, the machine operates with a poor steam rate and requires excessive steam flow per unit of output.

7. Improve the cycle efficiency by adding a regenerative feedwater heater to the 100 psig steam header (and possibly to the 400 psig header). It is an established fact that efficiency improves when boiler feedwater heaters condense steam that has passed through back-pressure or extraction turbines on its path from the boiler, and develops shaft power enroute.

8. Where substantial steam quantities pass through PRV's, such as the one operating between the 400 and 100 psig headers, consideration should be given to adding a steam turbine between these pressure limits. Properly sized, a turbine can expand nearly all of the steam that presently flows through the PRV, but it will produce power in place of the wasteful throttling of the steam that now exists.

The economics of this change will be dictated by the amount of shaft power that can be converted from the available enthalpy drop. Economics permitting, a good, efficient machine should be purchased rather than a wasteful turbine that would be little better than the PRV. Please note that until this PRV flow is minimized, the addition of the feedwater heater in Step 7 will not be prudent.

The turbine installed in this recommendation could be selected as a dedicated drive for some existing or new machine. However, the more appropriate load will probably be an induction generator that does not require steady turbine output. Recall that the PRV flow will vary in response to downstream pressure changes (load), and that the new turbine throttle flow might also be variable.

9. There is an obvious vent to the atmosphere from the 10 psig header -- fitted with a muffler, in fact. Where you find muffled flows to the atmosphere, you will undoubtedly find flows. **Steam venting to the atmosphere is avoidable and will be stopped in a properly managed steam system.**

In this particular case, steam demands from the 10 psig header, although fairly large, do not equal or exceed the flows that are let down to the header through turbines. The PRV has automatically and properly backed off to little or no flow. Therefore, the presence of vent steam simply means too many turbines, or inefficient turbines are operating.

After all 10 psig steam sources have been checked for leak-by or blow-through (and this includes PRV bypass valves), consideration must be given to replacing one or more turbines with equivalent companion (or purchased) electric motor drives. A reasonable alternative in this

case will be the replacement of units No. 7 and 8, the combined flows of which would stop the vent and open the PRV to minimal flow. Similarly, turbines No. 6 and 8 can be replaced with motors and elimination of their flows would stop the vent. It is likely that the economics of machine replacement would result in units No. 7 and 8 being removed.

Controlling exhaust flow from turbine No. 2 would represent a third option, but might be the least desirable because of the generator output requirements, and the excellent steam rate between the operating pressures of 850 psig and 10 psig.

10. During the data gathering phase (learning the system) of this project, all connected process loads were carefully examined. Several savings opportunities were discovered:

● One of the heaters that condenses 400 psig steam was found to be operating with the steam shell pressure at 150 psig. After additional study, it was agreed that the process fluid could be properly heated for normal throughputs with 100 psig steam. Plans were made to repipe the exchanger to normally be operated from this header, with the 400 psi connection being relegated to alternate source steam. Changes of this type will permit that mass of steam required in this heater to produce additional work between the 400 and 100 psig headers.

● It was also determined that one of the heaters taking 100 psig steam operated at 45 psig. An evaluation was made to verify that by doubling the heat exchanger surface area, the process stream could be heated by using 10 psig steam. Again, the additional increment of turbine work between the 100 and 10 psig headers would have to pay for the investment in the heat exchanger and a new turbine. This technically acceptable change might prove to be uneconomical.

● Some of the condensate from the heater units was collected and pumped back to the power house. Some of it was ditched. The wasted condensate was found to be more distant from the power house and would require extension of condensate piping, and the addition of collection tank/pump sets. With today's fuel costs, there is justification for returning large quantities of high temperature condensate.

● While examining the heater steam and condensate flow controls, it was found that the trap bypass line valve was open on one of the units from which condensate was recovered. Since this was a closed system, no steam was observed blowing to the atmosphere at the heater. However, when the condensate header was traced to the flash tank, steam was found blowing vigorously through the atmospheric vent. The trap had apparently failed and when the operator was unable to maintain temperature on the process fluid, he knew from past experience all he had to do was bypass the trap for a short while. As is most often the case, the bypass valve remained open.

- Drain systems on the other heat exchangers were checked to verify that noncondensible vent taps were properly located, and that the gases were being automatically vented.

11. Checks were made for process steam sparges. One case was found where a large steam flow was being injected into a distillation column. The flow was not metered, but by valve trim characteristics it was estimated to be about 50,000 pounds per hour. The need for this high flow was challenged.

The operating supervisor reluctantly cut the flow by modest increments over a period of time while he carefully monitored the separation and other parameters. He could no longer meet product specification if the steam flow was reduced below approximately 38,000 pounds per hour. He then slightly reopened the steam flow control valve, and operated satisfactorily while saving at least 10,000 pounds per hour.

In addition, he was able to reduce the cooling water flow through the overhead cooler/condenser, and cut the load on the cooling tower which had marginal performance at best.

12. It is not unusual to find substantial plumes of steam being vented from various flash tanks that receive hot condensate or blowdown waters. This one atmosphere vapor represents a loss of latent heat that should be recovered by some practical method. For example, a condensate flash tank can be fitted with cold treated water nozzles to provide quenching sprays to save the heat of the vapor. Cold makeup water heat exchanger coils can be fitted in the domes or vents from blowdown flash tanks. Of course, the piping runs and vessel appurtenances must be paid for by energy savings, with adequate return.

13. The example steam system was found to have an array of transport losses. These included a number of steam leaks with various levels of severity; there were a number of pipe sections, flanges, and valve bodies that had damaged or missing insulation. The steam line drip legs were fitted with a variety of steam traps, a few of which blew through steadily. Some of them discharged to waste, while others were piped to flash tanks in nearby buildings.

Condensate return lines were found to be bare throughout most of the system; a number of line and threaded joint leaks were found that resulted in a substantial loss of this liquid. It was reported that the use of return line corrosion

inhibitors had been started only recently. The economics of properly weatherized insulation systems were being evaluated.

This list of savings opportunities is not complete. As a result of this steam system analysis project, responsible supervisors and alert operators will intensify the search for significant avoidable losses and additional power potential, to further improve plant performance. Large financial savings resulting from these analyses far exceed the costs of the studies.

CONCLUSIONS

There are several inherent characteristics of an industrial steam power system operating in proper balance:

- The balance requires the efficient production of steam at the correct thermodynamic state, with mass flows equal to process demands.

- It requires the highest practical extraction of work from the steam as it expands through turbines enroute to valid heat loads.

- Flows through PRV's must be limited to rates that are essential for downstream pressure regulation.

- Steam energy must be applied to the processes at the lowest practical thermal state.

- There must be no venting of steam from your system.

Where properly applied, the simultaneous or sequential production of shaft power, and the supply of lower grade heat energy to valid processes is an excellent economic practice. Industry has employed these tactics for years with varying degrees of success. In recent years you have heard the term cogeneration applied -- usually with little agreement as to definition. However, it has always made good sense to knowledgeable industrial managers to cogenerate where conditions are favorable.

---

(1) SYNTHA III is a proprietary program of the SYNTHA Corporation, 41 West Putnam Avenue, Greenwich, CT 06830

Chapter 61

APPLICATION OF ENERGY CONSERVATION TECHNOLOGY TO STEAM BASED CENTRAL PLANTS

C. J. Kibert

Central Plants seek to take advantage of the economies of scale to provide heating and cooling for a group of buildings by concentrating all heating and cooling apparatus, except terminal air handlers, in a single location. The apparatus then provides conditioning to the buildings by pumping hot and chilled water through an underground piping network to the outlying buildings. The central location of the conditioning apparatus results in reduced operating costs because of a reduced requirement for maintenance personnel and reduced transportation costs.

Existing Central Plants which derive their primary energy from steam are prime candidates for the applications of energy conservation efforts and technology. The typical Central Plant is operated in a manner dictated by the experiences of the operators and will probably not be operated near optimal or least cost operating point. For Central Plants a number of considerations for energy conservation effort are:

1. On-line Optimization

2. Cogeneration

3. Pumping Modifications

4. Flue Gas Economizer

In order to give a concrete example of how these technologies can be applied, the Central Plant of the University of South Florida will be used as the basis (Figure 1.). This particular Central Plant has 4 boilers, each with 45,000 pounds per hour (PPH) capacity, two producing 235 psig (sat) steam and two producing 235 psig (100° superheat) steam. A variety of chillers provide chilled water for the campus. Two use high pressure steam. One is a condensing turbine which produces 2400 tons of refrigeration, the other a non-condensing or backpressure turbine with 1200 tons capacity. Three absorption chillers, each of 600 ton capacity, use low pressure (20 psig) steam as an energy source. The sixth chiller is a 1200 ton electrical centrifugal. The Plant also has four hot water heat exchangers, a 350 KW cogeneration turbine, and a number of large chilled and hot water pumps.

ON-LINE OPTIMIZER

The premise of an "on-line optimization

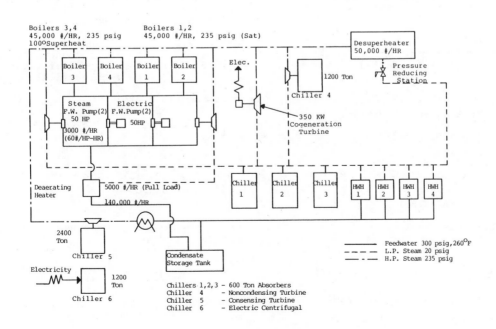

FIGURE 1. U.S.F. CENTRAL PLANT SCHEMATIC

system" or On-line Optimizer is that least cost operation of a plant can be had by creating models of important plant systems which faithfully describe the individual and integrated component operation and then analyzing the models in terms of plant conditions to optimize operations. The analysis is performed by means of a digital computer which contains the models, necessary mathematical analysis tools, and interfaces to sensors located throughout the plant. The sensors determine all critical temperatures, pressures, flows, and power consumption to allow performance analysis to be conducted.

For a typical boiler (Figure 2.) the temperature of the feedwater, the steam temperature and pressure, and fuel flow into the boiler are critical parameters to be measured. A model of the boiler system in terms of cost is then required. We begin by determining the amount of energy required for boiler operation for a given steam output:

$$E_i = \frac{h_i X_i}{N_i} \qquad (1)$$

where E_i = energy input, BTU/hr

   X_i = steam output, PPH

   N_i = efficiency at given steam output

   h_i = enthalpy input to steam, BTU/lb

   i = boiler number

N_i is the efficiency curve, one for each boiler which may be a polynomial determined by one of several fitting procedures. It will normally have the form:

$$N_i = a_i + b_i X_i + c_i X_i^2 + \ldots \qquad (2)$$

Figure 3 illustrates a set of curves for a 4 boiler system such as the USF Central Plant. The cost per hour of operation for boiler i is then:

$$K_i = \frac{q\, h_i X_i}{(a_i + b_i X_i + c_i X_i^2 + \ldots)} \qquad (3)$$

$$K \text{ in } (\$/hr)$$

where q = cost of fuel ($/BTU)

The total cost of boiler operation is

$$\hat{K} = \Sigma \frac{q\, h_i X_i}{(a_i + b_i X_i + c_i X_i^2 + \ldots)} \qquad (4)$$

$$\hat{K} \text{ in } (\$/hr)$$

subject to constraints:

$$\Sigma X_i = T_B \qquad (5)$$

$$M_i \geq X_i \geq 0 \qquad (6)$$

where T_B = total steam demand, PPH

   M_i = maximum boiler load, PPH

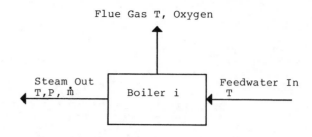

FIGURE 2. TYPICAL BOILER PARAMETERS

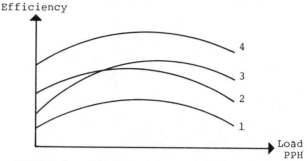

FIGURE 3. EFFICIENCY OF BOILERS FOR A FOUR BOILER SYSTEM

Each chiller is instrumented in a fashion similar to the boiler system to determine individual machine performance (Figure 4.). The energy input to chiller j, E_j (BTU/hr):

$$E_j = \frac{Y_j}{\beta_j} \qquad (7)$$

where Y_j = cooling load (BTU/hr)

   β_j = coefficient of performance

The cost of operating chiller j is C_j ($/hr). C_j depends on the type of chiller. If we use the following subscript system:

$j = 1,2,3$	:	absorbers
$j = 4$	:	noncondensing turbine
$j = 5$	:	condensing turbine
$j = 6$	:	electrical centrifugal

The cost functions are:

$$C_{1,2,3} = \frac{\hat{K}}{T_B h_{TC}} \qquad (\$/BTU) \qquad (8)$$

$$C_4 = \frac{\hat{K}}{T_B h_{NTC}} \qquad (\$/BTU) \qquad (9)$$

$$C_5 = \frac{\hat{K}}{T_B h_{TC}} \qquad (\$/BTU) \qquad (10)$$

$$C_6 = \frac{e}{3413} \qquad (\$/BTU) \qquad (11)$$

where h_{TC} = enthalpy change, condensing turbine, BTU/lb

h_{NTC} = enthalpy change, noncondensing turbine, BTU/lb

h_{ABS} = enthalpy change, absorbers, BTU/lb

The cost of operating chiller j is, Q_j ($/hr):

$$Q_j = \frac{C_j Y_j}{\beta_j} \qquad (12)$$

The total cost is, $\hat{Q}$ ($/hr)

$$\hat{Q} = \sum_j \frac{C_j Y_j}{\beta_j} \qquad (13)$$

with constraints

$$\sum_j Y_j = T_C$$

$$P_j \geq Y_j \geq 0$$

where T_C = maximum total chiller load, BTU/hr

       P_j = maximum chiller load, BTU/hr

The Coefficient of Performance, β_j, is similar to the boiler efficiency curve except that it is a function of both load and entering condenser water temperature (ECWT), T_k. That is:

$$\beta_j = f(Y_j, T_k) \qquad (14)$$

β_j can be expressed as a function of chiller load alone and an interpolating procedure for inexact ECWT's developed:

$$\beta_j(Y_j, T_k) = a_j + b_j Y_j + c_j Y_j^2 + \ldots$$
$$(\text{constant } T_k) \qquad (15)$$

A family of curves at various fixed T_k's is generated and used by means of a bias to produce the COP curve for the current ECWT as (Figure 5.). Another method of generating COP curves is by multiple nonlinear regression which treats load and ECWT as independent variables. The result is an equation of the form:

$$\beta_j(Y_j, \Delta T_j) = a_j F_j^2 + b_j \Delta T_j^2 + c_j F_j +$$
$$d_j \Delta T_j + e_j F_j \Delta T_j + f_j \qquad (16)$$

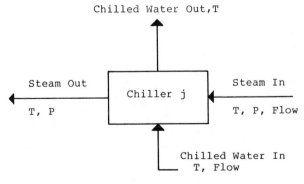

FIGURE 4. TYPICAL CHILLER PARAMETERS

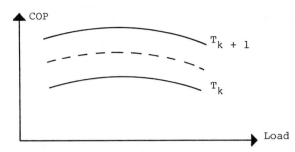

FIGURE 5. CHILLER PERFORMANCE VS. LOAD , ECWT

Once the major system models have been designed, analysis of the overall system can be conducted using various optimization techniques. These techniques must have the capability of determining the optimum of a nonlinear function subject to linear equality and inequality constraints. The techniques used in the On-line Optimizer for the USF Central Plant were:

1. Lagrange Multiplier
2. Nelder-Mead Simplex
3. Projected Gradient
4. Patterned Search

The Lagrange and Projected Gradient routines are mathematical analysis techniques which take advantage of the characteristics of the cost functions and differential calculus to determine the optimum point. The Nelder-Mead and Patterned Search techniques are "search" techniques which geometrically seek the optimum. Penalty functions which prevent the search from leaving feasible search regions must be incorporated into these latter two techniques.

On-line Optimizers usually incorporate one technique for analysis. The USF system used all four methods in order to provide a mean for cross-checking results. The Lagrange Method also has the capability for microcomputer implementation due to the compact size of the program which can be generated.

The actual optimization procedure must sometimes solve a coupled system of equations. This is necessitated by the tradeoff in steam costs versus electrical energy costs when steam driven machinery is optimized together with competing electrical machinery. This occurs in the USF Central Plant where steam driven turbines and absorbers are analyzed against an electrical chiller. Figure 6 illustrates a simplified schematic of the procedure for solving this "coupled" problem. A steam flow guess is made, the boilers optimized, and the resulting steam cost is fed into the chiller optimization routine together with the cost of electricity. The result of the chiller optimization is steam flow required, cost of chiller operation, and electrical requirements. The new steam requirement is compared to the initial guess. If the two are sufficiently close, the procedure terminates. If they are not within the preselected criterion for convergence, the procedure continues until satisfactory convergence is achieved.

Software to implement the On-line Optimizer includes the optimization routines, a routine to generate the boiler and chiller performance curves, a status reporting system, and an integrator system for determining periodic totals of flows, energy of usage, and operational costs (Figure 7.).

COGENERATION

Central Plants often have usages for high pressure and low pressure steam. If the high pressure steam uses are less than the low pressure requirement, a pressure reducing valve (PRV) may be required to throttle steam to low pressure to meet these needs. A turbine-generator may often be justified to fulfill the low pressure requirements, and may be retrofitted or installed in parallel with the PRV.

Analysis of plant logs can be utilized to determine a reasonable expected flow rate to size the turbine generator. The savings resulting in this retrofit can be stated as follows:

$$K_4 = K_3 + (K_2 - K_1)t \qquad (\$) \qquad (17)$$

where K_3 = demand savings
K_2 = electricity generated
K_1 = additional steam requirements

The demand savings to K_3 are:

$$K_3 = \frac{\dot{m}_{CG} P_o k_d}{3413} \qquad (\$/hr) \qquad (18)$$

where $\dot{m}_{CG}$ = steam through cogenerator, PPH
P_o = generator output, KW
k_d = demand charge, \$/Kw
3413 = converts BTU to KW

The worth of electrical energy generated is:

$$K_2 = \frac{\dot{m}_{CG} P_o k_e F}{3413} \qquad (19)$$

where k_e = electricity worth, \$/KWH
F = load factor

Additional steam required in terms of cost is K_1:

$$K_1 = \frac{Q_L k_g}{N_B}$$

where Q_L = steam energy , BTU/hr
k_g = fuel cost, \$/BTU
N_B = boiler efficiency

For a steam flow of 16,270 PPH, a cogenerator for the USF Central Plant provided a savings of \$6,528.00 per month. The simple payback for this system was 2.6 years. Figure 1 shows the location of the cogenerator in the USF Central Plant.

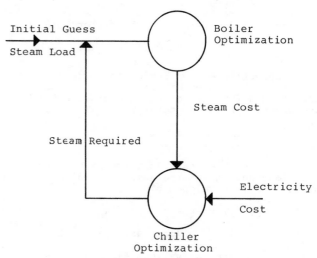

FIGURE 6.   SOLVING THE COUPLED BOILER-CHILLER OPTIMIZATION PROBLEM

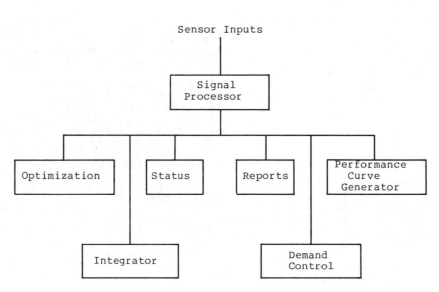

FIGURE 7.   ON-LINE OPTIMIZER SOFTWARE

PUMPING MODIFICATIONS

Expansion of a physical plant which is served by a Central Plant results in obvious changes in the circulation of hot and chilled water to the various buildings. As more buildings are added, substantial modifications to the Central Plant pumps are inevitably required to meet the changes in pumping head requirements which occur. Control valves in the buildings respond to the lack of pumping head at the Central Plant by remaining in a full open position continually. Degradation in coil performance also results because an insufficient differential pressure does not exist to allow design flows through the coils.

In the case of the USF Central Plant the area serviced had grown from about 2,000,000 square feet to over 3,000,000 square feet from 1970 to 1981. Analysis of the chilled water pumps indicated that they were operating inefficiently and not delivering proper flows to the campus. Consequently a 400 hp pump was installed to increase flow from 11,300 gpm at 136 feet of head to 15,000 gpm at 220 feet of head. The consequence was a reduction in throttling required on the larger pumps and a return to design conditions in the operation of the buildings. The addition of the 400 hp pump resulted in reduced pumping costs because the entire system could be operated in a region of maximum efficiency for the pumps. The total cost of pumps, motor starter, piping, pump pad, and wiring was approximately $43,000.

ECONOMIZER ANALYSIS

The USF Central Plant was analyzed for the installation of flue gas heat exchangers or "economizers" for combustion air preheating. With a stack temperature of 470 degrees F and a reduction in temperature to 265 degrees F a savings of over $33,500 per year was claculated. This computation was based on the following:

    4 Boilers with 160,000 PPH total
      capacity
    Air flow of 27,000 PPH per boiler
      at full capacity
    Annual operating hours: 6,210
    Combustion Efficiency: 90 %
    Excess Air: 2.5%
    Cost of Economizer,each: $ 70,000

The result of the analysis was a simple payback of 4.2 years. Due to high initial costs this particular measure was not selected for implementation.

SUMMARY

Central Plants lend themselves well to a number of energy conservation measures, some of which have been described in this paper. Knowledge of component and system efficiencies is critical to the ability to operate the plant at least cost. A number of optimization techniques have been described which are readily adaptable for solving the boiler-chiller system of cost functions. In cases where the Central Plant has high pressure and low pressure uses a cogeneration turbine may be appropriate for obtaining useful work at the point where throttling normally occurs. Pumping modifications to improve system performance are appropriate in cases where the pumps and system are not well matched due to expansion of the building area which the Central Plant serves. Economizers are appropriate in cases where there is a high flue gas stack temperature with a geometry appropriate for installation of a heat exchanger.

SECTION 19
SOLAR

Chapter 62

SHALLOW SOLAR POND PROGRESS

D. L. Burrows, D. R. Dinse, V. A. Needham III

ABSTRACT

Several lost-cost shallow solar pond projects are now under design or are being constructed throughout the nation. Four such systems are presently in design or pending construction for commercial and industrial applications with the Tennessee Valley Authority participation. These applications include: building service hot water for a hospital, process hot water for a photographic laboratory, process hot water for industrial textile dyeing, and TVA's own shallow solar pond performance test modules.

Design and construction techniques for these solar applications have produced system ranging for $10 - $30/ft² and yielding approximately .15 - .20 MM Btu/ft²yr. These systems are compared on differences in design, and actual or estimated construction cost. Economics are discussed. Some performance data comparison is presented on projects operational by June, 1982.

INTRODUCTION

In its search for practical options in energy sources for commercial and industrial applications, the Tennessee Valley Authority (TVA) is one of the first utilities in the Nation to test and demonstrate the effectiveness of shallow solar ponds for large volume water heating.

The shallow solar pond (SSP) is an offspring of flat plate solar energy collectors designed to supply large amounts of heat, ideally suited for commercial and industrial applications at a competitive cost with conventional fuel sources. Lawrence Livermore National Laboratory (LLNL), Livermore, California, has explored the SSP technology (See Figure 1) over the past several years in many different configurations utilizing many different materials in pond constructions[1]. The design which is discussed in the following now appears to offer a cost-effective advance in conventional flat-plate technology, where the load situation is such that the SSP can make a contribution in the normal, batch-mode operation. Typical initial cost at present for different sized SSP systems range from $10-$30 per square foot of installed collector area or approximately $70-$210 per MBtu. Performance projections by LLNL for a 16' by 200' (5m by 60m) SSP module employing a black plastic water bag filled to a 4-inch (100mm) depth are that under ideal summer conditions in the southwest portion of the United States, up to 0.3 MBtu per square foot-year will be collected and that the 16' by 200' module would then supply 8,000 gallons of water daily heated to 140 degrees Fahrenheit (assuming an inlet temperature of 60 degrees Fahrenheit). LLNL states further that a typical system may be expected to deliver 0.25 MBtu per square foot-year. TVA is now field testing its own SSP modules in order to establish an annual energy yield figure typical of the TVA area (some preliminary results of this testing are presented in table 2).

LLNL has been primarily responsible for the previous research and development work which has led to the present state-of-the-art in SSP technology. LLNL's work has focused on identifying low cost materials for pond module construction which yield a cost-competitive alternative source of energy without sacrificing performance to the point that cost effectiveness is negatively impacted. This process of convergence has yielded the present design for solar shallow ponds, a cross-section of which is shown in Figure 1. The pond module construction process is as follows:

o Clear land of all plant growth and treat pond bed with herbicide.

o Erect reinforced concrete curbing around pond bed.

o Remove curbing forms and screed in place (in the bottom of the pond bed) a layer of dry sand, sloped inward to the pond centerline.

o Place rigid foamglas insulation on top of sand.

o Place black plastic water bags and water manifold on top of insulation.

o Cover pond with corrugated fiberglass glazing strips, secured to curbing with key clamps and railing fittings. Glazing is held in place by conduit support bows and stainless steel band strapping.

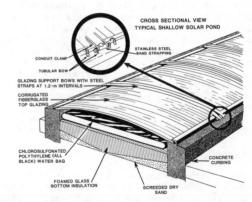

FIGURE 1. CROSS SECTION - TYPICAL SHALLOW SOLAR POND

Clearly, the shallow solar pond module itself is a simple, low-profile solar energy collector which presents no significant cost hurdle ($4-10 per square foot) to be overcome. It is, in fact, less expensive to build, per square foot, than the conventional flat plate solar collector. The SSP thus offers a significant solar technology to the commercial and industrial sectors, where the rapid recovery of cost traditionally required by these customers has not typically been available through investments in solar applications.

Through the projects now underway in the nation, and through TVA's own independent testing, the cost effectiveness of the SSP technology on the commercial scale is being determined with actual field data. The remainder of this paper will briefly overview the SSP projects in progress and TVA's efforts in both the laboratory environment and the field.

Overview of Projects

Although several projects have been initiated in the past (Sohio, Sweet Sue[2]), six ongoing project sites will be discussed: the California Forestry Service installation in El Dorado County; Union General Hospital in Blairsville, Georgia; Milan Army Ammunitions in Milan, Tennessee; U. S. Army installation in Ft. Benning, Georgia; the Michigan Dept. of National Resources installation near Kalamazoo, Michigan; and the TVA installation in Chattanooga, Tennessee. Estimated and actual system cost and performance are listed in Table 1.

The California Department of Forestry

The California Department of Forestry is finishing construction on a one-pond (16' by 100') SSP system at its Growlersburg Conservation Camp in El Dorado County. The camp is manned with prison inmates who work at various conservation projects which the Forestry Department plans, and these same inmates are constructing the SSP system at Growlersburg Camp. The pond is of the conventional LLNL design of approximately 18 inches high, steel reinforced concrete curbing containing a layer of screeded, dry sand, rigid foam insulation, black rubber water bags, and covered by a layer of corrugated fiberglass glazing. The pond will provide approximately 4,000 gallons-per-day of heated water which service the needs of the inmates in detention there. One minor design alteration has been necessitated at the site due to the occurance of heavy, wet snows during the winter months. This has been the addition of treated, redwood braces under the center points of the glazing support bows to prevent glazing collapse under snow loads. Except for the pond materials and the system controls, the Growlersburg Camp SSP is being constructed at minimal cost to the Forestry Department due largely to the facts that 1) the inmate labor charge is a fixed cost to the state, and 2) there is existing storage at the site from which supply water to the pond will be provided. It is not required that other dedicated cold water supply be constructed. Although additional shading problems at the site remain to be overcome, the project now stands at 95-percent complete, according to LLNL.

Union General Hospital

The Union General Hospital, located in Blairsville, Georgia, will soon be bidding a SSP system composed of one 16- by 150-foot pond, a 6,000-gallon potable water storage tank, associated valving and controls, and a 500-gallon capacity surge tank. The surge tank will bring the pond water to required pressure (approx. 65 psig) before introduction into the present oil-fired boiler, thereby displacing the need for an estimated 8,300 gallons of fuel oil annually (this figure includes boiler efficiency). The planned SSP system is of the conventional-type construction developed by LLNL and will receive funding support from the Appalachian Regional

Commission, the Tennessee Valley Authority, and the hospital. Upon completion, the SSP will be monitored by TVA in order to assess actual performance and cost effectiveness. Plans now call for an award date in early June, with construction to commence in late June or early July (exact dates remain to be fixed).

Milan Army Ammunition Plant Shallow Solar Pond

The Milan Army Ammunition Plant in Milan, Tennessee, is considering the construction of a shallow solar pond to provide process hot water for the plant's X-ray facility.

The plant uses approximately 5,000 to 6,000 gallons per day of 73-degree water which is presently heated by an oil-fired steam boiler through a hot water converter. This is the only load on the boiler in the summer; thus, it could be shut down in the summer if a shallow solar pond supplied the 73-degree water. This would allow for maintenance of the boiler as well as saving man-hours and energy because the existing boiler does not operate efficiently in the summer with the partial load. The ponds operate most efficiently during the summer when this application needs the energy.

The initial pond design was completed, however, construction estimates were substantially over budget. One reason for this is that the control and distribution system required is rather elaborate for only one pond module (16' by 100'). Multiple ponds can utilize the same control and distribution system, thus reducing the cost per square foot of pond area. Presently, the pond is being redesigned to try to bring it within budget.

The United States Army Corps of Engineers

The United States Army Corps of Engineers, Savannah District, is presently building the world's largest shallow solar pond (SSP) water heating system at the Fort Benning military installation near Columbus, Georgia. The final system will incorporate 80 16- by 200-foot shallow solar pond modules, an approximate 500,000-gallon above ground storage reservoir, and 2 pumps capable of delivering a total of 800 gallons-per-minute each to supply heated water for approximately 6,600 personnel and half of the post's laundry. The shallow solar pond modules are of the conventional construction developed by Lawrence Livermore National Laboratory (LLNL). As of May 1, 1982, the project was approximately 35-percent complete with roughly one-third of the 80 concrete curbs in place and trenching well underway for placement of piping. The Corps of Engineers now estimates a project completion date of February 25, 1983, which is reported as 10 percent ahead of the original construction schedule. A significant feature of the Fort Benning installation is its cost. At $15.50 per square foot of installed pond, the project will supply 6.4×10^4 MBtu annually (LLNL projections) for water heating. The corps projects an annual savings of $225,700 as a result of the shallow solar pond and anticipates 100-percent amoritization in just under 10 years. John J. Harte and Associates of Atlanta, Georgia, was employed by the corps as the design engineering firm for the project. The design will incorporate full onsite automation and is specifically intended for integration at a later date with a full-post energy management system.

The Michigan Department of Natural Resources

The department's Fisheries Division presently employs a shallow solar pond water heating system at its Wolf Lake Hatchery near Kalamazoo, Michigan. The SSP system is used to provide heated water which then passes through a heat exchanger for a fish breeding operation in which harvest is a function of water temperature up to an optimum point. Two unique features of this particular operation are that it is a summer only application, and that the pond modules are unglazed; a cover is put in place over the modules during periods of non-use and the Fisheries Division estimates the life of the water bags at roughly twenty years. In an additional effort to hold down cost, this SSP system uses compacted gravel as an insulating material under the water bags in place of the rigid foam insulation used in the conventional LLNL design. (System size and cost data were not available at the time of this writing.)

TVA Shallow Solar Test Pond

As part of a Tennessee Valley Authority effort to promote energy conservation, two shallow solar test ponds have been constructed in Chattanooga, Tennessee.

The ponds are 16' by 50', are extensively monitored, and seek to compare the performance of clear top/black bottom PVC bags and all-black hypalon bags. The construction is complete and the ponds are being manually operated with the automatic operation and data collection system scheduled to start in June of 1982.

The ponds will be filled to a two inch depth and operated for one week, with stagnation allowed over the weekend. This process will be repeated for a four inch depth the following week and then a six inch depth the next week. This entire sequence will be repeated for one year.

The data will be used to project pond performance for commercial and industrial applications by validating a TVA computer analysis and simulation of the ponds written around the TRNSYS computer program. This computer model was first developed using data from the Lawrence Livermore Laboratory shallow solar ponds in Livermore, California.

The TVA ponds will provide data to make accurate performance estimates for the Valley climate as well as allowing potential contractors and owners to view an actual pond installation.

RESULTS AND CONCLUSIONS

Performance

The results of the preliminary computer collector simulation and a limited amount of actual data at the TVA shallow solar test pond are shown in Table 2 below. A significant difference in the test pond data basis and the computer simulation is evident. This is because of differences in the inlet water temperatures. The computer simulation uses varying water main temperatures for Chattanooga (42 degrees in February to 74 degrees in August). The test pond uses a recirculating system from a holding pond. The average inlet temperature to the test pond was approximately 73 degrees Fahrenheit versus a water main temperature of 49 degrees for April. Pond performance is sensitive to inlet water temperature and pond depth; thus, the test data base pond performance will tend to be conservative. The test pond was run at a 4 1/2-inch depth, while the computer simulation was run at a 4-inch depth. Actual test data basis is presently limited to manually collected data in the spring of 1982. The testing will soon be fully automated with an extensive data acquisition system. The test data

basis numbers were calculated using the elevated inlet temperatures of the recirculating system. Typical pond depths are 2, 4, and 6 inches. Running the ponds deeper will provide higher efficiencies, but lower discharge temperatures.

Also shown in Table 2 are the horizontal plane insolation and a computer simulation of energy delivered for an actual system including transmission and storage losses.

Shallow solar ponds appear to offer a good alternative to conventional solar water heating systems because of the construction cost advantage. Annual performance is competitive with other flat-plate type collectors although it is heavily skewed to the summer months.

Cost Implications

LLNL's most recent cost projections for the SSP module itself are $4 to $7 per square foot of installed collector. The cost of associated installed mechanical interface delivery systems vary widely depending on several factors: physical location of pond(s), existing conventional water heating systems, required building water pressure, geographic characteristics of site, etc. While economies of scale cannot, at present, be verified to exist for all delivery systems, as was seen in the Fort Benning, Georgia, SSP installation, the cost per square foot of the installed modules certainly appears to exhibit economy of scale characteristics. TVA intends to more accurately identify and quantify, through its own SSP project involvement, where economics of scale exist for the SSP technology, if they do. Of course, this can only begin to be done after several distinct projects have been installed and cost data for each is analyzed.

What follows is a generic example of the economic analysis employed by TVA in its evaluation of proposed SSP installations. It will be clear from the case presented that in the absence of recognizable economies of scale, the important factors to be considered in a SSP economic evaluation focus on tax implications. Available tax credits, accelerated depreciation schedules, and financing scenarios can combine to make the SSP investment very attractive to the C&I business firm which has an adequate load. In the example used, the Fort Benning SSP project discussed earlier is treated as though the project owner were a profit-making commercial customer financing the full cost of the SSP system. The following factors are known:

Conventional Fuel Cost: $2.11/MBtu (Nat. Gas)
Boiler Efficiency: 60%
System Size: 256,000 Ft^2
System Energy Yield: 250,000 Btu/Ft^2-Yr.
Initial System Cost: $3.97 Million ($15.50/Ft^2)

The following assumptions are made to complete the cash flow analysis:

Applicable Tax Credits: 15% Energy Tax Credit;
10% Investment Tax Credit

Depreciation: 5 Yr. Accelerated Cost Recovery Method

Loan: $3,970,000 over 8 Yrs. at 16% Simple Interest

Fuel Escalation (Annual): 22%
Market Discount Rate: 22%
System Life: 15 Yrs.
Property Tax: N/A

Annual O&M Cost: 1/2% of Initial System Cost
Firm's Tax Bracket: 47%

The cash flow statement, then, may appear for the project as shown in Table 3. (A positive sign in front of the "Effective Tax" entries indicates a tax saving; a negative sign indicates a tax expense.)

It is clear that the large tax credit in year one and the accelerated depreciation schedule have a critical impact on the entire statement. Noteworthy is the fact that the cummulative present worth column starts with a positive figure and remains so throughout the life of the system. As is to be expected, the cummulative present worth reaches its lowest point, $124,466, at the end of year eight when the annual mortgage payment, $913,990 is 86.2 percent principal, and after all the depreciation has been exhausted. This analysis says that the investment is definitely favorable, given that the firm can accommodate the cash flow situation in the early years of the system's expected life.

GENERAL SUMMARY

o The effective energy collection of shallow solar pond modules per square foot is comparable to that of smaller modular flat plate collectors of conventional design.

o The installed cost of the SSP module is low relative to that of conventional collector modules.

o SSP system viability at a particular site is highly dependent on the type and amount of piping required (external to the pond modules), and the mechanical system chosen to integrate the SSP with the conventional building system.

o Current tax laws provide for rapid write-off SSP capital expense which favorably impacts the typical cash flow statement and can produce a cumulative present worth which is always positive.

For the present, the primary focus of TVA's SSP testing efforts is the determination of appropriate operational strategies to obtain optimal performance in the TVA area (i.e., pond depth, drain and fill times, corresponding energy yields, etc.). Future work with the technology by TVA may, however, evaluate other designs for SSP system configuration specifically for the purposes of cost effectiveness enhancement. These designs may include:

o Roof-mounted designs employing inflatable glazing materials and lighter-weight curbing (example: extruded aluminum).

o Non-batch (flow-through) SSP's.

o Non-batch SSP's in tandem with water-source heat pumps.

Table 1

Project Cost Summary

Site (square ft.)	Size (ft^2)	Cost[c] (000's)	Cost $/ft.^2$	Cost $-yr/MBtu
California Department of Forrestry	1,600	(a)	(a)	(a)
Union General Hospital	2,500	72	28.80	144.00
Ft. Benning Military Installation	256,000	3,970	15.50	62.00
Milan Army Ammunition Plant	1,136	59	51.94	207.76
Michigan Depart. of Natural Resources	(a)	(a)	(a)	(a)
TVA Test Pond	1,600	13,600[b]	8.50[b]	
Metro	44,000	352	8.00	44.44

NOTES: (a) - information not available
(b) - collector module only
(c) - Ft. Benning and TVA test pond are actually on construction bid cost.

Table 2

Shallow Solar Pond Performance Estimate - Chattanooga, Tennessee

Month	Q Solar Horizontal Surface (Btu/ft^2)	Q Collected- Simulation: (Btu/ft^2 @ 4" depth)	System Simulation- Energy Delivered From Storage (Btu/ft^2 @ 4" depth)	Limited Actual Data Evaluated Inlet Temperature (Btu/ft^2 @ 4 1/2" depth)
JAN	19,909	7,990	6,259	4,817
FEB	26,909	13,684	9,674	8,249
MAR	34,175	19,857	14,922	11,982
APR	48,721	32,146	23,475	19,404
MAY	51,478	33,148	25,805	20,005
JUN	56,127	38,963	24,798	23,506
JUL	50,923	32,805	22,547	19,790
AUG	49,752	27,290	18,567	16,452
SEP	40,180	23,477	16,900	14,160
OCT	33,954	17,926	12,794	10,818
NOV	24,409	9,993	7,277	6,024
DEC	17,611	5,429	4,667	3,273
TOTAL	454,148	262,708	187,685	158,480

Table 3

Year	Solar Savings	Depreciation	Effect On Tax	New Cash Flow	Present Worth Of Net Flow	Cumulative Present Worth
1	225,700	595,500	481,680	+ 766,039	+627,901	627,901
2	275,354	873,400	567,990	- 90,496	- 60,801	567,100
3	335,932	833,700	496,540	- 101,368	- 55,824	511,276
4	409,837	833,700	433,594	- 90,409	- 40,811	470,465
5	500,001	833,700	358,493	- 75,347	- 27,878	442,587
6	610,001		-123,006	- 446,846	-135,518	307,068
7	744,202		-230,114	- 419,753	-104,346	202,723
8	907,926		-358,144	- 384,058	- 78,256	124,466
9	1,107,670		-511,275	+ 576,544	- 96,293	220,759
10	1,351,357		-625,808	+ 705,699	- 96,610	317,369
11	1,648,655		-765,538	+ 863,267	- 96,869	414,239
12	2,011,359		-936,009	+1,055,500	+ 97,082	511,321
13	2,453,859		-1,143,984	+1,290,025	+ 97,257	608,578
14	2,993,707		-1,397,713	+1,576,144	+ 97,400	705,978
15	3,652,323		-1,707,262	+1,925,211	+ 97,517	803,495

(A positive sign in front of the "Effect on Tax" entries indicates a tax savings; a negative sign indicates a tax expense.)

REFERENCES

1. Casamajor, A. B., and Parson, R. E., "Design Guide For Shallow Solar Ponds," Lawrence Livermore Laboratory, Report UCRL-52385, November 1 (1979).

2. Hall, B. R., and Guinn, G. R., "Final Design Report For Solar Production Of Industrial Process Hot Water Using Shallow Solar Ponds," Teledyne Brown Engineering, May 26, 1978.

3. Silver, J. D., and Wessling, F. C., "Simulation Methods Developed For The Design Of Commercial Hot Water Systems Utilizing Shallow Solar Ponds." AS/ISES 1981 Proceedings of the 1981 Annual Meeting, Philadelphia, PA., Vol. 4.1 pp. 782-786.

Chapter 63

PROJECT MANAGEMENT OF A MAJOR SOLAR RETROFIT PROJECT

B. I. Benator, R. H. Wright, H. H. Briscoe, N. D. Stubblefield

INTRODUCTION AND OBJECTIVE

In Spring 1982, Floyd Medical Center, a regional hospital serving Northwest Georgia and Northeast Alabama, put on line a solar domestic hot water (DHW) system. The DHW retrofit project, taking 2½ years from energy audit to final inspection, was implemented in several distinct phases. It was designed to produce a practical energy savings project and not to be a solar research and development project. In the same light, the purpose of this paper is not to add to the copious technical literature currently available; rather, it is to discuss the real-world project management of a major solar retrofit project. It will discuss the actual techniques used, lessons learned, and provide guidance which may be useful to others in project management of similar projects. Some background information is helpful.

BACKGROUND

The project to be described is a 6500 sq.ft. flat plate collector solar domestic hot water system retrofit project for Floyd Medical Center, a major regional hospital in Rome, Georgia serving Northwest Georgia and Northeast Alabama. The 314 bed hospital uses approximately 10,000 - 15,000 gallons of hot water per day. The hospital has had an aggressive energy management program in place since 1973 as a result of top management's commitment to cost containment of controllable expenses. Mr. Ben Popay, Maintenance Director, has instituted numerous operational and maintenance energy conservation measures (ECMs) as well as capital investment ECMs such as solar screens, energy efficient lights, heat pumps, and a computerized energy management system.

The solar system (and computerized energy management system) was part of a comprehensive government grant program which is known generically as the "Schools and Hospitals Program" of the Department of Energy (DOE). Floyd Medical Center was the first hospital in Georgia to receive an energy audit under the program in 1979. Authors Benator and Stubblefield performed that audit. In January - February 1980, the hospital then directly funded a technical assistance program which met the requirements of 10 CFR 45, the controlling regulations of the grants program. This study (also conducted by Benator and Stubblefield) recommended the computerized energy management system and a solar domestic hot water system to be 50% funded by a DOE grant.

The initial analysis of the solar system suggested roof-based 15,000 sq.ft. flat plate collector array. However, one of the roof areas was to be saved for an additional two floors growth, and DOE rejected the initial payback analysis which was conducted by incorporating a 12% energy cost escalation into the payback analysis. This necessitated a recalculation of collector array size and use of the then current natural gas price. In the two month period since the application was initially submitted, natural gas prices rose from 24¢ to 29¢ per hundred cubic feet, thus supporting the concept of escalating gas prices in determining the payback period. However, the government regulation required current gas prices so that is what was used. The project payback period was computed to be approximately 14.5 years based on 29¢ per hundred cubic feet of natural gas.

In September 1980 the Hospital Authority of Floyd County selected BENATECH, INC. as the Project Management Consultant (PMC) for the solar system. Prior to BENATECH's selection as the hospital's PMC, Floyd Medical Center applied for and was granted a Certificate of Need for the solar project (and computerized energy management system) by the State Health Planning and Development Agency. This Certificate was valid for a period of six months from July 22, 1980, a fact which was significant because energy conservation measures had to be accomplished by the hospital before January 22, 1981 or an extension requested prior to the expiration date. If an extension request were not made prior to January 22, 1981 a new application would have to be submitted. This was a milestone date simply because completing a Certificate of Need (CON) application would be a time consuming process for both the hospital and their Project Management Consultant. Both organizations put that date into a tickler file for review in late December 1980 to request extension of the Certificate, if required. This turned out to be necessary since another hospital project delayed significant work on the solar project until December 1980. Accordingly, an extension request for the CON was made and approved.

Prior to looking at prespecification activities, it might be instructive to summarize the key facts about the project to date -- December 1980.

Client: Floyd Medical Center, Rome, Georgia
         Ben A. Ansley, Executive Director
         Bill G. Waters, Associate Executive
                         Director
         Ben Popay, Maintenance Director

Description: 314 bed regional hospital

System: 6500 sq.ft. flat plate array
         Solar Domestic Hot Water
         Active System

Project Management Consultant: BENATECH, INC.

PRESPECIFICATION ACTIVITIES

Because there had been a government solar demonstration program in progress for several years preceding this project, BENATECH contacted DOE, SERI, ORNL,[1] and the National Solar Heating and Cooling Information Center (now Conservation and Renewable Energy Inquiry and Referral Service) to obtain information about similar projects. The objective in doing this was to avoid reinventing the wheel and take advantage of lessons learned from other projects. It was decided at the outset by the client that this was to be a practical energy savings project, not a solar research and development project. BENATECH contacted several institutions and their solar engineers and/or contractors that were identified in the information provided by the above government agencies and laboratories. Questions were asked of the facility operators and their design professionals as to collector type (flat plate, parabolic, etc.), collector materials (aluminum, copper, etc.), heat transfer fluids (water, water-glycol mixture, synthetic liquid, etc.), construction problems, operating problems, lessons learned, etc. The results of this checking were that although no major technological complexities were involved in a flat plate solar energy DHW system, poor collector manufacturing processes, unskilled construction personnel, inadequate construction supervision, inattention to corrosion concerns, and lack of cooperation between the principal parties (i.e. owner, engineer, contractor) involved could individually or collectively act to cause problems in the project.

In order to avoid many of these problems, BENATECH recommended that Floyd Medical Center retain a solar consultant with expertise in the design and construction of commercial solar systems. In January 1981, Claude Terry & Associates, Inc. (CTA) was selected by the hospital to serve as their solar consultant. Mr. Neal Stubblefield of CTA was project manager for a retrofit solar installation in Atlanta, and worked with Mr. Benator to analyze numerous collectors, heat exchangers, and thermal insulation types. Of particular concern was the use of collectors with aluminum absorber plates and copper tubes with the potential for dissimilar metal galvanic corrosion. A simple solution would have been to spec the collectors so that only copper absorber plates and copper tubes would be permitted. However, this would have eliminated from consideration a solar collector manufacturer located in Rome, Georgia. And local servicing of the collector would be a positive benefit for the hospital. So BENATECH undertook an extensive two month (initially planned for two weeks) corrosion research project to see if the Rome manufacturer would be pre-qualified for the project. In addition to normal library research, BENATECH contacted SERI, DOE, Battelle Memorial Institute, the Georgia Office of Energy Resources, and the National Solar Heating and Cooling Information Center. BENATECH also used a professional information search firm, INFOSEARCH, to provide additional information about solar collector corrosion. The results of this research indicated that if collector loop liquid chemistry is properly controlled, and the collector is constructed so a liquid ion path between the aluminum and copper cannot occur, risk of galvanic corrosion was minimal. Just to be sure, however, in March 1981, we worked with the Rome, Georgia collector manufacturer to examine a collector that had been operating in a commercial laundry DHW system for three years. Messrs. Benator and Stubblefield physically examined the collector for deterioration compared to new collectors currently being manufactured. There were no visible signs of corrosion, but the polycarbonate glazing had turned slightly yellow from its original clear color. Research into polycarbonate glazing had indicated that this was normal and the percent transmission loss was minimal.

In February 1981, we received the Department of Labor (DOL) wage decision required by the Davis-Bacon Act to incorporate in the bid specs for the solar system. In addition, discussions were held between BENATECH and Armour and Cape, Inc. for the latter's potential involvement as project engineer for the solar system. Other collector manufacturers were reviewed for prequalification.

In March 1981, Mr. Rick Wright, P.E. of Armour and Cape submitted a proposal to be project engineer for the solar project. Although himself a registered professional (electrical) engineer in Georgia, Mr. Benator believed the hospital would best be served by having an experienced registered mechanical engineer serve as the design professional on the project with BENATECH serving as the customer's representative and project management consultant. In April, the hospital retained Armour and Cape as the project engineer.

During the period May - June, 1981, Mr. Rick Wright prepared a draft set of specifications and drawings for the project. These were reviewed by Mr. Benator and comments provided to Mr. Wright. The final specifications were issued in July 1981. Because of government regulations associated with the project, the following regulations and requirements were incorporated into the bid spec:

"The CONTRACTOR and any SUBCONTRACTOR he may employ shall comply with the following documents or provisions

a. Executive Order 11246, entitled "Equal Employment Opportunity," as amended by Executive Order 11375, and as supplemented in Department of Labor (DOL) regulations (41CFR Part 60).

b. Copeland "Anti-Kick Back" Act (18 USC 874) as supplemented in DOL regulations (29CFR Part 3).

c. Davis-Bacon Act (40 USC 276a to a-7) as supplemented by DOL regulations (29 CFR Part 5). See requirements of 29 CFR 5.5 included in this contract.

d. Federal Wage Decision No. 81 - GA - 239 dated 5-28-81 included in this contract.

e. Contract Work Hours and Safety Standards Act (40 USC 327-330), sections 103 and 107 as supplemented by DOL regulations (29CFR Part 5).

f. The OWNER, the PMC, Department of Energy, the Comptroller General of the United States or any of the United States or any of their duly authorized representatives shall have access to any books, documents, papers, and records of the CONTRACTOR for the purpose of making audits, examinations, excerpts, and transcriptions.

g. The provisions of Attachment O to OMB Circular A-110 (7/30/76) shall be deemed incorporated into this Agreement and made a part thereof.

h. All other applicable requirements of law."

In addition to meeting the requirements of government regulations, the bid spec also included the following information and requirements:

a. Compliance with "Recommended Requirements to Code Officials for Solar Heating, Cooling, and Hot Water Systems (DOE/CS/34281-01)"

b. Ten year full warranty collector guarantee

c. Three month and six month retainage provision

d. Relationship between the Project Management Consultant and the Project Engineer

e. Two heat exchangers for double barrier protection of domestic hot water from collector loop liquid

f. Identification of three acceptable collectors based on prequalification research and ASHRAE 93-77 test results.

g. Provision to allow owner to select best overall contractor, not necessarily the one with the lowest bid.

h. Provision for liquidated damages if the job was not completed in 120 days.

SPECIFICATION PUBLICATION

There were several important considerations associated with publishing the specifications once they were prepared. These included 1) the government requirement to encourage small and minority business participation, and 2) the need to provide broad and equitable distribution of the specification, yet not inordinately delay the project. To accomplish both of the objectives, the project was advertised several times in the Rome newspapers and local area builders exchanges. Plans were provided to contractors who provided a refundable $30 deposit if they returned the plans. Specific mention was made of the owner's request for small and minority business participation in all of these advertisements. The bid opening was set for 11:00 A.M. Monday, August 17 at Floyd Medical Center, Rome, Georgia. This date provided several weeks for the interested parties to visit the site, talk with the Project Management Consultant (PMC) and Project Engineer, and obtain subcontractor bids. A Monday opening was chosen to give bidders the weekend to review their bids before submitting. The 11:00 A.M. time was selected to allow out of town bidders to bring their bids without having to stay overnight if they didn't need to. It also provided the PMC and Engineer time to check last minute details such as room availability and letting the hospital personnel know where to send the bidders when they asked for the bid opening room, etc.

BID OPENING AND REVIEW

At 11:00 A.M. on August 17 the bids were opened and read. There were seven bidders. As you may recall, the specs did not require the hospital to select the lowest bid; other factors such as previous performance, professional reputation, collector warranty, and ability to service the system were considered as important as price in the overall scheme of things. We read the bids and prices to the assembled group as the bids were opened. It became quickly apparent that there was going to be some work involved to categorize and evaluate the options that were bid. There were three collector manufacturers approved for this project and most bidders submitted three bids, one for each type collector. In addition, because the specs asked for a 10 year collector warranty instead of the then industry standard 5 years, addi-

tional options were proposed in some cases to account for the additional five years above the manufacturer's standard five year collector warranty. Only one collector manufacturer offered a 10 year warranty with no additional cost. However, it was learned at the bid opening that this owner, who lived and had his plant in Rome, was in the process of selling his company to a person from Pennsylvania. Surprise! The obvious significance of this was that the prequalification of this firm and its owner that took place was not necessarily valid for the new firm and owner. A complete background check on this new owner would be necessary. Another wrinkle that occurred at bid opening was the non-inclusion of a bid bond by one of the bidders as requested in the bid spec. This was noticed by one of the bidders and was logged as a possible discrepancy for further review. During the detailed bid review process of the next several weeks, it became evident that two firms were close on dollars with other things being equal or comparable.

Mr. Benator then conducted reference checks on one of the two top candidates and Mr. Wright did the same for the other top candidate. Questions about on-site supervision, previous similar type jobs, bid clarifications, and a check of references was conducted. In addition, since both contractors were using the Rome collector manufacturer for their lowest bid, and since that manufacturer had recently changed ownership, a detailed reference check was conducted on the new owner. The new owner's reputation in Pennsylvania proved to be excellent, and he had committed to move to Rome which meant absentee management would not be a concern. When the smoke cleared from the reference checks on the two leading candidates, a firm in Rome-- Northwest Georgia Engineering Co., Inc. -- was the top choice. They had the lowest price, a 10 year collector warranty, and a superb reputation for quality work. The only thing remaining to do was to negotiate and sign a contract.

THE CONTRACT

Since the specification was written so as to be essentially the meat of the contract, this would seem to be a simple task. However, Murphy's Law is never to be taken lightly! When the hospital lawyer made the routine check with the Georgia Secretary of State to verify the contractor's corporate name, we learned that there was no such firm registered with the Secretary of State! Surprise, again. We checked with the president of the winning firm, Mr. Harry Briscoe, and were told that they had recently filed with their attorney a name change to Northwest (underline ours) Georgia Engineering Co., Inc. from North Georgia Engineering Co., Inc. since their mail was getting mixed up with another "North Georgia" named firm. However, their attorney had not yet filed the name change with the Secretary of State, and thus as far as the Secretary of State was concerned, there was no such firm. All contractual documents were changed to reflect the firm's name as it was at that time. With these changes made, the appropriate documents were executed on September 22, 1981, exactly two days prior to the expiration of the most recent Davis-Bacon wage determination for the project. Missing that expiration date could have delayed the project for up to six weeks while a new wage decision was being prepared.

PROJECT KICKOFF MEETING

With the contractor now under contract, a kickoff meeting was held at the hospital to establish groundrules for the conduct and management of the project. It was felt that the project would be most successful if each person associated with the project recognized that he/she was an integral part of the project team whose team objective was the best possible solar DHW system for the customer. This was emphasized at the kickoff meeting which was attended by the following people:

Mr. B. G. Waters, Associate Executive Director,
    Floyd Medical Center
Mr. Ben Popay, Maintenance Director,
    Floyd Medical Center
Mr. Barry Benator, Project Manager, BENATECH, INC.
Mr. Rick Wright, Project Engineer,
    Armour and Cape, Inc.
Mr. Harry Briscoe, Contractor, President,
    Northwest Georgia Engineering Co., Inc.
Mr. George Spayd, Collector Manufacturer, President, Solar Corporation of America
    (SOLCOA)

At this meeting it was emphasized by Mr. Benator that although he was the project manager for the hospital, Mr. Rick Wright as project engineer was responsible for the design, engineering, and technical monitoring of the project. In addition, it was emphasized that if problems should develop, Mr. Benator and Mr. Wright were to be notified immediately so they could be resolved quickly and not delay the project. Finally, it was stressed that open communication between the project participants was encouraged, and in fact, a necessity, if the project were to be a success for all concerned. During this meeting, details of contract administration were reviewed such as submittal and payment of invoices, posting of required equal employment opportunity and Davis-Bacon Act notices, and weekly review by BENATECH of contractor payments to comply with government regulations.

PROJECT MONITORING

The project proceeded smoothly after the kickoff meeting with only a few minor problems cropping up between the meeting and the end of 1981. In order to emphasize the necessity of working as a team, members of the project team were encouraged to work things out between themselves unless, of course, a significant design change was involved or the project schedule was in jeopardy. In this case, Mr. Benator and Mr. Wright were to be informed immediately.

One such instance occurred when an insurance inspector visited the site in December 1981 and advised the contractor that northern wind loads might exceed structural limits of the collector array as then designed and built. Mr. Briscoe immediately advised Messrs. Benator, Wright, and Spayd of the inspector's observations. Mr. Benator called a meeting of the principal parties to discuss this matter in January 1982. Although the insurance inspector could not say for sure the as-built structure would fail, it was felt by everyone at the meeting that his concerns should be addressed. To the high credit of both Northwest Georgia and SOLCOA, a sharing of costs arrangement to strengthen the collector array was worked out at the meeting. And within a month, structural modifications, approved by Mr. Wright, were added to the array which effectively dealt with the inspector's concerns. This willingness on the part of the contractor and collector vendor to resolve this issue without fighting over whose responsibility it was served both the customer and themselves in the long run.

The only other significant "glitch" in the project came at the very end of the project. The heat exchanger vendor which was supplying two heat exchangers delayed the shipment of one of the heat exchangers for over a month due to various quality control and documentation problems. Efforts by the heat exchanger representative to expedite shipment were met with numerous unkept promised ship dates (you know the problem). Finally, on March 17, 1982, Mr. Benator directly phoned the manufacturer and after several transfers reached the manufacturer's Customer Service Manager. After explaining the situation in detail as well as the impact on future jobs for the manufacturer this delay would have, Mr. Benator was assured that the heat exchanger would be in Rome the next day. It was delivered the next day. Although this could be considered a coincidence, and that it would have been delivered the next day anyway, further checking revealed 1) that the truck it was on was not a direct shipment (as had been previously promised by the heat exchanger vendor), 2) the truck had broken down in Virginia and the heat exchanger had been picked up for priority delivery on orders of the Customer Service Manager. The message in this vignette is that when normal channels fail to get the job done, the project manager must become involved to get things off dead center.

FINAL INSPECTION

On March 26, 1982, a final inspection of the solar system was held at the hospital. A punch list was prepared by the project engineer which included the requirement for appropriate system operating and maintenance instructions for the hospital. The contractor began work on the punch list items the same day. As a result of the individual and cooperative efforts of all concerned, this project was completed within budget for the hospital. Were it not for the heat exchanger delay, it also would have been completed on time. The system is now functioning to reduce the hospital's energy use and costs.

LESSONS LEARNED/RECOMMENDATIONS

The following lessons learned are presented so you can plan for them on your next solar project. The authors hope that by sharing these with you, your next solar project will go even smoother than this one.

1. Things take longer than planned.

   - Collector corrosion research was expected to take two weeks; it took two months.
   - Critical path equipment delays can disrupt an otherwise ideal, on-schedule, on-budget project. Be prepared to become personally involved if deadlines continue to be missed. Establish trigger milestone dates to step up efforts to meet project schedule requirements. A weekly status report of critical path items may be appropriate.

2. Be prepared for extra paperwork if the government is involved.

   - Ensure your specs incorporate all the government requirements your contractor and their subcontractors will be required to meet.
   - Be able to distill the requirements written in "governmentese" to English for your client and project team members. Meet with the cognizant government agencies to obtain interpretations if necessary.
   - Be prepared to submit periodic reports to the government for your client. For Floyd Medical Center, a semiannual implementation progress report was required. A final report has been submitted. Annual utility monitoring reports will be submitted for the next three

years.

3. Be ready for traps -- they are everywhere.

   - Recall the flap over the contractor's corporate name not being the same as the name under which the bid was submitted.
   - Recall the change of ownership of the collector manufacturer around bid time and the need for another round of qualification and reference checks before contract signing.
   - Have in place the team working relationships that will allow problems like the insurance inspector's concern over the collector array's structural integrity to be handled quickly, with minimal hassle to the customer and fairness to the firms that will respond to the problem.

4. Above all, try to anticipate probable problems and plan so they won't occur. When the unexpected does occur, involve your project team in the solution so the best overall resolution to the problem is obtained. The authors found that by insuring the customer's interests were top priority, the solution of problems and fairness to the participants generally took care of themselves.

----

[1] SERI -- Solar Energy Research Institute
ORNL -- Oak Ridge National Laboratory

SECTION 20
IMPROVING ENERGY
PRODUCTIVITY

Chapter 64

IMPROVING ENERGY PRODUCTIVITY THROUGH BETTER BUILDING OPERATIONS

J. D. Swetish

ABSTRACT

This paper is a summary of the energy conservation efforts asso ciated with offices and other commercial buildings within Deere & Company.

Concepts such as variable air volume HVAC systems, task ambient lighting, energy efficient lighting modifications, computerized energy management systems, etc. are reviewed and examples presented.

Deere & Company's first venture into large scale solar heating for a commercial building is described in detail.

The process of establishing an energy management program for the marketing and product distribution branch houses is discussed, together with the necessity of performing an energy audit and of measuring results.

The quantitative results of the Deere & Company energy conservation program are presented.

INTRODUCTION

The use of energy for manufacturing has been of significant importance throughout the history of Deere & Company. In February 1973, several months before the Arab Oil Embargo, an energy management program was initiated to meet the problems arising from the ever increasing energy costs, and from potential energy shortages.

The results of this program through fiscal 1981, based on a fiscal 1972 base year, have been a 37% decrease in the energy required to produce a finished ton of product and a 10% decrease in total energy consumed, despite the fact that there has been a 44% increase in the weight of product produced, and a 60% increase in manufacturing area.

This program takes on added importance because of escalating energy costs, and the need to switch to alternate energy sources because of reduced availability.

CONSERVATION MEASURES

Approximately 93% of all the energy consumed in Deere's North American facilities is used in the 13 manufacturing plants, and it is these plants that have been the major focus of our energy conservation efforts.

However, the office buildings at these manufacturing plants, and the other commercial buildings within the corporation have not been ignored. Several energy conservation measures have been implemented, or are being planned for these buildings, and the concepts utilized in these measures can be applied to other office and commercial buildings.

HVAC

Ventilation and air conditioning equipment comprise one of the major, if not the largest single energy consuming system in office and commercial buildings, and at Deere & Company, much has been done to reduce HVAC energy consumption.

Reduced Building Temperature

At the start of our conservation program, we embarked on a program of reduced building temperatures. Office temperatures were reduced from 72°F to 68°F, and then further reduced to 65°F in compliance with President Carter's Building Temperature Restriction Legislation. It is interesting to note that since President Reagan rescinded the Building Temperature Restrictions, we have continued to maintain our office building temperatures at 65°F, and we use 65°F as our inside design temperature for all new office buildings.

During the summer months, we now air condition our offices to 78°F instead of 72°F as was normal before the start of the conservation program. The results are that the employees dress accordingly both during summer and winter.

Night Setback

When the buildings are unoccupied such as at night, and on weekends, the interior temperature is reduced from 65°F to 50°F. The heating systems are switched over to the control of a 50°F set point thermostat by either time clocks or the building energy management system. At a predetermined time, the system fans shut down and the building temperature is allowed to drop to the 50°F set point. When this set point is reached the heating system is reactivated to maintain the night setback temperature. The heating system is switched back to the 65°F setpoint prior to the start of the next work period. During the cooling season, the air conditioning equipment is shut down when the buildings are unoccupied.

Variable Volume HVAC Systems

In the past few years, there has been a turnaround in the design of heating and air conditioning systems. For many years, constant volume reheat, multi-zone, and double duct systems have been very popular because they are "Forgiving Systems". That is, they have the capability of providing heating and cooling (or a mixture) throughout the year, regardless of the actual heating or cooling loads that might exist. Being constant air volume systems, it becomes necessary to temper the supply air to the conditioned spaces as the heating or cooling load changes.

With these systems, the supply and return fans may have to be in operation during the heating season when the building is unoccupied, either continuously or by means of a night thermostat to keep the building from freezing.

If perimeter radiation were used with these types of systems, and the radiation is sized for heat losses, then the supply and return fans would not have to operate during the heating season when the building is unoccupied. This reduces the operating costs to some degree.

However, the biggest savings in energy comes from the elimination of the reheat requirements. This can be accomplished on many existing installations by converting them to variable air volume systems.

By varying the air quantities to offset the existing cooling on interior zones, the reheat coils can be disconnected. If there is existing perimeter radiation, the reheat coils for the exterior zones can probably be disconnected as well.

Retrofitting existing double duct systems can also be accomplished by the addition of a VAV terminal in the hot and cold ducts for exterior zones (if no perimeter radiation exists) and by only using the cold duct for interior zones since only more or less cooling is needed for these zones.

The addition of inlet vanes on the supply and return fans (controlled by duct static pressure) will reduce the energy consumption of these fans when they operate at reduced air quantities.

Fig (1) gives a comparison of the amount of energy used by various HVAC systems, in terms of BTU/sq. ft./year for a typical office building located in the mid-west.

It can be seen that by changing from a constant volume reheat system without perimeter radiation to a variable air volume system with perimeter radiation, an energy savings of 62% is achieved.

It also indicates that the fans and reheat requirements in constant volume systems are the reasons for their high energy usage.

By switching to a variable air volume system with perimeter radiation, the fan energy is reduced by over 85%, and the reheat energy requirement is reduced to zero.

An example of upgrading HVAC systems occurred during the Engine Works office expansion project.

The John Deere Engine Works is a 1.1 million square foot complex manufacturing diesel engines that are used in John Deere agricultural and construction equipment.

The existing office building was a two (2) story structure with a total area under roof of 22,500 square feet, located on the north side of the factory building.

The project called for a 200% increase in office space, but the topographical constraints at the site prevented an expansion to the North.

The increase in square footage was made possible by doubling the length of the existing office building, and adding a third floor on top of the existing structure.

Because of building layout requirements, the existing 1st floor mechanical equipment room could not be expanded, and so a new penthouse was constructed on the factory roof, directly south of the office structure, to house the necessary additional HVAC equipment.

The existing VAV system that originally handled the 1st and 2nd floors was modified to include the expansion area of the 1st floor in addition to the areas already served. This was accomplished by modifying ductwork, and reducing the volume of air handled by the system to that point which is minimally satisfactory. The reduction of heat gain in the building by reducing lighting levels (see section on lighting) enabled the system to handle the additional square feet without increasing the fan motor horsepower or changing the fan. Static pressure controls were upgraded and, the discharge damper was eliminated, and inlet vane dampers were installed for more efficient fan operation at reduced air volumes.

Two new VAV systems were installed to handle the 2nd floor expansion and the new 3rd floor. One unit handles the 2nd floor expansion while another unit handles the 3rd floor. The electric perimeter radiation system was expanded to cover the entire office perimeter.

A study is underway to replace the inlet vanes with solid state variable frequency drives for the fan motors to further improve the HVAC systems efficiencies.

LIGHTING

Lighting, in office and commercial buildings can comprise a large portion of the building energy budget and much can be done to reduce the cost of lighting, both in installed costs and operating costs, without sacrificing the quality of illumination.

Correct Lighting Levels

As part of the Deere & Company energy management program, lighting levels in all buildings were studied, and then reduced to levels that conserved energy, but gave adequate illumination for the task being performed.

In our office buildings, general office areas were originally illuminated to 100 FC, the IES standard for many years and we have, in most cases, reduced this to 50 FC.

In contemplating reduction of lighting levels say from 100 FC to 50 FC, the removal of 50% of the lighting fixtures will give a lighting level of 50 FC as read by a light meter. But what the meter reads is not what the eye sees. The meter reads what can be called "raw footcandles" - some of which are good and some bad.

There are two types of "bad" light:

Direct glare - Light that enters the eyes when the worker is looking straight ahead. The results of this type of bad lighting are headaches and eyestrain.

Reflected glare - Or veiled glare. This is caused by lighting fixtures placed in the ceiling where the light reflects directly off the task into the workers eyes. The task acts as a mirror, and when veiling glare is present, it acts as a veil over the task to obscure it.

If glare can be controlled, the quality of the lighting system can be improved.

At the Engine Works office, the original lighting was designed for 100 FC maintained. We removed 50% of the lighting fixtures from the 1st and 2nd floors, and used these fixtures in the expansion areas of these floors. We purchased lighting fixtures for the 3rd floor only.

By carefully utilizing the contribution of natural light from the windows, and locating the light fixtures in the ceiling in relation to the work areas, we provided an acceptable lighting system covering an area 3 times as large as the original office with only a 50% increase in electrical usage.

Task-Ambient Lighting

Quite often, fixtures are placed in a ceiling to look good on a reflected ceiling plan: symmetrical and even. That is the easiest and quickest way to design a space, and ideal for "fast track" construction, but a disaster, often, for quality of illumination and energy savings. It is not always possible to find out what tasks will occur in a space, but when possible the savings can be huge. Task lighting concepts, not only save money, but improve esthetics. "One per module" design is boring and costly.

For example, some middle management offices are illuminated with four - 4 lamp glare producing lens fixtures. Two 6 lamp louver fixtures properly placed in relationship to the desk would produce more "good light" than the four - 4 lamp lens fixtures.

Even though louver fixtures cost much more than lens fixtures, the total cost of the installed system would be less, and it would use less energy. 12 lamps rather than 16 lamps, and 2 fixtures rather than 4.

Louver fixtures have been used with success at the Tractor Works office building. The reflected glare normally associated with 2x4 troffer type fluorescent fixtures is noticably absent.

Another concept of task lighting is to remove all the fixtures from the ceiling and utilize modular "work stations" which have task ambient lighting fixtures built into them. This concept has been utilized at a number of Deere & Company office buildings.

The work station modules that are used in the Administration Center West Wing have three fluorescent lamps mounted over the work area. Two of these lamps are pointed down to give adequate task illumination. The third lamp is pointed upwards to create ambient light. This ambient light is enhanced by the ceiling which consists of polished metal strips with sound absorbant spaces between them.

Similar modules are used at the Tractor Works office and the Industrial Training Center, the only difference being that the ceilings are the standard suspended acoustical type, and some conference modules have incandescent spotlights mounted in the ceiling to provide illumination on conference tables.

COMPUTERIZED ENERGY MANAGEMENT SYSTEMS

Central control and monitoring systems (CC&M) are installed at most of Deere & Company's manufacturing facilities, and consequently the office buildings at these locations are connected to them.

These CC&M systems typically monitor and control the operation of the HVAC, fire alarm, and security systems within the office buildings and give visual, aural, and typewritten notification of abnormal conditions or failures within these systems.

At the Corporate Administration Center in Moline, the CC&M system is used to control the lighting system in addition to the previously mentioned items.

The lights are turned on and off automatically depending on the occupancy of the building.

If people have to work during off shift hours, they can have the lights in their particular work area switched on by telephoning the CC&M operator and requesting that the lights be turned on for a certain zone and room.

The information giving the zone and room number for each work space is attached to the telephone unit located at the work space, along with the telephone number of the CC&M operator.

When the person leaves the building, he telephones the CC&M operator to have the lights turned out in his particular work area.

As these CC&M systems also control utility and process systems in the factories themselves, maintenance and reliability are of the utmost importance.

It is for these reasons that Deere & Company, to date, has only dealt with the major CC&M vendors such as Honeywell, Johnson Controls, and Trane Sentinal.

These vendors can provide a complete turnkey system fully integrated into the systems that are controlled by it.

A study is underway to investigate some of the smaller CC&M systems that could be used in our marketing and product distribution facilities. However, availability of spare parts, and vendors' maintenance service capability will continue to be of prime importance.

ALTERNATE ENERGY SOURCES

Let's take a look at Deere & Company's first large scale venture into solar heating - The Davenport Industrial Training Center.

The 55,000 square foot facility was constructed for international service and sales training for the Industrial Equipment Division. The Training Center offers 22 different training courses to over 3000 students yearly.

The facility's entrance opens into a unique lobby/cafeteria setting. There is a main office complex with landscape furniture and task/ambient lighting, five classrooms, an art department, and a complete audio-visual (TV) studio where training films, and presentations are prepared, together with more than 150,000 slides photographed per year.

The open shop area covers 20,000 square feet, and has 20 training stations.

The most visible commitment to energy conservation are the solar energy systems on the south face of the building. The south wall, which is completely covered by almost 10,000 square feet of 1/4" thick single pane glass comprises two separate solar systems. One to heat water, the other to heat air.

The building is divided almost equally with the solar water heating system for the office-classroom area, and the solar air heating system for the shop area.

Solar Water System

The water heating solar collector is mounted on the 1200 square foot south wall of the building's administrative area. Mats of synthetic rubber tubing are fastened to the wall to act as the solar collector. Radiation from the sun passes through the glass, impinges on the mats, and heats a mixture of ethylene glycol and water circulating through the tubes. In turn, this heated mixture is circulated through a heat exchanger to heat water. The 80°F hot water is subsequently used to supply energy to, and increase the heating efficiency of 12 water to air heat pumps used to heat and cool the administrative and classroom areas of the building. In addition, this solar system is used to heat domestic hot water for showers, dishwashers, etc.

The use of multiple heat pumps is also of interest, since the design takes advantage of the naturally-occurring heating and cooling requirements of the office and classroom areas. During the heating season, the interior areas of office type buildings typically require cooling to offset the heat released by people, lights and equipment, while the perimeter areas simultaneously require heat to offset losses through walls, doors, and windows.

By installing separate heat pumps to serve the various interior and perimeter zones, the simultaneous heating and cooling requirements are used to advantage. The heat pumps, operating in the cooling mode remove heat from the air and add heat to the circulating water. Conversely, the heat pumps that are in the heating mode take heat from the water to heat the space. If the water becomes too cold, there is a backup electric boiler, and if the water becomes too hot, there is a backup cooling tower. By operating the equipment in harmony, the efficiency of the heating and cooling functions are enhanced by approximately 10%.

Solar Air System

The south wall sections of the open shop and shop training areas total 8500 square feet and are comprised of a hollow core concrete wall, and a row of windows along the top of the wall. In this case, solar radiation passes through the outside glass, and is absorbed by the blackened surface of the concrete wall. As the solar energy heats the concrete wall, the air in the space between the glass and the wall also becomes hot.

Depending on the temperatures of the air between the glass and the wall, the wall, and the occupied space, air control dampers and fans are automatically operated so that air is drawn from the shop at floor level and circulated through the space between the glass and the wall or through the hollow cores in the concrete wall. The heated air is subsequently returned to the shop area.

During operation of the air heating system, after the first few hours of sunlight, the space between the wall and the glass heats up rapidly, while the concrete wall heats slowly. Under these conditions,

air is circulated through the space between the glass and the wall and used to heat the shop. After several hours of illumination, the temperature of the wall is sufficiently high so that air can be circulated through the hollow cores, and used for heating. Because of the large mass of the wall, it remains warm ie. it stores heat for use during periods of cloudiness or to heat the shop after sunset.

An electrically heated underfloor radiant heating system is the primary heating source for the shop areas. This system comprises electrical resistance heating mats buried under the 6" concrete floor slab and 12" of sand. The mats are energized only at night when the demand for electricity is lowest, and when the utility company is using their most efficient generating equipment.

When energized, a large mass of concrete, sand, and earth is heated, thus storing a large amount of heated energy. By conduction, the heat slowly migrates up through the sand to heat the floor slab and warm the shop space. The temperature at the surface of the floor slab will remain nearly constant at about 5°F above desired room temperature.

During the summer, for shop ventilation and to cool the solar chamber, the shop fans reverse direction, pulling air from the shop ceiling area, into the solar chamber, and out of large exhaust louvers located at each end of the solar chamber.

The solar contribution to the shop heating system is approximately 40%.

Overall, the Industrial Training Center was designed to save energy by taking advantage of naturally occurring phenomena. In addition to saving energy at this facility, the design of the center is important to future Deere & Company construction projects. It is interesting to note that of all the equipment related to the solar systems, the only specialized "solar" product is the synthetic rubber mat assembly. All the other equipment is standard construction material or HVAC equipment. The operation of the various technologies employed will be monitored so that a determination can be made on how best to apply similar energy-conscious engineering in the future.

MARKETING AND PRODUCT DISTRIBUTION ENERGY MANAGEMENT

PROGRAM

The previous portion of this paper has covered some of the energy conservation efforts at our manufacturing facility office buildings, and at the Industrial Training Center.

To increase Deere & Company's energy conservation savings, the existing energy management program is being expanded to include the marketing and product distribution locations in North America.

The 17 larger locations total 6 million square feet and had a total annual energy bill in excess of $2 million for 1981.

The goal of the program is to reduce the energy consumption (adjusted for climatalogical variations) by the end of 1983 to 80% of the energy consumed in 1980, the base year, broken out as follows –

    - 4% in fiscal 1981
    - 6% in fiscal 1982
    -10% in fiscal 1983

Due to current economic conditions, all energy reduction projects must have a 50% (minimum) return on investment with a 2 year payback. Consequently, only projects having a quick payback and low capital investment are being considered at this time.

Walk Through Audits

Walk through energy audits are being performed at each location. These surveys are <u>specifically</u> aimed at reducing the energy costs of the buildings' heating, cooling, and lighting systems.

As these distribution centers or "Branch Houses" do not have in-house engineering expertise, the audits are being conducted by a member of the Corporate Energy Management Division.

The types of projects being implemented are best described by looking at one particular Branch House.

Atlanta, Georgia, was the first Branch House surveyed back in early 1981. Projects installed since the walk through audit include a Honeywell time-leased computer energy management controller (BOSS) which optimizes temperatures and run cycles for the office and warehouse heating and cooling systems. The controller began functioning in August 1981. Energy saving fluorescent lamps were installed in January 1982 and a cooling tower expansion with a new 2 speed motor was completed during the summer of 1981 together with a few other smaller items.

Energy savings for 7 months (August 1981 through February 1982) have been almost twice those estimated.

The goal for Atlanta was 10% for 1982, and this has been reached 5 months early.

To enable us to keep track of the program, an energy data base has been set up, similar to that used to keep track of energy usage at the manufacturing facilities. This data base uses 1980 as its base year and enables rapid analysis of energy usage, costs, ratios, etc. At this time the data base is being upgraded to allow the energy usage to be correlated to building factors and climatalogical changes.

Projected Savings

Using 1980 as the base year, the composite cost of energy for the Branch Houses was $5.88/BTU x 10^6. This cost rose by 16% in 1981 to $6.81/BTU x 10^6. Looking forward 10 years, to 1991 assuming a 15% escalation of energy costs per year, the cost of energy in 1991 will be $27.55/BTU x 10^6 (305% increase).

The potential savings over this 10 year period will be $7.7 million, with a yearly savings of over $1 million by 1989.

Results

Comparing the first half of fiscal 1982 to the same period during fiscal 1980 in terms of energy usage. Natural gas usage increased 1%, fuel oil increased 5%, but electricity decreased 4%. At first it may seem that this net breakdown is poor because the 1982 goal of 10% is not being achieved, but this is not really true as the on site surveys are still in progress, the units that have been surveyed are just beginning to respond to recommendations, and the energy data has not yet been factored for weather. Which for Moline, Illinois, was 15%-20% colder in 1982 than 1980. Therefore, giving consideration for these factors, our previously set goals are attainable.

To illustrate this - The first quarter 1982 energy consumption was 4% over that required for the same period of 1980 <u>but</u> the second quarter 1982 consumption was 3% below the 1980 level to equal the breakeven point now. The things that are being done at the units are just now beginning to take effect.

Conclusions

Yes, energy conservation is a challenging arena. The wise selection of energy mix, and the efficient use of energy throughout the Corporation is a goal to which Deere & Company is committed.

In conclusion, it is our message to encourage the adoption of vibrant energy management and conservation programs, and to work together to achieve wise use of energy.

HVAC SYSTEMS ENERGY CONSUMPTION COMPARISON
$BTUx10^3$/SQ FT/YEAR
FOR BUILDINGS LOCATED IN THE MIDWEST

SYSTEM	FANS	COOLING	REHEAT	HEATING	TOTAL
VAV With Perimeter Radiation	10	25	0	71	106
Induction VAV With Perimeter Radiation	10	25	0	71	106
Induction Vav With Perimeter Air	18	25	0	65	108
VAV Reheat With Perimeter Radiation	11	27	20	87	145
VAV Reheat - No Perimeter Radiation	31	29	24	95	179
Induction Reheat - No Perimeter Radiation	23	25	10	80	138
VAV Double Duct - No Perimeter Radiation	31	25	17	87	160
CAV Reheat - No Perimeter Radiation	74	43	90	74	281
CAV Double Duct - No Perimeter Radiation	75	25	50	70	220

FIGURE 1

Chapter 65

IMPROVING ENERGY PRODUCTIVITY THROUGH BUILDING SIMULATION

W. E. Henry, Jr.

ABSTRACT

Approximately 500 million square feet of new and renovated commercial space will be added to the nation's building stock annually. Much of this space will be designed by architects and engineers. Many decisions in the design process will affect building energy use; the evaluation of these decisions and the understanding of their interaction is not intuitive. Computer energy simulation of buildings will allow the designer to evaluate the maximum number of variables. The most energy efficient building which meets all requirements will result from the process.

INTRODUCTION

While statistics on the existing stock of buildings are understandably hard to come by, we do know that approximately 40% of the square footage of non-farm buildings is office, bank, hotel, and large retail type structures. The larger of these structures are likely to be architect/engineer designed. In 1976, the most recent year for which statistics are readily available, there were 451 million square feet of floor space of these types constructed or renovated in major fashion. If these buildings use energy at the average rate of 84,000 BTU's per square foot, then they will consume 3.79×10^3 BTU annually. This is the equivalent of 6.3 million barrels of oil annually and may not seem like a lot when one realizes that in November of 1981, 5 million barrels of residual and distillate fuel oil were consumed in the nation.

The table in Figure 1 contains some recent forecasts for new space construction over the next twenty years. Over two <u>billion</u> square feet of space will be added to the nation's stock of buildings in this period. At the same time, older and often less efficient space will be demolished to make room for these new buildings. Clearly, we have a significant opportunity to reduce the energy the nation uses in its buildings.

Building Type	Year	Total
Office & Bank	1976	108.0
	1980	130.8
	1990	149.5
	2000	178.6
Educational	1976	119.7
	1980	94.7
	1990	89.9
	2000	99.3
Hospitals & Health	1976	74.2
	1980	61.0
	1990	75.5
	2000	81.0
Retail & Other	1976	516.1
	1980	610.2
	1990	683.5
	2000	809.0
Nonhousekeeping	1976	33.6
	1980	33.4
	1990	29.1
	2000	29.1
TOTAL:	1976	851.6
	1980	930.1
	1990	1027.5
	2000	1197.0

NOTE: Totals may not add due to rounding.

Source: Arthur D. Little, Inc., estimates.

FIGURE 1. NONRESIDENTIAL CONSTRUCTION FORECASTS
(Million Square Feet)

FACTORS AFFECTING BUILDING ENERGY USE

Many design decisions in the building process will affect energy use in a significant way. Consider the following:

Basic Form of Building

A study by Victor Olgyay[1] concluded that there were optimum shapes for houses in different climates. In addition to determining the optimum shape, Olgyay specified limits to the variance from optimum. Figure 2 shows the results of Olgyay's research.

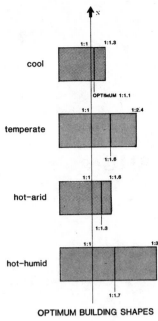

OPTIMUM BUILDING SHAPES

FIGURE 2.

Orientation

As a result of the interaction of building with sun and wind, its orientation can have up to a 10% impact on energy use. This may be even greater in buildings with a larger proportion of glass to wall than normal. Figure 3 shows the effect of alternate shapes and orientations.

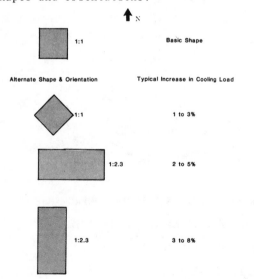

EFFECT OF ORIENTATION ON COOLING LOAD

FIGURE 3.

Envelope Mass

The mass of the building structure will have a measurable impact on both the magnitude and the timing of peak cooling and heating loads. In a building of greater mass, the cooling load will be lower and

will occur longer after the instantaneous heat gain than with a building of lower mass. The effect of mass on heating requirements is well known. Winter sun will provide energy for a building at a rate four to eight times as rapidly as the building can use that energy. Properly designed, the mass of the building can store that energy until it is required (when solar energy is not available). Figure 4 shows the impact of mass on the magnitude and timing of cooling loads in a building.

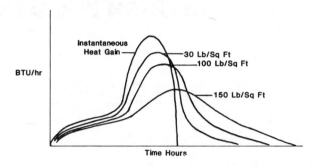

ACTUAL COOLING LOAD VS. TIME
FOR LIGHT, MEDIUM, AND HEAVY CONSTRUCTION

FIGURE 4.

Daylighting

We are once again discovering a "new/old" technique for lighting our buildings: daylight. There are many advantages of daylight, not the least of which are good quality and low operating costs. These must be balanced against the introduction of unwanted glare and heat gain into the space. Benjamin Evans[2] shows calculations of the benefit of daylight in a school lunchroom to be a net annual saving in energy. Figure 5 shows the results of that calculation.

	BTU/Year
Lighting Energy Saved	7,390,000
Increased Cooling Load Due to Sun (Net)	3,042,000
Additional Heat Energy Required Due to Loss of Light Energy	1,430,000
NET ENERGY SAVINGS:	2,918,000 BTU/Year

FIGURE 5. ANNUAL ENERGY USE FOR A SCHOOL LUNCHROOM

(Assumes Oklahoma City latitude and lights off for 7-3/4 hours per day.)

Other factors, such as glazing placement (8% effect), choice of lighting systems (12% effect) and choice of HVAC systems (10% effect), will have a great influence on energy use in buildings.

Of course, it must be realized that new buildings are making more provision for copy machinery, word processors, and computers. All of these devices result in greater internal heat gains, with the net result that buildings are less influenced by exterior climate and more influenced by interior climate. This places further emphasis on selection and specification of HVAC systems which can make effective use of heat gains when they are needed and economically remove them when they are not.

We must also realize that many factors which we have represented as having impact on energy use are not in fact within the designer's control: lot lines or zoning ordinances will dictate orientation, aspect ratio will be dictated by floor space needs, mass and glazing placement and size will be dictated by the owner's desires or by historic district requirements, etc.

Nonetheless, the designer has considerable freedom to make choices which will affect the energy use of the buildings by ten or twenty percent or more.

HOW MAY A BALANCED SOLUTION BE ACHIEVED?

The designer must balance the client's program with building codes and his own input to produce a structure that meets all needs, energy included. An energy efficient design need not cost more than a "conventional" one, but some aspects of construction will necessarily raise the cost:

- o   double glazing
- o   Thicker or more massive structure
- o   Extra insulation
- o   Economizers
- o   Heat recovery
- o   Etc.

The effect of these measures on a building's cost can be accurately estimated. However, their long term effect on energy use is often more difficult to calculate, and in some cases, even the trend of their effect is not intuitive. An example will serve to illustrate the point.

The author was recently involved in an energy simulation of a new administration/engineering building for Pratt and Whitney Aircraft of Canada, a division of United Technologies. Among other questions, the management of Pratt and Whitney wanted to know the effect of increasing the occupancy of the building from 1200 people to 1900 people. Even in Montreal, a building of this type is cooling dominated, and the assumption was that additional people would add to the cooling load and thereby increase energy usage.

Instead, predicted energy use with the additional persons was nearly the same as that with the lower number. Inspection of the simulation print-out indicated that the increased heat gain in winter had not, in fact, offset the increased summer cooling load as had been at first suspected. Resisting the conclusion that the simulation was in error, a closer inspection of the numbers revealed that increased occupancy had reduced the humidification load by a greater amount than it had increased the cooling load. Net energy use had gone down

slightly. This conclusion, which was not intuitive, led to money saving changes in cooling and humidification equipment.

The balanced solution desired by the owner and the designer may be best achieved when they are able to evaluate a number of possible solutions to the design problem. This evaluation must include as accurate a quantification of costs and savings as possible.

WHAT TOOLS ARE AVAILABLE TO ASSIST IN THE EVALUATION?

There are several levels of building energy simulation available to the architect or engineer. The most commonly used is probably the degree day method. Energy use (E) is predicted by the relationship:

$$E = \frac{(\text{Design Heat Loss})(\text{Degree Days})(.24)}{(T_{od} - T_{id})(\text{system efficiency})(\text{fuel heating value})}$$

where: T_{od} = outside design temperature
          T_{id} = inside design temperature

This method is not commonly used for cooling calculations. Originally designed to simulate residential energy use where heating is assumed to be required below 65°F, the method has been adopted for simulation of other facilities by using degree days to bases other than 65°F. Obviously, this method must be limited to simple buildings and is not useful in analyzing the building design options previously discussed. Nothing in the equation takes orientation, building mass, or other variables into account.

The second technique in use is called the Bin Method. This method requires calculation of heating and cooling loads at each of the temperature bins for which the hourly frequencies of outdoor temperatures are tabulated. These bins are 5° in width. This technique, which requires more calculation than the degree day method, will generally produce more accurate results and will allow the architect/ engineer to take interheat gains and varying system efficiencies into account.

A recent study[3] shows agreement between these two methods and the DOE-2 computer simulation for simple buildings for predicting energy use.

The third and most sophisticated method uses a computer to simulate the energy performance of the building. A recent ASHRAE bibliography lists 43 such programs. Some are simply computerized versions of the degree day or bin methods; other simulate building energy use on an hour-by-hour basis for all 8760 hours in a year.

These simulation tools are generally available through timesharing systems such as CDC CYBERNET, G.E. Information Systems, or those offered by the Boeing Company. In addition, if one has computational equipment of sufficient capacity, then the programs may be purchased. Time share is the most cost-effective approach for the majority of designers.

Many simulations offer a "layered approach" to the process. It is possible at early stages of the design to input only simple information, such as building size, shape, and orientation. The program will then use "default values" to fill in the blanks. A number of preliminary design concepts can be compared without the need of detail. More computer runs can then be made as additional decisions on wall sections, fenestration, and systems are required.

Final runs will be limited to one or two choices of systems because of the level of detail required. This includes:

o   Operating schedule
o   Temperature schedules
o   Building equipment and performance data
o   Weather data/location
o   Building construction data
o   Internal load data and profile
o   Window data and overhangs
o   Zone data
o   Energy source heat content
o   Required submetering

The simulation input will require an architect or engineer familiar with the building in question and a simulation specialist familiar with the program in use. With experience, (two to three simulations), these persons can be one and the same. For buildings of 6-10 zones (5 is the useful minimum), it will take about one man week to prepare the input. Some examples of the uses of energy simulation are contained in Figure 6.

Application Area	New Bldg.	Existing Bldg.
Evaluate Envelope Options	X	
Orientation	X	
Daylighting	X	X
System (Re)sizing	X	X
Placement of Glass	X	
Alternative HVAC Systems	X	X
Evaluate EMS Systems	X	X
Evaluate Deadband Thermostats	X	X
Evaluation of Passive/Active Solar	X	X

FIGURE 6.   SOME EXAMPLES OF ENERGY SIMULATION
APPLICATIONS

Costs of simulation programs can be divided into hardware and software costs. Hardware for a timeshare system is a suitable terminal and a modem. These are usually leased and, depending on speed and print quality, will cost from $200 to $1000 per month. To purchase a computer suitable for running these programs, one must be prepared to spend over $100,000. Software costs on a timeshare system are usually broken out as follows:

Setup $400 per zone
Run $ 30 per zone
Storage   Varies with system

You can easily see that once a building is simulated in the computer, the incremental cost of additional simulations is relatively low. Of course, if a firm has chosen to purchase the necessary hardware and become a licensee for a simulation program (an additional $5-10,000 charge), then the cost of additional runs is quite low. Note that labor costs are not included in the above estimates.

As a reward for the somewhat tedious task of entering the required data, the designer receives more than he gave, and often more than he wanted! Outputs include:

Summary of input data

Building load data
    by hour
        month
        year

Energy use data

System sizing
    auto sizing
or check designer's sizing

The whole effort may be expected to cost between $6,000 and $10,000 for one simulation and three additional runs.

Is it worth it? Despite the fact that energy simulation will add a small percentage to the design fee, we think its value can be demonstated. Consider the benefits:

Evaluation of more alternatives than can be done manually

Accurate comparison of alternatives

Hour by hour loads

Accurate system sizing

Prediction of the interrelationship of large numbers of variables

The exercise has only to result in the accurate sizing of the chiller boiler or air handlers to have an immediate first cost benefit to the owner; it has only to affect the annual energy budget by a few percent to pay for itself in the first year or two, and when one thinks of the millions of new square feet of building under design each year, the benefit to the nation is clear as well.

REFERENCES

1   Olgyay, Victor, "Design with Climate", Princeton University Press, Princeton, New Jersey, 1963.

2   Evans, Benjamin E., Daylight in Architecture, McGraw Hill, New York, NY, 1981.

3   Kusuda, Sud and Alereza, "Comparison of DOE-2 Generated Residential Design Energy Budgets with those Calculated by Degree Day and Bin Methods", ASHRAE Transactions, 1981, Volume 87, Part 1.

Chapter 66

IMPROVING ENERGY PRODUCTIVITY THROUGH BETTER MANUFACTURING PROCESSES

B. D. Norian

Industry consumes 43 percent of all the electric power generated in this country, and about 47 percent of the total energy needs of the economy. It is therefore reasonable to expect that a conscientious effort to conserve energy on the part of the industrial sector will have a more significant impact on the total energy consumption of the nation than will a similar effort on either the commercial or residential fronts. As representatives of this major segment of energy use, it is our moral and fiscal responsibility to combine our cumulative knowledge, experience, and expertise to maintain or increase productivity while minimizing the unitary energy input of the facility. Others have addressed in detail the state-of-the-art techniques of proven concepts to reduce energy consumption in the areas of the building envelope, combustion efficiency and steam generation, HVAC, lighting, and domestic hot water. Heat recovery devices, steam trap maintenance programs, cascaded cooling water systems, and even refuse incinerators with integral waste heat boilers are becoming commonplace as fuel costs command an increasingly higher fraction of total manufacturing costs. This paper, however, addresses some of the more obscure strategies by which energy and energy dollars can be conserved in industry. Although each problem, rationale, and solution described has been successfully implemented in a specific application, the generality of each is such that with perhaps only minor modification, they can be employed in other industries.

COVER OPEN PROCESSES
WHICH USE HEATED FLUIDS

There are many industrial processes which use a tank of heated liquid into which parts are dipped. Salt baths, chemical washes and rinses, and electroplating are typical examples. Because, for reasons peculiar to the process, the fluids must be maintained at elevated temperatures, thermal losses from the open surface of the tank can be excessive. Further, the evaporation of aqueous liquids can create an abnormally high latent heat load to be removed by comfort cooling equipment, while corrosive vapors require a high ventilation rate using specialized materials for the wetted parts to ensure a reasonable service life.

The most cost-effective strategy to use in a particular application depends on the size of the tank, the temperature of its contents, the chemistry of the fluid, and the frequency with which the tank must be accessed. Common solutions include the use of an insulated cover to retard heat loss between uses or during the hours that the plant is shut down; alternatively, a layer of plastic spheres can be floated on the surface of the tank. These strategies are particularly viable in the case of fluids which must remain heated continuously.

A hot-dip galvanizing processor formerly maintained the vat of zinc in a molten condition using 24 gas burners. To ensure that the metal would not solidify, the burners were fired around the clock, although the plant was essentially a single-shift operation. After fabricating a cover from heavy angle iron and firebrick, which is dropped onto the large tank at the end of the shift, only two burners were required to offset the thermal losses of the closed vessel. Savings in natural gas during the first year were estimated at $12,000.

An electroplater was contemplating the installation of an elaborate exhaust system over several of his tanks, because the vapors emanating from them were seriously corroding the steel frame of the building as well as the piping and electrical conduits. One can only imagine the effect on the employees. Alternatively, a consultant recommended floating a double layer of polyethylene spheres on the surface of the errant tanks to retard the evaporation of the corrosive liquids. Not only did the environmental conditions within the plant show a marked improvement, but the fuel oil consumption also decreased significantly. Investigating further, it soon became apparent that the energy savings was directly attributable to the reducton in the evaporative losses from the tanks. Recalling that every pound of water that is evaporated requires the addition of nearly 1,000 BTUs of heat to the air to maintain the same temperature, plus the addition of a pound of make-up water to the tank which must be heated to the normal temperature of the contents, the source of the savings becomes apparent. The energy savings amounted to $1,300 during the first year, not to mention the avoided cost of installing and operating the exotic ventilation system.

CRYOGENICALLY TREAT TOOLING AND
WEAR PARTS FOR LONGER LIFE

Although still relatively unknown in industry, it has long been recognized that subjecting cutting tools, welding electrodes, and other wear parts to cryogenic temperatures under carefully controlled conditions will result in a dramatic increase in the service life of the components. Typically, tool life can be extended by 300 to 500 percent at a cost which is on the order of 20 percent of the cost of the item. Simply stated, the thermal excursions to cryogenic temperatures change the microstructure of ferrous alloys. The last few percentage points of retained austenite, an unavoidable result of commercial heat treating, are converted to martensite, a tougher, more wear-resistant configuration of the crystalline structure of the alloy. While an obvious advantage is a significant decrease in the tooling budget, increased productivity per unit time as a result of fewer tool changes will effectively reduce the gross energy requirement per unit of output. More direct energy savings are possible by using cryogenically treated copper welding electrodes. Although the mechanism is not understood, the use of treated electrodes requires that the current be reduced by approximately ten percent; an additional benefit is that the service life of the electrodes is effectively doubled.

A manufacturer of photographic films found that by using cryogenically treated slitter blades, the interval between resharpenings was significantly extended, reducing the downtime required to change blades and reset the machine. Productivity was subsequently increased by over $50,000 worth of film annually. Similarly, an automotive assembly plant, using treated copper spot welding electrodes, was able to reduce electricity costs by an estimated $10,000 per year, halve their budget for replacement electrodes, and, because the cryogenically treated electrodes produce a smoother surface, eliminate a secondary operation of disc-sanding the craters formed by conventional electrodes in semi-visible locations such as the door sills.

IMPROVE THE POWER FACTOR
OF THE PLANT

A conventional electric meter measures the current component that is in phase with the applied voltage, which is the current flowing through the resistance in the circuit. This component, measured in kilowatts (kw), records the real work performed by the electrical circuit. The out-of-phase component, the kilovolt-amperes reactive (kvar), provides the magnetizing force necessary for the operation of the work-performing device. The vector sum of these components, which are graphically at right angles to each other, is known as the apparent power, and is measured in kilovolt-amperes (kva). All components of the distribution system must be sized to carry the kva of the system. The angle between

the kw and kva vectors is known as the phase angle; the power factor is the cosine of this angle, or kw/kva. As the kvar decreases, so does the phase angle, and the magnitude of the kva approaches that of the kw. When the phase angle is identically zero, the power factor becomes 1.000 (the cosine of zero degrees), or 100 percent. At unity power factor, the kva is equal to the kw, and all of the heat developed within the system is a function of the current that is performing useful work.

An inordinately low power factor effectively decreases the capacity of the distribution system to handle work-performing electrical loads. It also results in currents that are higher than required for a given task, creating an excessive voltage drop and contributing to the system losses. Finally, the tariff schedules of many utility companies specify severe financial penalties for low power factor. Many utilities that do not penalize users for poor power factor may specify a minimum acceptable level in their industrial power contracts.

Power factor improvement is simply a mechanism whereby the kvar of an electrical system is reduced. This can be accomplished in one or more of four ways:

- The use of "high power factor" utilization equipment, such as lighting ballasts, and ensuring that induction motors are operated at or near full load will significantly improve the power factor of the facility.

- A synchronous motor, which has a leading power factor can be used to drive a piece of equipment that is operated constantly (such as an air compressor) to offset the effects of a number of smaller induction motors, which have a lagging power factor. Because synchronous motors are practical only in large sizes -- several hundred horsepower -- this approach is necessarily expensive.

- A synchronous condenser is similar to a synchronous motor, in that it is a large piece of rotating machinery. Its function, however, is purely that of power factor improvement, and it does not drive a load. Synchronous condensers are rarely a practical solution to low power factor in industrial plants; they are used primarily by utility companies.

- The installation of a power factor improvement capacitor is accepted as the simplest and most effective strategy to improve the power factor in an industrial plant. Because the current through a capacitor leads the applied voltage by 90 degrees, capacitors have the effect of cancelling the inductive or lagging kvar on a one-for-one basis. Installed in banks to provide the required amount of reactance, capacitors can be mounted in any convenient location, with the wiring run to the point of electrical connection.

Reducing electrical costs through power factor improvement presents an interesting contrast to the approach of most other facilities projects. In general, the level of effort required to effect an economical solution to a problem is proportional to the severity of the problem itself. Conversely, the complexity of solution to a poor power factor is inversely proportional to the severity of the problem. In the case of an extremely poor power factor, the installation of any amount of capacitors anywhere in the system will provide a very favorable payback.

A large midwestern manufacturing plant had a punch press operation that consisted of equipment typical of the industry in which small motors drive large flywheels continuously, and the actual punching operations are essentially powered by the inertial effect of the wheels. Thus, the motors are essentially unloaded most of the time. The aggregate effect of hundreds of induction motors running unloaded created a low power factor for which the plant was penalized by the utility rate structure. When a new air compressor was needed, the decision was made to drive the unit with a synchronous motor. Power factor correction was assured because the compressor was dedicated to the punch press department and would be operating any time the presses were in production. Compared to the savings in the monthly electric bill, which were projected at $9,500 per year, the incremental cost of the synchronous motor was almost insignificant.

RECIRCULATE AIR IN WEB DRYER

Many continuous-web coating, laminating, or dyeing operations employ a heated oven to evaporate the water or other solvents used in the process before rolling the finished product onto a core. The length of the dryer is determined by the production rate (feet of product per unit time), the amount of solvent used per unit area, the permeability of the substrate, and the rate at which the solvent can be evaporated, which, in turn, is a function of the volatility of the diluent and the maximum temperature that can be tolerated by the material. Typically, dryers are heated by a series of steam coils and blowers, or, alternatively, by an integral fossil-fueled furnace. Gas-fired units can be direct or indirect fired. The most common air flow pattern is for the system to ingest plant air, heat it to the required temperature, pass it over the product in a counterflow configuration, and then exhaust the air to atmosphere. Thus, every cubic foot of air that is exhausted from the dryer must be introduced by a make-up air system or, more likely, infiltrated through the inevitable cracks and crevices

found in virtually all buildings. During the heating season, this air must be tempered to ensure employee comfort while, at all times, the air must be heated within the dryer.

As mentioned above, the rate at which a given solvent is evaporated is a function of the velocity of the air across the web, the temperature, and, in the limit, the relative concentration of the solvent vapor in the air stream. In commercial dryers, the relative vapor levels are typically on the order of ten percent, even in systems which contain several zones and cascade the heated air from the final, or exit zone, to the primary, or entrance section.

The energy-saving alternative is to install a section of ductwork from the dryer discharge to the make-up air entrance, including face and by-pass dampers which allow the percent of recirculated air to be continuously varied while maintaining a constant flow of air across the product. In the case of water-based processes, it is possible to automate the fresh air/recirculated air balance by installing a proportional humidistat in the exhaust duct which controls a modulating motor connected to the damper system. By noting the quality of the product as the setting of the humidistat is gradually increased, the optimum operating point in terms of energy economy can be readily established. With other solvents, optimum damper settings can be determined by trial and error. If flammable solvents are used, explosion-proof electrical components are required. Further, it may be necessary to use electronic sensing equipment to override the optimum damper setting to maintain the vapor concentration below the lower explosive limit (LEL).

A paper converter applied a water-based clay coating to a heavy paper stock. The in-line dryer was steam-heated in four zones, each using a typical once-through air flow pattern. All four independent air paths were fitted with face and by-pass dampers to allow a variable degree of recirculation, which was controlled by a dedicated humidistat in each exhaust duct. Several days of experimentation yielded the most aggressive settings that would produce an acceptable product. Although it was difficult to accurately measure the savings (this equipment required less than two percent of the steam output of a central boiler plant), it was calculated to be approximately $2,800 per year.

ALTERNATE LOADS TO REDUCE DEMAND

The industrial rate structures of most utility companies comprise two elements, not including the ubiquitous fuel adjustment charge. The energy consumption, measured in kilowatt hours, and typically billed on a classic declining block rate structure, is simply the aggregate amount of electric power consumed by the facility during the billing period. The second factor, the demand charge, is based on the highest average amount of power, measured in kilowatts, that the complex requires, or demands, during a brief interval, commonly 15 or 30 minutes. Because the demand charge can amount to a significant fraction of the total bill, and also represent a substantial sum of money in the absolute sense, numerous strategies have been devised to reduce the billed demand in an effort to trim operating costs. While a microprocessor-based energy management system, properly designed and programmed, can provide the mechanism to achieve this goal, the initial investment is formidable, particularly if there are not many loads that can be deemed sacrificial.

Alternatively, many manufacturing processes use equipment that is employed sequentially to produce the finished product. Often, particularly in the case of electrically heated apparatus, power must be supplied during the stand-by modes to maintain working temperatures. Because at least once during the course of every billing month it is reasonable to expect that all the equipment is energized simultaneously, the effect on the billed demand is additive. If the temperature of one or more of the devices is not critical, it may be possible to effect a significant savings in the demand charge by rewiring the contactors such that the non-critical elements are sacrificial to those that require more precise temperature regulation. It may even pay to increase the wattage of selected heating elements so that adequate temperatures may be maintained although the real time available for heating is decreased.

A firm specializing in dip-brazing aluminum assemblies for the microwave industry pre-heats the parts in an electrically heated oven to evaporate any residual moisture. A secondary reason is to prevent a rapid local temperature drop in the molten salt bath in which the parts are heated further to effect the desired brazing action. For years, both units had cycled according to need to maintain the desired temperature. Because the temperature of the preheat oven is not critical, while the salt bath must be maintained within extremely close limits to prevent crystallization of the salt or melting of the components to be brazed, and each had a time-averaged duty cycle on the order of 50 percent, a relay was wired into the control circuits of both devices so that when the salt bath called for heat, the preheat oven was inhibited. Although there was no significant difference in the temperature of the oven, the demand charge dropped by over $50 per month, for an annual cost reduction of $625. Note that there was no energy savings; the kilowatt-hour consumption remained essentially unchanged.

MINIMIZE ROOF REPAIR COSTS WITH
INFRARED SCANS

Infrared scanners are portable devices capable of rendering heat energy in the sub-visible spectral range quantifiable in the hands of a trained operator. Because the equipment is necessarily expensive, and considerable experience is required to accurately interpret the results, it is generally more advantageous to retain the services of an infrared specialist than to lease or purchase the scanner. Common industrial applications for infrared detection equipment include:

- Identification of infiltration leaks in building envelopes

- Location of defective insulation or leaks in underground steam distribution lines

- Detection of loose electrical connections in feeder panels

- Verification of the operation of steam traps

- Isolation of heat-related problems in regard to product output or quality

It is only recently, however, that infrared scanning equipment has been used extensively as a tool to optimize roof repairs in conjunction with minimizing heating costs. Typically, the flat roofs common to most large industrial buildings are weatherproofed with five layers of asphalt-impregnated felt, alternated with applications of molten asphalt which serves as a binder as well as contributes to the watertight-integrity of the membrane. The layer of pea gravel generally spread on the top surface is primarily to provide a more durable surface for foot traffic to facilitate the maintenance and repair of roof-mounted mechanical equipment. Hence, the common term for such a built-up roof: tar and gravel. A secondary function of the stones is to reduce the surface temperature of the roof by reflecting the incident solar radiation; lower temperatures retard the evaporation of the volatiles in the binder, increasing the life of the roof. Between the weatherproof surface and the roof deck is generally a layer of insulating board, such as "Fescoboard", which can be from one to three inches thick.

Because the primary function of roofing systems is to keep the rain out of the building, they are considered serviceable if they do not leak. The onset of the first leak is certainly cause for concern, but human nature and high repair costs being what they are, several months may elapse before the indicated repairs are effected, and then, only the area surrounding the apparent leak is rebuilt. In truth, a typical roof leak will admit water for weeks or even months before it appears as a drip from the ceiling. In the interim, the water is absorbed by the porous insulating board. Thus, even after the leak is repaired, a significant area of soaked insulation remains. In addition to the obvious long-term effects of corroding or rotting the roof structure, the wet insulation provides a more direct path for the transfer of heat from the inside to the outside of the building than if there were no insulation at all. The resultant heat loss during the years that the condition will probably exist (until the entire roof is ultimately replaced) will cost many times that of contracting for the repair of the leak.

Using an infrared scanner to measure the relative surface temperature of the roof during cool, cloudy weather will definitely outline areas of wet insulation, indicating a leak, or a pre-existing leak which has been repaired without replacing the insulation. The affected areas can be outlined directly on the roof surface with an aerosol paint for later identification. The ability to detect even small leaks before they become apparent in the customary manner will ultimately reduce not only the roof repair budget, but also the cost of maintaining comfort conditions in the occupied space below.

A large manufacturing company housed in an old New England "mill building" complex could not afford the total replacement of the roof that was imminent, and was interested in obtaining the maximum benefit from a limited repair budget. By contracting for an infrared scan of the entire roof, with the problem areas outlined in yellow paint, they were able to secure competitive bids from qualified roofing contractors based on the precisely defined areas requiring replacement. Although it is difficult to quantify the effect of the strategy on either the long-term roof repair budget or the resultant energy savings, the latter was calculated by the Facilities Engineering Department to be approximately $1,100 during a typical winter for the 4,000 square feet of roof that was rebuilt.

INSULATE HIGH TEMPERATURE EQUIPMENT

It is reasonable to assume that by now, most of us have packed as much insulation as possible into every accessible crevice of our homes in an effort to minimize fuel consumption and costs for comfort heat. We are convinced that the concept of adding insulation between two surfaces that are maintained at different temperatures will reduce the rate of heat transfer between them. If we must pay for an outside energy source to maintain this temperature difference, it is a simple matter to calculate the cost-benefit ratio of purchasing and installing the insulating material.

Curiously, this thinking has not pervaded industry with the same zeal and enthusiasm that abounds in the residential sector. While it is true that most firms dutifully insulate steam distribution mains, great expanses of sheet metal and tangles of process piping continue to radiate heat energy into the occupied space. Even during the winter months, the magnitude of the radiated heat more than offsets the heat loss of the building; doors and windows are often opened by the workers for relief. In summer, the oppression seriously affects productivity as the efficiency of the workers is impaired. Throughout the year, the energy and dollar waste continues.

In attacking the problem, it makes sense to insulate first where the heat loss is greatest. Because heat flow is proportional to the temperature difference and the affected surface area, the hottest and the biggest machine elements should be insulated first. Admittedly, in many cases this is easier said than done. Odd shapes or specific environmental conditions may render common insulating materials or standard methods of application virtually useless. In such cases, it is necessary to be innovative. Even an installation that is less than perfect is better than no insulation at all. Most manufacturers of insulating materials will work with an engineer to determine the optimum product and application technique to effect the desired result.

A manufacturer of heavy paper stock used two continuous Fourdrinier machines with in-line dryers consisting of 96 steam-heated drums three feet in diameter. The drums were constructed with chromium-plated steel sleeves, and cast iron end plates. The hubs contained the necessary bearings as well as the rotary joints to introduce the steam to the interior cavity and remove the accumulated condensate. Although the end plates, with an aggregate area of over 2,700 square feet, were never insulated (probably because it would have been a formidable task, and not cost-effective when the equipment was built), they were maintained at a temperature of 265°F, and collectively radiated sufficient heat energy to maintain excessive temperatures during the winter months, and necessitated the use of large exhaust fans in the summer to permit reasonable working conditions in the area. The application of insulation to the hemispherical drum ends presented many problems. The material had to be shaped to provide a reasonable fit to the parts, it must withstand the anticipated physical abuse from tools and workers attempting to maintain the machine output, and it must provide a reasonable measure of resistance to the passage of heat to the surroundings.

The solution was to cut discs of 1-1/2 inch polyisocyanurate, cutting them across the diameter so that they could be fitted to the end caps without removing the bearing assemblies, and cementing them in place with a high-temperature adhesive. Hardware cloth was then wired over the insulation and fastened to protruding studs. Finally, the entire assembly was plastered with a plastic refractory material to provide mechanical protection against the hazards of normal production activity.

Based on the readings of the dedicated steam flow meter, and the internal charges for steam from the central boiler, the annual savings were projected to be nearly $28,000. A secondary benefit was that the comfort level within the affected area of the plant was considerably enhanced.

SIZE ELECTRICAL MOTORS TO THE LOAD

Most commercially available production equipment is designed to perform its function on a variety of similar products. Because different products, or different sizes of the same product, do not have equal horsepower requirements, the machine is generally equipped with an electric motor of sufficient capacity to handle the heaviest load in terms of product size or line speed. Further, electric motors are furnished in discrete sizes. Thus, if the calculated load is 3.5 horsepower, a five HP motor must be provided.

Most industrial facilities utilize three phase electric power. The common induction, or "squirrel cage", motor is notoriously inefficient at partial load; the power factor also decreases as the load is reduced, further increasing the cost of operation.

To determine the load factor of a given motor, it is only necessary to isolate one of the power leads, accessible by removing an electrical junction box, and use an inductive ammeter (such as an Amprobe) to compare the actual consumption to the name plate current, usually expressed in "full load amperes" (FLA). In the case of dual-voltage motors, it is important to know the characteristics for which the motor is wired. Finally, if the machine is subjected to variable loads, the measurement should be taken at the maximum power requirement.

If the measured FLA is significantly below the maximum name plate rating, there is a possibility that a smaller motor can be substituted effectively. The class of service will be a factor (the ambient temperature, or the fact that the motor drives a fan and is located within the air stream, for example), as will the rated "service factor" of the motor, which indicates the tolerable overload expressed in percent. The plant Electrical Engineer can provide valuable assistance in the determination of the proper motor size to minimize the consumption of electricity per horsepower-hour of work performed.

A cosmetics manufacturer was using a standard bottling line to produce small samples to be used as give-aways. Because of the variety of cosmetics produced, it was decided to apply the labels manually. Thus, the labeling operation, the slowest element of the production line, determined the speed at which the automatic filling and capping machinery would run. The combination of the small product size and the slow line speed meant that the two 20 HP motors supplied with the equipment could be replaced by five HP units. Although the plant was not penalized by the utility company for low power factor, the direct savings in consumption and billed demand was calculated to be $900 per year.

RESCHEDULE HIGH-DEMAND PROCESSES TO THE SECOND SHIFT

Demand charges have become an integral part of virtually all industrial electric bills. As a further incentive to reduce the maximum demand, many utility companies increase their demand charges during the summer billing months as a financial incentive for customers to reduce their peak demand at a time when the generating capacity of the power company is strained due to air conditioning loads. Others impose a "ratchet" clause in their tariff schedules which states, typically, that the billed demand each month will be the higher of the actual metered demand, or 90 percent of the highest metered demand recorded within the preceding 11 months. Either practice can be extremely costly to a facility with a characteristically high summer demand peak.

The concept described in this section will effectively reduce the monthly metered demand, and can be applied if there is a peripheral process which is not directly related by time or function to the primary plant output, and which itself constitutes a significant element of electrical demand. Simply stated, the strategy involves rescheduling the subprocess to the second (or third) shift. Even for a three-shift operation, the peak demand is typically recorded during the first shift, coincident with the occupancy of the office staff. Thus, eliminating the additive effect of the high-demand operation to the first shift base load will result in a potentially significant dollar savings. A fraction of the apparent cost reduction may be lost to the salary premium, or shift differential, commonly paid to workers on shifts other than the first, or to support security personnel required as a result of the change. Clearly, the sum of all the effects must be considered to determine the feasibility and cost-effectiveness of the rescheduling.

A meat processing plant in New York City, historically plagued by high electricity costs, buys frozen sides of beef, and cuts them into individual portions for the restaurant trade before refreezing them. Because it is necessary to have the beef at a temperature of 28 °F to facilitate the cutting process, the cartons of frozen beef at 0°F are run through a large microwave oven. The thawed meat is then stored in a holding refrigerator until it is cut. In an effort to reduce the total electric bill, it was suggested that the microwave process be performed during the second shift, when only a cleanup crew was working in the plant. A total of two men were transferred from the first shift to thaw and store a sufficient amount of beef for the cutters on the following day. The result was an immediate reduction in the billed demand of 70 kilowatts which, at the time the change was instituted, saved $10,900 annually.

VERIFYING IMPROVED ENERGY PRODUCTIVITY

The ability to verify expected energy productivity savings is essential to any program to reduce energy costs in an industrial facility. As mentioned above, many of the most cost-effective solutions require innovative strategies whose savings may not be easy to predict. For this reason, it is often desirable to try an improvement on a limited basis and to monitor the energy savings over a period of time before fully implementing the change. In this way, the energy savings attributable to the change can be calculated and its cost effectiveness determined.

In order to properly interpret energy bills, it is necessary to separate those factors affecting energy costs that cannot be controlled, such as weather variation and fuel price increases, from the effects of any energy productivity improvements. Such uncontrollable factors can obscure the effects of improvements unless they are considered in the savings calculations.

Energyworks uses a microcomputer-based computer program, called Energy Bill Analysis, to isolate the impact of energy productivity changes from other, uncontrollable energy cost factors. Originally developed for in-house use to produce reports for commercial and industrial clients, it is now available for use on popular microcomputers and as well as on a service bureau basis for users without access to microcomputers. This program enables managers to monitor and verify savings due to energy productivity improvements as well as to interpret energy bills to assure continuing performance in industrial, commercial, or residential applications.

CONCLUSION

It is not always necessary to spend thousands of dollars in implementation costs to reduce energy costs. The diversity of equipment and processes in most manufacturing plants affords the engineer the opportunity to analyze and innovate. The concepts and strategies outlined here represent but a sample of the potential for conservation that still exists in industry, even after nearly a decade of escalating energy costs.

While energy costs may be only a small fraction of the total cost of producing, marketing, and distributing a manufactured item in many industries, the relative proportion, and therefore the financial impact, are likely to continue to increase in the years to come. It is therefore incumbent upon those responsible for such matters to reduce process energy requirements to the lowest overall cost. Firms that are successful in this endeavor will have a decidedly competitive advantage in the years to come.

SECTION 21
POWER GENERATION
AND FUELS UPDATE

Chapter 67

SHORT RUN ALTERNATIVES TO CURRENT ELECTRIC UTILITY FUELS

W. W. Westcott, F. L. Merat

Since economic development goes forward only with the use of energy, we are now at the beginning of a great change. As the 75% of mankind which has been living on an agricultural base is beginning to buy machines and manufactured consumer goods, the direct and indirect pull on existing energy sources will increase so that in the year 2000, which is not 20 years away, it is reasonable to compute a world energy usage of 2 to 2.5 times the current consumption with resulting increases in cost [1].

A rapid change of an existing price structure has happened at least once before [2]. In England during the period from 1575 to 1650 when it came out of the so-called Dark Ages and into commerce, there was an increased need for fuel. The basic manufacturing fuel at that time was wood. England then became deforested, and the only economic alternative was coal. Considerable engineering was required to adapt this alternate to such processes as smelting, glass manufacture, and the textile processing industry.

The problems faced in England at the turn of the 17th century were no less serious than those faced by us today. They did some ingenious and difficult research, design and development to solve their basic problem. In the process, they achieved a head start on mainland Europe in commerce and manufacturing which was to last for two centuries. [3] Surely, we too, will develop alternates and make the necessary adaptions.

An alternative is only a viable way of doing something in a way other than it was done before and alternatives to current electric utility fuels do not have to be fuels at all. They must be ways of making available power follow the load curve. Usually the alternative is a method which costs less and fulfills all other requirements. Today, as in England of the 1600's, there are alternatives, set in a much more complicated economy. Some are well known, some less well known, and some can only be dimly seen. One reason we are all here today is that there are a number of alternatives, none of which seems to work economically in all circumstances.

As the largest user of solid fuels, and one of the largest users of petroleum fuels, the electric utilities are in a position to carefully investigate alternatives.

After the authors' examination of available alternatives, there are some which stand out in light of several or all of the following criteria:

1. Their costs are well-defined. That is, they are currently in use and have been shown to result in lower overall costs.

2. They are widespread enough in their potential to be of interest to more than one area or condition.

3. They are fairly simple technologically, acceptable environmentally, and incorporate no problems that cannot be solved with existing technology.

4. They have long lives, so that their cost will result in a solution over a period of decades.

Ideally, they should last until some non-fuel consuming method of production is developed, and is shown to be technologically, economically, and environmentally favorable.

5. To be short run, they should be available within the short run planning period usually used by utilities. To pick a number, say eight years from concept and initial investigation, to installation and initial running.

6. Their scale should be large enough to be of interest for utility operation.

One example of an alternative is pumped hydro, which uses available nuclear or high-efficiency coal base stations to generate semi-peaking power. Here, fuel is used, but in an innovative and more cost-effective manner than if the alternative were not used. Pumped hydro will be discussed in more detail later, as will methods which follow from this concept.

The alternative may be one which has nothing to do with power generation per se, but results from successful management practices by the utility. The almost universal demand charge and the European practice of time-of-day rates are examples. Purchased power is another non-generation alternative. Interesting things can be done with purchased power, especially in reducing capital expenditures. It is safe to say that no solution is without new problems, and, therefore, that no solution is ideal.

There are a number of papers at this conference which deal in one way or another with alternatives. These include: co-generation, which from the utility viewpoint is the sale of waste heat; load management; and heat recovery from waste materials. It is the purpose of this paper to cover some of the more promising electric fuel alternatives but without going into great detail. Most all of the alternatives given are covered in varying degrees of depth in the literature, and have operating experience which can be investigated.

The first alternative to be examined is burning of municipal waste. This alternative is attractive because of a decrease in landfill sites, increased amounts of refuse being generated, and higher costs of other fuels [4]. The two examples most relevant to electric utilities appear to be the Wheelbrator-Frye/Boston Edison operation and the Lakeland, Florida municipal utility plant. Both can generate about 35MW from municipal refuse. The Wheelbrator plant uses mass burning [5]. the Lakeland plant prescreens and mixes the refuse with coal [6]. Both are shown as viable, but limited by the available refuse within economic hauling distance. The cost structures are quite different, and require enough explanation to be outside the scope of this paper [7]

A similar alternative is the burning of peat. Peat and municipal refuse are somewhat similar in that they are solid fuels with a lower B.T.U. content per pound than coal, do not have to be mined, and require somewhat special equipment to burn them. They are both limited to about 35-50 MW generating capacity due to the cost of transportation [8]. This limitation is of a general, not a specific, nature. As an example: a location net to a river or other body of water with barge transportation will ease costs and allow larger generating capacity. The B.T.U. range of peat is 6,000-almost 7,000 B.T.U./pound dry weight. Peat is of interest not only because it is competitive with coal, but because of its widespread distribution, including areas now relying only on petroleum for electric generation [8].

The general restrictions and advantages of burning refuse and peat are also found in the burning of another alternative, the so-called biomass products such as bagasse (sugar cane stalks), wood, and wood waste [9]. It is worth noting that both bagasse and wood waste have been used for years on a small scale (by utility standards) for electric generation by sugar refineries and sawmills, respectively.

Natural gas wells are viable peaking plant energy alternatives in cases where there are proven reserves in the area. This is because both risks and costs are readily determinable.

Another alternative to be considered because of recent developments is "dry geo-thermal" generation. This is the method whereby hot areas relatively close to the earth's surface are drilled, the rock is fractured with high-pressure water, feedstock water is pumped down the well, and the steam, or heated water, is brought up through a separate set of wells some distance away. Los Alamos National Laboratories has done the most recent work in this area. Their recovery temperatures are between 250-300 degrees Centigrade (480-580 degrees Fahrenheit). A small 10 MW plant is being investigated by a utility. In this example, the drilling depth is approximately 14,000 feet; construction costs are calculated to be $1,500/kW of generating capacity, with drilling costs representing 50-60% of overall cost [10]. This process does not have the high degree of corrosive alkali metal contaminants which have been a problem in the development of wet geo-thermal deposits, and has a life-span which should carry the plant indefinitely into the future.

There are two aspects to dry geo-thermal which, in the authors' opinion, make it an interesting alternative for development. One is that the geo-thermal potential is some 10-15% of the world's land mass [11]. In a few areas the source comes close enough to the surface to be economically tapped now. The other aspect is that since 50-60% of small plant cost is in the drilling, and the process scales well, improvements in drilling methods would lower overall costs significantly, allowing a much wider area to be suitable for generation using a $1,500/kW criteria [12]. A revolutionary change in drilling, or some not yet conceived alternative to drilling, would allow geo-thermal generation in 15% of the world's land mass [13]. Improved geo-thermal drilling methods are currently being worked on. It must be remembered that not only is this method a fuel-less method of generation, but is base plant. That is, it operates 24 hours per day, making it possible for use in conjunction with compressed air or steam storage for peaking operation. The kWh cost overall appears to be in the 40 mil range and is protected against rising fuel costs.

In summary, geo-thermal is not only practical in specific locations today, but has the short term potential to become much more widely usable [11]

Another alternate is the general classification of pumped storage installations, which shift usage of fuels in time and allow for the substitution of one fuel for another.

There is at present a 290 MW capacity compressed air storage facility at Huntsorf, Germany near the North Sea [14], and a projected facility to be built in Edgan County, Illinois. URS Engineering of San Mateo, California has aquafer rights to build yet another facility of similar capacity to the Huntsorf station. Compressed air energy storage plants (CAES) compress air using available capacity from nuclear or efficient coal base stations to run air turbines for semi-peaking and peaking periods. The storage space for the German plant is salt caverns which were leached hydraulically with the effluent being discharged into the North Sea. The Illinois location would use a natural aquafer, or underground void, currently filled with water, as the storage space. Voids such as abandoned mines and depleted gas wells may also be usable. The general idea is that if power can be obtained for 10 mils/kilowatt-hour and part of that is lost in conversion, it is still better than generating or buying peaking power at 50-60 mils. Construction costs for CAES are on the order of $400-$500 per kW capacity [14]. Among the advantages of CAES are that the process uses virtually no real estate and the area can be flat. The main disadvantage is that such underground voids are not universally available. However, it has been shown beyond reasonable doubt that underground compressed air energy storage is economic in areas where such voids occur and can be purchased at reasonable cost.

In the general area of pumped storage is underground pumped hydro, a combination of pumped hydro and CAES which also appears to fill the criteria of a short term alternative. With this method, water is impounded on the surface and the pump/turbines and motor/generators are in a cavern a short distance above the underground water storage site. Projected costs are in the general range of those for CAES [15]. There are some problems foreseen due to the contamination of the water by underground minerals.

In 1955 pumped storage was in its beginnings [16]. Today, there are a number of pumped storage methods available for selection.

Whereas, all of the mentioned alternatives give more freedom to management in determining alternate ways of supplying their service areas, they raise new questions. Here have been presented some of the methods which are simple and economic enough to present viable prospects.

The problem today is not so much a lack of alternatives, as it is too many possible ones. If the energy situation in which we find ourselves is indeed a crisis, it presents us with not only dangers, but opportunities.

References

1. Sassin, Wolfgang, "Energy," Scientific American, September, 1980, pp. 128-129
2. Nef, John U., "An Early Energy Crisis and its Consequences," Scientific American, November, 1977, p. 140.
3. Ibid., pp. 142-151
4. McGowin, C. R.,"Municipal Solid Wate: A Utility Fuel?", NCRR Bulletin, Vol. 11, No. 4,National Center for Resource Recover, Inc., Washington, D.C., December, 1981, pp. 101-108
5. Broglio, Ronald, Wheelabrator-Frye, Inc., Danvers, Mass., May 28, 1982. private commun.
6. Conference Proceedings, Municipal Solid Wate As a Utility Fuel, Miami, Florida, May 13-14, 1982. Available thru Electric Power Research Institute, Palo Alto, Calif.
7. Ibid
8. Manfred, Ralph, Electric Power Research Institute, Palo Alto, Calif., May 25, 1982. private commun.

9.   "Bio-Energy," The Bio-Energy Council, Washington,
      D.C., pp. 1-2.

10.   Smith, Morton, Los Alamos Laboratories,Los Alamos,
      New Mexico, May 24, 1982. Private communication.

11.   White, D.E., et al., "Assessment of Geothermal
      Resources of the U.S., 1975," Geological Survey
      Circular 726, U.S. Geology Survey, U.S. Depart-
      ment of the Interior.

12.   Pollack, Henry, N., and Chapman, David, S., "The
      Flow of Heat from the Earth's Interior," Scienti-
      fic American, August, 1977, pp. 74-75.

13.   Smith, Morton, "Progress of the U.S. Hot Dry Rock
      Program," The International Conference on Geo-
      thermal Energy, Florence, Italy, May 11-14, 1982,
      pp. 310-319.

14.   Maass, Peter, and Stys, Z. Stanley, "Operation Ex-
      perience with Huntorf," American Power Conference,
      Chicago, Ill., April 21-23, 1980.

15.   Tam, S.W., et al., "Assessment of High Head Turbo
      Machinery for Underground Pumped Hydro Electric
      Storage Plants," Second Miami International
      Conference on Alternate Energy Sources", Coral
      Gables, Florida, December 1979.

16.   Ibid

Chapter 68

AN OUTLOOK FOR LIQUID FUELS FROM SHALE OIL

V. K. Gupta

INTRODUCTION

Recent trends in oil consumption in U.S. indicate that approximately half of our nation's energy comes from petroleum. The United State's need for liquid fuels will continue well into the future, but the current and future availability of crude oil has become this nation's most pressing problem. Based on the recent trends in technology, no other fuel sources can be expected to power our transportation industry. These include fueling our cars, trucks, tractors, planes, ships, and our defense equipment. It will certainly be past this century, before other technologies will be developed and commercialised. Two principal alternatives are available to offset present and predicted deficits between liquid fuels demand and their suuply. One is to increase our reliance on imported crude oil from sources such as OPEC; the other is to develope alternate liquid fuels from domestic sources other than crude petroleum. Continously increasing dependence on imported oil has made U.S. more susceptible to foreign political events, and the gravity of the situation is apparent from the effects on our economy even without interruption of the oil supply. Synthetic fuels, or synfuels as they are sometimes called, are the liquid and gaseous hydrocarbons derived from coal, oil shale, tar sands, and other organic substances such as agricultural and urban wastes. These liquids are usually comparable to natural petroleum, that may be refined into gasoline, diesel oil, jet fuels, or other fuels no different in performance than natural products. Prospects of synfuels production from coal described by the author (1), offers a great promise. The alternative fuel shale oil is ready now and could be providing us with significant quantities of liquid fuels in five years. U.S. has large resources of oil shale.

For many years, the commercial production of shale oil in U.S. has been, "just around the corner", but the commercial production of liquid hydrocarbons from shale has, as yet, made little or no impact on the energy scene. A note appeared in American Chemical Society (ACS) publication in 1927, which read, "Shale Oil obtained in the operation of the Federal Government's experimental oil shale plant near Rulison Colorado, is now available for distribution by the Bureau of Mines to the laboratories that might be interested in conducting tests with such oils."(2) Fifty years later, an announcement that appeared in another ACS publication stated, "products from shale oil may be available as early as 1981."(3) The progress in half a century has been virtually nil except for the refinement of some experimental processes and the issuance of hundreds of patents largely related to extraction technology.

It still remains to be seen whether the predicted temporary oil glut, and World and U.S. economic recession, together with the dire predictions by no growth groups of environmental, social and economic disasters if a commercial shale oil operation is established in Western United States, will stop the present impetus of oil shale development or whether now is the time. However, if one examines the energy problem as is evident from the data on increasing dependence on petroleum and natural gas(see Fig. 1), and one wants to obtain energy independence, it is necessary for us to proceed with the development of oil shale and coal conversion industries. There are many problems yet to be overcome.

The purpose of this paper is to better inform the public about this huge domestic resource and our nation's need to utilize it. Oil from rich oil shale deposits of Rocky Mountains alone can provide significant quantities of oil at a reasonable price for the near term. This paper will concentrate on the availability of oil shale, the

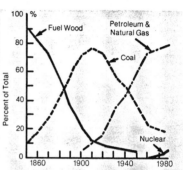

Figure 1: Current Energy Trends.

nature of the oil shale, the comparison of various technologies of extracting oil from the rock, upgrading of the oil, and its environmental and social impact. An effort has also been made to point out the future trends and prospects of Shale Oil industry in U.S.

POTENTIAL AND DISTRIBUTION OF SHALES

Most of the workable shales in this country are found in the Green River Formation, which is about 16,000 square miles area in the states of Colorado, Wyoming, and Utah (see Fig. 2). There is an estimated 1.9-2.0 trillion barrels (bbl) of crude oil equivalent present in this formation. About 85% of it is found in the Piceance Basin in Colorado, which is the primary target area of interest to the most of companies. Within the 600 billion barrels of oil equivalence of high grade shale (25-100 gal./ton), an 80 billion bbl of oil equivalence can be readily recovered using current technology. This alone would replace the domestic supply of oil for 20 years. From the data in Table 1, it is obvious that the U.S. oil shale source is huge. It is estimated that potential resource from shales other than the Green River Formation may yield on the order of 130 trillion bbl of oil. Evidently reserves of this general magnitude represent a liquid fuel supply of at least 200 years.

COMPOSITION OF OIL SHALE

Its composition is complex and varies both with strata, and in case of some saline constituents with radial distance from the disposal center. Excluding

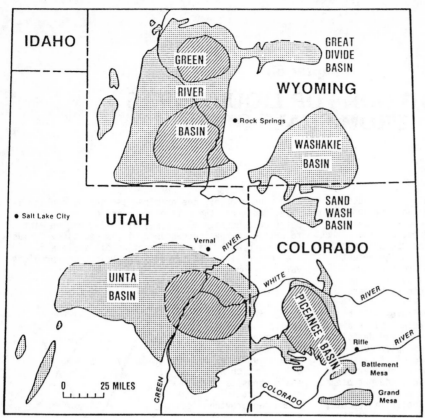

FIGURE : 2

GREEN RIVER FORMATION

 UNAPPRAISED OR LOW GRADE SHALE

 SHALE YIELDING AT LEAST 25 GALLONS OF OIL/TON

Table 1. Oil Shale Resources of U.S.(4)

Deposits	Oil Equivalent		
	25-100[a]	10-25	5-10
Green River Formation	1,200	2,800	4,000
Devonian and Missispian Shale	-	1,000	2,000
Alaskan Marine Shale	250	200	large
Shale associated with Coal	60	250	210
Other Shales[b]	500	22,000	134,000

a. oil yields in gal./ton.

b. These include Ordovician Black Shale, Permian Phosphoria Formation, Monterey Shale, Cretaceous Shale beds, and Nonesuch Shale etc.

these saline minerals, the qualitative composition is stratigraphically constant over wide areas. The organic matter in shale is composed mostly of kerogen with a few percent bitumen; the two can be differentiated by the solubility of bitumen in benzene and similar solvents. The elemental analysis in table 2 is typical representation of the composition of kerogen. The structure of kerogen is not definitely known but it is probably a multi-polymer with molecular weight of about 3200. Upon pyrolysis, it is broken into various organic fragments, forming a viscous oily liquid, oil shale. A pragmatic index of oil shale value is the Fisher Assay, a standardised laboratory pyrolysis, which measures the oil obtainable from a given weight of shale under controlled conditions. This is expressed in gallons of of oil per short ton of shale. The Fisher Assay of kerogen from different grades of Shale is given in table 3.

Table 2. Elemental Analysis of Kerogen (5)

Component	Amount Wt%
Carbon	80.52
Hydrogen	10.30
Nitrogen	2.39
Sulfur	1.04
Oxygen	5.75
C/H Atomic Ratio	7.81

Table 3. Conversion of Kerogen by the Fisher Assay(6)

Grade of Shale gal/ton	10.5	26.7	36.3	57.1	61.8	75.0
Conversion of kerogen by Fisher Assay to:						
Oil wt%	51	65	69	66	69	71
Gas wt%	14	12	11	12	12	11
Organic residue wt%	35	23	20	22	19	18
	100	100	100	100	100	100

The inorganic portion of the oil shale comprises a complex and variable mineral. The host rock is dolomite marlstone whose major non-saline constituents are analcime $Na_2O.Al_2O_3.4SiO_2.2H_2O$, quartz SiO_2, K-Feldspar $K_2O.Al_2O_3.6SiO_2$, Na-feldspar (albeite) $Na_2O.Al_2O_3.2H_2O$, dolomite $MgCa(CO_3)_2$, calcite $CaCO_3$, and illite $K_2O.3Al_2O_3.6SiO_2.2H_2O$. Pyrite and marcasite

FeS$_2$ are also found. Table 4 shows a typical listing
of inorganic minerals present in a 25 gal/ton shale.
Minor constituents of the shale such as As, Pb, Mn, Ba,
Cr, V, B, Se, Te, Zn, Mo, Sr, P, and Cu are also important because of environmental and refining considerations. Their concentrations vary in the range of
0.001 - 0.1 (wt%).

Table 4. Typical Mineral Content of 25 Gal/Ton Shale
(5)

Mineral	Formula	Wt %
Dolomite	CaMg(CO$_3$)	33
Calcite	CaCO$_3$	20
Plagioclase	NaAlSi$_3$O$_8$.CaAl$_2$Si$_2$O$_8$	12
Illite	K$_2$O.3Al$_2$O$_3$.6SiO$_2$.2H$_2$O	11
Quartz	SiO$_2$	10
Analcime	NaAlSi$_2$O$_6$. XH$_2$O	7
Orthoclase	KAlSi$_3$O$_8$	4
Iron*	Fe	2
Pyrite or Marcasite	FeS$_2$	1

* In combined state as ferroan limestone.

EXTRACTION TECHNOLOGY

The recovery of syncrude from oil shale comprises,
basically opening up the ore bed (fracturing or mining,
and crushing) and applying heat to the ore to release
the organic material from the mineral matrix. There
are three basic technologies: open pit mining, underground room and pillar mining with surface retorting,
and in situ fracturing and extraction. The option of
the basic extraction technique depends primarily on the
depth of the ore bed, the bed thickness and assay of
the deposit. For Devonian Shale, suggested guidelines
are: 1) do not mine if the overburden is 200 ft. thick
or the seam is less than 10 ft. thick, 2) do not mine
if the overburden ratio is 2.5, and 3) do not mine
shale containing less than 10 percent organic carbon(7).
Similar guidelines, but with different trigger numbers
would have to be applied to other ores depending on
depth, assay, thickness, extraction process, accessibility, and so forth. A major problem in dealing with
oil shale is that on the average a ton of shale yields
less than a barrel of oil (8), hence massive quantities
of spent shale must be disposed of,and the spent shale
is highly alkaline making land reclamation and protection of aquifers extremely difficult.

RETORTING TECHNOLOGIES

The oil retorting process can be classified as
either ex situ (surface) and in situ (within the existing formation). The advantages of both the processes
may be summarised as:

Ex Situ Recovery

Advantages

1. Recovery efficiency is high, up to 80-90% of
   the organic content of the shale can be retorted.
2. The process variables can be easily controlled.

Disadvantages

1. Operation cost is high, due to the cost of
   mining, crushing, and transportation of oil
   shale.
2. Spent shale disposal followed by land revegetation is a severe problem.
3. Pollution of environment both air and water
   are significant problems.
4. The process is limited to only rich shale
   resources that are accessible for mining.

In Situ Recovery

Advantages

1. Environmentally desirable since all operations
   are carried underground. Neither mining nor
   spent shale disposal is needed.
2. Oil can be recovered even from deep deposits
   of oil shale formation.
3. Even leaner shale deposits containing less than
   15 gal/ton oil can be utilized.
4. The process is economic due to reduced or no
   cost of mining, crushing, and transportation.

Disadvantages

1. Process control is difficult due to insufficient permeability and communication within
   the oil shale formation.
2. Recovery efficiency is generally lower.
3. Possible contamination of the sub-surface
   waters.

To compare the advantages versus disadvantages,
the in situ methods for production of oil from shales
optimize recovery while minimizing recovery impact.
As a result of this greater emphasis has been placed
on in situ processes.

Ex Situ Processes: In these processes, shale rock is
mined crushed, and transported to the retort where it
is subjected to temperatures in the range of 500-550°C
in which the chemical bonds joining the organic
compounds and the rock matrix are broken. The products
of the retorting process in the gaseous state are,
collected, condensed, and upgraded into liquid products
that are comparable to petroleum products. The oil
so obtained is sent to the refinery by pipeline,
where it is refined into the final product.

Retorting processes are generally of two types:
those in which gas is used to transfer heat and those
in which heated solids are circulated to transfer heat.
The retorting process is strongly affected by the
quality of the feedstock, becomes less efficient when
the shale is lean. The advantage of low cost in mining
lean shale is lost in high retorting costs. Some gas
combustion retorts do not perform well for rich shales
due to plastic deformation and compaction of the shale
or liquid flooding   of the retorting column leading
to excessive pressure drop and consequent malfunction.
Thus an appropriate balance has to be achieved, between
the type of mine and grade of shale, and the type of
retort used. Some retorts use combustion of residual
carbon from the processed shale to supply heat for
retorting purposes. In such cases, light hydrocarbon
gases along with hydrogen and carbon monoxide are
diluted when combined with combustion produced flue
gas, resulting in a low Btu gas. Since the particle
size cannot be too small in a gas flow retort to avoid
excessive pressure drop, retorting of kerogen takes
longer at high temperature, the conversion process
results in higher yield of gaseous products and carbon

residue on shale. Therefore, air combustion retorts produce relatively large volumes of low Btu gas. Three internal combustion retorts that have been used are Gas Combustion, Paraho, and Union A.

Some other retorts, such as Tosco II and Lurgi-Ruhr gas use indirect heating by heat transfer from hot solids. Solid-heat carrier retorts did not develop until reliable equipment was available for heating, transporting, and separating large quantities of hot solid material. This type of retorts are still the most complex of the various retorting systems due to the difficulties in soids handling. These processes produce more oil and lesser volumes of gas, which is of high Btu content.

Other heat transfer methods include microwave, laser, and circulation of a substance which releases heat by chemical or nuclear reaction. These retorting systems are laboratory curosities and will remain so for some time to come. Since all retorts are of limited flexibility, it is possible that a large commercial plant will use retorts of different kinds. Selection of an optimal combination of different retort designs will need considerations of overall plant thermal efficiency, overall resource utilization, and overall processing costs. For example, crushing, and screening costs may be minimised by using an internal combustion or gas-recycle retort to handle coarse shale and a solid-heat-carrier retort to handle fine shale. Combining several retort designs in a single plant will also permit greater flexibility in the composition of the product mix from the plant. The design and operation of these retorts has been well described in a report by Colorado School of Mines(9).

Development of the immense shale oil reserves of the Green River formation will require solutin of many environmental and social problems. Potential environmental problems include disruption of the land, disposal of large quantities sulfur containing low Btu retort gas, particulate emissions from mining, solids-handling, and retorting operations, and disposal of spent or retorted shale. Although other problems also very serious, but the disposal of spent shale is the greatest environmental roadblock to oil shale development. A good grade of Green River shale will yield about 30 gallons of crude shale oil per ton of feed shale provided it is pyrolysed under carefully controlled conditions of the Fisher Assay. A realistic figure of 90-percent of Fisher Assay for internal combustion retorts, with 30 gal/ton shale, will require 1.6 tons of raw shale, producing a barrel of crude shale oil leaving behind 1.3 tons of solid residue. Uncompacted shale has a bulk density about equal to that of water. The 1.3 tons of spent shale for every barrel of shale-oil product will occupy about 42 cubic feet. A commercial size 50,000-barrel-per-day shale-oil plant will produce 65,000 tons(over 2 million cubic feet) of retorted shale each day. If it were spread over the surrounding countryside, a single day's production of spent shale will cover an area of 40 acres to a depth of 1 foot. Disposal of this volume of spent shale is a colossal environmental problem. Compaction of disposed shale is an expensive process, requires water, which is a scarce natural resource in the shale areas. Prolonged irrigation of the surface will be needed to promote sprouting and growth of natural grasses, to prevent escape of fugitive dust, and to stop erosion of the surface by wind and water. Creation of the stable non-eroding surface over the disposal area is critical since retorted shale contains soluble salts and organic compounds which will be leached by water flowing over and through the eroded shale. Contaminated surface runoff must be stored in evaporating ponds until plant growth in the revegetated areas is extensive enough to inhibit further erosion.

The properties of shale oil derived from each of the processes are quite different. Data from a number of processes is listed in Table 5. For comparison, data on Arabian crude has also been included in the table. Data on some in situ processes has also been included in the table.

In Situ Processes: In situ processing is an alternative approach to oil shale development which offers less surface disturbance, fewer surface installations, and smaller labor forces. The requirements for manpower and equipment are less because there is little or no mining involved. Surface retorts are not needed, and spent shale disposal is not required, thus reducing material handling problems which plague surface processing by leaving the shale underground and transporting to the surface only gaseous and liquid fuels.

Although some proposed in situ processes feature low-temperature separation of hydrocarbons from mineral matter by donor solvents or by bacterial action, most rely on thermal treatment where shale is heated to 800°F and kerogen is decomposed to oil, gas, and coke like residual carbon. An in situ retort is analogous to a huge aboveground retort in that it contains a quantity of broken oil shale which is permeable to a heat carrier fluid. The heat carrier is forced into the retort at one point and the products of pyrolysis are withdrawn from the retort at other points. The carrier may be heated before it is injected, or it may be heated while inside the retort by combustion of a portion of the kerogen pyrolysis products. In the latter case heat is carried through the retort by the combustion gases as in an internal combustion retort.

Whatever is the method of heat supply, shale in the in situ retort must be highly permeable to fluid flow. Efforts for heating impermeable shale deposits by conduction have concluded that shale's low thermal diffusivity make externally heated in situ retorts uneconomical, thus in situ retorts must rely on convective heat transfer modes. This mechanism requires that shale be broken into chunks so that continous gas pathways will be available through the entire bed of shale. Permeability is the major problem with in situ processing of most of the oil shale beds with the exception of saline zones of Green River Formation.

A recent review of the literature indicates that all in situ oil shale processes can be classified into the following categories:

    I.   Subsurface chimeny
         a. Hot gases
         b. Hot fluids
         c. Chemical extraction.
   II.   Natural fractures
         a. Unmodified
         b. Enlargement by leaching
  III.   Physical induction-no subsurface voids.

The above classification is based on the geometry of subsurface retort, and the methods used to degrade and extract exposed kerogens and bitumens. A detailed description of these methods can be found in several reports (5,9-11). Most techniques presently patented utilize a subsurface chimney formed either by explosive fracture of the rock or by chemical spalling of rock from the sides of boreholes until a subsurface retort is formed. The rubblized material is pyrolyzed using external heating sources - either hot gases or hot

Table 5. Shale Oil Properties From Different Processes(5)

Process/Property	API Gravity	Pour Point °F	Viscosity @ 100° F	Carbon Wt%	Hydrogen Wt%	Nitrogen Wt%	Oxygen Wt%	Sulfur Wt%	Ash Wt%	C/H Ratio
Arabian Light Crude	33.4	-30	47	-	-	-	-	1.8	-	-
Fisher Assay	20.7	75	113	85.23	11.38	1.80	-	0.98	-	7.49
NTU 150- Ton	25.2	70	79	84.58	11.76	1.77	-	0.76	0.01	7.19
Gas Comb. 150-Ton	21.2	85	95	-	-	2.11	-	0.68	-	-
Union A	20.7	90	121	-	-	1.9	-	0.81	0.04	-
Tosco II	21.1	80	106	85.1	11.6	1.9	0.8	0.9	-	7.34
Paraho(DM)	19.3	85	285	84.9	11.5	2.19	1.4	0.61	0.66	7.38
Union B	21.2	60	150	85.3	11.15	1.77	1.12	0.61	0.03	7.3
Circular Grate DF	18-25	70	100-200	-	-	1.8-1.2	-	0.7-0.9	-	-
Hytort	-	60	29/70	-	-	1.68	-	0.57	-	-
Oxy Vert. MIS	22.5	70	70	84.86	11.80	1.50	1.13	0.71	-	7.19
Oxy Geokinetics	24/25	50/65	16	-	-	1.50	-	0.64	-	7.2
True In Situ	25.4	80	42.6@100°F	84.89	11.82	1.62	1.09	0.42	0.26	7.18
Equity* True In Situ	41.2	0	-	85.3	13.4	0.53	0.56	0.49	-	6.37
USBM* True In Situ	28.4	10	45	84.88	12.02	1.69	0.81	0.60	0.01	7.06
Sinclair* True In Situ	30.6	35	-	-	-	1.4	0.49	1.28	-	-
LETC* True In Situ	-	20	48	-	-	1.4	-	0.94	-	-

* All these were experimental and had very low yields.

pyrolytic fluids, or combustion of organics initiated by the addition of an internal ignition material. A number of methods have been proposed that make use of natural fractures and zones of soluble minerals that can be opened and enlarged by solution, eliminating the need for an explosively generated chimney. In these methods gas pressures and aggressive aqueous fluids are used to expose internal surfaces for secondary pyrolytic extraction of oil precursors. One proposal is to use ultrasonic destruction of mineral matrix and breakdown of kerogen to produce shale oil along natural fractures(12). Woods Research and Development Company (13) have used lasers to generate voids and at the same time heat for pyrolytic destruction of organics in the kerogen. All methods in the literature require generation of greater degree of porosity and permeability than is initially present in the rock. Matrix breakdown is usually accomplished thermally, though those processes that do not require a subsurface chimney may include a limited amount of chemical destruction of mineral and organic matter prior to retorting process. Other recovery methods include any method for oil recovery other than retorting. Due to limited work on these methods, only information on Bioleaching of oil shale and Solvent processing is discussed.

Bioleaching Of Oil Shale: This ia a process in which microbially produced sulfuric acid is used to enhance the recovery of kerogen from shale. Microorganisms capable of oxidizing sulfur to sulfate are employed to produce sulfuric acid, which in turn dissolves kerogen thus increasing the matrix porosity, facilitating kerogen removal. Laboratory experiments have shown an enhanced kerogen recovery following bioleaching. The process can be used both in ex situ and in situ oil shale processing. The basic process steps include the following:

1. Fracture the targeted resource shale with explosives or by mechanical means to allow access to increase the surface area.

2. Either inject an acid broth produced from surface cultivated microbes, or introduce Thiobacillus thiooxidans directly to the formation containing sulfur rich medium.

3. Recover kerogen or bitumen released by floatation on the medium surface, or mine and retort the shale after leaching process has been completed.

4. Recover sulfur for recycling from the spent medium using Desulfovibrio desulfuricans.

Yen et al.,(14-15) showed that the carbonate fraction of the mineral matrix was dissolved almost completely in sulfuric acid. The process improved shale quality from 25gal/ton to 40 gal/ton oil. A large amount of water is required for the process. In addition, the bacteria that can withstand chemical and thermal environments at 1000 feet depth may be a problem. Any strain of bacteria developed will have to be carefully screened for its potential environmental impacts.

Solvent Processing: This process consists of heating the shale in the presence of a solvent at low temperature 320-430 C(600-800 F), with or without a high partial pressure of hydrogen. In this case the products are separated from the mineral matrix, by gravity flow, sedimentation, filteration or other methods. Solvent treatment followed by retorting has also been used. The experience so far indicated that 95-100% of the organic matter can be recovered by solvent extraction method(16). The advantages of solvent extraction are: simpler and more rapid transfer of heat, higher yield of liquid, and less refractory nature of liquid. The main disadvantage is that shales below 20-25 gal/ton can not be easily extracted since the mineral portion is continous. The product at low temperature tends to be viscous liquid and separation by sedimentation is difficult unless high solvent to organic ratio is used. There is a tendency to crack extracted material to light products even at low temperature, thus solvent losses remain a technical and environmental problem. A stable aromatic solvent is preferred to minimize cracking losses.

OIL CHARACTERISTICS AND USES

Crude shale oil can be utilized in several different ways. It can be burnt in its raw form in the boiler or it can be refined into liquid fuels or it can be processed into synthetic crude oil for later conversion in a refinery or petrochemical plant. The oil obtained from shale is low to medium gravity, low in light ends high in distillate fractions and produces little residue in conventional crude oils of similar gravity. It contains small quantities of arsenic and iron, and has a high pour point, and is high in nitrogen and intermediate in sulfur content. The data already given in Table 5 shows the characteristics of several shale oils from different retorting processes. The properties of the oil so produced reflect the effects of processing conditions at that scale of operation, thus there is no assurance that oil produced from a commercial plant will have exactly the same properties, the determination of accurate values will depend upon the properties of oil from the commercial unit itself.

The distillation data shows that shale oil contains only small amounts of low boiling and high boiling materials, mainly rich in middle distillates making it an excellent source of turbine fuels and high quality diesel fuels. The presence of unsaturated hydrocarbons in the shale oil shows that some cracking of the oil occurs during retorting processes.

Conversion To Synthetic Crude Oil: Three properties of crude oil need to be modified for it to be used as synthetic crude. These are lowering the pour point, reducing the nitrogen content, and removing the trace quantities of arsenic and iron, which can be accomplished by using conventional refining steps with operating conditions adjusted to the process requirements for shale oil. The basic approach is to use severe hydrotreating of the crude shale oil. The trace metals will be removed prior to hydrotreating step, usually in guard bed reactors in hydrogen environment. These metals are poinonous to hydrotreating catalysts, and would be adsorbed on to the surface of the catalyst causing frequent catalyst replacement. Several refiners have tested and proven technology for removal of these metals(17-21).

Hydrogenation under severe conditions(high pressure high temperature, and low space velocity) has been tested on several crude shale oils and found to provide synthetic crude with superior properties(17,22). If hydrotreating were carried out at the production site, a less severe, minimal treatment may possibly be used. The major purpose is to remove the trace metals, and to reduce nitrogen and sulfur content as well as pour point, so that the oil could easily be pipelined to be acceptable to refiners and be sold at prices comparable to high quality crude oils. Pretreatment will reduce the hydrogen consumption and processing costs at the refinery. On the contrary, if the hydrotreating were carried out at the refinery, it could be adjusted to meet the quality of products desired, and the overall refining capability of the refiner. Both approaches, production site hydrotreating or refinery hydrotreating have been used.

Pretreated or upgraded shale oil is a highly desirable refinery feedstock. It is high in saturates content(paraffins and naphthenes), low in residue, and produces an excellent middle-range distillate. Downstream refining is similar to that for raw shale oil, but without the pretreatment steps required to remove nitrogen and heavy metals. The process yields a high quality oil and should command a premium price. The most beneficiary use for such a product will be either

a refinery or a petrochemical plant. Direct use as a fuel would have to be justified at the refinery feedstock price.

Refining To Liquid Fuels: The ultimate goal for developing these alternative fuels is to produce more liquid fuels. Thus it is imperative that the technology is capable of providing all the needed liquid fuels. Many refiners have performed experiments designed at fitting shale oil into existing process facilities, but all of this information is not public, however, the work supported by government is in the open literature (17,23). The most recent and complete evaluation of conventional processing of oil shale has been reported by Chevron(17). Amoco, Ashland, Suntech, and UOP are conducting studies on producing jet fuel from shale oil under funding from Air Force, and their work has been reported in recent technology reviews(20,21). The data for processing shale oil into motor gasoline was presented at a recent API meeting in Chicago(24).

Chevron identified three procedures for producing fuels from shale oil, which are: hydrotreating whole shale oil followed by hydrocracking; hydrotreating followed by fluid catalytic cracking; and coking followed by hydrotreating. Each of these procedures yields different quality product. The most flexible scheme used for jet fuel production is hydrotreating whole shale oil after removal of trace metals. The product is then distilled, providing three fractions. The naphtha is reformed, the middle distillate is further hydrotreated, and the vacuum gas oil is hydrocracked. Hydrotreating the crude shale oil is also the first step after trace metals removal when high octane gasoline is the desired product, followed by fluid catalytic cracking, and subsequent reforming of the light naphtha fraction. The middle distillate is further hydrotreated. The least expensive approach is to coke the raw shale oil, then hydrotreat the naphtha and distillate fractions from the coker, making this approach an attractive route for the production of diesel fuel, although the total yield of liquid fuels is low due to initial coking step. A typical flow diagram for the refining of crude shale is given in figure 3.

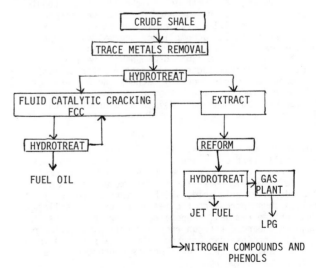

Figure 3: The Extractacracking Process

For example, an analysis of raw and hydrotreated Occidental shale oil given in table 6 shows that 97% sulfur and 65% nitrogen removal are attained in the main hydrotreater. The presence of water in the hydro-treated product may account for its relatively high content of oxygen. The analysis of fractions from hydrotreatment are shown in table 7. Nitrogen, sulfur, and aromatic content increase with increasing boiling range of the fractions.

Table 6. Analysis of Raw And Hydrotreated Occidental Shale Oil(21).

	Dewatered and Filtered Feed	Hydrotreated Product
API GRAVITY @ 60° F	23.0	30.9
Distillation, ° F		
IBP	413	169
10 vol.%	511	422
50	730	679
90	959	983
EP	-	1000
Recovery	90	97
Aromatics	42.5	37
Polar Aromatics	23.6	-
Asphaltenes	2.4	-
Chemical Analysis, wt%		
Carbon	84.82	86.25
Hydrogen	12.04	12.77
Sulfur, PPM	6200	159
Nitrogen, Total	1.46	0.50
Oxygen	1.18	0.11
Arsenic, PPM	33	<1
Ash, wt%	<0.005	NIL

Table 7. Analysis of Fractions From Hydrogenated Occidental Shale Oil(0.50 wt% Total Nitrogen)(21).

	NAPTHA	LT. GAS OIL	HVY.GAS OIL
Yield, vol%	14.8	12.2	63.4
Boiling Range, ° F	1-450	450-535	535-1000
API Gravity @ 60° F	44.0	37.5	28.9
Distillation, ° F			
IBP	296	460	535
30	365	477	670
50	385	481	740
70	403	487	817
EP	448	536	1000
Pour Point, ° F	-	-	+43
Aromatics, %	19.5 vol	30 wt%	41 wt%
Elemental Analysis, wt%			
Carbon	85.91	86.16	86.25
Hydrogen	14.04	13.40	13.17
Sulfur, PPM	46	51	72
Nitrogen, Total	0.08	0.36	0.52
Oxygen*	0.03	0.17	0.14

* High oxygen content is due to moisture in the sample.

Based on the results of hydrocracking and hydro-treating operations, it may be concluded that raw shale crude can be modified to obtain liquid fuels, specially the military turbine fuels and diesel fuels. The fuels not only meet military specifications, but exceed military specifications particularly in combustion value and nitrogen and sulfur content. Refining of raw shale oil will require some additional costs due to capital investment for hydrotreatment and in the cost of providing for hydrogen. Initial estimates placed the total processing cost at about $ 10 to $ 12 per barrel (1978 dollars) $4 to $6 per barrel above the cost of refining traditional crudes. These costs have been reduced due to reduced hydrogen requirements in recently developed processes by Chevron and Pace Company.

ENVIRONEMNTAL AND SOCIAL CONSIDERATIONS

Very few projects have undergone the environmental scruitny that shale oil development has. Eighty percent of this nation's shale reserves are on public lands held by the government. The potential utilization of these lands for energy production has subjected them to close State and Federal review. According to a recent report of the Colorado Research Institute: "Potential oil shale developers are engaged in perhaps the most extensive environmental program in the history of mineral develoment. Even before the leasing of a federal oil shale lands in 1974, a great many environmental study and planning effort were active, supported by private, government, and university related oil shale ventures. Today each potential operator has gathered an impressive set of baseline data in an effort to develop this resource in an environmentally responsible manner."

Water Usage: Water availability and quality are the major concerns in the development of shale oil. In the Green River Formation, the major water sources are the flowing waters of the Upper Colorado River Basin, supplemented by local supplies of groundwater. It seems that sufficient water is available to support a substantial shale oil industry. In a surface retorting facility, disposal of spent shale accounts for about 40% of the water needs. Modified in-situ production requires relatively little water. The recent report issued by Colorado Energy Institute estimates that surface retorting processes will require about two barrels of water for each barrel of shale oil produced, whereas in-situ processing will require only about one barrel of water for each barrel of oil produced. The projected water demand is well within the Colorado Compact allocation even for a two-million barrel-a-day shale oil industry. A Department of Interior environmental study prepared in 1973 found that 341,000 acre-feet of water per year from the Upper Colorado Basin could be made available for oil shale development, which is sufficient to support major shale oil production capacity.

Water Quality: Oil shale processing also raises serious questions concerning the protection of water quality. These are: leaching of mobile salts from spent shale in aboveground processes into groundwater supplies. Thus shale disposal must be handled so that groundwater does not come in contact with the spent shale. Backfilling of underground mines will be acceptable only if the permeability of the backfill is low enough to prevent significant salt release, or it can be shown that mobile salts are permanently bonded in the host rock.

Modified in-situ processes offer substantially less of a problem as to the water quality due to higher temperatures and longer retorting times. Extensive laboratory studies by Occidental Research Corporation and the U.S. Bureau of Mines indicate that carbonates present in shale are converted to insoluble silicates when the shale is subjected to temperatures in excess of 1300° F over long periods of time. If leached the spent shale produces a leachate whose pH is comparable to that of the ground water in the Colorado Basin. Therefore in-situ processes are better than ex-situ processes as to the water quality.

Air Quality: One of the major concern in shale oil production is particulate emission. According to the Department of Interior environmental study, only limited local air quality is to be impacted. The main sources of this pollution are dust emissions from vehicles and construction operations, dust produced during crushing and retorting of shale, solid particulates from blasting, mining, overburden spent shale disposal, and gases from retorting and oil upgrading operations(hydrogen sulfide, sulfur dioxide, nitrogen dioxide, and carbon monoxide). Federal and state agencies expect most shale oil operations to meet existing emission standards. Technology for recovering sulfur from retorting and output gases is already known. Trace hydrocarbons and carbon monoxide emissions are expected to be within required limits. Modified in-situ processes have advantage over surface retorting with respect to particulate emissions.

Land Use: Spent shale disposal is the principal concern regarding land use, as two tons of rock is needed for every barrel of oi production by surface retorting. For a 100,0000-barrel-per-day plant, 30 acres of land per year are needed to dispose of the spent shale. If backfilling of mineshafts were not practical or allowed, the land requirements could increase to as high as 75 acres per year. Once again modified in-situ proceeses cause least land disturbance and require none spent shale disposal. Only disruptions are roads, mines, and site construction needs. Federal and state regulations require that intensive programs of land restoration and revegetation be undertaken. Research in this area has demonstrated that vegetation on spent shale can be successfully established. Occidental has experimented with wheat and barley crops, and the yields were comparative to other normal soils of Colorado.

In addition, the oil shale industry would bring some societal disruptions associated with shale mining such as: reduction in property values near the mines due to dust, noise, and aesthetics, and damage to roads. On the overall scale of things, there is the probability that shale oil production would be even more disruptive of the environment and society than would coal production(25). This difference may be attributed to the fact that shale rock on per ton basis contains much less oil as compared to coal.

CONSTRAINTS TO COMMERCIALISATION OF SHALE OIL

There are number of uncertainities facing the industry such as, economic, legal, technical, environmental, and regulatory. Many of these uncertainities are caused by the fact that the construction of any shale oil plant will be a first of its kind. Some of these uncertainities are:

- The present uncertainties about the future costs of shale oil.
- Uncertainity in future price of world oil.
- Large capital investments required for each project.
- Large capital risks relative to company assets.
- Risks of major project delays.
- Lack of governmental support in providing finances, regulations, and necessary tax incentives.
- Ownership of shale oil lands.
- Large number of legal permits are to be obtained before any work on any project can be started.
- Rules for obtaining these permits change constantly.
- Scale up of processes from pilot plant to commercial plant, the largest oil shale plant in U.S. has a capacity of 1,200 tons of shale oil per day, and the commercial plant will be at least 6 to 10 times larger than existing plants.

- Requirement of an environmental impact statement EIS in advance of any major federal action that may significantly affect the quality of environment.
- Constant revision of environmental regulations and standards both at state and federal level.

All of this leads to increased uncertainity for oil shale development which discourages any commitment towards development of a large scale project. Many of the constraints are beyond the control of oil shale industry. On the other hand, positive governmental action in most of these areas would eliminate many of the uncertainities facing the development of oil shale industry.

PROSPECTS OF SHALE OIL INDUSTRY IN U.S.

The current status of oil shale projects has been described in a recent report prepared by EPRI(10). An estimate that was developed from a poll of the oil shale industry taken by EPA in Denver region, and published in the March, 1980, Cameron Synthetic Fuels Report quarterly indicates a figure of 883,000 barrel per-day shale oil industry by 1996. These production goals tend to be optimistic. The 583,000 barrels per day production by 1990 may not be achieved, keeping in mind the recent activities in regard to Colony Oil Shale Development Project. A goal of 400,000 barrels per day may be a more reasonable expectation. The goals set for 1985 seem to be out of question at the present rate of development and at present levels of governmental support. Another view on the future of shale oil was expressed at a recent symposium and drew almost universal support from those in attendance. Simply stated, it was the feeling that shale as a source of hydrocarbon fuels and feedstocks, would never get off the ground, in large majors because of the imposing environmental problems and expected low net yields of useful products(26).

SUMMARY

It ia apparent from the present energy consumption patterns and available resources that there is an urgent need for new major sources of primary energy, specially liquid fuels for transportation, residental, and commercial sectors of our economy. Coal and shale appear to be ideal sources- available in large quantities and technologically convertible to fluids. However both sources suffer from similar deficiencies such as economic, ecologic, and logistic. The technologies being developed are ex-situ and in-situ processing of shale rock. Both the processes have problems associated with them, but in-situ processes seem to be favorable due less environmental impact. The future of shale oil industry in U.S. will depend upon several factors including technological, environmental, social, and economic.

REFERENCES

1. V.K. Gupta, 'Synthetic Fuels From Coal - The Energy in Transition,' Proceedings of the 4th World Engineering Congress (1981).
2. Ind. Eng. Chem. 19, 338(1927).
3. Chem Eng. News 55(30),14(July 25, 1977).
4. D.C. Duncan, and V.E. Swanson, 'Organic-rich shales of the United States and World Land Areas,' Geol. Survey Circ. No. 523(1965).
5. 'Oil Shale Data Book,' TRW Energy Systems Group June (1979).
6. K.E. Stanfield,'Properties of Colorado Oil Shale,' USBM Report of Investigation, 4825(1951).

7. B.S. Lee, Energy User News 3(8), 22-23(1978).

8. W.D. Metz, Science, 184(413) 1271-1275 (1974).

9. 'A Practical Approach to Development of a Shale Oil Industry in The Unites States,' Colorado School of Mines Research Institute, Project L 50701 (1975).

10. "Shale Oil Potential for Electric Power Fuels," EPRI AP- 2186 Contract TPS 80-710(1981).

11. T.F. Yen,'Science and Technology of Oil Shale,'

12. M. Prats,'Method for Producing Shale oil From an Oil Shale Formation,' U.S. Patent 3,593,789 (Shell Oil Company).

13. 'Proceedure for In Situ Recovery of Minerals and Derivatives Thereof with the aid of Laser Beams,' Neth. Appl.6,905,815(Woods R&D Co.).

14. T.F. Yen et. al., 'Feasibility Studies of Bioleaching Production of Oil Shale Kerogen,' NSF-RANN GI 35638 Final Report (1975).

15. T.F. Yen, et. al.,'Method of Converting Oil Shale into a Fuel,' U.S. Patent 3,982,995 (1976).

16. H.B. Jensen, W.I. Burnet, and W.I.R. Murphy, 'Thermal Solution and Hydrogenation of Green River Oil Shale,' Bulletin 533, Bureau of Mines (1953).

17. R.F. Sullivan, and B.E. Strangeland,'Converting Green River Oil Shale into Transportable Fuel,' 11th Oil Shale Symposium Proceedings, Colorado School of Mines, Golden Colorado, 120-134(1978).

18. A.H. Frumkin, E.J. Owens, and R.B. Sutherland, 'Alternative Routes for Refining Shale Oil,' Chemical Engineering Progress 64-72(1979).

19. E.D. Burger et. al.,'Pre-refining of Shale Oil,' Paper presented at ACS meeting in Chicago Ill. August 24-25(1975).

20. 'Jet Fuel Looks to Shale Oil,' 1980 Technology Review, Proceeding of the Symposium, November 19-20(1980).

21. 'Jet Fuel From Shale Oil,' 1981 Technology Review, Proceedings of the Symposium, November 17-18 (1981).

22. B. Baral, Union Oil Company, Personal Communication, Jan. (1978).

23. E.T. Robinson,'Refining of Paraho Shale Oil into Military Specification Fuels,' 12th Oil Shale Symposium, Colorado School of Mines, 195-212 (1979).

24. P.E. Lovell, M.G. Fryback, H.E. Reif, and J.P. Schwedock,'Shale Oil in a Good Source of Mogas,' Oil and Gas Journal 79, 92-101(1981).

25. E.H. Thorndike,'Energy and Environment: A Primer for Scientists and Engineers,' Addison Wesley, Reading, MA 167-168(1976).

26. A. Kaufman,'Two Views of Resource Scarcity: Engineering vs. Economic,' presented at Natl. Symp. Crit. Strategic Mater.; Amer. Chem. Soc., Washington, DC June (1978).

Chapter 69

SOME THERMODYNAMIC ASPECTS OF DESALTING THROUGH STEAM BOILER PLANTS

S. S. Stecco, A. Galletti

1) INTRODUCTION

During the last years combined heat and electricity power plants have rapidly developed, reaching important performance conditions: among them desalting and electricity production systems deserve a very special mention owing to growing fresh water needs for industrial purposes.

In different occasions such item has been presented and discussed /1,2,3,4/, showing also a possible improvement of thermodynamic performance, by means of a direct-contact heat exchanger /5/. The various possible solutions for sea water desalting are shown in table 1.

energy increase.

A semplified approach, referring to a negligible amount of fresh water production with respect to sea water mass flow (and then for constant saline concentration) leads to

$$E = 0.64 \text{ kWh/m}^3$$

in reversible conditions.

In practice very different figures are experienced, so that Clearfayt /7/ suggests a more realistic minimum at 3.5 kWh/m^3, while existing plants show the following average values

| multiple effects | $E = 105 \text{ kWh/m}^3$ |
| multiflash | $E = 60 \text{ kWh/m}^3$ |

SEPARATED COMPONENT	PHASE	PROCESSES	DRIVING POTENTIAL
WATER	STEAM	DISTILLATION	ENTHALPY OR PRESSURE
WATER	LIQUID	REVERSE OSMOSIS	PRESSURE
WATER	SOLID	FREEZING	PRESSURE OR ENTHALPY
SALT	STEAM	------- NOT KNOWN ---------	
SALT	LIQUID	ELECTRO DIALYSIS	ELECTRICITY
SALT	SOLID	ION EXCHANGE	CHEMICAL POTENTIAL

TABLE 1

The Office Saline Water (OSW) considered and tested various solutions from 130 m^3/day to 3800 m^3/day. Details on these plants can be found in literature /5,6,7/.

2) THERMODYNAMIC CONSIDERATION FOR DESALTING

There is a minimum of energy, from a theoretical point of view, for desalting the unit mass of water, corresponding to Gibbs free

for the most important applications.

Other solutions can be significant for low salinity applications (fig. 1).

In flash desalting process the following energy considerations hold

$$Q + W = A + (T_2 - T_0) + D(T_1 - T_0)$$

where (fig. 2)

Q = heat power entering the system

W = power for fluids circulation

A = rate of refrigerating circuit losses

D = fresh water mass flow

B = brine mass flow

T_0 = feeding water temperature

T_1 = distilled water output temperature

T_2 = brine output temperature

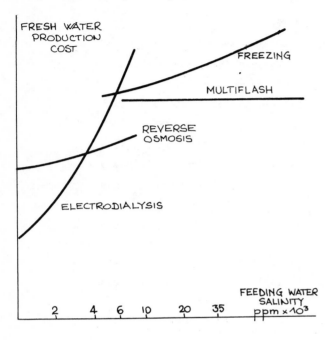

FIG. 1

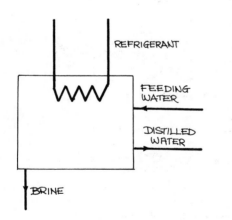

FIG. 2

in any coherent unit systems so that the specific energy consumption is

$$E = \frac{Q + W}{D}$$

in which, in the following analysis the ratio

A/D is considered negligible and $T_2 \cong T_1$, so that

$$E = \frac{B+D}{D} (T_1 - T_0)$$

3) STEAM POWER PLANTS WITH DESALTING FACILITIES

The temperature of desalting multiflash process, considered the most suitable for high production, make it convenient to couple it to a steam turbine. The consequent higher steam specific consumption depends on

a) fresh water to electricity production ratio

b) thermodynamic characteristics and efficiencies of separate processes.

The solution to be taken into consideration have then to be "compared" from these two points of view.

Referring now to a specific and existing power plant (LIVORNO, 155MW electric output) the different possibilities are here examined. The plant under consideration has the following thermodynamic characteristics:

HIGH PRESSURE TURBINE INLET

Temperature	T = 535°C
Pressure	p = 14.6 MPa
Enthalpy	h = 3416 kJ/kg

MEDIUM PRESSURE TURBINE INLET

Temperature	T = 535°C
Pressure	p = 3.7 MPa
Enthalpy	h = 3527 kJ/kg

CONDENSER INLET

Pressure	p = 0.00485 MPa
Enthalpy	h = 239 kJ/kg

MASS FLOWS

Boiler feeding	132 kg/s
Reheater	118.6 kg/s
Condenser	83.5 kg/s

3.1) Direct-contact solution

This solution represents a variation of classical multiflash system, with higher heat transfer characteristics, by means of elimination of metallic surfaces. Fig. 3 shows that steam condenses by direct contact with di-

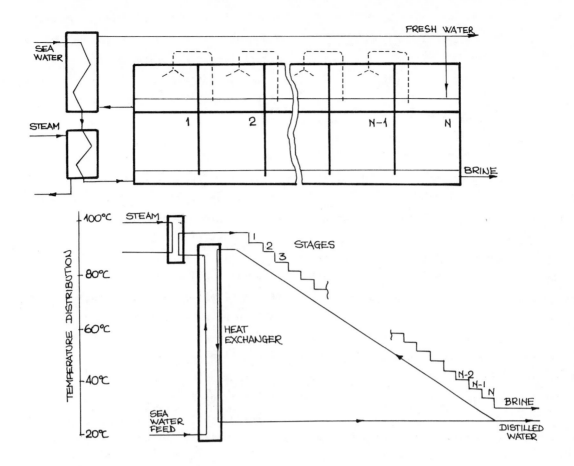

FIG. 3

stilled water, flowing in an open channel.
Two water flows cross the whole stage system
in opposite direction. The fresh water coming
from lower temperature stages to higher temperature ones, preheated by condensation heat of
steam, increas its mass flow at each stage.
The brine partially vaporizes and decreases
its temperature at each stage.
Feeding sea water is heated by means of two
heat exchangers: in the first one heat recovery from distilled water is performed, in the
second one the temperature is further increased by steam flow. Different arrangements are
possible as it will be shown later.
This solution, as described in /5/, has the
following advantages

a) reduced cost with respect to surface heat
   exchangers (the decrease can reach 40%)

b) no deposits on tubes and then reduced cor-

rosion

c) lower energy consumption.

The condensed steam flow, reduced of fresh
water production flow, enters desalting plant
stages, where it is preheated by brine, and
reintegration of mass flow by means of fresh
water there produced is performed.
From desalting plant the water flow enters
the steam power plant for a normal thermodynamic cycle to the condenser.
It is important noticing that fresh water tem
perature is related to condenser output tempe
rature T_C, and maximum temperature of sea water $T_{SW} = 95°C$, for corrosion and sealing phe
nomena, while the temperature drop is $\Delta T = 5°C$
to increase the transformation ratio T_R, defi
ned as the ratio between fresh water production to brine feeding.
We need then a stage number of order of 20,

while

$$\Delta T > \Delta T_s + \Delta T_p$$

where

ΔT_s = stage temperature drop

ΔT_p = temperature loss due to real conditions effects[*]

For T_c = 32°C T_{sw} = 95°C   T = 5°C   we get

$$T_R = 9.84\%$$

i.e. a fresh water production of 9.84% of water flow coming from condenser.

Three possibilities can be technically significant for brine heating, namely:

a) Brine heating from ambient temperature to T_{sw} by means of exhaust gases from steam generator flowing through a suitable heat exchanger.

b) Brine heating from ambient temperature to a temperature of 60°C by means, of first extraction condensing steam. The remaining temperature gap is overcome by exhaust gases.

c) Brine heating from ambient temperature by means of the first and second extraction condensing steam.

A thermodynamic analysis of solution a) b) and c) follows.

3.1.1) Exhaust gases heating

We refer now to fig. 4 a), in which the solution is summarized in its most important thermodynamic aspects. It is important to point out that two low-pressure exchangers are dropped out and exhaust gases undergo a temperature decrease of 135°C.

For a medium-high steam power plant the decrease of sensible heat recovering in Ljungstroem

_____

(*) Among the various causes of temperature loss the following seem to be the most significant ones:

a) lower vapour pressure of brine with respect to distilled water

b) water level effects

c) filtering losses

heat exchanger, together with the problem of reduced regeneration factor, can lead to a strong increase in specific fuel consumption, so that the thermodynamic optimization should in any case be matched to a significant cost-benefit analysis.

3.1.2) Regenerative and exhaust gases heating

In this solution brine heating is performed by means of two different and successive fluids: the first low pressure heat exchanger can reach in the plant under consideration, a temperature of 60°C, whilst the second step (to 95°C) is obtained through exhaust gas heating (fig. 4b).

The inconvenient of reduction in efficiency for the steam generator is here then appreciably reduced.

3.1.3) Regenerative heating

Two steam extractions are here employed, the first of which is unchanged in terms of mass flow and temperature with respect to original plant, while the second one needs major modifications (fig. 4c). The thermodynamic computations result into a very little condenser flow variation, with an increased extractions flow of only 18%, so that the specific consumption is altered only by a 0.05%, with a fresh water production of 9.09 kg/s.

3.1.4) General considerations

The possible solutions, examined in some details, lead to table 2, in which the most important parameters are summarized.

There is no doubt that the last example is very appealing, specially referring to specific energy consumption for desalting.

However the transformation ratio has here a fixed value, namely 9.84%. Applications in this case refer to high electricity productions coupled to a limited need of fresh water, whose production is in any case in a fixed ratio to condenser steam flow.

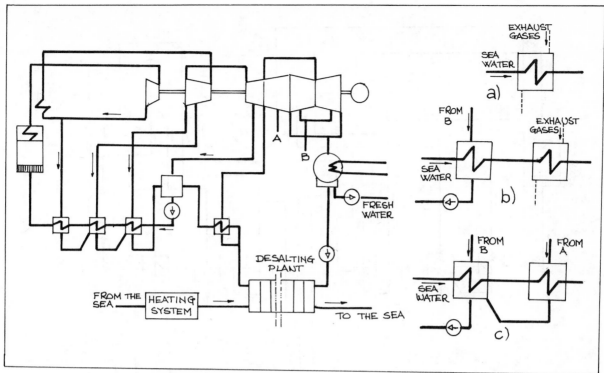

FIG. 4

PLANT SOLUTION	MASS FLOW AT CONDENSER kg/s	SPECIFIC CONSUMPTION kcal/kWh	FRESH WATER MASS FLOW kg/s
ORIGINAL PLANT 155 MW	89.160	2079.58	0
EXHAUST GASES HEATING	97.396	2191.97	9.54
REGENERATIVE PLUS EXHAUST	97.396	2132.39	9.54
REGENERATIVE HEATING	92.352	2080.76	9.09

TABLE 2

3.2) <u>Coupled plant with brine heater</u>

When fresh water production needs to be considered independent from electricity production
we can consider a preheating process for brine, by means of a special steam extraction
from low pressure turbine (fig.5), which can
be matched to different requirements.

The steam power cycle has only the modification of a new extraction circuit, which is in
most cases only an adjacent branch of an alrea
dy existing pipeline.
Calculations are performed on an optimization
basis following the scheme of fig. 6: the
fresh water production can be easily changed

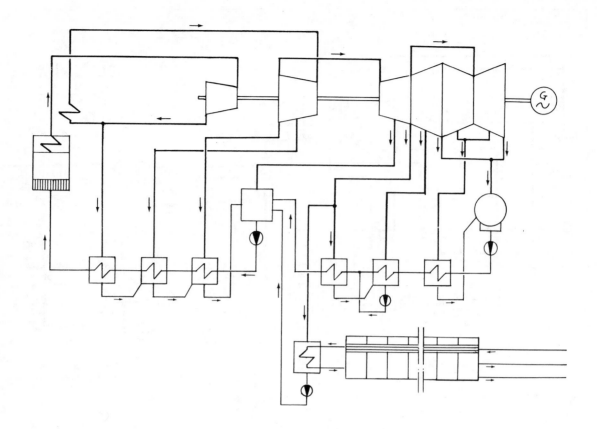

FIG. 5

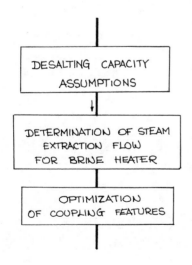

FIG. 6

varing the steam flow to brine heater itself, while the new thermodynamic characteristics of the circuit produce a new equilibrium resulting in a different distribution of extraction flows.

For a maximum temperature variation in the brine heater of 6.5°C, and entering steam conditions of 110°C and 0,095 MPa, a transformation ratio of 12% can be reached (always referring to the same power plant of Livorno). The computer program has considered a fresh water production from 100 m^3/h to 800 m^3/h: fig. 7 and 8 show the mass flow variation in the last five heat extraction, numbered from lower pressure to higher pressure ones, and the mass flow at condenser.

3.3) Exergy analysis of brine-heater solution

Since the brine-heater solution, previously described, is by far the most important coupling system between steam power and desalting plants, some further comment is needed, from a thermodynamic point of view.

Economic costs of combined power plants can be compared by means of the "second-law analysis" /8,9,10/.

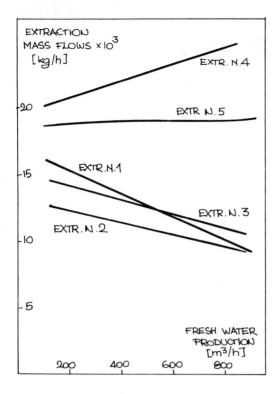

FIG. 7

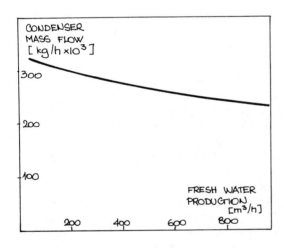

FIG. 8

For the examined solution (but the considerations can be considered of general interest) the desalting process energy consumption is, referring to unit production of fresh water,

$$E_c = 224 \text{ kJ/kg}$$

which is, in any case, a very realistic and interesting value.

The second-law analysis let us go further on in these comments: after easy calculations the

following figures can be established; with respect to a production of 800 m³/h of fresh water.

PROCESS	EXERGY VARIATIONS kW
Steam extraction for brine heater	12.291
Sea water heating	11.259

When considering that, for such production the loss in electric output is 7410 kW, the following conclusions arise:

1) the exergy efficiency of desalting process when coupled to steam power station is 91,6%, a very high value so that the process is certainly the most interesting one from a thermodynamic point of view.

2) Comparing the exergy gain in the sea water flow to the power loss we can conclude that the exergy efficiency of the whole process is increased by some 3850 kW, so that corresponding costs are sensibly reduced.

4) REFERENCES

/1/ S. Stecco - Un evaporatore monostadio a flash per ricerche sulla dissalazione - XXVIII Congresso Nazionale ATI, l'Aquila (Italia), 1971.

/2/ S. Stecco - Towards a geometrical and thermodynamic optimization in a single stage flash evaporator - 4th Intern. Conference "Fresh Water from the Sea", Heidelberg (Germania), 1973.

/3/ S. Stecco, A. Bosio - Sull'influenza del battente liquido nelle prestazioni di un dissalatore monostadio a flash - XXX Congresso Nazionale ATI, Firenze (Italia), 1974.

/4/ S. Stecco, G. Chiatti - Improvement in dual purpose desalination with flash evaporators - 1st Inter. Conf. "International Desalting and Environmental Association", Puerto Rico, 1975.

/5/ S. Stecco - Steam power and desalination:

some coupling possibilities - "Energy
for Industry", Ed by P. O'Callaghan, Per-
gamon Press, 1979.

/6/ Maffei - Deduzioni delle relazioni inter-
correnti fra i diversi parametri di un
impianto di dissalazione tipo flash ed
ottimazione della superficie di scambio
termico - Com. Pers.

/7/ Clerfayt - La production d'eau potable
par dessalement - CEBEDOC Sprl, 1967.

/8/ A.B. Campbell - Editor in Chief - ENERGY:
Second law analysis of Energy devices
and process - 5,8-9, august-september
1980.

/9/ S. Stecco, G. Manfrida - Second-law analysis of composite power plants - 17th
IECEC Conference, Los Angeles,august
8-13, 1982.

/10/ G. Manfrida, S. Stecco - Exergy-loss mo-
delling of power plants and industrial
processes - 17th IECEC Conference, Los
Angeles,august 8-13, 1982.

SECTION 22
ENERGY MANAGEMENT

Chapter 70

ENERGY MANAGEMENT

R. J. Trieste

I. INTRODUCTION

Energy, uncontrolled, moves in all directions; but with effective control maximum utilization can be achieved thereby reducing its overall use.

In order to have an effective program all energies utilized must be accounted for and controlled. To accomplish this end statistical data on energy use must be collected and put into some kind of order, its use must be reported periodically and its use must also be analyzed to determine where and how energy reductions can be accomplished. If the energy utilized is managed effectively, conservation will follow automatically.

This is the program developed at the Brooklyn Union Gas Company. It accounts for 100% of the energy used internally by the corporation and its primary purpose is to put Brooklyn Union on an energy diet.

II. ENERGY MANAGEMENT SYSTEM

At Brooklyn Union, all of the energies consumed are utilized in five areas which comprise buildings, gate stations, holder stations, the fleet and the plants.

The energies used in these areas are electricity, natural gas, naphtha, methanol, gasoline, diesel fuel, fuel oil, and two additional energy related materials, nitrogen and water. The annual cost of these nine items was almost five million dollars in the 1981 operating year.

The base year selected for the program was fiscal 1979. Brooklyn Union's fiscal year begins October 1st and ends September 30th. All energy use is compared for the present period, cumulative to date, to the same period in fiscal 1979 and it is broken down into use rates and total use by area. The use rate will account for weather where it applies or average uses that are not weather affected.

Four programs make up Brooklyn Union's Energy Management System. These being a Statistical Analysis Program, a Recommendation Program, an Awareness Program and a Reporting Program.

The following is a discussion of each of these programs in further detail.

A. Statistical Analysis Program

1. Buildings: The buildings utilized in the corporation are subdivided into groupings depending on their use. In this way one can analyze how an individual building is consuming its energy as compared to other buildings in the group and the group as a whole. Seven different building groups have been developed which cover the General Office, District Offices, Satellite Stations, Service Stations, Canarsie Service Center and two areas of the Greenpoint Energy Center. The seven building areas contain 47 individual buildings.

These buildings use natural gas, electricity and fuel oil for heating, air conditioning, lighting and machine use. The system that has been developed covers a use rate for these energies and a total use. The use rate adjusts for weather and provides a means of determining if the energy use is decreasing or increasing. The use rates are based on the energy consumed per square foot of building floor space per heating and cooling degree day. In this way if the use rate is decreasing we know the total use will decrease depending upon the weather. If only total use is monitored a conclusion may be reached that energy is decreasing for example, but in reality the weather is only warmer.

2. Gate Stations: Gate Stations are handled similarly to buildings. These stations, which number 17, consume natural gas and electricity for space heating, process heating, controls and security. Since the energy consumed in the gate stations is affected by the weather, use rates take this factor into account here too.

Gate Stations are the means of receiving natural gas from the long distance transcontinental pipeline systems and delivering that natural gas into the companies distribution systems that supply it's customers. As a result then, the affect of weather is to increase or decrease the throughput of the Gate Stations. When the weather is colder the stations as a whole will handle more natural gas. The use rate developed then, is the energy consumed per heating and cooling degree day.

Some Gate Stations are more energy intensive then others and all stations must be on the line during the winter period to satisfy customer demands. During the low load periods in the summer our game plan is to maximize the use of the less energy intensive Gate Stations with the result being an overall reduction in energy use.

3. Holder Stations: There are two holder stations and they provide a means of helping to balance the hourly swings in customer natural gas demands. Each contains a pumping station and two gas holders. These stations are handled similar to the Gate Stations with one additional wrinkle.

The Holder Stations consume natural gas and electricity for space heating, controls, lighting and pumping. Pumping is handled with electric driven and gas engine driven compressors. The system developed for this area includes a use rate and total use similar to the Gate Stations and in addition to use rate and total use for running the compressors.

4. Fleet: The fleet at Brooklyn Union consists of trucks, heavy construction equipment, sedans, vans and compressors which utilize both gasoline and diesel fuel.

The Energy Management System that has been designed for the fleet is broken down into responsibility areas and also provides for a further breakdown by sections within each department so that the fuel utilized in each section can be compared to other sections within a department and department to department within the corporation.

The fleet statistics cover the rolling stock, compressors, diesel fueled equipment, compressed natural gas vehicles and propane fueled vehicles. The number of units covered in these four areas total over 1,200.

5. Plants: Brooklyn Union has both a Synthetic Natural Gas Plant and a Liquid Natural Gas Plant. These Plants are utilized to supplement the peak day requirements as well as some base requirements.

As was done in the other areas an Energy Management System was developed for each Plant to account for the total energies consumed in the Plants. Both use rates and total use are tracked for natural gas, electricity, methanol and naphtha in these Plants.

In the LNG Plant the system covers the full twelve month period while in the SNG Plant the system only covers the portion of the year when the Plant is not producing. During production a complete energy balance is run from the time the first cubic foot of gas is produced to the last cubic foot. The Energy Management System discussed in this article covers the Plant start-up prior to production, the Plant shutdown after production and the remaining period of the year when the Plant is completely shutdown.

The LNG Plant contains two storage tanks that have a combined capacity of 1.6 BCF, vaporization facilities and liquifaction facilities.

The SNG Plant has a capacity of 60 MMCFH per day and utilizes naphtha as a feedstock.

6. Total Energy Use: The total energy use in the five consuming areas is converted to equivalent BTU's so that the total corporate energy use can be tracked. This is especially important when switching from one type of fuel to another type since it is only in this way the net gain or loss can be seen.

In addition to energy units the cost of each energy and the total cost is also compared. This cost is then allocated to the twenty-four user departments within the corporation in proportion to the building space they occupy, the fleet they utilize and the facilities they operate.

B. Recommendation Program

By analyzing the statistical data opportunities for improvement can be selected. Areas that show trends out of line must be looked into to determine why the energy use levels are where they are. This

is one way of looking for energy opportunities. Once they are uncovered they must be persued from the recommendation stage through the construction stages to insure they become operational in a timely manner.

C. Awareness Program

Many different avenues are being used to make all employees aware of the energy use and how to conserve it. These avenues compose but are not limited to reminder stickers, bumper stickers, hard hat stickers, articles in company newspapers, speaking engagements, films in conjunction with normal training, posters and any other means to keep the energy conservation message in the limelight without going stale.

In addition to the above avenues, employee motivation programs are being used. Our first program called S.A.V.E. (Save Americas Vital Energy) ran last year. It utilized a number of contests that strived to bring energy conservation from the employee's home to the work place. This program covered all energy uses and proved to be successful. Through recommendations developed by the program, the cost of the program had a payback of under two years.

This year we are into our second motivation program. Where S.A.V.E. looked at all energy uses, our current program G.R.I.P. (Gasoline Reduction Incentive Program) is zeroed in to gasoline only. The program began in March, 1982 and will end in February, 1983. At this point in time it is too early to evaluate its effectiveness.

Our objective in the awareness program is to make energy conservation habit forming.

D. Reporting Program

In the energy consuming areas all of the statistics are displayed as tabulations and in graphical form. The tabulation provides a statistical means of comparing the energy use and the graphs provide a visual means. Both presentations have advantages and disadvantages but combined they provide a good means to effectively demonstrate how the program is proceeding.

The graphical and statistical areas are done on a cumulative basis and reported on monthly. This report is distributed to all departments and officers within the corporation striving to put the energy use responsibility on the user departments.

Audits on thermostat set points and room temperatures are performed in all seven building areas several times during the heating and cooling seasons. Reports on these audits are forwarded not only to the departments that are responsible for the maintenance but to the other departments that utilize these areas as well.

II. CURRENT PROGRESS

A. Overview

This report covers the six month period ending March, 1982 compared to the same six month period in the previous year and fiscal 1979 our base year. All reports generated on our Energy Management System are reported in this manner in order to document cumulative energy savings from the previous and base years.

1. Weather

The weather for this period is less severe as

compared to the same period last year but more severe than the base year. Thus far the number of heating and cooling degree days are 1.9% less than last year and 4.5% higher than the base year and as a result our energy use should be lower this year compared to last year and higher this year compared to the base year. This is especially so in the Buildings, Gate and Holder Stations areas. The use rates, which adjust for weather, truly indicates how efficiently the various energies are being utilized and this point will be covered further on in this section.

The cumulative heating and cooling degree days to date are indicated in Figure 1.

2. Natural Gas Throughput

As the weather gets colder, the amount of natural gas sold to our customers will increase. The change in volume handled has a direct affect on the energies consumed in the Gate and Holder Stations areas. For the period 20.9% more gas was handled as compared to the base year which is 5.3% more than last year. This affect of increased volumes handled will be covered under Gate and Holder Stations.

Figure 2 indicates this throughput as compared to the base year.

3. Total Energy Use

The total energy use is a composite of all the energies consumed within the company operations. This use has decreased 14.3% for the period as compared to the base year but is 4.6% higher than last year as indicated in Figure 1. The current replacement cost of all energies saved to date amounts to $976,000.

Adjusting the total use for weather this reduction equates to the lowering of the overall use rate by 18.1% as indicated in Figure 2.

B. Buildings

These areas are most sensitive to weather due to the operation of the individual building HVAC systems. As the current weather is warmer or colder than the comparison periods the energy use will swing accordingly and it is only through the use rates that the energies efficiencies can be seen.

1. Electricity

As indicated in Figure 3 total electric use for the period has increased 7.3% as compared to last year and 26.7% as compared to the base year. The use rates track the total use since they indicate an increase of 9.7% and 21.5% for both comparison years as indicated in Figure 4. Many of the recommendations made for this energy use have not been implemented yet. As they are implemented we expect a drastic reduction in electric use.

2. Natural Gas

Natural gas use and use rates in the Buildings Areas has shown considerable improvement as indicated in Figures 3 & 4. For the period as compared to last year the use is down 5.4% and as compared to the base year the total reduction is 13.0%. The use rates for the same comparison periods are down 0.7% and 14.4%.

3. Fuel Oil

The fuel oil use tracks similar to the natural gas use and use rates and are indicated in Figures 3 & 4. Total use for the two comparison periods is down 24.6% and 34.3% and the use rates are down 23.4% and 37.2%.

C. Gate & Holder Stations

Like the Buildings area, these stations are also affected by weather. Weather changes have a direct affect on the individual stations HVAC systems and an indirect affect on the amount of gas these stations handle. As a result the amount of energy consumed in the stations for both reasons will usually increase as the weather gets colder.

As indicated previously, the throughput for both comparison years is higher but the weather is slightly milder as compared to last year but more severe as compared to the base year.

1. Electricity

Total use and the use rates are both reduced for electricity in this area as can be seen in Figures 5 & 6. Total use is down 24.5% and 22.6% for the two comparison periods and the use rates are down 23.0% and 25.9%.

2. Natural Gas

Natural gas, like electricity, has also shown considerable improvement as indicated in Figures 5 & 6. The natural gas use as compared to last year is down 4.7% and as compared to the base year is down 47.8%. The use rate is unchanged as compared to last year but is down 48.6% as compared to the base year.

D. Fleet

The Fleet's energy use is a function of the work load and how efficiently the vehicles are utilized. In order to improve the utilization of each vehicle we are stressing planning, scheduling and idling of engines as examples of ways to reduce fuel consumption. The workload has increased since we have increased the miles being traveled by the fleet. The energy use in this area for gasoline, diesel and CNG vehicles is indicated in Figure 7.

1. Gasoline

Gasoline is utilized in all vehicles and most compressors except for the heavy construction equipment. As compared to last year, gasoline use has increased 11.7% but is down from the base year by 4.3%. A specific conservation program on gasoline use has been implemented this year. We expect this program will have a considerable impact on gasoline use.

2. Diesel Fuel

Diesel fuel consumption has decreased this year as compared to last year by 26.5% but is higher than the base year by 1.7% as can be seen in Figure 7.

3. CNG Vehicles

At the present time we have nineteen vehicles on

CNG, sixteen more than last year and none in the base year. The CNG use is indicated in Figure 7 and no comparison is being made due to the increase in the number of vehicles each year.

E. Plants

The two Plants at our Greenpoint Energy Center are the LNG and SNG Plants. Each of these Plants consumes electricity and natural gas while only the LNG Plant uses methanol and the SNG Plant naphtha.

1. Electricity

All of the electricity is purchased for the SNG Plant while only a portion of the electrical requirement for the LNG Plant is purchased with the balance being self generated in a total energy plant. In addition to reducing our electrical use we are also striving to maximize the amount of electricity we produce in the LNG Plant. The combination of both have shown considerable progress as indicated in Figure 8. Purchased electricity has decreased in both Plants 32.2% as compared to last year and 36.6% as compared to the base year.

2. Natural Gas

As with electricity, natural gas is utilized in both Plants too. In the SNG Plant the boilers were fired with naphtha up until 1980 when they were converted to natural gas firing. In the LNG Plant the total energy plant utilizes natural gas fueled engines and we are producing more of our own electricity. As a result of both Plant changes, natural gas use is up as can be seen in Figure 8. As compared to last year natural gas use is up 17.9% and as compared to the base year natural gas use has increased 289.7%.

3. Naphtha

Naphtha is only utilized in the SNG Plant and its use is for feedstock and heater fuel. As previously explained it had also been utilized as boiler fuel. As a result of this conversion to natural gas firing and other conservation measures, our naphtha use has been considerably reduced. As can be seen in Figure 9 our naphtha use is down 25.2% as compared to last year and 94.1% as compared to the base year.

4. Methanol

Methanol is only utilized in the LNG Plant for the production of liquid. Since its use is a function of liquid production its use will vary and the important factor in this energy use is its utilization rate. For the period this rate has decreased 32.4% as compared to last year and 41.8% as compared to the base year. Liquid production for the two comparison periods was higher by 98.6% and 47.2% while total methanol use was up 35.2% as compared to last year and down 45.0% as compared to the base year.

F. Recommendations

As explained in Section II identifying energy saving opportunities is one phase of our Energy Management Program. Since the conception of the Energy Management area 115 recommendations have been made on conserving energy in specific areas. Of these only 56 are in operation thus far and the balance are in various phases of completion. The potential savings from these 115 recommendations per year are 827,142 Kwhrs of electricity, 26,941 Mcf of natural gas and 31,146 gallons of gasoline. At current costs the projected annual savings due only to these 115 recommendations amounts to $223,000.

CUMULATIVE ENERGY USE

(NATURAL GAS, ELECTRICITY, NAPHTHA, METHANOL, GASOLINE, DIESEL FUEL & FUEL OIL)

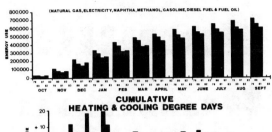

CUMULATIVE
HEATING & COOLING DEGREE DAYS

FIGURE 1

CUMULATIVE
NATURAL GAS THROUGHPUT

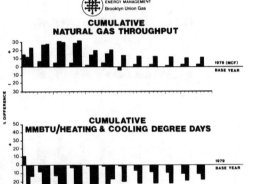

CUMULATIVE
MMBTU/HEATING & COOLING DEGREE DAYS

FIGURE 2

BUILDINGS
CUMULATIVE ENERGY USE

NATURAL GAS

ELECTRICITY

FUEL OIL

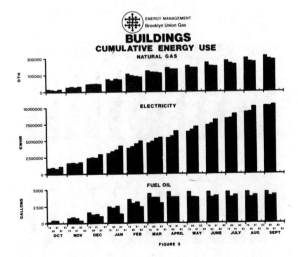

FIGURE 3

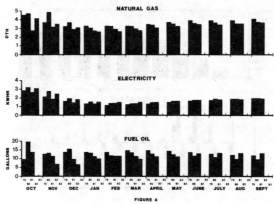

BUILDINGS
CUMULATIVE ENERGY USE RATE

FIGURE 4

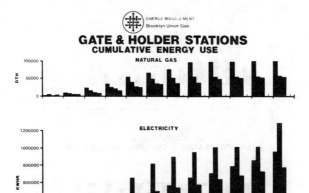

GATE & HOLDER STATIONS
CUMULATIVE ENERGY USE

FIGURE 5

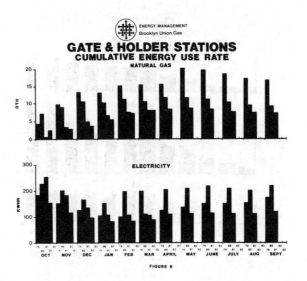

GATE & HOLDER STATIONS
CUMULATIVE ENERGY USE RATE

FIGURE 6

456

ENERGY MANAGEMENT
Brooklyn Union Gas

FLEET
CUMULATIVE ENERGY USE

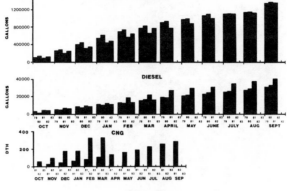

GASOLINE

DIESEL

CNG

FIGURE 7

ENERGY MANAGEMENT
Brooklyn Union Gas

PLANTS
CUMULATIVE ENERGY USE

NATURAL GAS

ELECTRICITY

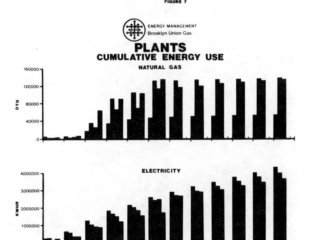

FIGURE 8

ENERGY MANAGEMENT
Brooklyn Union Gas

PLANT (SNG/LNG)
CUMULATIVE ENERGY USE

NAPHTHA (SNG)

METHANOL (LNG)

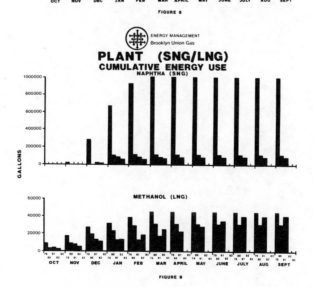

FIGURE 9

Chapter 71

A MOTIVATION MODEL FOR CORPORATE ENERGY CONSERVATION

L. A. Janicke

During the period 1973-1974 when world oil prices soared, Americans received a clear message that the days of plentiful and relatively cheap energy were over. Since those years, we have suffered through spot shortages and have witnessed steadily rising prices, largely fueling a long and severe period of inflation.

How have we responded? Let's look at some of the facts:

* Our import level of oil has remained fairly constant in terms of percentage, yet risen in terms of volume.

* The nuclear alternative, once thought our best hope for the future, is proving quite expensive and unpopular with a large percentage of the population due to concerns about safety.

* Solar energy, still in its infancy, is too expensive and cannot be used reliably in many sections of the country.

* Environmental concerns and the high cost of capital have combined to halt full utilization of our coal resources.

* Once shocked by skyrocketing prices, Americans have now absorbed the cost of energy into their budgets.

* The present market glut of fuel is having a negative effect on conservation practices. Although demand and prices have fallen as the result of a worldwide recession, this is largely interpreted as a signal that the energy crisis is over.

THE NEED FOR A MOTIVATION PROGRAM

With decades of inexpensive and abundant energy a part of our history, we have continued to view a wasteful lifestyle as a right. During the oil embargo and subsequent shortages of gasoline, many Americans, including government officials, recommended a seizure of foreign oil fields. Not until Detroit downsized most vehicles did we grudgingly trade-in our oversized and inefficient V-8's.

This unwillingness to sacrifice is not surprising. Living in a wasteful environment for a significant length of time has spawned wasteful habits which are difficult to overcome. We have enjoyed comfortable lifestyles, largely as a result of both plentiful and affordable energy. And any deterioration of those lifestyles is viewed as sacrifice.

Our technological innovations and engineering refinements have addressed this problem by the development of energy-efficient designs, along with the installation of computerized management of energy consumption. This is not always viewed favorably by many individuals who view the increased use of computer control as a harbinger of George Orwell's prophecy.

Until every facet of energy consumption is managed by computer, people will continue to waste, sometimes out of ignorance and, at other times, out of rebellion against lifestyle curtailment. An important corollary is that many people never believed an energy crisis existed, charging big business with a cooperative machination to drive prices upward, a belief reinforced with each price decline during this temporary market glut.

If we are to enjoy a spirited, continuing effort to conserve energy in America, we must first accomplish several difficult, but obtainable, tasks:

* We must motivate, rather than move, people toward conservation on a <u>voluntary</u> basis. Forcing regulation, no matter how well intentioned, will always lead to some degree of rebellion.

* People must be educated to the facts concerning fossil fuels and our presently limited alternatives. They must understand the finite character of our resources, and receive education in the methods they can employ to conserve.

* Conservation must be viewed in terms of personal savings, instead of personal sacrifice.

* We must teach the use of energy resources is a privilege which we all share, not a right of birth.

Once again, the burden of accomplishing these goals realistically falls on Corporate America. With its business expertise and financial resources, relevant motivation programs addressing large segments of the population in manageable groups can be designed and implemented successfully.

The sections which follow on the structure of

such programs are not theoretical in approach, but rather are based on our experiences as consultants in eliciting and obtaining voluntary employee cooperation. Whether the productivity need centers about increasing sales, improving job safety and health, or conserving energy, the same principles of motivation apply: elicit voluntary acceptance of goals through education; then reward good behavior through the use of incentives.

SELLING TOP MANAGEMENT

Even a finely conceived motivation program will die a premature death unless Top Management accepts the need and agrees with its objectives.

The easiest type of motivation program to sell is one which is designed to increase sales volume or market share. With some notable exceptions, corporations perceive growth as a function of revenues, obviously increased through greater volume. This type of program, therefore, is generally congruent with on-going corporate goals. Additionally, the results obtained lend themselves to measurement, i.e., rewards are generally made for obtaining or exceeding a volume target.

On the other hand, the bottom line is an arithmetic subtraction of expense from revenues, and controlling expense often has equal impact on net profit, particularly in highly competitive situations.

The control of energy use results in an expense reduction which can be substantial, depending on the nature of the business and the amount of energy employed. Yet, the application of an incentive program to the expense side of the business is a concept foreign to many otherwise knowledgeable executives.

In our experience, the person who will most readily embrace a motivation program designed to reduce expense is the Chief Executive or Chief Operating Officer. Although extremely busy with a variety of responsibilities, the CEO or COO will quickly recognize the value of such a program, provided it is well-conceived and sold on a cost-justified basis.

While the indirect benefits of motivation programs are often difficult to justify, energy is quantifiable and lends itself to direct measurement. In our designs, and recommendations for designs, we usually outline cost justification for our clients based on some form of direct measurement. In this way, a payback period can be determined and compared with alternative investments.

Armed with a properly designed program and implementation strategy, along with a projected payback schedule, middle level managers will have little difficulty in securing serious consideration for such a program by Top Management.

THE PLANNING PHASE

In the exploratory phase of program development, you will find it helpful to construct a formal, written plan which should include at least the following elements:

1. State Your Objectives

What do you wish to gain from the implementation of an energy conservation program? To whom should it apply? Does it make more sense to run a pilot program or apply it to the entire group? Your objective(s) must be clearly delineated prior to beginning the program's design.

2. State How You Will Measure Results

Results can be divided into financial (quantifiable) and non-financial (qualitative). Do not make the mistake of implementing any program before the machinery for measurement is installed. Outlining your expectations prior to implementation not only allows you to determine objectively its success or failure, but also enables you to make adjustments, as needed, while the program is operational.

3. Solicit Managerial Input

Provide a rough outline of your concept to all levels of management and solicit their opinions and advice. Do not expect to obtain unanimity of thought at this point. What you are seeking is valuable input and considerations which are not apparent to you. Also, by openly soliciting their opinions, managers are preconditioned to the necessity of their active involvement and support. As a final note, managers providing input are also making a commitment to sharing in the success or failure of the program ultimately implemented.

4. Determine Your Resource Pool

Successful programs are intricately detailed and best left to professionals with experience and a successful track record. This does not suggest you need hire an outside consultant. Many companies have excellent personnel experienced in corporate communications and graphic design. When checking into the viability of using internal resources, be sure to obtain an estimate of costs. Contracts with internal departments, in terms of accountability, should not differ from those made with outside consultants.

5. Calculate An Appropriate Level of Investment

In modern industry, neither machinery nor motivation programs are patchwork, unless you anticipate early failure. A well designed program will have many elements interweaved and each costs money. Having calculated your projected payback, you should determine an acceptable period for the recapture of program cost. This determination will yield your appropriate level of investment.

PROGRAM RECOMMENDATIONS

Following is an outline of contents for a typical energy conservation incentive program. It is designed to educate employees in practical energy conservation techniques and to reward them for positive behavior. Similar elements were used in programs designed for Brooklyn Union Gas Company and Merck and Company.

1. Design a Logo and Theme

A program theme and dedicated logotype should be developed for use in all program communications. This element lends visible importance to the program, and provides an integration of all subsequent program materials and communications.

2. Use Teaser Posters

The purpose of a teaser poster is to generate curiosity and interest in a program prior to its actual announcement. The first teaser should be extremely nebulous, referring to the program name and little else, other than a commencement date. The second teaser should hint at rewards which can be won in the program. No details, however, should be announced in either teaser poster.

Both posters should feature strong graphic design and multi-color print on good paper. Since this is the first communication about the program, it is important to convey a quality image.

We recommend the placement of teaser posters in many areas to gain recognition through repetition. Timing is also important. The first poster should precede program implementation by approximately one month. The second poster then replaces the first about two weeks later.

In order to gain maximum results from the use of these teasers, ask managers for their cooperation in not revealing program details prior to the official announcement.

3. Involve the CEO and COO

An important element in the implementation of any motivation program is both support and involvement from the Chief Executive and/or Chief Operating Officer. Regardless of the vehicle chosen for formally announcing the program, we recommend each employee receive a one-page letter from one of these officers, announcing the rationale for the program, its objectives, and outlining, in a broad way, the program's contents.

4. Use Educational Promotions

In a typical one-year program, we recommend the use of four educational events, one each quarter. These events combine the features of a newsletter with those of a promotion. Typically, employees receive a list of energy-saving tips which they can easily implement in their homes or automobiles in order to enjoy a personal and measurable savings through conservation.

We usually combine the newsletter with a promotion in which the employee is asked to participate in some form of game or quiz on the subject of energy conservation. A pre-determined number of prizes are awarded via random drawing from all employees who participate and obtain a perfect score.

In this particular design, we look for a direct payback from the third quarterly event in which we ask employees to submit their best tips for saving energy on the job. Prizes are then awarded on the basis of merit and potential savings, judged by someone who is qualified at the client company.

5. Choose Incentives Carefully

When possible, we like to tie-in the incentive to the subject. In this particular design, we recommend the use of energy-saving or energy-related incentives. In the former group, we have used bicycles, solar calculators, and tire gauges; in the latter, insulated jugs, portable gas grills, and knitted wool caps.

We have three other recommendations concerning incentives:

a. Test Popularity: Through surveys and informal feedback, we test the popularity of a premium with similar groups. Items proving unpopular are quickly replaced.

b. Buy Quality: Off-brands and poor quality incentives will gain nothing but negative comments and a poor level of cooperation from participants.

c. Use Dual Value Prizes: In each promotion, we offer a few large prizes and many smaller ones. Even though only a few "Grand Prizes" are awarded in each contest, they tend to work as an important carrot, attracting maximum participation.

While on the subject of prizes, we do not recommend the use of cash as an incentive. Although we will not take the time in this paper to detail all of the reasons, our experience indicates it simply does not work as well as merchandise.

6. Announce Winners Promptly

Building and maintaining credibility throughout the program is integral to its success. Immediately following the drawing, all winners should be announced. We prefer the printing and distribution of a sample poster which details all winners' names.

EVALUATION AND MAINTENANCE

Prior to the expiration of the program, you should conduct an objective evaluation of actual results against expectations. While a financial evaluation is relatively simple, qualitative measurement is often quite difficult.

At Brooklyn Union Gas, we chose to secure feedback from all participants through the use of a formal survey. We designed the instrument to measure the value and popularity of most program elements. Questions were reviewed and modified by a qualified Educational Psychologist in order to eliminate obvious areas of bias. Additionally, the survey forms

were secretly encoded to measure data from
various departments.

Data from the survey was run through a standard statistical package which included
t-Tests for data comparisons, and a report
was prepared for the company.

Largely as a result of the survey, many modifications were made in a second-year program
which is currently operational. At the time
of this writing, the new program is already
exceeding financial expectations, a trend
which both we and company officials expect to
continue.

Habits are not modified or replaced quickly.
Although a one-year program can be cost-justified, it is a serious mistake to believe it
will effect permanent change. Maintenance
programs must be designed and implemented to
continue the process of awareness and conservation.

Incentive programs require time, forethought,
and an investment, but they are designed to
modify human behavior. Only by convincing
employees to conserve energy today can we ensure adequate supplies to meet the challenges
of tomorrow.

SECTION 23
ENERGY POLICY

Chapter 72

OIL PRICE FORECASTING

H. N. Morris

Introduction

Historically, OPEC inherited the pricing structure that had been devised by the major oil companies and the international oil market.

This system revolved around the crude oil known as Saudi Arabian 34⁰ light, which was also called the marker crude. All other crude oils were priced in relationship to the marker crude, with bonus prices being paid for lighter oils, oils with less sulfur, and oils that came from North Africa due to lower freight rates when shipped to Europe or the U.S. Initially, these premiums were valued at a few cents per barrel, and usually only the freight advantage proved to add materially to the cost of oil, and only then in times of stress, e.g., the closure of the Suez Canal.

However, for the last few years, OPEC had been attempting to come up with a price structure with "reflected the market" value of crude oil. The Western countries were constantly arguing at international trade meetings such as the North/South conference, that OPEC as a cartel was arbitrarily setting the price of crude in a manner which had no relationship to the market. OPEC, on the other hand, argued that as fast as they made adjustments in price, the real value was immediately eroded by the "exported inflation" of the Wester world and the continuing decline in the value of the dollar in which they were paid.

In order to overcome these two "problems"

(1) Establishing the "market" price of crude.
(2) Protecting revenue against inflationery erosion .
OPEC realized that the existing pricing system was inadequate.

For a great many years they had been considering the idea of being paid with a basket of currencies, e.g., the International Monetary Fund's Special Drawing Rights (SDRs), and/or some other multi-currency system, particularly one that introduced gold into the equation.

Obviously, this was an extremely complex matter and a monumental decision to make, with risks of a completely unkown nature. There was no way to be sure that such a move would not only collapse the Western monetary structure, but also take down the OPEC members' reserves with it. Therefore, although OPEC has made threatening noises to drop the dollar from time to time, it has never bitten the bullet. It seemed simpler on the other hand, to devise a pricing mechanism that "reflected the market". The system that would fill OPEC's need was the European Monetary parity method, which had already been tried and tested, and for simplicity's sake is called "The Snake in the Tunnel". Again, this would be a radical departure from the existing structure and a move in an unknown direction. Therefore, the more conservative members of OPEC resisted the change.

However, the advent of the Ayatollah Khomeni in Iran, and the removal of Iran's oil from the market for a short period of time, allowed the more adventurous members of OPEC to force through the "Snake in the Tunnel" system.

The Snake in the Tunnel.

The concept of the "Snake in the Tunnel" system is quite simple. A bottom price and a ceiling price are set, i.e., "the tunnel". The prices that the various members charge for their crude are whatever they can get between the bottom and ceiling prices. Theoretically, these are supposed to vary between the two lines of the tunnel. As time goes on, this fluctuating level of prices draws a squiggly line, i.e., "the snake". The system was introduced in July, 1979.

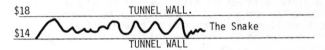

If the prices had stayed at the bottom, the Western world could have argued that bottom level was too high. But in fact, OPEC's Strategists knew that the market position was extremely tight, and prices not only gravitated to the $18 level, they broke through on the spot market to about $23. OPEC then revised the tunnel markers in October, 1979, setting the lower level at $18 and the upper at $23.50.

In the past, the spot market had played a balancing role between supply and demand have been both an instrument for and dominated by, the major international oil companies and a few large brokers.

In the original "Snake in the Tunnel" exercise in July, the spot market was disregarded by OPEC. Suddenly it became the focal point of attention, and it did not take some OPEC members long to exploit its potential. By hold crude off the normal market, a shortage was automatically created, therefore driving prices up on the spot market. The oil that was witheld from the normal market was now sold at the new higher prices on the spot market.

With literally thousands of new brokers in the field, coupled to the uncertainty of Iran's supplies, this strategy was so successful that prices of up to $45 per barrel were paid on the spot market.

In fact, the spot market became _the_ market, and the Saudis, with their $18 crude, began to look rather foolish when other OPEC members were obtaining double this price for their oil. In any event, the Saudis had to increase their price to bring it more into line. In fact, prior to the Caracas meeting in December 1979, they raised their price from $18 to $24 per barrel.

The Caracas Meeting   December 1979

When the Caracas meeting took place at the end of
December, there was no doubt that the system was
working as far as OPEC was concerned. Saudi Arabian
crude 34° light was priced at approximately $12 per
barrel entering 1979, entering 1980 it was $24 per
barrel, and other OPEC members had seen their crude
leap from $13 per barrel to as high as $40 per barrel
on the open market.

Therefore, the only thing left to discuss was how high
to set the tunnel ceiling. Here it was argued that
there was no reason to set a ceiling as long as the
spot market could be exploited. The minimum price
could always be manipulated upwards to suit any
improving situation, but more important, it could be
moved upwards to retard any serious wholesale erosion
of price should this occur.

The Saudis objected to this mechanism insisting that a
ceiling price should be established, particularly if
they were going to hold the line on the floor price,
which they claim could not be adjusted rapidly enough
to keep them in line with the spot market fluctuation.

Unfortunately, once again it was reported that OPEC was
in disarray because they could not reach a uniform
price for crude. The reason that OPEC did not reach a
uniform price was quite simple, they were not trying
to do this as the uniform price was an acronysm which
had been disbanded in favor of the new system.

OPEC of course, regularly gives the impression that it
is striving in one direction, when it is actually
moving in another, and this gives rise to elaborate,
but totally misinformed comments on what actually
takes place at OPEC meetings and the implications on
world oil prices and supply.

Post Caracas

Although there was an immediate sense of euphoria
amongst some of the OPEC members, Libya, Algeria,
Nigeria et al, the Saudis were not overjoyed with
the rate of change, and in particular, the fait
accompli of price increases which have taken place
outside the mechanisms set up by OPEC through the
exploitation of the spot market.

The Saudi objection basically revolves around the fact
that because they were the "good guy" holding the
bottom of the tunnel price, they also looked foolish
when other members sold their oil openly at twice the
price. They also objected to the fact that OPEC
agreements were being broken as soon as the resolution
to adjourn the meeting had been passed.

However, Saudi Arabia had benfited strongly from the
"Snake in the Tunnel" and there is no doubt that the
manipulation of the spot market by the other members
has worked in their favor. Irrespective of this, the
Saudis were being hurt internally on the political
front, and had lost a certain amount of face inter-
nationally as they were predicting slow movements in
prices through 1979 not the quantum leaps that were
experienced.

Saudi Arabia had always prided itself on the fact that
it was a "very good member of OPEC" who abided by all
OPEC agreements and therefore could not tolerate this
situation for very long for not only were they losing
prestige, but pricing agreements made at OPEC meetings
were being exploited to their detriment. Even with a
high OPEC output, we would probably have seen the
Saudis making some move in 1980/1981 to counter the

influence of the spot market, probably by under-
selling the other members who were exploiting it.
However, the Saudis did not have to do this as things
started to move in their favor as the world market for
oil started to collapse.

How much of this collapse was due to conservation, and
how much was due to the recession is not the subject
of this paper, however, we do believe that the
recession has played an overriding role in the drop
in world demand.

Once this demand dropped below 27 million barrels per
day for OPEC oil, the Saudis real power became
apparent. They could use their vast oil output of
about 10½ million barrels per day to create a surplus
thereby bringing the spot market into line together
with the members who were exploiting it. Throughout
this period the Saudi Arabians gradually raised their
prices in line with the agreement from the Caracas
base of $24 per barrel to $32 per barrel, but at the
same time refused to ease production unless the
other members agreed to abide by the "Snake in the
Tunnel" limits on price.

Apart from the fact that the Saudis were over-
producing there was another major factor which helped
to squash demand for OPEC oil. The oil industry in
order to protect itself from cutoff and major price
fluctuations, particularly with the advent of the
Iraq/Iran War, had established large stock piles and
reserves. When it became apparent that the demand
for oil was dropping rapidly, the industry moved to
liquidate these stocks.

Various figures have been voiced on the size of these
stocks and the rate of liquidation, in fact the
Saudis complained at the beginning of 1982 that the
oil industry was overstripping at the rate of 4
million barrels per day in order to bring undue
pressure upon the weaker members of OPEC such as
Nigeria. However, the oversupply of oil certainly
had the desired effect as far as Saudi Arabia was
concerned, for it did collapse the spot market prices
and bring the other members into line. The Saudis
then agreed in a compromising vein to raise the
bottom of the tunnel to $34 per barrel from $32 per
barrel, and raise the top of the tunnel to $38 per
barrel where it stands now, although at the time of
writing, even the better crudes are only selling in
the $35/$36 per barrel range.

1982 and Beyond

Throughout the period of adjustment in the 1980/1982
period, speculation and uninformed analysis resounded.
The two major points which were voiced were -

> oil prices would collapse almost to their
> pre-Embargo level, and

> OPEC would therefore collapse with it.

As previously discussed, Saudi Arabia is a "very good
member of OPEC", and would not make a move to see
this happen, therefore in spite of all the wisdom
being espoused about the collapse, Saudi Arabia has
maintained her bottom of the tunnel price of $34 per
barrel and drastically cut production. However,
throughout this period OPEC has managed to achieve
something that she could never achieve or agree on
in the previous 20 years of her existence. She is now
operating a producting sharing scheme. If this
production sharing scheme is actually maintained,
then OPEC will become a true Cartel, and will be
able to manipulate the market and tune it to a far

greater degree than she has ever done in the past.

The question now arises, "will demand start to
increase again?" In this respect, the vast overhang
of oil company stocks are nearing the end and will
probably disappear by the end of 1982. The other major
criteria is when and if we come out of the recession,
and this recession is not just a United States
phenomena it is worldwide. When we come out of the
recession the demand for OPEC oil will radically
increase and although we would not expect to see
quantum leaps as we have in the past, there will be
an unrelenting pressure on prices and this time
OPEC may be even more sophisticated in its economics
to deal with this than in the past. Far from being
pronounced dead, we believe that OPEC will have come
out of this period more sophisticated than when it
went in, and should the Iran/Iraq War develop into a
Persian Gulf war, then the pressure on prices could
become enormous.

The danger here of course lies in the fact that such
action would once again make price considerations
irrelevant and supply considerations paramount and
with stocks depleted, it would be impossible to
guage what would happen to prices.

Chapter 73

ENERGY GENERATION: NUCLEAR VS. COAL

N. S. Parate

ABSTRACT

1. INTRODUCTION

2. ENERGY CHOICE

3. NUCLEAR AND COAL

4. UTILITY MANAGEMENT

5. OBSERVATIONS

6. CONCLUSIONS

ABSTRACT : The importance of economic and cheap energy to a society like ours, which is geared to an increasing energy use and which is shaken by the oil crisis of late 1973 is great. There will be continuing increase of electric energy use inspite of various conservation measures, environmental restrictions and limitations. The energy crisis and self sufficiency policies of almost all nations has compelled us to realise and appreciate the finiteness of fossil fuels. These fossil fuels supply 90 % of world energy need. The supply problem is closely linked with the population growth, increased industrialisation and rising aspirations of the people worldwide. Various experts differ dramatically on the type of strategy and solutions to solve the energy needs and power generation requirements. The major choice of fuel resource today and in near future is without doubt Nuclear and Coal. Solar and Other sources will be the choice fuel in the beginning of next century. They have limited value at present in the generation mix for electric energy. Various studies indicate that the projected increase in world electric energy demand could not be met without a major contribution from Nuclear Power. Nuclear industry has been clouded with uncertainties during recent years particularly in United States. Most of the developing world is deprived of sufficient reserves of coal. More than 80 % of world's coal is in USSR & USA. The real advantage of the nuclear power is the low cost of fuel and clean environment. The various experiences of electric utilities in USA are presented and discussed in the paper. The Nuclear industry advocates argue that the nuclear energy is clean, cheap and safe with unlimited available fuel source but the opponents argue about the environmental damage, riskfactors, waste disposal and decommissioning issues aswellas the recent power plant operational & design issues echoed by the Nuclear Regulatory Commission.

1. INTRODUCTION

Energy is the backbone of modern society. The importance of economic energy is great to a society like ours, which is geared to a continuous energy use. Very few realised the gravity of our dependence before the great blackout of Northeast in 1965.

The oil crisis of 1973 has shaken the society of the industrialised as well as developing nations around the world. Alternate energy resources and technologies have attracted since then more and more attention. There will be continuing increase in the use of energy in spite of various conservation measures and pattern of life styles. The energy crisis and awakening of self-dependence has compelled most of the nations to appreciate the finiteness of fossil fuels and hence the development of nuclear and other alternate energy fuels and electric power generation technologies. The electric supply is the major part of the energy requirement of the present day society. It is closely linked with the population growth, rising aspirations and increased industrialisation of the nations worldwide. It is an extremely complex subject that encompasses social, economic, political and technological consideration. Various experts differ dramatically on the types of solutions to solve the energy needs throughout the world.

2. ENERGY CHOICE

The absence of ideal and perfect choice between various alternatives of energy fuels and resources at economically affordable level in each nation makes a generation mix from different types of energy resources, fuels a desirable and appropriate choice. Coal and Nuclear are the prominent electric generation choice today and in the near future particularly for the industrialised nations. Solar and Biomass seems to be the alternate future energy choice particularly in the developing countries. This is mainly because of the decentralised nature of these resources and the rural population of these countries and the cottage industry developments and needs for the supply of the energy.(17,18)

2.1 ELECTRIC GENERATION : Various studies indicate that Coal and Nuclear power are clearly capable of providing the major share of the amount of electric generation requirement at affordable and competetive price in the near future. The primary energy requirement to deliver a unit of energy from Coal or Uranium to the user is substantially greater(11) in terms of mining, transport, processing & generation than is the case for oil, gas. The higher cost is partially offset by the much higher enduse efficiency achievable with electric generation.(5,7)

3. NUCLEAR AND COAL

The review of the literature and analysis of datas concerning Nuclear and Coal generation, related costs and issues has been recently discussed by the author(19,20) in a paper for the UPADI-82, Aug. Peurto Rico and other internal report for the Pennsylvania Public Utilities Commission, Harrisburg (Please see tables 1 to 7 and Fig. 1 to 7 attached)

Table 1.   WORLD ENERGY DEMAND   AND   SUPPLY 1985-2020 (Exajoules)

Primary energy (EJ)	1985	2000	2020
Coal	77-88	122- 171	230-405
Oil	147-169	163- .224	145-248
Natural Gas	50-55	71- 92	84-118
Nuclear	22-24	85-111	246-380
Hydro	23-23	34-34	56-56
Wood and Solar	30-33	44-57	78-115
Total primary energy demand.	349-392	519-689	839-1323
Potential total energy production	350-420	580-680	820-1010

DISTRIBUTION OF POTENTIAL ENERGY SUPPLY

		COAL	OIL	GAS	NUCLEAR	RENEWABLE	TOTAL
OECD	2000	50-70	34-44	20-35	50-80	20-30	190-240
	2020	70-110	26-44	10-20	100-160	35-70	280-360
Centrally Planned	2000	70-100	20-40	20-40	20-40	20-35	160-230
	2020	110-150	20-40	25-45	50-100	30-60	280-360
Developing	2000	15-30	100-140	25-45	7-15	20-35	180-240
	2020	20-50	70-130	30-70	30-60	30-60	220-330
World	2000	145-190	165-210	70-110	85-125	65-90	580-680
	2020	220-290	135-200	70-120	200-300	105-175	820-1010

Shares of world primary energy demand.

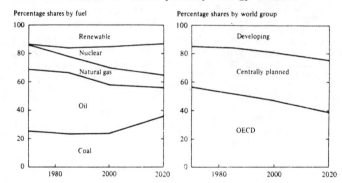

Percentage shares by fuel   Percentage shares by world group

Table 2.   Energy Demand

1. North America: End -2000   2000-2020

Electricity growth	4.2		2.9-3.4
Energy demand in year	1985	2000	2020
Primary energy demand(EJ)	97-108	134-167	193-262
Potential Total energy production.	90-110	110-160	140-220

2. USSR and East Europe: End -2000   2000-2020

Electricity growth	3.8		3.3-3.9
Energy demand in year	1985	2000	2020
Primary energy demand(EJ)	72-84	119-162	213-319
Potential Total energy production.	60-90	110-190	150-300

3. China and CP Asia: End -2000   2000-2020

Electricity growth	9.2		4.8-6.3
Energy demand in year	1985	2000	2020
Primary energy demand(EJ)	32-38	45-67	73-148
Potential Total energy production.	30-50	35-70	70-120

North America.

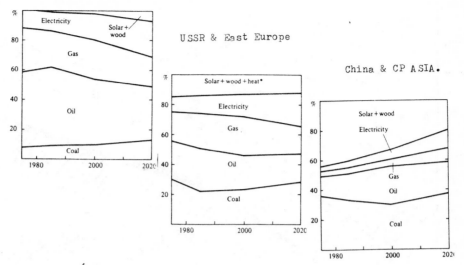

USSR & East Europe

China & CP ASIA.

Table 2.(CONTD) Energy demand.

4.Japan,Australia and Newzealand:	End 2000		2000-2020
Electricity growth	5.1		2.7-3.1
Energy demand in year	1985	2000	2020
Primary energy demand(EJ)	25-28	35-48	55-78
Potential Total energy production.	7-11	20-30	35-70

5. West Europe:	End 2000		2000-2020
Electricity growth	4.5		2.7-3.2
Energy demand in year	1985	2000	2020
Primary energy demand(EJ)	61-69	91-110	124-170
Potential Total energy production.	30-40	45-70	55-110

Secondary Energy by source

Japan,Australia,Newzealand.   West Europe.

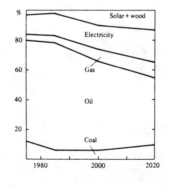
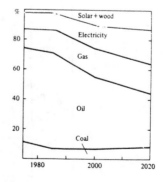

France :Primary energy 1990

The French energy plan has it's aim to achieve a balance of
1/3 Nuclear,1/3 Oil and 1/3 Coal plus other resources.
By 1986 the installed nuclear capacity with annual addition
of 5000 MW Nuclear power will supply 50 % of the total French
electric generation. Germany has been aiming at substantial
contribution of Nuclear energy but the recent resistance from
the public has played a drawback in nuclear program.

471

ESTIMATED WORLD URANIUM PRODUCTION CAPABILITY TO 1990

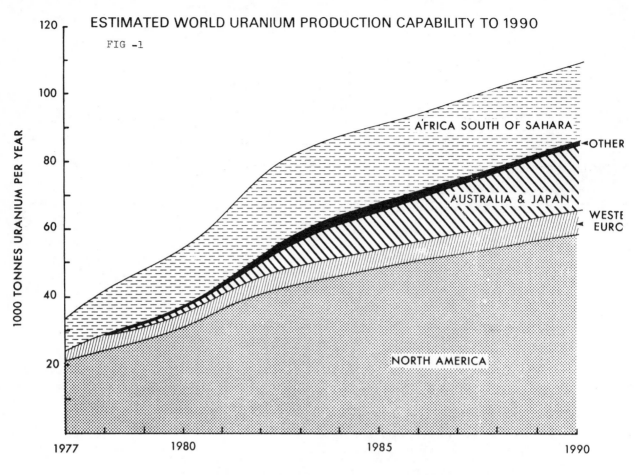

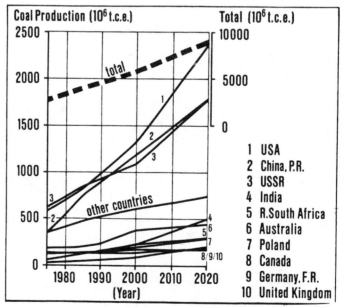

Fig. 2 : Survey of the Future Trend in Production Figures for the Main
Coal-Producing Countries

472

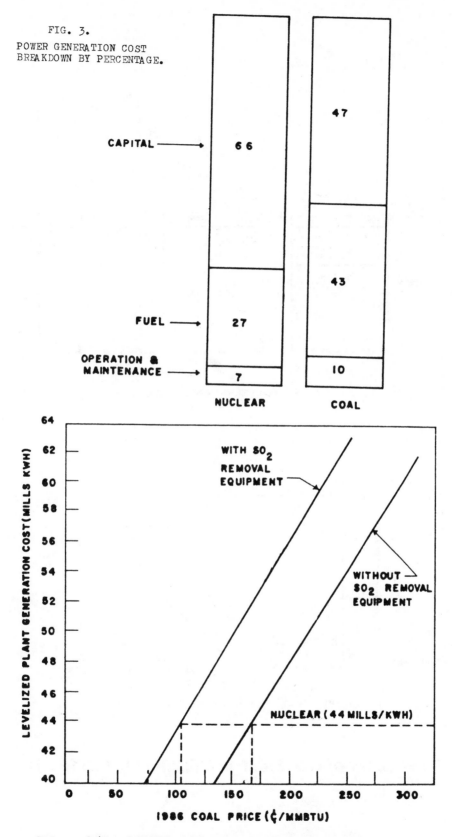

FIG. 3.
POWER GENERATION COST
BREAKDOWN BY PERCENTAGE.

CAPITAL — 66

47

43

FUEL — 27

OPERATION &
MAINTENANCE — 7

10

NUCLEAR   COAL

WITH SO$_2$
REMOVAL
EQUIPMENT

WITHOUT
SO$_2$ REMOVAL
EQUIPMENT

NUCLEAR (44 MILLS/KWH)

LEVELIZED PLANT GENERATION COST (MILLS KWH)

1986 COAL PRICE (¢/MMBTU)

FIG. 4   COAL PRICES AND BREAKEVEN ECONOMICS

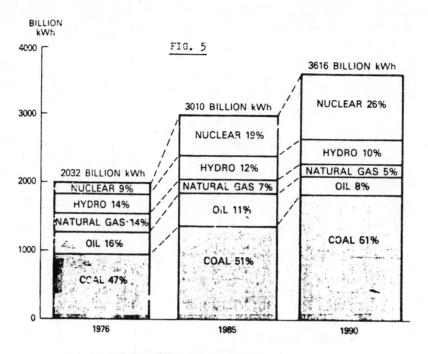

FIG. 5

ELECTRICITY GENERATION BY FUEL TYPE
1976 AND PROJECTED

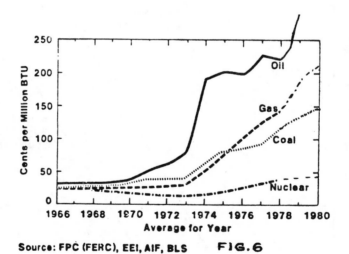

Source: FPC (FERC), EEI, AIF, BLS   FIG.6

Fuel Costs of U.S. Electric Utilities (1966-1978)

3.1 NUCLEAR : Nuclear fuel offers the world virtually unlimited source of clean, reasonably priced energy according to it's advocates. The opponents argue about the environmental damage, risk factors in case of accidents or leakage, lack of experience of utilities, particularly in developing countries, waste disposal and decommissioning issues and a billion dollar investment for 1000 MW nuclear power plant project. Nuclear industry has had a stormy history in USA particularly during these years although most of the utilities can easily illustrate the savings that nuclear power has meant for their customers. The licensing, environmental requirements and various permits during construction and operations of a nuclear unit results in a 10-15 years leadtime for newplants. The combinations of economic problems in financing new units, uncertaining of load forcasting(reduced from 6 percent to 2-3 percent throughout the USA at present) 10-15 years ahead, highlevel nuclear waste disposal or an enrichment of nuclear fuel policies reprocessing and breeder reactor development issue and various actions (site investigations, earthquake safeguards) by the Nuclear Regulatory Commission, State Public Utility Commissions and courts are frustrating for the Nuclear Industry and Power Generation at competetive price/costs. In terms of air quality, nuclear power is the cleanest source of energy. No atmospheric pollutants characteristic of fossil plants are emitted. The radiation emanating from the nuclear plants is less than from many fossil plants as per opinions expressed during some recent studies(1,16).

Strong research and development programs are being conducted to develop nuclear waste disposal repository candidate sites in USA by the department of Energy. The program is being managed in the past by Union Carbide and at present by Battelle Memorial Institute. The repository candidate sites are for the ultimate disposal of spent nuclear fuel from the various commercial power plants. Various studies conclude that the nuclear power will continue to produce the lowest cost of energy except from Hydroelectric sources.Once a nuclear plant is built, the fuel cost is relatively insensative(4,10) to inflation and is a very small part of the total generation cost.

Most of the European countries particularly France and Germany appear to be more dependent on nuclear power for their energy needs than the United States(Table 2).

3.2. COAL : Coal is clearly the most abundant domestic fossil fuel resource in USA with identified sufficient reserve(9) as shown in fig.2 to meet the nations demand for an estimated 200 years. Coal has been a major contributor to the United States energy for 100 years. Coal has been at the center of national debate and as a fuel it is abundant, cheap and provides many jobs in the area. But coal is dirty, dirty and risky to mine, dirty to handle and dirty to burn. The difficulty is aggravated with the environmental regulations and policies of the governments from time to time (8, 12,13).These regulations related to Coal use & mining add to the cost of burning coal to generate electric energy. Some of the major detrimental effects due to coal use are the huge scars left after large strip mining operations, acid mine drainage, black lung disease for the miners, respiratory and heart disease due to sulpher and nitrogen oxides, acid rain due to coal burning, subsidence and other environmental damage(mine waste disposal, fly ash and incinerator residue, pollution) issues. With the current technology and progress,

much more coal can be mined and burned in a clean and economic way. Equally important are the healthy climate for investment to develop large mines and continuing government supporting policies. Starting a new mine is an expensive and time consuming process. To develop a new deep mine producing two million tons per year with a 20 to 30 year life span require five to seven years and an investment of 60 million dollar or more. Mine(14) expansion is similarly time consuming and costly. Reasonable surface mining regulations and reclamation practices, development and improvement in coal utilisation technology having lower emmission release and environmental impacts, with the use of scrubbers and fluidised bed combustion technique are being adopted and improved. Environmental concerns limiting sulphur pollution and hence the scrubers and or precipitators on coal fired plants may hurt the economics of coal and possibly lead the utilities to use other fuel resources.

3.3 FUELS : In United States, Coal is forcasted to provide over 56 percent of the energy for the electric generation by 1990. Oil and Natural Gas consumption will decline under the most likely forcast but still remains the dominent fuel in several areas of the United States, particularly so in the gulf coast states.and coastal areas. Recent oil cartel shake up and continuing glut of oil supply in the market due to supply demand & political situation has introduced a new factor for the revision of the forcasts and policies. Nuclear power is (Fig.5) projected to provide approximately 20 percent of the total electricity requirement in 1990 versus 15 percent today in USA. There is enough supply of Uranium(9) and the trend is towards upward estimates throughout the world.

The developing countries have been showing an increased rate of energy consumption and yet the developed world nations consume almost 15 times more energy. This inequality would naturally decline as the process of development progresses and as the economic gap slowly narrows down. Barring a relatively small number of countries in Central and North America and the Middle East, Most of the developing countries are deprived of sufficient reserves of fossil fuels. More than 80 percent of world's coal is concentrated in the USSR and the USA. Peoples Republic of China and Europe(Fig.2) account for another 15 percent and the remaining 5 percent is distributed among the rest of the countries. The energy demand in Latin America will be satisfied by the Hydro and Conventional energy source for a decade or two(6) according to some studies and the authors observations while stationed in South America for two years recently. Nuclear energy is being developed in Argentina, Brazil and Mexico . Number of construction and research development activities are under progress with collaboration from different developed countries like Germany, USA, France.

The Geothermal, Wind, Tidal energy sources, wherever they exist, are of limited value and can at best serve as supplemental sources in a comprehensive energy program. Solar energy is the only other source which holds the possibility of some day becoming a major source of energy. Being clean, nondepletable and abundant, it can provide sufficient and infinite amount of energy and can prove the ultimate solution of the world's energy problem. Several other forms of this solar energy (wind, waves, wood, biomass, thermal gradient) are technically capable of providing and producing energy but at costs that substantially exceed the present conventional energy costs.

Table 3

ENERGY GENERATED BY FUEL TYPE (%)
1970

	Coal	Nuclear	Light Oil	Heavy Oil	Hydro	Gas
Vepco	55.42	0.00	0.40	41.82	1.87	0.49
CP&L	87.06	0.02	1.85	0.00	3.61	7.46
Duke	88.68	0.00	2.04	0.00	4.14	5.14
Alabama Power	79.21	0.00	0.04	0.00	15.77	4.98
Appalachian Power	98.33	0.00	0.36	0.00	1.85	0.02
Baltimore Gas and Electric	60.51	0.00	1.89	32.43	0.00	5.18
Cleveland Electric	100.42	0.00	0.00	0.00	0.00	0.00
Commonwealth Edison	62.91	6.33	2.27	3.33	0.05	25.10
Consumers Power	87.10	1.97	0.21	0.00	2.33	8.34
Detroit Edison	83.05	0.04	2.19	4.84	0.00	9.88
Florida Power	0.00	0.00	0.00	56.25	0.00	43.75
Georgia Power	75.88	0.00	0.18	0.00	5.64	18.30
Gulf States						
Illinois Power	85.31	0.00	0.13	0.00	0.16	14.41
Jersey Central	14.02	48.51	0.41	33.71	0.00	6.23
Long Island Lighting	0.00	0.00	1.84	90.14	0.00	8.02
Louisiana Power	0.00	0.00	0.06	0.00	0.00	99.94
Niagara Mohawk	57.67	10.36	0.90	13.96	17.10	0.02
Ohio Edison	100.00	0.00	0.00	0.00	0.00	0.00
Ohio Power	98.55	0.00	1.45	0.00	0.00	0.00
Pennsylvania Power	89.99	0.00	5.16	0.00	4.86	0.00
Philadelphia Electric	38.30	0.70	3.53	58.32	0.00	2.69
Public Service	34.65	0.00	1.30	56.49	0.00	8.30

Computed from information contained on Page 432a, Form 1 (FERC)

Table 4

ENERGY GENERATED BY FUEL TYPE (%)
1979

Company	Coal	Nuclear	Light Oil	Heavy Oil	Hydro	Gas
Virginia Electric & Power Company	33.25	21.49	3.33	37.78	3.42	.73
Carolina Power & Light Company	60.13	35.65	.83	0.0	3.36	.04
Duke Power Company	68.40	26.06	.06	0.0	5.16	.34
Alabama Power Company	73.46	5.97	.69	-	17.29	2.60
Appalachian Power Company	97.09	-	.44	-	2.65	-
Baltimore Gas & Electric Company	22.21	50.60	.29	23.84	-	3.06
Cleveland Electric Illuminating Company	87.21	9.48	.33	4.37	-	-
Commonwealth Edison Company	43.65	42.27	1.61	10.15	.02	3.19
Consumers Power Company	63.21	16.13	1.10	18.07	1.94	1.74
Detroit Edison Company	88.75	-	1.97	8.59	-	1.98
Florida Power & Light Company	-	25.96	1.59	53.17	-	19.28
Georgia Power Company	88.85	5.78	.46	.17	4.51	.24
Gulf States Utilities	83.97	-	.19	15.83	-	-
Illinois Power Company	96.30	-	1.10	1.61	.05	.94
Jersey Central Power & Light Company	19.11	59.29	4.78	12.57	-	5.90
Long Island Lighting Company	-	-	1.31	90.30	-	8.39
Louisiana Power & Light Company	-	-	.53	17.59	-	81.88
Niagara Mohawk Power Company	33.40	13.80	.03	38.79	13.70	.28
Ohio Edison Company	94.58	3.17	2.25	-	-	-
Ohio Power Company	96.80	-	3.20	-	-	-
Pennsylvania Power & Light Company	81.36	-	.76	17.88	-	-
Philadelphia Electric Company	35.36	38.91	2.44	26.33	-	.13
Public Service Electric & Gas Company	15.10	28.58	1.78	34.82	-	20.30

Comparison of Fossil Coal-Fired* Units 400 MW and Above
to All Nuclear Units

Table 5   UNIT YEAR AVERAGES
1969-1978

EQUIVALENT AVAILABILITY

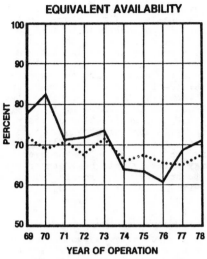

CAPACITY FACTOR

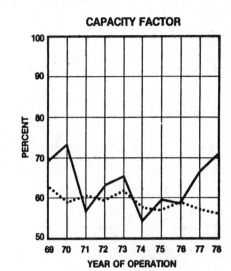

FOSSIL ··········
NUCLEAR ———

*COAL PRIMARY FUEL

Equivalent Avail	69	70	71	72	73	74	75	76	77	78
Fossil	71.9	69.2	71.1	67.2	71.9	65.9	67.2	65.8	65.0	67.3
Nuclear	78.5	82.5	71.1	71.5	72.8	63.7	63.1	60.6	68.1	70.5

YEAR

Capacity Factor	69	70	71	72	73	74	75	76	77	78
Fossil	63.1	59.2	60.2	59.7	62.4	57.5	57.2	59.3	57.4	55.7
Nuclear	69.9	73.3	56.9	63.1	65.1	53.6	59.4	59.0	65.9	70.3

YEAR

Source: "Ten Year Review - 1969-1978 Report on
Equivalent Availability," NERC.

Table 6   TOP PLANT ENERGY PRODUCTION EXPENSE Mills/KWH

Company	HEAVY OIL 1970	1973	1975	Coal 1970	1973	1975	1979	Nuclear 1979
Virginia Electric & Power Company	3.44	6.80	20.09	3.66	5.03	11.85	16.35	9.56
Carolina Power & Light Company	12.74	6.82	-	3.58	4.55	11.08	14.21	7.29
Duke Power Company	-	-	-	3.60	4.46	8.68	14.03	6.88
Alabama Power Company	-	-	-	2.74	4.24	9.79	16.11	18.40
Appalachian Power Company	-	-	-	2.39	3.65	8.59	14.03	-
Baltimore Gas and Electric Company	4.02	7.64	19.68	3.08	5.25	8.46	13.15	8.33
Cleveland Electric Illuminating Company	-	-	-	4.59	5.38	12.05	18.77	9.30
Commonwealth Edison Company	5.67	8.85	15.32	3.33	4.90	10.89	18.49	5.95
Consumers Power Company	13.89	8.22	19.18	3.26	4.85	9.27	15.95	12.05
Detroit Edison Company	13.15	13.08	22.64	3.64	5.28	8.93	16.39	-
Florida Power Company	3.78	6.12	14.32	-	-	-	-	5.89
Georgia Power Company	-	7.54	22.89	3.56	4.27	10.31	13.59	9.69
Gulf States Utilities	-	-	-	-	-	-	-	-
Illinois Power Company	-	4.64	NA	2.14	2.95	NA	11.74	-
Jersey Central Power & Light Company	5.54	10.83	27.66	3.67	5.31	8.55	13.16	6.09
Long Island Lighting Company	3.18	NA	16.37	-	-	-	-	-
Louisiana Power & Light Company	-	-	17.80	-	-	-	-	-
Niagara Mohawk Power Company	4.76	6.57	17.36	5.16	6.42	14.01	16.73	NA
Ohio Edison Company	-	-	27.63	2.84	4.96	11.98	14.28	16.86
Ohio Power Company	-	-	-	2.63	3.75	6.70	9.49	-
Pennsylvania Power & Light Company	-	-	24.11	3.08	5.23	8.46	13.16	-
Philadelphia Electric Company	5.75	11.28	24.31	3.08	5.25	8.46	13.16	6.26
Public Service Electric & Gas Company	4.11	8.53	21.59	2.90	5.25	8.46	13.16	6.89

Barring the promises held by fusion and hydrogen, nuclear fission ,simple and fast,seems to be the best alternative that commend itself for immediat e future development for meeting the needs of the developing and technologically capable countries like India,China,Japan,Brazil and Others.
In spite of well publicised grounds on which nuclear energy expansion is opposed by well meaning critics, there is much that can be said in favour of developing nuclear energy as the energy base for the economic development of the developing countries. Coal and Petroleum technologies were thresholds to the new era of economic develop ment. Nuclear technology is another of such milestones in the history of human endeavour of putting science and technology to work for the benefit of mankind. It represents another industrial revolution which the developing nations should not and cannot afford to miss. The real advantage obtained by the nuclear option is in the matter of fuel costs and also the technology development. The nuclear fuel costs are almost in the ratio of 1:3 for nuclear and Coal/Oil (4)(Tables 7 & 8.)

4. UTILITY MANAGEMENT

The Electric Utilities originally served the local areas but were drawn into consolidating & making large operating units to obtain the benefits of scale economies and superior management,control. Same reasoning followed in the building up of larger units(Nuclear 1000 MW & Coal 450) and centralised clustered power plants in United State. However difficulties of raising large amounts of new capital,steeply rising operating costs,public opposition to successive rate increases,lower growth rate than expected in the earlier planning, many new planned construction have been delayed postponed or cancelled.
Power pooling arrangement, among the various electric utilities covers the entire USA with sizeable network of interconnections. Power pooling was devised to enhance(15) service, reliability and to reduce the operating costs. This helped to to realise the utilities to gain financial benefit with better,safe management. Financial benefits are often realised with staggered construction of large generating units,short term capacity transactions(during outages and severe weather conditions) and economic dispatch. Reduction of installed reserve capacity is made possible by mutual emergency assistance arrangement and associated coordinated transmission planning. It is now eminently practical for numerous individual companies within a broad service area to share the benefits of nuclear generated power. Joint ventures in fuels management(Coal & Uranium Mining Fuels Supply,Fuel Cycle Operations,Transportation and Waste disposal & Decommissioning) seems to be another variant of the pooling arrangement.

5. PERFORMANCE OBSERVATIONS

The actual performance observations of number of US electric utilities has been studied and compare ed by the author during his Public Utility Commission experiences and testimonies. The results and the tables are complied from FERC (Gray Book) information and Utilities(6,13). These tables 3-7 indicate the comparative informations about the capacity factors,capital costs,fuel costs, energy production expenses and percentage generation according to fuel types, explaining the change in the generation mix and increase in nuclear generation since 1970 throughout the United States with particular effects on emphasis on nuclear generation and reduction in oil generation.

6. GENERAL OBSERVATION

The fundamental problem that we must cope with is the fact that the United States and other economies are shifting from dependence on energy provided principally by oil and gas to a more diversified mix of supplies with a major emphasis on Nuclear and Coal. This transition will likely lead during the 21 st century to a substantial reliance on energy from renewable and nondepleting source(Solar and Hydrogen). The effect of the improvement in energy efficiency in USA has been to hold energy consumption well below the levels that would have been reached if energy price and use pattern had followed the pre-embargo trends. Non depleting and renewable sources for exemple are not expected to meet substantial portions of energy demand until well into 21 st century. Energy from hydroelectric and Geothermal sources is limited by the availability of suitable sites. Realisation of the potenti al of solar energy is contingent upon technologica l breakthrough (photovoltaic cells, solar panels) that will bringdown it's cost. Today converting the sun's energy directly into heat/electricity with existing technology is typically four to eight times as expensive as obtaining the same from conventionally generating sources.

7. CONCLUSION

The above discussion clearly indicates the following general conclusions and the policy issues.
1. The nuclear generation is economically cheaper now and will remain so in the near future.
2. The kind of generating stations between now and future years will be nuclear and coal. Increas ingly the generation mix will be the practical and desirable choice for any nation or utilities.
3. More and more efforts will be put to develop and make it cost competetive the renewable and non depleting energy sources and the appropriate technologies.

ACKNOWLEDGEMENT : The author would like to express his sincere thanks to colleagues at Pennsylvania Public Utilities Commission,Harrisburg and North Carolina Public Utilities Commission,Public Staff Raleigh.
Special mention is of Mr. Robert Pennell,Energy Advisor, Pa. PUC, Harrisburg.

REFERENCES

1. Atomic Energy Forum,Washington D.C.'Licensing Design and Construction problems:Priorities for solution' Jan. 1978.
2. Mr. Corey Gordon 'Cost Comparison of Nuclear and Conventional Electric Generation' Public Utilities Fortnightly,April 22,1976.
3. Mr. Crawley & Kokolski, United Engineers and Constructors,Philadelphia'Testimony presented before the Pa. Public Utility Commission, Philadelphia Electric Co. Rate Case RID 00000438 June 78.
4. EPRI PS-45-SR 'Coal and Nuclear Generating Cost April 1977.
5. Hardie R.W. & Chamberlin J.H.'Analysis of Electric Power Generation Cost' Nuclear Technology, Vol 33,April 77.
6. Ismael Escobar & Mario Ibacache' Contribution al analysta del desarrallo de fuentes no convencionales de energie en America Latina' Mayo 1977,InterAmerican Development Bank,D.C.
7. Lee, T. H.,Manager,Strategic Planning Operation for Power Generation,Business Group,G.E.C.' Testimony presented before Connecticut P.U.C. 'Power Generation Economics' Jan.22,76.

8. Parate Nath. S. 'Testimony presented before the Pennsylvania Public Utility Commission,Harrisburg,Pa.Rate Case-Coal Surcharge(Emergency),West Penn. Power R-7801546,Dec. 1978

9. Parate Nath. S. 'Testimony presented to Pa. PUC Harrisburg,Philadelphia Electric Company,Rate Case RID 00000438 (Decommissioning & Nuclear Power Plant Life,Depreciation rates), July 1978.

10. Parate Nath. S. 'Testimony presented to the Pa. PUC,Metropolitan Edison Electric Co. (Met Ed) Rate case R-78060626, Nov. 1978.

11. Parate Nath. S. 'Testimony presented before North Carolina Public Utility Commission,Raleigh Duke Power Company,Docket No. E-7 Sub 295,Apr.80

12. Parate Nath. S. 'Testimony presented before N.C.PUC,Raleigh,Carolina Power and Light Co. Docket No. E-2 Sub 383, March 1980.

13. Parate Nath. S. ' Testimony presented before the N.C. PUC,Raleigh, Virginia Electric Power Company, Docket No. E-22,Sub 383, April 1980.

14. Parate Nath. S.' Testimony presented before the West Virginia Public Service Commission,Charleston, W.Va. for Appalachian Power Company Fuel Clause case no. 79-322-E-GI, Charleston,Sept.79.

15. Reem F. Herbert ' Nuclear Power pooling by Electric Utilities' Memorandum,Federal Energy Ad Administration, June 18.75 and Private discussion, March 1982.

16. Rao Waman ' Nuclear Energy and Economic Development, Energy Communication ,Vol.2(4)377-406,77.

17. World Energy Resources 1985-2020,Publication of World Energy Conference,1978,IPC Sc. & Tech.

18. Sood D.K. and Parate Nath.S. ' Alternate Energy Resources and Technologies for rural third world countries' Paper to be presented at XVII UPADI Congress of the Pan American Federation of Engineering Societies, August 1982,SanJuan,P.R.

19. Parate Nath. S. and Sabnis G.M.' Some view on Economics of Power Generation: Nuclear vs Coal. Paper to be presented at UPADI, XVII Congress of the Pan American Federation of Engineering Societies, August 82,SanJuan, Puerto Rico.

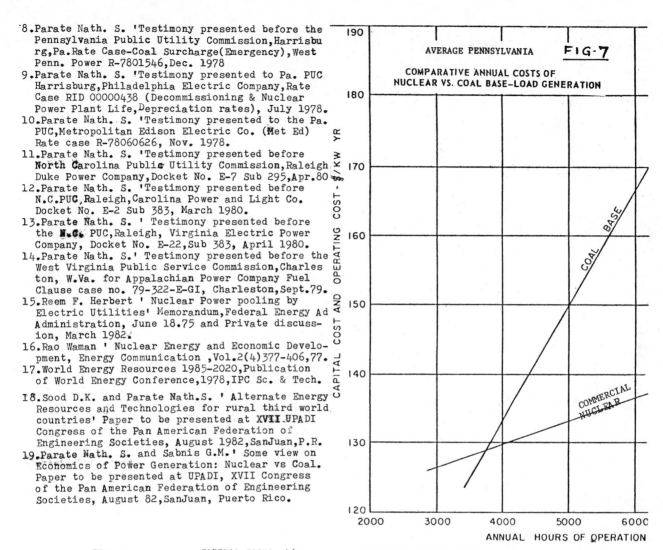

Table 7

Region	CAPITAL COSTS $/KW		LEVELIZED BUSBAR COSTS - MILLS/KWH	
	Coal	Nuclear	Coal	Nuclear
Northeast	698.50	829.00	48.10	41.60
Southeast	569.00	711.50	44.25	37.60
East Central	663.00	787.00	46.85	40.20
West Central	654.00	754.50	42.55	39.10
South Central	649.50	734.50	41.05	38.75
West	714.00	823.00	46.20	41.30
National	664.00	791.00	45.15	40.20

Table AVERAGE COST 1986.(USA)

NOTE: Capital cost estimates were developed for the Electric Power Research Institute by the Bechtel Power Corporation for coal-fired generating plants and by United Engineers and Constructors for nuclear generating plants.

479

SECTION 24
BUILDING OWNERS AND
MANAGERS FORUM

Chapter 74

ENERGY SAVINGS COMPUTATIONS FOR CONTROL BY ENERGY MANAGEMENT SYSTEM

D. M. Fowler

INTRODUCTION

The energy savings computations discussed below were done for a 49 floor all-electric high rise office building of 1.4 million gross square feet in downtown Dallas. The building exterior consists of double pane reflective glass, 37% of which is vision glass. The cooling system includes two 1520 ton chillers and one 760 ton chiller, serving variable air volume air handlers every other floor. Heating is by means of resistance duct heaters and fan powered mixing boxes with electric heating coils downstream.

The owner, a major developer, needed assistance in deciding whether or not to install an energy management system (EMS). They also wanted assistance in deciding which loads were economical to control if they decided to install an EMS.

As is normally the case for a new building, the basic building design was essentially complete when these studies were started. For each separate computation below, the local controls as provided in the initial design documents and specifications are outlined, followed by a brief discussion and the EMS energy savings calculations. Before this energy savings study was undertaken, a computer building simulation study had been completed for the purpose of establishing the design energy budget or goal. This earlier study will be referred to from time to time.

POSSIBLE SAVINGS

Energy Use

An existing all-electric high rise office building in Dallas with double pane reflective glass that is operated on a rigid schedule of 12 hours per day and 6 hours for Saturday of HVAC uses energy at the rate of 87,000 BTU/Sq. Ft./Yr., or 25.49 KWH/Sq. Ft./Yr. It does not have a microprocessor or computer based energy management control system installed, but the building operator does manually adjust the chilled water temperature with the outdoor weather conditions. Lighting is 4 watts/square foot, whereas our previous energy studies of your building have used 2 watts/square foot as the lighting level. If the 4 watts/square foot of lighting is reduced to 2 watts/square foot, the existing building energy would be reduced to 17 KWH/Sq. Ft./Yr., or 58,000 BTU/Sq. Ft./Yr. Such an actual energy operating budget seems appropriate for consideration for your building as a conservative estimate on which to calculate savings. Annual energy costs (KWH and fuel costs) at current rates (DP&L Rate G) amount to $665,700.

17 KWH X 1,412,657 GSF X $.02772 = $665,700/Year

For demand use, we will use the monthly peak demand calculations from the first energy study of your building. These demand use figures are for electric cooling with centrifugal machines and for electric resistance heat.

Cooling months of Jun, Jul, Aug, Sep, Oct (DP&L Rate G)

Month	Demand	Cost (@ $7.56)
Jun	7462	$ 56,412.72
Jul	7461	56,405.16
Aug	7464	56,427.84
Sep	7001	52,927.56
Oct	7083	53,547.48
	TOTAL	$275,720.76

Heating months of Nov, Dec, Jan, Feb, Mar, Apr (DP&L Rate GH electric heating rider)

Billed demand (winter months) = (.5625 X A)+(.25 X B)

Where A = Summer peak demand (Aug = 7465)

B = Actual or recorded demand

Month	Actual Demand	Billed Demand	Cost (@ $7.56)
Nov	6907	5925	$ 44,793.00
Dec	6593	5847	44,203.32
Jan	6591	5946	44,951.76
Feb	6229	5756	43,515.36
Mar	5985	5695	43,054.20
Apr	5745	5635	42,600.60
	TOTAL		$263,118.24

The month of May had a demand of 6671		50,432.76
Yearly Demand Total		$589,271.76
Yearly Energy & Fuel Total		$665,700.00
TOTAL		$1,254,972.00

This total amounts to an annual cost of 89¢/Sq. Ft. for the 1,412,657 gross square feet.

Energy Savings

Energy Management and Control Systems in general will save 10-15% of annual electricity costs when only start/stop scheduling, optimized start/stop, duty cycling, demand monitoring and limiting, and night set-back are used. In other words, these savings result without either lighting control, chilled water reset, condenser water reset, or chiller optimization. But these are savings which normally result in existing buildings even where the building operators have been very energy conscious. These systems can improve upon normal manual operation and can save on demand charges where manual or time-clock operation can not.

	Annual	10% Saving	15% Saving
Energy and Fuel Charges	$665,700	$66,570	$99,855
Demand Charges	$589,272	$58,927	$87,941
Total	$1,254,972	$125,497	$187,796

Pay-Back

A pay-back criteria of three years would permit an economic expenditure of three times the above savings.

$$3 \times \$125,497 = \$376,491$$
$$3 \times \$187,796 = \$563,388$$

DETAILED SAVINGS CALCULATIONS

Since there are approximately 50 air handlers in your building, and since it is necessary to control each one of these for after-hours usage at nights and on week-ends, a microprocessor which can handle a maximum number of 40 loads is adequate for this building. Savings estimates which follow, will be made then only for micro-computer or computer-based systems. In each of the savings estimates which follows, the local controls as currently specified are described, followed by a discussion of these local controls and how an EMS would function to save energy. The discussion is followed by the energy savings estimated with installation of a computer-based EMS.

HEATING, VENTILATING AND AIR CONDITIONING (HVAC) AIR SIDE SYSTEMS

Start/Stop of Air Handling Units

Local Controls: A single 10-step cam action 7-day time clock on the Engineer's Central Control Panel controls office space air handling units. Ventilating units, exhaust fans, and the like are controlled manually. Four outside air supply units are interlocked with the air handling units which they serve. Overtime use is enabled by a courtesy panel for floors one through four and by manual override above the fourth floor. A second time clock is provided for chiller plant equipment.

Discussion: Time clock control is adequate for a single tenant building of smaller size with reasonably fixed operating hours. With time clocks, air handlers can be started at the appropriate time in the morning, but must run until the latest tenant leaves in the evening or must be enabled by a manual override if all are turned off at the end of the day. The manual override system does not permit the tenant to be easily billed for after hours usage, unless someone from the operating staff performs the override function. For 50 air handlers this does not seem practical from a personnel and economic viewpoint.

A microprocessor is not suited for this building because it is limited to 40 loads, whereas this building requires a capability for 50 just for the air handlers.

Energy Savings: If all fans were permitted to run until the last tenant and the janitors were completed with their work, say at 10 PM, an additional four hours of run time would be needed over what would be required if the fans could be controlled individually by an EMS. If the fans are all turned off at 6 PM and tenants and janitorial crew have access to a manual override, at least two hours of additional run time would be involved.

A saving then of 2-4 hours should be possible for the 1000 KW of fans, with electricity cost at $0.02772 and no demand saving. Since the variable air volume fans will be operating under part load conditions during these periods, a load factor of .6 will be used for them.

(2-4 hours)(5 days/week)(52 weeks/year)(1000 KW)(.60)($.02772/KWH) = $8,649 - $17,298

"Optimum" Start/Stop and "Set Up" Temperature

Local Controls: The single time clock on the Engineer's Central Control Panel can be adjusted to start units at a fixed time before occupancy to permit pre-cooling of the building. During the heating season, the fan powered mixing boxes remain on around the clock to maintain minimum occupancy temperatures.

Discussion: This program saves energy by starting the heating or cooling system only as early as is necessary to achieve desired indoor comfort conditions, with the start time based on either outside or inside temperatures.

Energy Savings: For the 1000 KW of fans, the optimum start/stop program will save an estimated one to two hours for each operating day of the year, for a savings of $4,324. - $8,648.

(1 hour)(5)(52)(1000)(.6)($.02772) = $4324

Control of Heating - Night Set-Back and Summer Lock-Out of the Heating System

Local Controls: The building is heated by electric resistance coils in the fan powered mixing boxes (FPMB). Heating is controlled in sequence by the same pneumatic thermostat system which controls the cooling. The FPMBs will cycle on at night during the heating season to maintain the temperature set point of the thermostats; and unless disabled, they will also cycle on in similar fashion during the cooling season.

Discussion: There are about 500 heating units in the building, so control of them by an EMS is costly in terms of the relays and number of points involved. But the savings are correspondingly high. The fan powered mixing boxes are circuited electrically separately from the electric heating coils. This allows both to be controlled separately.

Energy Savings: Energy can be saved through a form of night set-back and through locking-out the FPMBs during the cooling season.

Night Set-Back and Summer Lock-Out Heating Savings: Savings here can be achieved by locking-out heating above a fixed outside temperature, or inside temperatures if an adequate number of temperature sensors are provided in the EMS. An estimated 1/6 to 1/3 of required building heating can be saved in this manner. From our earlier computer building simulation, 4,439,622 KWH of heat were required for the year.

(1/6-1/3)(4,439,622)($.02772) = $20,511 - $41,022

Summer Lock-Out of the Heating System Fan Savings: The fans in the 500 heating units (295 KW) could be locked out for an estimated 200 hours during the cooling season for an annual savings of $16,355.

295 X 2000 X $.02772 = $16,355

Outside Air Control - Supply and Exhaust

Local Controls: Four outside air supply units are interlocked at the Engineer's Central Control Panel with the air handling units which they serve. Inlet vanes on each unit are controlled by static pressure. Exhaust fans are not tied in to the Engineer's Central Control Panel.

Discussion - Supply Air Fans: The supply air fans are not required except during occupied hours. The extra hours of operation during morning cool down of the building are useful only if the outside air is assisting in the cooling of the building. At least four hours per day of operating time for the supply fans can be saved.

Cooling Requirements

Wet Bulb °F*	Enthalpy BTU/#	Enthalpy To 63°WB (28.5 BTU/#)	#Hours Per Year*	BTU/# Dry Air Hours Per Year
75	38.6	10.1	54	545
74	37.7	9.2	229	2107
73	36.8	8.3	404	3353
72	35.9	7.4	566	4188
70	34.1	5.6	808	4525
68	32.4	3.9	995	3880
65	30.0	1.5	966	1449
				19,947

* Engineering Weather Data NAVFAC P-89 1 July 1978

Energy Savings

Cooling

19,947 BTU/# Hours X 103,600 CFM X

$$\frac{60\text{Min./Hr.}}{13.5\text{ CF/\#}} \times \frac{4\text{ Hrs.}}{24\text{ Hrs.}} \times \frac{5 \times 52}{365\text{ Days}}$$

   = 1,090 Million BTU/Year

$$\frac{1,090\text{ Mil.BTU/Yr.}}{12,000\text{ BTU/Ton Hr.}} \times 1\text{ KWH/Ton Hour} \times \$0.02772/\text{KWH}$$

   = $2,519/Year

Fans

4 Hours/Day X 260 Days X 103,600 CFM X

$$\frac{4.5''\text{ S.P.}}{5745} \times \$.02772 = \$2,339$$

Total Annual Savings = $4,858

Discussion - Exhaust Air Fans: Exhaust air fans normally operate 24 hours per day and usually are not controlled in any way. If controlled by the EMS they can be turned off for at least 10 hours of each operating day and 24 hours for each non-operating day.

   10 X 5 X 52 = 2600 Hours

   24 X 2 X 52 = 2496 Hours

   5096 Hours

Energy Savings

Cooling

$$\frac{5096\text{ Hours}}{8760\text{ Hours}} \times 23,500\text{ CFM of controllable exhaust}$$
X 19,947 BTU/# Dry Air X

$$\frac{60\text{ Min./Hr.}}{13.5\text{ CFM}} \quad \frac{1\text{KWH/Ton Hr.}}{12,000\text{ BTU/Ton Hr.}} =$$
   100,997 KWH/Year

Heating

23,500 CFM X 2,400 Heating Degree Days X 24 Hrs./Day
X 1.08 BTU/CFM °F Hour X

$$\frac{5096\text{ Hours}}{8760\text{ Hours}} \quad \frac{}{3413\text{ BTUH/KWH}} = 249,174\text{ KWH/Year}$$

Fans

$$\frac{23,500\text{ CFM X 1''}}{5745} \times 5096\text{ Hours} = 20,845\text{ KWH/Year}$$

Total Annual Energy Saving   371,016   KWH/Year

   371,016 X $.02772 = $10,285/Year

CHILLER CONTROL

Local Controls

Lead or first chiller with chilled water pump, condenser water pump, and cooling tower is started with a time clock. Additional chillers are started or stopped manually by the building operators. Building operators normally leave the chilled water temperature at the design setting. Condenser water temperature is controlled to the lowest permitted by the chiller manufacturer.

Discussion: The building operator will have to start and stop chillers for both day and evening use, based on the judgement of the operator. Specific starting and stopping sequences in an EMS program can save a large number of hours of chiller operation, compared to manual procedures. Many building operators begin the day by automatically starting 2 or 3 chillers at the same time on the hottest summer days and let them run all day. So the start/stop/selection procedure is important for savings.

Second, building operators normally do not change the leaving chilled water temperature of each chiller, even though that temperature is only required for the few days of the year that meet or exceed design conditions. An EMS can continually adjust the leaving water temperature, for a saving of over 1% for each degree the temperature is raised.

Third, condenser water temperature can be continually reset by an EMS in the same manner as the chilled water temperature, with a similar saving. An EMS can also reset the condenser water temperature by the ambient or ourside wet bulb temperature and cycle off tower fans as the condenser water temperature approaches the outside wet bulb temperature, saving additional fan horsepower.

Fourth, an EMS can be designed to provide demand control for chillers. In addition to the savings possible through proper chiller selection to meet the instantaneous load, as already discussed, the demand limiter on each chiller can be used to set up digital outputs for stepped load shedding of each chiller.

Automatic chiller start/stop controls are available from Carrier, Chillitrol, CESCO, Johnson, Barber-Colman, and others, in addition to being available as a part of EM systems. Some controls are limited to handling one chiller and most can handle only two. The logic options in handling more than two obviously get more detailed.

Energy Savings

Chiller Start/Stop: Experience indicates that a good building operator will start and leave on the 2nd and 3rd chiller 25% more time than is necessary with automatic controls. And from our earlier computer building simulation, we estimate that chiller #2 (1520 tons) will need to run 1500 hours and chiller #3 (760 tons) for 500 hours. The EMS would be able to save the added 25%, which we estimate would be at a minimum energy input of 20% of peak chiller input.

Chiller #2

(1520 tons)(.8 KW/Ton)(1500 Hours)(.25)(.2)

   = 91,200 KWH

CHW and CW Pumps

   74 KW X 1500 X .25   = 27,750 KWH
   118,950 KWH

Chiller #3

(760 tons)(.8 KW/Ton)(500 Hours)(.25)(.2)

   = 15,200 KWH

CHW and CW Pumps

   37 KW X 500 X .25   = 4,625 KWH
   29,825 KWH

TOTAL   =   138,775 KWH

Savings = (138,775)($.02772/KWH)   = $3,847

Chilled Water Reset: From our earlier computer building simulation we find that chillers for your building will use 2,965,715 KWH per year. For this savings calculation, we assume from experience a weighted average chilled water increase for the entire year of 6° for a 6% saving.

2,965,715 X .06 X $.02772   = $4,933

Cooling Tower: Based on experience, we assume that cooling tower fans can be cycled off when the condenser water temperature is within 3° of the outside wet bulb temperature. We estimate that this can be done 50% of the time, when chillers are less than 40% loaded.

(2000 Hours)(.5)(180 KW) X $.02772   = $4,990

Total energy savings then for chiller control are, very conservatively:

Automatic Start/Stop/Selection	$ 3,847
Chilled Water Reset	4,933
Cooling Tower	4,990
	$13,770

DEMAND MONITORING AND CONTROL

Local Controls: There is no way to provide demand monitoring and control or limiting through local controls. This function can only be provided through a microprocessor or computer-based EMS.

Discussion: Peak demands may be classified as being of three general types:

Morning start-up; Monday morning is usually the most pronounced, if the building cooling system has been shut down over the week-end.

Daytime peaks

Random peaks

The manner in which these may appear on a graphical demand meter are shown in the example shown on the next page.

Morning Start-Up: The only effective way of controlling this type of peak is through use of some type of chiller control system which provides for a soft start. Among these types, of course, if that provided by a computer-based EMS.

Daytime Peaks: These peaks are caused by the natural simultaneous solar, transmission, outside air and internal load peaks and may last for several hours. In large office buildings, these peaks can be controlled only by limiting chiller loading and allowing some drift upward in space temperature.

It should be recognized that when a cooling system is designed and sized based on 1% design weather conditions, this means that, on the average, there will be only 30 hours during the year that are above these conditions. For the Dallas area the 1% summer design conditions include a 102°F dry bulb temperature with a mean coincident wet bulb temperature of 75°F.

Random Peaks: These peaks occur in most buildings and are caused by simultaneous operation of miscellaneous equipment: hot water heaters, fans, elevators, and the like. These peaks can be controlled by an EMS with demand monitoring and load shedding. Often a load will have to be shed for only one 15 minute demand interval. If the load to

be shed is one step on a chiller, there will not be any discernable change in comfort conditions for such a short period of time.

All the earlier energy savings calculations were based on current 1981 rates for energy (KWH) and fuel (per KWH), plus 5% sales tax. The figure of $.02772/KWH includes no demand increment.

Summer Demand Savings

The loads that would be shed for summer demand limiting are the small chiller, one step of 20% on one of the two large chillers, and the electric domestic hot water heaters.

small chiller 608 KW + 37 KW of Pumps =	645 KW
large chiller (20%) 1216 KW X .2 =	243 KW
hot water heaters 96 KW X .5 =	48 KW
(diversity of .5)	936 KW

With proper setting of demand targets for each month, it should be possible to save 936 KW for each of the summer months (5) and May.

936 X 6 X $7.56 = $42,457

And since the mathematics of the rate structure provide a minimum bill or ratchet, there is also a saving for the six winter months.

Winter billing demand = .5625 summer peak + .25 actual winter

Thus, .5625 X 936 or 526.5 KW of demand costs are avoided for the winter months.

526.5 KW X 6 X $7.56 = $23,882

A reduction of 936 KW for each of the five summer cooling months plus May, results in an annual saving of $66,339.

Winter Demand Savings

The heating rider GH is in effect for this electrically heated building. The saving of demand costs is accordingly less than in the summer months, since the actual demand for the heating months is reduced. From the formula above, we see that winter billing demand is reduced only by an amount equal to 1/4 or .25 of each KW of actual demand for the billing period.

The loads which would be shed for the winter months include 40% of the small chiller (or the equivalent if a large one is operating), the electric hot water heaters, and 10 floors of electric heat for 12 minutes each hour.

small chiller (40%) 608 X .4 =	243 KW
hot water heaters 96 X .5 =	48 KW
10 floors @ 55 KW each =	550 KW
TOTAL =	841 KW

841 KW X .25 X 6 X $7.56   = $9,537

Total annual demand cost savings then amount to $75,876.

ELECTRIC DOMESTIC HOT WATER HEATERS

Local Controls: The sixteen electric hot water heaters (52 gallons and 6 KW each) and sixteen 1/12 HP circulating pumps make up the hot water system. Each normally cycles on as required to maintain a set leaving water temperature, and the circulating pump runs continuously unless turned off.

Discussion: Controlling these water heaters for energy savings alone is not economical. But when demand savings, discussed earlier, are included, such control does become feasible.

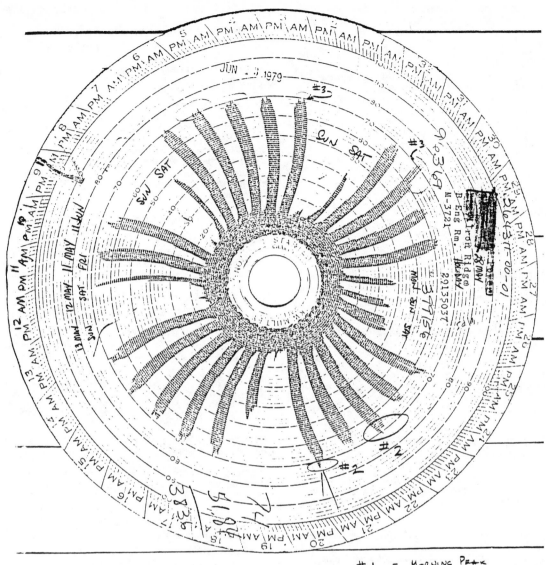

#1 = MORNING PEAK
#2 = DAYTIME PEAK LOADS
#3 = RANDOM PEAK

LIGHTING CONTROL

Local Controls: The basic design involves a modular wiring system with local switching in individual offices and floor control, if desired, using the regular circuit breakers.

A bid alternate provided low voltage relays for each pair of lighting circuits (one per 1000 square feet of floor space, approximately) in the event connection with an energy management system would be required. The contractor priced this alternate at $166,170.

Discussion: Ideally, lighting circuits should be:

- "enabled" just prior to the first occupant arriving.
- switched on in each individual office or space by occupant of that office when and if they arrive for work.
- switched off in each individual office or space by the occupant at any time during the day that the occupant(s) leave.
- "disabled" as soon as all occupants have departed.

- "enabled" a floor at a time by the night-time cleaning crew when they begin cleaning that floor.
- "disabled" by the night-time cleaning crew when they complete their cleaning on a floor.

With currently provided local controls, those actions entitled "enable" and "disable" involve the switching of floor circuit breakers in the electrical closet on each floor. Switching of these circuit breakers on 50 floors is obviously a time consuming task, even if only two trips a day by building operators or security personnel are necessary.

The provision of local switching provides added energy savings, particularly in perimeter offices and unoccupied offices. A recent article in the Illuminating Engineering Society (IES) magazine stated that energy savings in lighting of as much as 40% and demand savings as much as 15% resulted when local switching was added where only floor circuit breakers had existed. Thus, any good energy saving lighting system should provide local switching to enable the occupants to effect these savings.

With the bid alternate low voltage relays, a computer based EMS can perform those functions above defined as "enable" and "disable".

Energy Savings

With the ability to "enable" a floor remotely through the EMS instead of having a building security employee do so at the floor circuit breaker, a major saving is possible in hours or use. Whereas the hours of use with circuit breaker control may be 7 AM - 9 PM, with EMS remote control, it should be possible to reduce the hours of use of "on time" to 7 AM - 6 PM or 7 AM - 7 PM. Anyone needing light after 6 PM can call the central after-hours security office on the first floor and have the lights for that floor "enabled" through the EMS for a specified period.

Such a 3 hour saving for the 2560 KW of lighting in the building would save

$$3 \text{ hours X 5 days/week X 52 weeks/year X 2560 KW}$$
$$= 1,996,800 \text{ KWH}$$

When lighting loads are reduced, cooling loads are also reduced, and heating requirements are increased since the heat of light helps to heat the building.

The added heating requirement will be that heat of light which must be replaced by resistance heat for the heating season of 15 weeks.

$$3 \text{ X } 5 \text{ X } 15 \text{ X } 2560 = 576,000 \text{ KWH}$$

The added cooling saving will result from the reduced heat of light which does not have to be removed for the cooling season of 35 weeks.

$$\frac{3 \text{ X } 5 \text{ X } 35 \text{ X } 2560}{12,000 \text{ BTU/Ton-Hour}} \text{ X } 3413 \text{ BTU/KWH}$$
$$\text{X } 1.0 \text{ KWH/Ton-Hour } = 382,256 \text{ KWH}$$

The net energy saving is 1,803,056 KWH, which at a cost of \$.02772 is equivalent to \$49,981 per year. If the daily saving amounted to only 2 hours, the annual saving would be \$33,320.

Payback

If the 50 points for lighting control cost \$500 per point, the total cost of lighting control will be \$191,170.

\$500 X 50	=	\$ 25,000
Bid Alternate	=	166,170
		\$191,170

The payback is less than four years if 3 hours of lighting use are saved each day, and over 5 years if only 2 hours are saved.

Future Savings

If the bid alternate relays are purchased now with single point control (one digital output per floor), there are future potential savings to be had.

Each of the 50 floors is now designed for about 6 circuits for the common core areas and 36 circuits for the tenant areas. Let us assume that each floor of your building will average 3 hours of after-hours light use per day, or 18 hours per week, and that only two of the 36 tenant circuits are required for that use. Each circuit averages about 1.5 KW.

If the EMS is converted at a later date by installing a multiplexer on each floor so that each relay (2 circuits) may be controlled individually, only 2 circuits would need to be "enabled" after hours, instead of 36. The energy for 34 circuits would be saved, for an annual saving of \$66,162.

$$34 \text{ X 1.5 X 18 hours/week X 52 weeks X 50 floors}$$
$$= 2,386,800 \text{ KWH}$$

At \$.02772/KWH, this is an additional saving each year of \$66,162.

Most building operators bill tenants for after-hours lighting use based on the lease agreement, and most base those billings on a record kept at the security desk on the first floor. If the relays which are installed have an extra pair of contacts, these can be used to return status to the CRT of the EMS; that is, whether the circuit is "on" or "off". The calculated point program of the computer can then be used to accumulate after-hours use, calculate, and print out a bill for the tenant.

SUMMARY OF ENERGY SAVINGS

The individual savings computations in this section were based on computing savings based on estimated reduced "run time" or "on time" and on reduced outputs. These savings are each listed below, together with the total.

KWH and fuel savings	\$78,752 - \$112,236 per year
Demand Savings	\$75,876 per year

These figures can be compared with the 10-15% savings of projected annual energy use discussed at the beginning of this study.

	10%	15%
KWH and fuel savings	\$ 66,570	\$ 99,855
Demand Savings	58,927	87,941
TOTAL	\$125,497	\$187,796

It can be seen that the sum of the detailed savings, excluding lighting, falls generally within the range of 10-15%.

If lighting control is included as a part of the EMS, additional annual savings could range from \$33,320 to \$49,981.

	Energy and Fuel Savings	Demand Savings
1. HVAC Air-Side		
Start/Stop	\$8,649 - \$17,298	
Optimum Start/Stop	\$4,324 - \$8,648	
Night Set-Back and Summer Lock-Out		
Heat	\$20,511 - \$41,022	
Fans	\$16,355 - \$16,355	
Outside Air		
Supply Air Fans	\$4,858 - \$4,858	
Exhaust Air Fans	\$10,285 - \$10,285	
2. Chiller Control		
Start/Stop	\$3,847 - \$3,847	
Chilled Water Reset	\$4,933 - \$4,933	
Cooling Tower	\$4,990 - \$4,990	
3. Demand Monitoring and Control		
Summer Demand		\$66,339
Winter Demand		9,537
TOTALS	\$78,752 - \$112,236	\$75,876

Total Electricity Cost Savings (KWH, fuel, demand, tax)
\$154,628 - \$188,112

4. Lighting Control	
one point per floor	\$33,320 - \$49,981

Chapter 75

THE AUTOMATED INTEGRATED DATA (AID) CONCEPT

N. R. Gifford

ABSTRACT

Today's modern large-scale or high-rise building contains various control and monitoring packages which are associated with the building's systems such as elevators, life-safety, and HVAC.

Although the trend has been toward greater sophistication in technology for the various systems, it has only been through the advent of the Division 17 Specification that there has been an attempt to integrate some of these elements.

This paper addresses the need for the integration of these systems but not in the traditional manner which has been attempted to date. Due to the greater intelligence associated with each individual control and monitoring system, the alternative to having a single computer responsible for the various building elements is to create an Automated Integrated Data (AID) Center which can be utilized by all the departments responsible for each building system.

The goals of the AID concept are to: establish a common technology level for all hardware and software for the building's systems; provide a communications network for all these systems; and implement an operational concept which can reduce costs, improve efficiency, and address the building as a complete entity for any given situation.

BACKGROUND

Integrating building systems evolved with the addition of Division 17 where energy management, life-safety and automatic temperature controls were specified as a single bid package. A number of problems evolved with this concept which have caused mixed reactions from the various industries affected.

Problems associated with utilizing a single computer and a single point of control raised many operational considerations from the life-safety industry. Additionally, questions arose regarding reconfiguring the central computer software and the amount of additional effort it took to get these large complex systems fully operational. Some integrated installations were never successfully completed and were eventually replaced with smaller stand-alone systems.

But the integrated systems did provide what was perceived as an advantage to the owner by minimizing the number of bidding vendors and/or contractors and gave them a single source of responsibility with common hardware and software. The key word is "perceived" because the situation did not always work to the owner's advantage mainly because of hardware and software problems. But the concept of common hardware, common software, single source responsibility and the ability to communicate with multiple building systems to effect control without redundant hardware is still a valid concept which would be a benefit to the owner.

DIVISION 17 ALTERNATIVE

If disagreement as to the validity of single computer responsibility with the advantage of centralized monitoring and ease of operation continues, perhaps there is now an alternative which eliminates the jeopardies and provides even greater functional capabilities. For example, an Automated Integrated Data (AID) Center could be the solution which was not available 3 to 5 years ago. If so, how does the AID concept differ from the integrated system concept and how would the jeopardies be minimized or eliminated?

The hardware and software architecture is the biggest difference between the AID concept and the integrated system. All of the various control and monitoring packages throughout the building would be stand-alone intelligent "sub-systems" which would be data linked to the AID Center. Consequently, the AID Center is no longer a central controlling computer but is the data manager for all the systems in the building.

Before expanding on the AID concept, it is necessary to reevaluate the type of building control and monitoring system which could be integrated into the concept. Additionally, the benefit to the user for integrating each system requires discussion.

BUILDING SUB-SYSTEMS

Heating, Ventillating, Air Conditioning (HVAC)

The Variable Air Volume (VAV) systems being designed into today's buildings are attempting to provide the greatest energy and operational efficiency.

In a high-rise building, the VAV system is being designed around a central air shaft concept which runs vertically through the core of the building. The exact number of shafts depends upon the VAV design but there are a number of different configurations. Common to all these designs are very large fans which can range from 150 to 300 horsepower with CFM ratings up to 200,000 or more. The fans typically are vane axials set up in tandem or centrifugals with vortex dampers, but, in either design we are concerned with the importance of maintaining, monitoring, and controlling these systems.

The importance of monitoring these systems cannot be underestimated and should include various points such as: motor current, phase loss, static pressure, coil temperatures, damper positions, fan run times, and all critical points in the associated mechanical systems such as the chilled water plant.

Downstream of the central fan system are the floor VAV control systems which are becoming more sophisticated in design and control aspects. They range from a simple modulating damper to control air flow to terminal units which may have reheat coils, fans and dampers. Additionally, floor dampers are being incorporated for air control on a floor-by-floor basis for applications such as life-safety smoke control, or occupied/unoccupied air control.

The particular system being utilized will dictate the control and monitoring aspects to be incorporated into the AID network. Typically, floor damper control, floor thermostat zone control, and reheat coils would be included in the design. The greater the degree of monitoring and control dictates how much could be saved in energy, operational costs, and maintenance costs which, over the life of the system, could be greater than energy savings alone. In any event, the need for data analysis is very real and reinforces the need for a sophisticated automation system.

Automatic Temperature Controls

Associated with the HVAC package is the ATC sub-system or the automatic temperature controls. The state-of-the-art is now moving toward solid state direct digital control and away from the traditional pneumatic system for many good reasons.

First, the DDC system offers greater energy savings potential through software based control loops which can hold temperatures to close tolerances. Second, the maintenance associated with a DDC system is minimal compared to a pneumatic system. This results in material and labor cost savings to the owner. Third, the DDC system offers greater monitoring capabilities as an integral part of the control system as opposed to a basic monitoring system.

The DDC sub-system incorporates distributed intelligence in utilizing microprocessor based stand-alone units which are data linked to the AID Center. It is through the DDC/AID network that operations personnel will have complete interaction capabilities of not only the HVAC package, but all the ATC associated with it.

Part of the new technology currently being explored is solid state control of the floor VAV units in an attempt to gain tighter control of the entire building. The control capabilities being investigated include utilizing telephone wiring to transmit temperatures back to the DDC system via digital messages over the phone circuits. Control of the floor VAV unit would be affected through its factory mounted solid state control box instead of the traditional pneumatic sensor.

The DDC technology and the ability to integrate the DDC sub-system into the AID network is available today. The cost of the data integration capability is an integral part of the DDC. A higher initial cost will be realized with the DDC systems over the traditional pneumatics, but the life cycle cost of the entire package will show a favorable payback through a combination of energy, material and labor savings.

Elevators

Elevators are one of the major building systems to be integrated into an AID concept. They are another example of the development of microprocessor technology in control and monitoring packages. Elevators should be monitored for proper operation during normal and emergency situations.

During normal operations it is important to both the building occupants and the owner to have elevators positioned on the proper floors to service the traffic during the day. From the AID Center via the technology of color graphics display, the elevators can be monitored for proper position, speed of travel and mode of operation. From an operational viewpoint, it is important to immediately respond if an elevator malfunctions, if it stops between floors or if it fails to close or open its doors. Through a talk/listen circuit incorporated into the elevator microprocessor control system, communications from the AID Center could immediately assure the elevator occupants that the emergency is being handled and that they are not in danger.

Additionally, during normal operations, other elevator functions could be monitored and problems diagnosed. Information could be utilized for maintenance, load balancing, or even energy management in relation to electrical demand in the building.

During emergency situations associated with the life-safety system, elevator monitoring becomes even more critical. Codes define a two-phase operation for emergency operation of elevators. First, the elevators must be taken out of normal service and returned to the lobby or some other floor. Secondly, firemen can take over complete manual control of the elevators to evacuate occupants from the building and shuttle additional firemen to emergency areas.

Through the color graphics displays, the AID Center will give firemen invaluable information on the relationship of the emergency zones in the building to the elevators. If the situation changes from the time the elevator leaves the lobby until it arrives at the designated floor, that information can be given to the fireman in the cab before he gets there.

This technology is available today and there is little premium that must be paid to integrate the capabilities into the AID concept.

Life-Safety

The new Life-Safety Codes have caused dramatic changes in the application of life-safety systems. These changes have resulted in higher costs for these systems.

The emergence of the microprocessor based multiplexed fire alarm systems have been a major step forward in hardware development. Where the original hardwired fire alarm systems required expensive wiring runs for all circuits, the multiplexed systems have digitized the signals over communications lines which are installed at a fraction of the cost.

As part of the new code requirements, voice communication is now part of any high-rise building. Voice communication, in fact, has replaced bells and horns with other types of signals including types of pre-recorded messages for directing occupants during an emergency situation. But, the voice package system is an expensive feature of the life-safety system. In certain parts of the country, local codes still require that the voice package be hardwired rather than take advantage of multiplexing technology. Multiplex systems can fully digitize all life safety signalling in compliance with all codes and offer effective alternatives to the owner.

Technology of tomorrow, which is available today, could put an emergency voice communications station on each occupant's desk instead of scattered throughout the building. Telephone systems, which supervise

490

communication lines within the building, could be utilized with the life-safety system. The telephone, with a speaker in each desk-set, could be used for emergency messages just as the life-safety system does. But even more importantly, they could be "talk/listen" sets which could give firemen the ability to listen for occupants on the floor before having to dispatch a fireman from the lobby.

The integrity and supervision of the life-safety system can be maintained while eliminating the expensive wiring required in the building. The telephone/life safety system combination can also provide the maximum in protection and flexibility with no additional installation costs. This is especially beneficial to high-rise buildings with frequently changing tenant requirements.

Another technological feature, available today, is the application of color graphics display for life-safety functions. By integrating a microprocessor-based life-safety system, with or without the telephone network, the AID Center will generate color graphic displays of the building riser, floor plans, evacuation routes or any other graphics needed by firemen. Once integrated into the AID Center, the firemen could take advantage of other graphics from the elevators and HVAC systems to correlate the relationship of the emergency situation and to quickly determine actions to be taken. As the conditions in the building change, the graphics dynamically change also. This will give firemen a pictorial or graphic means of tracking the smoke or fire in the building.

All the technology that has been addressed is readily available and the cost factors for these capabilities are basically inherent in each control package.

Data Communications

Once the stand-alone sub-systems are in place in the building, the data link to integrate all the sub-systems into the Center needs to be established.

Typically, multi-conductor cables are pulled through the building for any type of communication. This is extremely expensive and, in many cases, makes a centralized monitoring budget so high that it is deleted from the project. By using the AID concept, however, greater capability can be provided within the communications link. By employing low cost multiplexers, for example, the relatively low data rate signals from many sub-systems can often be economically combined into a single data stream and transmitted over single pair wiring at significant cost savings. Of even greater potential, is the integration of building control signals with other communications traffic in the building. This is particularly true as digital data traffic in general becomes more commonplace and as communications, such as voice telephone, becomes digital. In this case, not only can redundant wiring be eliminated but multiplexers can be shared resulting in dramatic savings.

The transmission medium over which the data travels is also under development. A few years ago, the emphasis was toward carrier current signal transmission over the A.C. voltage lines in the building. Carrier current signaling has not been developed to accommodate high data transmission rates and does not have the capability of multiplexing multiple data messages.

There is, however, a new thrust toward fiber optic data links which offer tremendous potential in high-rise buildings. A Fiber Optic Nerve rising through the core of the building has the capability of providing the transmission medium to all building systems

and additional data for other building elements. The Optic Nerve, with its associated hardware can transmit multiplexed digital signals at the speed of light. This provides the capability to handle a far greater amount of data messages simultaneously.

The economics of fiber optics is becoming more attractive to the developer for a number of reasons. Although the material costs are initially high, the labor and materials to install the optics is less than traditional wiring. Owner's can offer their tenants greater savings through additional hardware capabilities linked through the optic nerve. Telephone systems which require a great deal of wiring space, vertically and horizontally in the building, can be data linked through optics to reduce these space requirements.

THE AID CONCEPT

The following is a review of the basic premises for the AID Center.

First, a common technology level of hardware and software capability has been established. All sub-systems in the building are solid-state and microprocessor-based and must have the capability to transmit data to the AID Center. This provides the user with the latest state-of-the-art systems, common architecture and low maintenance associated with each system.

The integrity of each system is maintained because each is a completely intelligent stand-alone system. A means is provided by which the user can monitor and control from a central point through a communications link established to provide data management to all building systems. New technology will continue to decrease costs and increase capabilities associated with additional data systems in the building.

New technology also is the key to providing greater operational potential for the user from and through the AID Center. The AID concept will provide the user with ways to address situations practically anywhere in the building. A more efficient building can be realized from an energy utilization and manpower requirement standpoint. A safer building is provided because of greater emergency capabilities which are all monitored through the AID Center.

Through higher technology, the AID Center can provide the best of both worlds: greater reliability through distributed intelligence and centralized monitoring and reporting through data integration.

Chapter 76

IMPROVING HVAC SERVICE THROUGH REMOTE MONITORING OF ENERGY MANAGEMENT SYSTEMS

J. P. Kettler

HVAC service contractors are becoming more knowledgeable in EMS. They have to be to survive in today's market! However, when called in by a building owner to survey his system and recommend EMS, these contractors often overlook features of EMS systems that can make life easier for both the contractor and their customers.

In the same way, an HVAC system improperly maintained can waste money, an energy management system improperly programmed or maintained, is of little value.

We are all familiar with the simplest of energy management controllers, the simple seven-day time clock. And, anyone familiar with commercial HVAC jobs can cite instances of seeing these seven-day time clocks on jobs with all the trippers removed so the systems run 24 hours a day, 7 days a week, totally bypassing the device installed to save energy. The very same thing is happening to many of the sophisticated energy management systems being installed. Why? Because energy management systems are not something to be installed, turned on and forgotten. Most energy management controls must be programmed and "tuned" to match the characteristics of a building. This is especially true as systems become more sophisticated. However, if the operator is not totally familiar with the energy management system and its interface with the HVAC system; When in doubt, the easiest thing to do is bypass the automatic control system.

There are basically four levels of energy management systems:

1. Time clocks.
2. Duty cycling and demand limiting devices.
3. Combinations of 1 and 2, plus remote analog read and reset capability.
4. Combination of 1, 2, and 3, plus system color graphics and computer algorithms for chiller/boiler optimization, etc.

This paper is not going to go into the individual payback periods and energy saving comparison of the various EMS products because this varies from system to system. Rather, it is intended to emphasize the benefits remote monitoring can offer in addition to start-stop, load shedding, etc., functions that most energy management systems provide. These benefits being the potential to reduce service costs, increase tenant satisfaction, provide more effective troubleshooting and reduce a contractor's risk in providing full service contracts.

If a contractor just listens to the first appeal of an owner to provide "something" that reduces energy consumption and costs almost nothing, the knowledgeable contractor begins to think "time clock." This includes some of the very sophisticated clocking devices that include a full year's program including holiday and daylight saving scheduling for a year or more.

A step up in clocking functions may include features to provide optimal start. This is the ability to delay the HVAC system start-up on mild days, and start systems earlier as outdoor temperatures and, therefore, building mass temperature varies further from the neutral or desired operating temperature.

Depending on the building and its occupancy, the clocking function devices, or an EMS control that incorporates clocking function plus demand limiting and/or duty cycling may be the ideal device. However, experiences of system owners, and this was pointed out in an article in a recent issue of "Energy User News," shows that a major error can be made by investing in a system that does not include monitoring capability.

Monitoring permits the owner to keep closer tabs on the operation of his building than he could with a basic on-off/alarm type energy management system. It permits a knowledgeable operator to determine if the HVAC system is indeed operating properly.

For example, an energy management system may be installed to optimally start and stop an air handling unit. The energy management controls may start the equipment at the ideal time to get the building up to temperature, and may energize the outdoor air damper to provide ventilation perhaps one half hour after the building is occupied in the morning. The damper may open further to provide free cooling if outdoor air conditions are below 75°F. But, what happens if the controls fail and that damper remains open when the outdoor temperature reaches 90 degrees? Depending on design conditions and the installed cooling capacity, the owner may never know it. Depending on the type of HVAC service the owner has, a serviceman, if he is thorough, may find it the next time periodic maintenance is performed on the building's mechanical system. Depending on outdoor conditions at the time the next service is performed, the problem may go undetected. In the meantime, the owner has been paying for unnecessary cooling of outdoor air whenever the temperature is above 75°F. However, if the energy management system offered monitoring capability in addition to the optimal start function, a knowledgeable operator would be able to determine that the above controls were faulty without actually going out and inspecting the air handling unit. A comparison of outdoor air, return air, and mixed air temperatures will tell an operator the exact position of these dampers. The operator can do this without climbing over ductwork, inserting thermometers at various points, waiting for them to settle out, etc. The key factor here is that the operator must be knowledgeable in HVAC systems, and most commercial buildings today do not have that type of expertise on their staff. This can be corrected, and building owners can effectively have this type of expertise available to them if they have their energy management system remotely monitored by a knowledgeable service contractor. Energy management systems are available today that can be operated as stand alone or "remote monitored" systems. That is, the unit can be operated and monitored on-site by a knowledgeable operator, but it can also be monitored remotely if a knowledgeable operator is not on duty. Remotely monitored EMS controllers vary from simple devices that strictly perform time clock functions and totalize energy consumption to complex systems that permit remote read and reset of analog points. It is the ability to read and reset analog points that can make a major impact on system operation.

An example of the latter type of system would have a micro processor based computer and video screen on-site. This is the operating system with a menu driven program that permits an operator with no computer knowledge to enter desired start times and operating temperatures or alarm limits. When an operating parameter exceeds the preset limit, the alarm condition is indicated at the central console. This system comes equipped with a telephone modem so that it can be monitored by a remote "Service Computer." Typically, the Service Computer would be located in a servicing contractor's office. This Service Computer would automatically call various "on-site" computers at scheduled intervals, perhaps hourly, to determine their operating status. If all control points are within preprogrammed limits, the "Service Computer" proceeds to call the next "on-site" computer. If any point is in an alarm condition, the alarm point and time will be printed out at the "Service Computer." This permits the contractor to immediately respond to a trouble situation. However, rather than just send a serviceman to check the point that initiated the alarm, the Service Computer allows the service contractor to analyze the situation from his shop. For example, if the alarm indicates a high space temperature, the contractor, through the Service Computer, can monitor the temperature of the air being supplied to the space. If the supply air temperature is high, he may remotely check and reset the chilled water temperature, for example, before sending a serviceman to the site. When he does send a serviceman, the serviceman will be more efficient because the basic "troubleshooting" analysis will have already been done and the serviceman can go right to the source of the problem, or at least have a list of specific items to check rather than just respond to a complaint that "it is too warm in the west wing."

If the service contractor does not "man" the remote Service Computer 24 hours a day, it does not mean that his customer's buildings are not monitored 24 hours a day. The Service Computer can be programmed so that during off hours, on receipt of information that certain critical alarm points have exceeded their limits, the Service Computer will start calling either servicemen's homes, or an answering service and relay a message through a computer voice synthesis system that an alarm condition exists. For example, when the telephone is answered, the Service Computer will relay the message, "Building X, Point Y, is above high limit."

In an actual case where a contractor is providing this type of service to a developer with multiple tenants in multiple commercial buildings, this type system has provided additional benefits in customer and tenant satisfaction. For example: Prior to installing this type of system, the contractor was providing full HVAC maintenance to the buildings. This full maintenance contract included monthly inspection of the HVAC

systems. In the event of a system malfunction that resulted in overheating or subcooling of spaces, the tenant would call the developer who in turn called the contractor to correct the situation. Since installing the system, the contractor usually receives the alarm indication before the tenant is fully aware there is a problem. The contractor then analyzes the situation, takes whatever action he feels is necessary, and informs the developer that the alarm was received and what action is being taken. Then, if the tenant calls, the owner can inform the tenant that they are aware of the situation and corrective action is being taken. In a specific instance, an alarm was received that the suction pressure on a compressor was too low. By remotely reading suction and head pressure, current draw on the motor, and discharge air temperature, it became apparent to the contractor the unit had lost its refrigerant charge. A serviceman with refrigerant was at the job site before the tenant was aware there was a problem. Because the contractor can observe the operation of the building from his shop, general on-site inspections are now performed every other month rather than monthly. In spite of the reduction in monthly inspections, the contractor feels he has a better chance of preventing catastrophic failure of equipment now than he did before the monitoring system was installed.

This same contractor remotely monitors other commercial buildings in which he does not have full service contracts. In these cases, the owner operates the building systems from his "on-site" computer, and calls the contractor when specific service is required. However, the contractor has the ability to remotely monitor the on-site computer, and with his HVAC expertise will make recommendations to the owner on changes that could be made to upgrade the system to improve its operation.

Remote monitoring offers the owner capability to improve tenant satisfaction, extend the life of his building's mechanical systems, and reduce energy consumption. It offers the service contractor the opportunity to offer improved customer service and increase the efficiency of his employees.

As an example of some points that should be monitored in a building and how the information gained from monitoring these points can be used, let's take a look at a typical 50 ton air conditioning system.

CONDITIONS TO BE MONITORED:

A. Refrigeration System:

1. Head Pressure:

Continued excessive head pressure can be caused by a restriction in a refrigerant line, dirty condenser fan motor or belt failure, or other reasons. Lowering head pressure extends the life of the equipment and increases efficiency.

2. Suction Pressure:

Low suction pressure can be caused by a loss of refrigerant, bad expansion valve, dirty evaporative coil, failure of compressor unloaders, or other reasons. Increasing suction pressure reduces chances of compressor failure due to liquid slugging or poor oil circulation. High suction pressure can be caused by a bad expansion valve or bad compressor valves. Correcting the expansion valve problem increases efficiency and prevents winding failure due to insufficient cooling. Replacing compressor valves increases capacity and efficiency and prevents having to buy a replacement compressor.

B. Air Handling Unit Measurements:

Air temperatures (outside, return, mixed, and discharge) discharge velocity.

1. Excessive outside air increases energy consumption and affects total system capacity.

2. Low airflow across the coil could cause freeze up and decreases efficiency of the refrigeration system.

3. Temperatures drop across the coil and CFM shows total heating capacity of the unit. This can be used to determine efficiency of the unit. This can quickly provide information for the serviceman since a determination can be made remotely whether the problem is with the central unit or at the zone control. Service time will be reduced.

4. No airflow can be used as an alarm indication when there is a call for the unit to be on. This indication is more precise than an auxiliary contact in a starter.

5. Measuring all conditions at the AHU before sending a serviceman allows better instructions to be written for the serviceman and reduces callback. The overall effect is that it saves time for the building manager and improves tenant relations, resulting in renewed leases.

C. Equipment Voltage and Amperage:

1. Excessive amperage draw on the compressor motor could be caused by shorting in windings, moisture in the refrigerant, or excessive head pressure, or slugging of liquid on the suction side. All of these can lead to early motor burnout and lower the optimum operating efficiency.

2. Amp draw on the fan motor but no airflow could be an indication of broken belts.

3. Monitoring voltage on all phases would point out low voltage or imbalance between phases, which could cause motor windings to fail.

4. Monitoring volts and amps combined with run time permits tabulation and accumulation of KWH consumption of a piece of equipment on a monthly basis.

5. Totalizing compressor operating hours and monitoring changes in performance permits better scheduling of routine maintenance and overhaul services. This may well mean the difference between rebuilding and replacing a compressor due to failure.

Is remote monitoring capability cost effective?

We understand how this remote monitoring capability can be desirable, but can it be cost effective? Each case must be analyzed separately, however, as stated at the beginning of this article, the possibility of including this feature should never be overlooked when considering an energy management system.

The following are some incremental savings and costs for monitoring some of the functions in the previous example. Realize that these are incremental costs only - and are used to demonstrate the thought process to analyze what can be gained from remote monitoring rather than show total system costs. The base on-site computer may have installed costs from $15,000 up, depending on capacity and function. Payback analysis of energy management systems should be performed for a specific project rather than trying to compare one project to another because buildings, their systems, and operation are all different, and seemingly small differences can have major effects on payback. Additionally, each owner must place his own value on occupant satisfaction and comfort.

Therefore, the following is offered as an example only. It is based on numbers used by a contractor who has experience with mechanical service and monitoring of EMS installations, and relates to the 50 ton system previously discussed.

ITEM	FUNCTION/DEVICE MONITORED	RESULTS	YEARLY SAVINGS (EXAMPLE) * **	INCREMENTAL COST (MAT'L & LABOR) ADDITION *
A	Suction Pressure Head Pressure	Extended Equipment Life & Efficiency	(1) Life 8 yrs. vs. 6 = $417.00 (2) Efficiency = $594.00	(3) $ 830.00
B	AHU Temperature (ODA, MA, R.A. Disc. A) + Air Velocity	Assume 5% ODA Reduction Thru "Fine Tuning." Reduced Time Troubleshooting & Callback	(4) Energy Savings = $626.00 (5) Reduced Service = $200.00 (6) Reduced Callback = $100.00	(7) $1,680.00
C	Amp & Volts	Extend Equipment Life & Improve Efficiency	(8) Extend Life = $ 60.00 (9) Increase Efficiency = $ 40.00 (10) Reduce Load = $225.00 (11) Rebuild vs. Replace = $375.00 (12) Reduce Service Time = $100.00	(13) $2,120.00

CHART 1. INCREMENTAL SAVINGS AND COSTS FOR REMOTE MONITORING 50 TON SYSTEM

* SEE DETAILED CALCULATIONS AT THE END OF THIS ARTICLE.
** NOTE THAT THESE SAVINGS ARE IN ADDITION TO ENERGY COST SAVINGS EXPECTED FROM OPTIMAL START, DEMAND LIMITING, ETC., FUNCTIONS OF THE ENERGY MANAGEMENT SYSTEM.

Chart 1 is only a partial listing to demonstrate added savings and benefits remote monitoring of energy management systems can offer both contractor and owner. After establishing the points to be monitored and normal setting and alarm limits, remote monitoring makes "fine tuning" of a system easier. For example, if monitoring areas served by a variable air volume system on a warm summer day indicates that all areas can be satisfied, the system static and/or fan speed should be reduced to determine if additional energy can be saved without affecting occupant comfort. Monitoring the cooling equipment and space temperatures on a design day in summer will indicate whether the cooling equipment is oversized. If it is, consideration should be given to permanently unloading part of the machine to reduce electrical demand on start-up.

There are but a few of the simple checks that can be done to fine tune a system to match a building.

The point is that an energy management system is a tool to be continually used and monitored by a knowledgeable operator. If installed and forgotten, or turned over to an operator that does not fully understand HVAC systems, the systems soon become useless. If a knowledgeable operator is not available at the building site, an energy management system designed for remote monitoring should be selected. For an owner, the benefits of remote monitoring are improved service and tenant satisfaction and lower energy bills. To the service contractor, remote monitoring means more efficient use of servicemen and equipment, reduced risk in providing full maintenance service contracts, and an opportunity to upgrade the customer's system.

Calculations and assumptions used for energy savings and system cost in Chart 1. These should be modified to suit each specific job.

1. Assumed replacement cost $10,000. Life extended from 6 years to 8 years.
   $(1/6 - 1/8) \, 10,000 = \$417$ savings.

2. Assumptions: 6¢/KWH, 1.1 KW/ton, 1200 FLH/year, 50 PSIG average lower head pressure resulting in a 15% decrease in operating cost.
   50 ton x 1.1 KW/ton x 1200 hrs. x 0.06/KWH x 0.15 incr. effic. = $594.00

3. Sensor cost 2 @ $120 = $240.00
   Misc Mat'l.   = 40.00
   Labor 6 MH @ $25.00/hr. = 150.00 $830.00
   Monitoring Panel   = 400.00

4. Assumed 5% reduction in outdoor air, 1000 FLH cooling, 1000 FLH heating.
   Summer savings = 50 ton x 450 CFM/ton, x 1000 hrs. x .57 Btu/CFM x .06 KWH x 8.33 x 10^{-5} ton/Btu x 1.1 KW/ton x .05 = $353.00
   Winter savings = 50 ton x 450 CFM/ton x 1000 hrs. x 1.08 x 75°F x $3.00/10^6$ x .05 = $273.00

5. Reduced on-site troubleshooting 8 hr./year x $25.00/ hr. = $200.00

6. Callback reduced 4 hrs. x $25.00/hr. = $100.00

7. 4 sensors @ $60, 1 @ $200 = $440.00
   Misc. Mat'l   = 40.00
   Labor 16 hrs. @ $25.00 = 400.00
   Monitor Panel   = 800.00

8. $3,000.00 cost to rewind motor, 2% increased life
   .02 x $3000 = $60.00

9. 1% increased efficiency. 50 ton x 1.1 KW/ton x 1200 hrs. x $.06/KW x .01 = $40.00

10. Fine tuning: Assume 1¢/sq. ft. reduction in utility cost = $50 tons x 450 sq. ft./ton x $.01/ft.^2 = $225.00

11. $3,000 savings to rebuild rather than replace a compressor. Rebuild unit after 8 years operation. 1/8 x 3000 = $375.00

12. Monitoring results in service mechanic having right tools/parts (fan belts) 4 hrs./yr. x $25.00/ hr. = $100.00

13. Sensors 3 @ $140.00, 3 @ $60.00 = $ 600.00
    Misc. Mat'l.   =   120.00
    Labor 8 hrs @ $25.00   =   200.00
    Monitor Panel   = 1,200.00
                         $2,120.00

SECTION 25
WATER ENERGY MANAGEMENT

Chapter 77

ENERGY/WATER NITROGEN CONSERVATION AT A PHOTOGRAPHIC CHEMICAL MANUFACTURING FACILITY

F. A. Piccolo, T. H. Riker

During the latter part of 1979 when energy costs started rising at an accelerated rate and resource shortages were posing a threat to the economy, Polaroid's Chemical Manufacturing Facility in Waltham, Massachusetts decided it was time to revitalize the Conservation Program that was initiated after the 1973 oil embargo.

The Polaroid Chemical Manufacturing Facility occupies three of nineteen buildings at the Waltham Site that covers over 200 acres. It produces dyes, developers, opacifiers, polymers, sensitizers and various other chemicals necessary for photographic film manufacturing. Steam, electricity, water, natural gas, nitrogen and on occasion CO_2 are utilized for various functions throughout the facility. In general, Polaroid Corporation is primarily in the photographic field and is considered to be a high technology company with relatively low energy intensity, costs approximately 2% of sales. However, energy costs in our specialized chemical operations utilities play a much more dominant role reflecting 4% of the cost of sales.

We felt the most effective and expedient way to carry out this program was to form a Task Force comprised of three disciplines (Production, Engineering and Maintenance). Meetings were scheduled on a monthly basis. The minutes of the first meeting stated a message loud and clear --- We the Task Force would establish an intensive game plan; however, in order for it to work, we would need serious participation by operating personnel and strong support of upper management. We received both of these essentials and before too long the Program started to show results. In 1980 and 1981 we saved $1,000,000 in energy costs vs. a budget of $5,500,000. Actual unit consumption during that period for each utility was down significantly; Electrical 20%, Steam 28%, Nitrogen 57%, Water 64%, Natural Gas 93%. The bottom line --- an increase in production coupled with a decrease in unit utility consumption.

After the second meeting, we adopted the charter --- AWARENESS - PARTICIPATION - RESPONSE. At each meeting we would review and update the status of each action item and equipment modification project. We would also devote part of the meeting toward our awareness public relations activity, which we felt was the key to success of the Program. Some of the slogans we used were displayed on an Energy Message Board; "Don't be a drip, save a drop" and "The watt you save may be

your ohm" are examples of slogans that we used. A poster with a separatory funnel simulating the energy budget dollars for each utility with the balance remaining shown each month attracted attention. Handouts with energy saving ideas were made available on a regular basis. There are about 250 people in this three-shift, seven-day week operation and before too long energy conservation was a daily topic of conversation.

Having a means of measuring actual consumption for each area was also important. We installed at least one meter/recorder for each utility in each building/area. This helped establish a competitive atmosphere between buildings and served as a Conservation Program in itself.

Monthly Reports showing current actual consumption data vs. budget, as well as for the same period in the previous year, acted as a report card that people anxiously awaited to be published to see how well they had done. The Monthly Report was good feedback, but we felt we had to do more to justify actual consumption. So we decided to issue a Monthly Efficiency Report. We had our Facilities Engineering Department calculate the fixed occupancy (HVAC and Lighting) consumption/cost for each building and used this as our non-process load. We then asked each Production Engineer to determine utility requirements to manufacture each product and used this as our process load. These two documents made it possible for us to compute a theoretical load each month. This value was compared to the actual monthly consumption and reported as efficiency. Efficiency values for each utility were plotted on an annual graph to indicate trends that were taking place. Our production requirements change frequently involving a variety of chemical products each with different utility requirements; therefore, this report turned out to be extremely effective. Any time efficiency values fell to "embarrasing" levels, management was quick to ask -- Why?

Another element which helped us considerably was to familiarize as many division people as possible in the function and design of each utility system. Training and having responsible people paid handsome dividends when it became necessary to follow up unusual conditions of high consumption --- These were referred to as "spikes". The response to "spikes" was immediate, and correcting the problem was now a matter of hours rather than days or weeks.

Charts indicating potential loss values for various size openings at different pressures also helped emphasize the importance of response time. It all adds up to:
AWARENESS + PARTICIPATION + RESPONSE =
ENERGY CONSERVATION RESULTS.

UTILITY SPECIFICS

Steam

Steam, which represents approximately 25% of our Energy Budget, is used for heating (comfort and process), humidification, vacuum jets and an emergency hydraulic pump. Steam is supplied from a central boiler plant on site that is capable of providing 200,000 lb./hr. of 250 psig saturated steam at 406°F. The largest of our three boilers is equipped with an economizer that increases efficiency to 83%. The three boilers burn #6 fuel oil with a sulfur content of ½%. 75% of all condensate is collected and returned to the steam plant. Only condensate that is suspect of contamination is not returned. Our first move was to issue a memorandum reminding everyone that the government temperature-humidity guidelines of 65°F and floating humidity in the winter and 78°F - 65% RH in the summer would be enforced for non-process areas. We also stressed that temperature would be lowered to 55°F in unoccupied areas. Any area that stored chemicals that required temperatures above 55°F were obviously exempt from this guideline.

Steam Conservation Projects that were identified as ones with the greatest savings were:

A) Seal building cracks and pipe penetrations to minimize heat loss.

B) Reduce number of air changes in process area to bare minimum. Our plants require frequent air changes to minimize solvent vapor concentrations.

C) Heat reclamation.

D) Establish and implement a steam trap preventive maintenance program.

E) Repair insulation on steam and hot water lines.

In addition to tightening up the buildings, we replaced two motorized sliding doors on outside walls that were constantly failing in the open position with the hinged type. It was evident that the change was effective when operators on the first floor stopped complaining of being cold and unit heaters did not come on as frequently.

A comprehensive air balance study of the ventilation system was done in two of three of our process buildings. Results indicated that the rate of air changes in some areas can be reduced without violating any safety requirements. Funds have been approved for this project and all equipment modifications should be completed before the '82-'83 heating season. Estimated annual savings are $18,000 -- a payback of one year. Once the air change frequency has been established, we can then address the feasbility of heat reclamation.

It seems like every journal I've picked up over the past three years has an article regarding the lucrative savings that were being realized from a Steam Trap PM Program --- Were these articles written by steam trap manufacturers? I didn't think so. Our steam costs were twice what theoretical calculations indicated, and some of our traps were originals, 1965 and 1971 vintage. We decided it was time to get serious about getting our Program off the ground. We started by breaking down the categories by service in each building; e.g. One of our buildings has a 15, 70, 140 and 250 psig steam service. Concurrently, we updated engineering drawings made up for each service and assigned tag numbers for each trap. Initially we replaced the drip traps on service headers feeling that these disc traps had seen the most severe service. Each trap will eventually have a test valve to allow us to test traps more frequently without interrupting the production operation. Traps on process equipment heating/cooling loops will be checked for appropriate service as well as operating efficiency.

Nitrogen

Nitrogen gas, an essential utility in the Chemical Manufacturing Industry necessary for safety purposes, is used to inert and purge vessels and pipelines. This inert gas is supplied from an 11,000 gallon liquid nitrogen storage vessel. The -320°F liquid is vaporized through atmospheric vaporizer and distributed to each plant through "miles" of copper and black iron pipe with many valves and fittings. Obviously, the leak sources are many. During pre-conservation times, nitrogen consumption was between 10 - 13 million ft^3/month. At that time the nitrogen rate was $.265/100 ft^3 making the daily cost for nitrogen $1,000/day! This exorbitant cost was posted on bulletin boards in the process areas and in the main lobby. Once the word got out, everyone made a conscious effort to do something about it. In the three years that the Task Force has been in effect, nitrogen consumption cost has been reduced to one-third of what it was in 1979.

The success of the Program can be attributed to the extensive maintenance effort in tightening up of the system and tours conducted with production personnel familiarizing them with piping routes and valve arrangements. Bypass valves were painted white so that they were noticeable and purge rates were reiterated to ensure that nitrogen was used efficiently.

The solvent tank farm at our site has nineteen underground tanks and all are blanketed with nitrogen to ensure a safe inert atmosphere. At one point consumption was 10-20 times the theoretical value. Two-inch caps removed to determine level by "sticking" tanks were occasionally left off and discs on conservation vents that were not seating properly were the main causes of this waste. Level is checked on a daily basis, and one, two-inch cap left off could waste as much as 50,000 ft^3/day. Spring loaded caps replaced the screw caps to resolve this probelm. Guides were installed in the conservation vents to prevent the discs from seating in a cocked position. In addition, rotometers were installed on nitrogen purge lines to indicate flow that was caused by leakage rather than displacement. This assignment was given to the individual responsible for checking tank level. Consumption in this area is currently down by a factor of four.

At the present time each operating department has its own meter and nitrogen is charged out by a measured total rather than the previous allocation method. This assures every operation that it is charged for what they actually use. It also constructively forces a competitive atmosphere between users which subsequently lends itself to nitrogen conservation.

Overall, the Program is working well and we feel confident that we can maintain the current consumption rates and set realistic goals towards further reductions without jeopardizing the effectiveness of our safety procedures.

Water

The natural resource that is so essential to life has been in critically short supply for the past three years in the Northeast and current demand projections predict the situation will worsen in the future. Conservation efforts have been statewide and much public attention has been given to the shortages that exist and programs that have been implemented to help alleviate the problem.

We at Polaroid Corporation have made a firm commitment to actively find ways to reduce consumption to the bare minimum necessary for our manufacturing operation. At the start of our Program, we held a symposium that included State and Community Officials. The session was extremely informative and enlightening. A 30 minute film on the entire water distribution system for Eastern Massachusetts was shown and a panel of speakers talked about various programs that were underway to resolve this serious problem.

Water stewards were assigned to each building with the responsibility to tailor programs that would fit their specific area. A common goal throughout the Corporation was to eliminate all "once-through" systems. At our Chemical Manufacturing Facility vacuum pumps and condensers fell into this category.

We took on the vacuum pumps first and re-circulated 80% of the water required for sealing and cooling in seven pumps installed in three process buildings. Water savings totaled 100,000 gal./day. On condensers that used water for cooling, we converted most to a closed cooling tower water loop where possible, and on others installed flow restrictions reducing flow to the bare minimum.

Water hoses left running at our service stations were another source of heavy waste. Spring loaded deadman valves that were installed initially had been removed because of the "inconvenience". We installed automatic shut-off valves controlled by timers

on some of our service stations and it has proven effective and economically feasible. Payback calculates to 2½ months. On service stations where auto-shut-off had not been installed, we implemented a "tagging system"; if a hose was left running for a specific reason, a tag stating such would be required, otherwise it would be turned off.

Operating mode for deionized water systems with two resin beds was modified. Instead of operating in a standby-service mode, we operated both resin beds in parallel. We found that regenerations were required less frequently and water savings came to 1,000 gal./day. Additional meters are continually being installed to help account for and justify water consumption. Readings are taken on a weekly basis at our Facility, and at the present time we can account for 85% of our actual consumption.

Figure 1 is water consumption data for the Waltham Chemical Facility for the period 1974 - 1981.

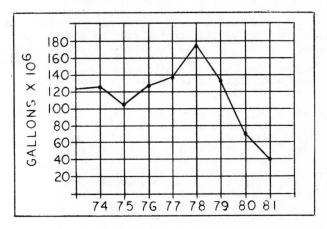

FIGURE 1. ANNUAL WATER CONSUMPTION

Overall I look at 1979 as the year that we turned the tide. Usage has become a daily concern and involvement by everyone has made a favorable impact on reducing consumption. We feel confident that an atmosphere has been created where operating decisions are based on actual needs rather than a better-to-be-safe-than-sorry attitude. The ultimate goal is to make this attitude a way of life.

Natural Gas

Exclusively used by two Thermal Oxidizer units for burning waste solvents. Back in 1978 and 1979 these units were consuming a total of 60 - 70 million ft^3/year.
Since that time our Chemical Recovery and Disposal Department has refined the technique of segregating high and low BTU content waste streams reducing gas consumption significantly. They have also reduced considerably the volume of waste solvent by increasing solvent recovery activity -- a practice that pays a double dividend.

The Program has been so successful that in 1981 natural gas consumption was down to a mere 4.3 million ft^3 and the larger of the two Thermal Oxidizers has been declared obsolete and is being taken out of service.

Natural gas cost savings realized in 1980 and 1981 are approximately $500,000.

Electrical

Cost of this utility is greater than all of the others combined. For 1982 it represents 62% of the total budget. In 1979 electrical consumption for process refrigeration was 32% of the total electrical load. We decided that this is where we had to concentrate our conservation efforts. Process refrigeration in building W6X is provided by two, 300 ton Carrier centrifugal units equipped with 700 HP hermetically sealed motors. At one time the 42% brine solution temperature was maintained at -17°F. After reviewing the processes closely with production personnel, we concluded that we could raise the brine temperature shut-off point and still provide adequate cooling. We did this on two occasions. Initially, we raised the cut-off to -13.5°F and the second time to -11°C where it stands at the present time. Close monitoring of the entire system by our air conditioning mechanics also contributed to the energy savings that were realized. In 1980 and 1981 consumption was reduced to 21% of electrical cost. Approximate energy cost savings attributed to temperature differential are $100,000/year.

Our nitrogen system was equipped with electrical vaporizers that were consuming 30,000 - 40,000 KWH/month. We replaced these with atmospheric types that gave us equivalent rate and capacity and took advantage of the $2,000/month savings. Rental cost for the atmospheric units was less, compounding the savings. Another feature was that power failure had no effect on the operation of these units.

One of our vacuum systems was using a Nash CL-2003 Liquid Ring pump that was driven by a 125 HP motor. We found that we could adequately provide this service with an existing Nash CL-1003 equipped with a 50 HP motor -- savings 300,000 KWH/year.

In the meantime, we worked at promoting efficient use of other occupancy and process related needs such as:

. Reduce light levels to minimum standard by installing phantom tubes when lamps fail.

. Assign responsibility to turn off lights in unoccupied areas during off-standard hours.

. Monitor process utilities at least twice per shift to ensure that bare minimum to provide necessary service only if left running.

. Provide photo-electric cells for all outside lighting.

Our experience with respect to conserving electricity is that it is most challenging. Since the bulk of the electrical equipment falls into a "fixed" category, the controllable portion is limited. Results from the first year of the Conservation Program reflected a 20% decrease in consumption. Obviously, much of this resulted from the major equipment modifications that we had implemented. Second year results were in the 2 - 3% range. This has been somewhat frustrating, but when you can account for and justify 95% of the actual consumption, we realize that a significant change in operating mode and/or better control capability is required to further reduce consumption. We hope to achieve this with the proposed Energy Management System that is currently under consideration.

The achievement reached by the collective efforts of many is certainly gratifying. The fact that we have been able to increase productivity while decreasing actual unit consumption is a good barometer of the success of our Program. However, we do realize that while succeeding can be difficult, maintaining/improving that level of success can be more demanding.

Where do we go from here? Our goal is to continue to work hard at keeping awareness on a high note and encourage participation by everyone in the plant.

Chapter 78

THE IMPACT OF WATER CONSERVATION PROGRAMS ON AMERICAN INDUSTRY

S. Zeman

Water, like other natural resources, has long been regarded as a commodity of infinite supply. Our assumption of continued abundance has often led us to
squander and pollute water with little awareness of
the consequences. And supply per se is not the sole
concern when we consider that approximately 4 trillion
gallons of rain fall upon the United States each day,
and that we use but one tenth of that for agricultural,
industrial, and consumer needs.[1] The key to understanding the dilemma of our water shortage is in recognizing
several important facts relative to our most precious
natural resource:

1. Water is not a predictable resource. The West,
   which comprises 60% of the area of the U.S. and is
   one of the fastest growing sections of the country,
   heavily agricultural and increasingly industrial
   and commercial as well, receives about 25% of the
   country's rainfall.[2] The East, regarded as amply
   supplied, is subject to periods of drought (such
   as are presently being suffered).

2. Pollution and hazardous waste disposal have significantly reduced potable water supplies in many areas.
   Despite improvement and more far-reaching government regulation, this remains an area of critical
   concern.

3. Many of the nation's water supply systems are hope-
   lessly inefficient and antiquated. New York City
   and Boston, for example, have systems which are
   estimated by some experts to leak as much water as
   they provide. This is by no means an isolated problem, and the concern here is that these systems
   will require enormous expenditures of time and money
   to be made efficient.

The thesis of this paper is that the relatively recent
movement toward industrial water conservation and water
management programs in this country has made a major
threefold impact upon American industry.

First, American industry is learning and beginning to
recognize that water conservation and management can
reap enormous savings, both in terms of gallons of
water consumed and in terms of dollars spent for acquisition, processing, usage, and disposal. Additionally,
the inception of water management programs has created
additional indirect but important benefits, such as
increased productivity, streamlined procedures, elimination of other sources of waste within the production
process, and community involvement and acceptance of
the necessity of industrial and municipal co-operation
in managing water resources.

Second, water conservation and management programs
instituted by industry have been extremely valuable in
educating employees - the homeowners and residential
consumers of this country - in the whole realm of water
conservation problems and solutions. Most industrial

programs, in fact, rely heavily upon employee input and
suggestion for their successful implementation, and
this employee participation not only aids the employer
by improving employee satisfaction, but also substantially decreases water consumption on the residential
level, thereby greatly improving the overall water conservation picture.

Third, industry can now, as a result of its pioneering
efforts and successes in water conservation and management, lead the way in the nationwide movement to conserve water. The growing concern in this country over
dwindling supplies of clean water, and the government's
commitment in recent years to regulating consumption,
led industry to face a critical question - how can we
continue to produce at an essential level with less and
less water to use? The programs and technological
innovations have been extremely encouraging, and will
undoubtedly serve as models for even more extensive and
innovative water saving programs in the future, as their
viability becomes more evident.

Faced, then, with an enormous problem of national proportions, American industry has begun to turn to water
conservation and management programs. While this is a
relatively new field and the technological expertise
is still growing, the key ingredient to date has been
conservation - using less water to do the same amount
of work. How is American industry coping with water
conservation and management? Let's examine more thoroughly the three major areas of impact.

I. Realizing Savings Through Water Management Programs

The two most important factors in hastening industrial
adoption of water management and conservation programs
have been rising costs and government regulation.
While the United States has the capability and the
technical expertise to renovate and restore the many
antiquated water supply systems in the country, the
cost of such rehabilitation is extremely high. Industry, at the mercy of increasingly overtaxed municipal
water supply systems, is becoming increasingly aware
that the cost of water is rising in a spiral parallel
with energy costs, that the two are in fact inexorably
entwined and are going to continue to escalate.

In addition, industrial use of water in this country is
exceeded only by agricultural use, and this water usage
is increasingly subject to stringent governmental
regulation which is expected to reduce the water available to industry substantially by the year 2000 (by as
much as 62% as compared to 1975 usage figures, according
to an estimate of the Water Resources Council).[3]

Given the inevitability of rising water costs and decreased supply, how has water conservation and management affected American industry thus far? The results
to date suggest that, while water management practices
will vary greatly depending upon the size and specific

nature of the industry in question, one fact is emerging quite obviously from the efforts of the industrial water conservation programs in place at this time: simple water conservation, when applied in a thorough and organization-wide water management program tailored to the specific site in question, delivers enormous savings in water used and dollars spent.

According to Associated Industries of Massachusetts, for example, the Gillette, Honeywell, Polaroid, and Raytheon companies operating in that state have reduced water consumption by 630.8 million gallons per year, or 29.4%, while increasing industrial output at the same time.[4]

The Gillette Company, recognized as the industrial water conservation leader, has been committed to water conservation since 1973 as an integral part of their overall energy conservation program. Gillette recognized at this time that water was going to be the next critical world resource in jeopardy, and has continued to recognize that the cost of water and sewage processing has risen at an even faster rate than that of energy.[5] The implication here is clear; any industry which does not incorporate a sound and practicable water conservation and management program into the framework of its overall operational make-up is losing money to increasing water expenditures at an even faster rate than if it were paying no heed to the rising cost of its other energy needs.

Gillette has, since the inception of its water saving efforts, achieved water savings ranging between 34% and 70% per location, a savings which can be translated into $1,200,000 worldwide. In 1972 the Gillette Safety Razor Division consumed 750 million gallons of water; in 1982, projected figures show that this division will use 200 million gallons, or an annual savings of some 550 million gallons, enough to supply 10,000 homes with water.[6] Figure I shows the annual water usage of Gillette's Safety Razor Division. If water and its related costs are rising at a rate which exceeds other energy costs, the potential for savings in the future is enormous, and likely to grow.

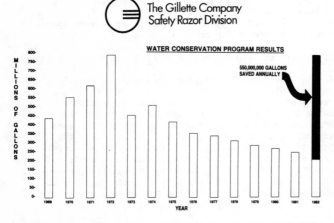

FIGURE I. WATER CONSERVATION PROGRAM RESULTS, GILLETTE

In 1981, the Waltham, Massachusetts Division of Polaroid Corporation was requested to institute water conservation measures, a response from corporate headquarters to pressure from the various municipal suppliers in the greater Boston area attempting to cope with record drought conditions. In that year, the Waltham site reduced water usage by 33.8% as compared to 1979 consumption. In 1982 Polaroid is intending to reduce the 1981 usage by another 15%; if successful, the division will have realized a $300,000 savings over water expenses

incurred in 1979 (See Table I).[7]

Year	Total Water Consumed	Increase or Decrease From Previous Year
1977	5.08×10^8 gal.	
1978	5.10×10^8	+ .4%
1979	5.31×10^8	+ 4.1%
1980	4.69×10^8	-11.7%
1981	3.50×10^8	-25.4%

TABLE I. WATER CONSERVATION PROGRAM RESULTS, POLAROID CORPORATION, WALTHAM SITE

The Sprague Electric Semiconductor Division in Concord, New Hampshire, began to examine water use in 1978. Their water conservation efforts are paying great dividends to date. In 1978, Sprague consumed 7,446,500 cubic feet of water. In 1981, water consumption dropped 21.3% from 1980, to a total of 5,246,700 cubic feet. Water consumption at the site is the lowest it has been since 1974 (See Table II).[8]

Year	Total City Water Consumed	Increase or Decrease From Previous Year
1974	5,219,100 ft.^{3}	
1975	5,411,200 ft.^{3}	+ 3.6%
1976	6,200,100 ft.^{3}	+12.7%
1977	7,152,000 ft.^{3}	+13.3%
1978	7,446,500 ft.^{3}	+ 4.0%
1979	8,432,600 ft.^{3}	+11.7%
1980	6,668,600 ft.^{3}	-20.9%
1981	5,246,700 ft.^{3}	-21.3%

TABLE II. WATER CONSERVATION PROGRAM RESULTS, SPRAGUE ELECTRIC, CONCORD, NEW HAMPSHIRE

Clearly, water conservation and management for industry is a concept whose time has arrived, and is already reaping substantial savings in water and money for those companies, large and small, with the foresight and dedication to recognize the opportunity and necessity for saving. Industrial water conservation has been shown to be both cost effective and economical in facing the challenge of rising costs and dwindling resources faced by all of American industry.

II. Employee Participation and Education

The resounding echo which rises from the reports of water conservation success in such large companies as Polaroid, Raytheon, IBM, and Gillette is that the key element in a successful water saving effort is the involvement of employees. This key ingredient is stressed in particular by the pioneering Gillette Company's Energy Conservation Coordinator, Cameron D. Beers, who states simply that the conservation of energy is an integral part of life on the job at Gillette, and at Gillette water and energy are in fact synonymous.[9] He refers not only to an isolated few in upper level management; he speaks of all of Gillette's employees. Policy which depends heavily upon the cooperation and commitment of employees has provided the striking results which Gillette and those companies which have followed its lead have achieved. Awareness of the problem is high in these companies, and the employees, because of their involvement and education in the issue of conservation, have taken this concern into their homes and residences and accepted the responsibility of spreading the word and seeing that the program continues to benefit them both at home and on the job. While technology has had its impact upon industry in providing efficient equipment, hardware, and design, it is the employees who are involved in the day-to-day processes, who see additional potential for elimination of waste and opportunities to save even more. The commitment of top level management to this goal of employee education, participation, and awareness is the foundation upon which successful water conservation and management programs are built, more important and crucial to success than enormous expenditures for new equipment and sophisticated design changes.

III. Leading The Way Toward Water Conservation And Management

Of necessity, industry in this country will have to provide the leadership in attacking headlong the reality of conserving and managing water supplies. Because municipal systems cannot and should not be relied upon to provide infinite supplies of water, because government regulation of water consumption and disposal is stringent and constrains industrial usage, and because water costs are escalating; industry must rise to the challenge of confronting water shortages in two significant ways:

First, a growing number of companies will be forced to initiate water conservation and management programs tailored to individual sites. These programs will essentially be water saving measures which attempt to reduce water consumption in existing sites by thoroughly auditing the entire framework of the location - building(s), procedures, and hardware - for water waste. Retrofitting and steamlining must then be undertaken. These programs require employer commitment and employee participation to be fully successful in restructuring and maintaining water conservation measures. The technology and hardware involved in this type of overall water conservation commitment need not be elaborate or expensive to reap significant results, as can be seen in the Gillette, Polaroid, and Sprague examples alluded to in section 1. While a good deal of retrofitting was made in these cases, the companies attest to the underlying importance of an effective audit, on-going employee participation, and company dedication to the spirit of conservation as the real key ingredients in the overall success of their programs.

Second, developing hand-in-hand with the committed water conservation efforts of the entire framework of the company - there must be a marked increase in the change over to water conservation technology and hardware specifically designed to combat the problems of water shortage, discharge of pollutants, and excessive use of potable water; and an upsurge in the re-examination and application of existing but underutilized knowledge. Each company's employees, when properly educated and encouraged, have the knowledge of the specific manufacturing procedures and can make the suggestions that will have the largest impact on savings.

By the year 2000, more than 40% of the nation's daily water usage, or nearly 100 billion gallons per day, will be utilized by industry. The primary industrial functions served by this water are cooling, boiler feedwater, and processing. Of these, cooling water is the largest consumer and the candidate for greatest savings through reclamation.[10] Reclamation is by no means a new concept - it refers to the use of wastewater and reclaimed water for nonpotable industrial water supply needs. The Bethlehem Steel Plant in Baltimore, Maryland, perhaps the nation's largest user of municipal wastewater, supplies 15% of its total water demand in this way. Amarillo, Texas, has for years (since 1954) been successfully reclaiming water and reselling it to individual consumers for nonpotable use.[11] Innovation in reclamation is developing today in the area of establishing large scale regional wastewater planning and facility development. Its feasibility as an alternative to using the dwindling supplies of fresh water less and less available throughout the United States makes this kind of large-scale planning more attractive to industry and municipal suppliers alike, and there are a growing number of co-operative efforts currently

in the works, which, despite the expected difficulties of establishing proper management and supervision, show enormous potential for success. There are even "plural" water supply systems in the works which will supply wastewater of varying degrees of treatment to individual purchasers. Approached on this kind of regional basis, reuse of wastewater makes possible large-scale planning for source substitution and the use of lower quality reclaimed water to meet industrial demands.[12]

Other technological innovations in water conservation are being developed in the realm of water saving machinery and hardware. Recirculating cooling towers which eliminate the wasteful "once-through" technique are increasingly prevalent, and their problems in terms of scale buildup, slime and algae formation, and corrosion are under constant study with new solutions and innovations continually emerging.

Design changes and improvements will be an important element in the industrial conversion to water saving measures which is in progress today. These design changes and improvements fall into the following categories of water consumption:

1. Reducing excessive flows for cooling and rinsing wherever possible.

2. Reusing of relatively clean spray water and in dip rinses, and adopting modern valves, nozzles, and related water hardware which has been specifically designed to be efficient.

3. Installing water saving hardware in all possible locations, including sanitary facilities, production and processing sites, and septage and indoor and outdoor plant maintenance facilities.

The rapid technological improvements and the growing availability of computer hardware, however, has provided industry with perhaps its most significant water conservation tool. Since water resource management problems may involve a great many variables which can affect the solution to a given problem, the computer's capacity to store and analyze information quickly and efficiently is the engineer's avenue to exploring a wide variety of alternatives to any given water usage situation. In Florida's Cross Bar Ranch well field computers are used to accomplish accurate well siting, monitoring of the aquifer, and logging, storage, and analysis of all important well field data.[13] Computers will increasingly be used to address the problems of water pollution as well. Computer systems are being used today which describe pollution loads and screen many pollution control alternatives. Essentially, the computer can and will be utilized in the battle to conserve water by finding the least expensive and most environmentally sound ways of managing water resources.

Perhaps the finest example to date of the impact of water conservation and management programs upon industry in the United States in terms of technological advancement is embodied in the IBM plant located in Tucson, Arizona. As a result of its water conservation program there, IBM will utilize no municipal water and it will add no water to the municipal sewage system. The program was specifically designed for a new plant and it includes on-site wells, a closed, chilled water storage system, a hot water storage system for heating, and a complex self-contained system for waste water treatment, reclamation, and reuse.[14] While such a system would not be economically feasible for the retrofitting of an existing facility, the strides made by industry are showcased in this plant, and point clearly to the importance that the growing implementation of water conservation programs can have upon industry in this country.

Some technological and design advancements are still in
the "drawing board" stage, including water desalination
and massive diversion of existing water supplies; these
may be economically impractical at present, but have
future potential. The fundamental question for industry has already been answered, however, in the growing
number of facilities adopting technological and design
advancements. Clearly, industry can learn to cope
with the water supply crisis and must establish itself
as the leader in instituting the task of ensuring that
our supplies are ample and that they continue to last
into the future.

Industry in the United States, then, is facing the
challenge of water shortage and reaping tremendous
benefits by involving employees in the industrial water
conservation process and by devising innovative and
efficient water saving hardware, technology, and design
plans. This allows productivity to remain high while
consumption decreases significantly. This in no way
indicates that industry has overcome the impact of the
crisis in water supply in this country. On the contrary, in the field of water conservation the task now
at hand for American industry is greater than any previous one. Simply, despite the substantial efforts
already in place, water shortages continue to plague
industry, and will do so until conservation is the
watchword of all industrial consumers, rather than a
pioneering few willing to assume the responsibility of
managing water supplies.

FOOTNOTES

1. Thomas Y. Canby, "Our Most Precious Recource: Water," <u>National Georgraphic</u>, August, 1980, p. 148.

2. Armstrong, Ronald N. and Taylor, Robert S. "The Computer: Water Management Tool," <u>Consulting Engineer</u>, September, 1981, p. 97.

3. William W. Durrell, "Water Conservation: Protecting A Vital Resource For Industry," <u>Enterprise</u>, June, 1982.

4. Ibid.

5. Ibid.

6. Ibid.

7. Ibid.

8. Ibid.

9. Ibid.

10. Westerhoff, Garret P. "Water Cost Encourages Reuse and Conservation," <u>Consulting Engineer</u>, September, 1981, p. 88.

11. Ibid., p. 88.

12. Ibid., p. 89.

13. Armstrong, Ronald N. and Taylor, Robert S., op. cit., p. 98-99.

14. William W. Durrell, "Water Conservation: Protecting A Vital Resource For Industry," <u>Enterprise</u>, June, 1982.

SELECTED BIBLIOGRAPHY

Armstrong, Ronald N. and Taylor, Robert S. "The
    Computer: Water Management Tool." Consulting
    Engineer, (September 1981), pp. 97-100.

Canby, Thomas Y. "Our Most Precious Resource: Water."
    National Georgraphic, (August, 1980), pp. 144-179.

"The Browining of America" Newsweek, (February 23,
    1981), pp. 26-37.

"Water Conservation: Protecting A Vital Resource For
    Industry." Resource Management Associates, (1981),
    pp. 1-7.

"Water Supply and Conservation." Plant Engineering
    Library, (1978), pp. 22-23, 26-28, 29-30, 31-38.

"Water: Will We Have Enough To Go Around?" U.S. News
    and World Report, (June 29, 1981), pp. 34-38.

Westerhoff, Garret P. "Water Cost Encourages Reuse and
    Conservation." Consulting Engineer, (September,
    1981), pp. 87-90.

Wolman, Dr. Abel. "Water: Too Much, Too Little, Too
    Bad?" Consulting Engineer, (September, 1981),
    pp. 83-86.

Wheeler, William. "Updating Antiquated City Water
    Systems." Consulting Engineer, (September, 1981),
    pp. 101-105.